HETEROGENEOUS
REACTOR DESIGN

BUTTERWORTHS SERIES IN CHEMICAL ENGINEERING

Enlargement and Compaction of Particulate Solids
Nayland Stanley-Wood

Fundamentals of Fluidized Beds
John G. Yates

Heterogeneous Reactor Design
Hong H. Lee

Liquid and Liquid Mixtures
J. S. Rowlinson and F. L. Swinton

Phase Equilibria in Chemical Engineering
Stanley M. Walas

Solid Liquid Separation
Ladislav Svarovsky

HETEROGENEOUS REACTOR DESIGN

Hong H. Lee
Associate Professor of Chemical Engineering
University of Florida

BUTTERWORTH PUBLISHERS
Boston • London
Sydney • Wellington • Durban • Toronto

Library of Congress Cataloging in Publication Data

Lee, Hong H.
  Heterogeneous reactor design.

  Includes index.
  1. Heterogeneous catalysis. 2. Catalysts. 3. Chemical reactors.
I. Title.
TP156.C35L428   1984   660.2′995   84–1717
ISBN 0–409–95073–4

Butterworth Publishers
80 Montvale Avenue
Stoneham, MA 02180

10 9 8 7 6 5 4 3 2 1

Printed in the United States of America

TO MY MOTHER

CONTENTS

PART III   REACTOR DESIGN AND ANALYSIS

PART IV   TRANSPORT PROPERTIES

PREFACE

This book is on the design of reactors for heterogeneous, solid-catalyzed reactions. As such, the catalyst phase and bulk-fluid phase are treated separately as opposed to a homogeneous treatment, and all subjects are discussed with reactor design as the focal point. This text is intended primarily for instruction at the graduate level, but it is also intended for practicing engineers who have had a preliminary course in reactor design. Parts of the material can be used for an undergraduate course when the emphasis is on heterogeneous reactor design.

The material in this text is presented in the order of the logical steps that one would follow in proceeding from catalyst selection to reactor design. Accordingly, this book is divided into four parts. Part I (Chapters 1 through 3) provides an understanding of catalysis, support and catalyst preparation, and surface reactions. It also provides basic characterization techniques useful in choosing a catalyst system and in providing basic information on the chosen system. Part II (Chapters 4 through 8) deals with the catalyst pellets and particles, in which the reaction takes place, using global rates as the focal point. It provides a quantitative capability for obtaining global rates in terms of the intrinsic kinetics and those parameters pertinent to diffusion and catalyst deactivation. Part III (Chapters 9 through 13) treats the design and analysis of various types of reactors. It provides specific means of relating the reaction taking place in the catalyst pellets to the overall conversion in the reactor. Part IV (Chapter 14) contains a detailed treatment of transport properties.

The general tone and the common thread in all subjects treated are such that students can recognize reactor design as the focal point. The approach allows the students to acquire a rather self-sufficient capability for handling the design of heterogeneous reactors. The mechanics of carrying out the reactor design, whether they involve numerical integration of the reactor conservation equations or construction of a design diagram, are detailed so that students can put into practice what they have learned.

In presenting the material, emphasis is placed on fundamentals and generality. The origin of and the rationale behind the basic relationships used are explained in terms of the fundamentals of physics and chemistry. These fundamentals are used to describe quantitatively the macroscopic behavior of the physical and chemical rate processes underlying the reactor design. Arbitrary kinetics are used throughout the text so that the results can be applied to any given kinetic scheme. The treatment of catalyst deactivation by both physical (sintering) and chemical means is rather complete, as is the application of the results to the design and analysis of a reactor undergoing catalyst deactivation. An approach based on "reactor point effectiveness" is introduced in this book and is utilized to simplify the design and analysis of heterogeneous reactors. Reactor point effectiveness, which is the ratio of global rate to intrinsic rate for fresh catalyst expressed in terms of the bulk fluid, makes it possible to transform seemingly untenable complex design problems into manageable ones. Since the global rates can be expressed explicitly for arbitrary kinetics of the main reaction and catalyst deactivation, almost all heterogeneous reactor problems can be treated as a homogeneous reactor problem whether the reaction is affected by diffusion, catalyst deactivation, or both. This leads to a design practice in which catalyst deactivation is taken into account during the design phase, as opposed to the usual practice of designing a reactor based on the activity of fresh catalyst and then operating the reactor so as to account for the declining activity with time.

The material in this text has been used at the University of Florida for the last five years at the graduate level, and for two years at the undergraduate level. At the graduate level, for which this book is primarily intended, it has been found to be much more effective to emphasize concepts and reasonings and to maintain the continuity of contents and tone rather than to dwell on number-plugging examples. Typically, two lectures a week are followed by a recitation in which problems relevant to the material presented in the lectures are solved. Accordingly, examples do not appear in the text as such, but rather as a part of the text whenever such illustrations are necessary to help maintain the thread of concepts and reasonings. In a typical 15 week semester, Part I is covered in four weeks with major emphasis on Chapter 1. Chapters 2 and 3 are usually combined in presenting the material. Part II is covered in five weeks with major emphasis on Chapters 4 through 6. Chapter 7 is usually covered in conjunction with Chapter 12. Part III is covered in five weeks. Chapter 13 is not covered. Part IV is covered in one week in the form of a summary with emphasis on experimental methods. Not all material can be covered in one semester and therefore some sections are usually left for students to read. At the undergraduate level, emphasis is placed on the use of results and number-plugging examples so as to ensure that students can utilize the results given. Work was mainly confined to simple kinetics and plug-flow reactors and only parts of chapters were covered. Senior students who took the course felt very comfortable with the design of heterogeneous reactors and were very enthusiastic, as indicated by many who voluntarily solved the problems on their programmable calculators. By the time graduate students take this course, they usually have had courses in kinetics and transport phenomena. The undergraduate students will have had a course in homogeneous kinetics and reactor design.

Problems at the end of each chapter are listed in the order that the material is presented. Suggestions for solving problems are given when it is appropriate. Notation is given for each chapter to avoid confusion due to changes that occur in the notation. Each chapter starts with an introductory section in which the subject matter to be covered is outlined and each chapter ends with a summary section. The chapter on detailed treatment of transport properties is reserved for the end of the book so that better continuity for reactor design can be maintained.

The author is indebted to J.B. Butt for his direct and indirect involvement in this endeavor. He is also indebted to J.M. Smith, whose encouragement set the endeavor into motion and who showed continuing interest in its progress. The suggestions and criticism of E.B. Nauman and R.L. Kable are gratefully acknowledged. Past and present students of the author also contributed in a significant way. They are: L.M. Akella, D.J. Miller, J.C. Hong, I.A. Toor, K.J. Klingman, and E.C. Stassinos. To Jeanne Ojeda, Linda Padgett, and in particular Melissa Michaels, who did most of the typing, I express my thanks. Finally, I wish to express my special gratitude to my wife, Sukie, for help and encouragement.

Hong H. Lee

PART I

Catalysts, Reactions, and Kinetics

CHAPTER 1

Catalysts and Characterization

1–1 INTRODUCTION

A catalyst is a substance that accelerates (or sometimes decelerates) the rate of approach to chemical equilibrium. In so doing, it is neither consumed nor is its effectiveness reduced unless it is deactivated in the course of reaction. Only heterogeneous solid catalysts are considered in this book. Catalysts are usually metals or metal compounds. The catalyst surface exposed to fluid reactants is responsible for the catalytic effect. It is natural then that the catalyst be made to have a high exposed surface area per unit weight. On the other hand, the reactor that contains the catalyst should be as small as practically possible. Therefore, the catalyst is usually spread on a substance of high surface area. Such a catalyst is called a supported catalyst.

Ultimately, one would like to tailor a supported catalyst for the reaction under consideration in a reproducible manner. Unfortunately, however, this capability does not yet exist and even the characterization techniques necessary for this capability are far from satisfactory. However, there are some basic tools that are rather well established for the characterization of catalysts and supports. These will be treated in some detail.

This chapter gives a description of catalysts, reactions, and catalysis, categorized in terms of the type of catalyst. Some detailed procedures for catalyst preparation are described qualitatively to provide some insight into the variables that can affect the performance of the prepared catalysts. An approach for quantifying the preparation procedures is presented for impregnation to provide a basis for further studies in this important and yet unexplored area of catalyst preparation. Treated in detail are those well-established basic tools for the characterization of catalysts in terms of catalyst dispersion and specific activity. Catalytic activity and correlations for it follow as a natural consequence of the discussion. Total surface area, pore volume, and pore size distribution essential for the characterization of the support are then treated in detail. The premise on which engineers deal with reactions and reactors is invariably "given a catalyst." This chapter provides a basic level of understanding required in moving toward eliminating this restriction.

3

1–2   CATALYSTS AND REACTIONS

Solid catalysts can be grouped into three types: supported metal catalysts usually used for reforming reactions, acid and zeolite catalysts for cracking reactions, and metal compounds (oxides and sulfides) for partial oxidation reactions. A list of typical functions and examples of catalysts is given in Table 1.1.

It is well understood that a catalyst cannot change the ultimate equilibrium dictated by thermodynamics but accelerates the rate of approach to equilibrium. Why such an acceleration takes place can be explained with the aid of the energy diagram (Satterfield 1980) shown in Figure 1.1. The diagram shows the changes in energy associated with the different steps in a simple exothermic reaction. The dotted line is for the homogeneous reaction with its activation energy of E_{hom}, while the solid line is for the heterogeneous catalytic reaction. The activation energy for the limiting step of heterogeneous reaction is usually lower than that for the homogeneous reaction[*] as shown in the diagram. In such cases, the observed rate of a catalyzed reaction should be faster than the corresponding homogeneous reaction, each of the steps in the catalyzed reaction requiring a smaller activation energy than E_{hom}. The diagram also helps understand why no significant increase in rate is achieved by any catalyst when high temperatures are required, as in some endothermic reactions, in order for a substantial amount of the product to be present at equilibrium.

The mechanisms of catalysis for each group of the catalysts listed in Table 1.1 are difficult if not impossible, to generalize. However, a few statements can be made on each group of catalysts. Almost all the metal catalysts are transition metals having d orbitals. These orbitals are not completely filled due to band overlap or electron overflow from the d to the s band (Gates, Katzer, and Schuit 1979). While it is not clear how this bulk atom behavior is related to the surface atom behavior, it is clearly plausible that a singly occupied surface orbital can interact with a molecular orbital of an approaching gas species, resulting in a bonding at the metal surface. This chemisorption can be regarded as the formation of a surface compound. In order for a catalyst to be effective, the stability of the compound should not be too high, since this would lead to complete coverage of the surface by the stable compound, and not be too low, since this would lead to very little surface coverage by the compound. When bonding level is adequate for a given reaction, the reaction can proceed at a satisfactory rate.

Metal compounds (mostly metal oxides) are usually used for partial oxidations. The catalysts are typically transition metal oxides in which oxygen is readily transferred to and from the structure. These compounds are usually nonstoichiometric. Other types of oxides are those in which the active species is chemisorbed oxygen, as molecules or atoms. The behavior of most oxidation catalysts (Satterfield 1980) can be interpreted within the framework of a redox mechanism. This postu-

---

[*] Schlosser, E.G. (Heterogeneous Katalyse, Verlag Chemie, Wernheim, Germany, (1971)) showed theoretically that the catalytic rate can be faster than the homogeneous rate even when E_{hom} is less than the activation energy of the limiting step.

Table 1.1   Classification of Solid Catalysts (after G.C. Bond 1974)

Group	Function	Example
Metals	hydrogenation dehydrogenation hydrogenolysis* (hydrocracking)	Fe, Ni, Pd, Pt, Ag
Metal Compounds		
Semiconducting oxides and sulfides	oxidation dehydrogenation desulfurization	NiO, ZnO, MnO$_2$ Cr$_2$O$_3$, Bi$_2$O$_3$ · MoO$_3$ WS$_2$
Insulator oxides	dehydration	Al$_2$O$_3$, SiO$_2$, MgO
Acid and zeolite	cracking	SiO$_2$ · Al$_2$O$_3$ (amorphous or crystalline)

* Addition of hydrogen across a single bond to cause splitting into two molecules, e.g., $C_2H_6 + H_2 \longrightarrow 2CH_4$.

lates that the oxide catalyst is first reduced by the gas molecule being oxidized and then the reduced catalyst is in turn oxidized by oxygen from the gas phase. The form of the oxygen in the catalyst can be either chemisorbed or lattice oxygen. In the case of lattice oxygen, the active species is presumably the O^{2-} ion.

One important fact about the acid catalysts is that neither silica nor alumina nor a mechanical mixture of the two dry oxides is an active cracking catalyst, but a cogelled mixture of silica and alumina containing mainly silica is highly active. The incorporation of alumina into silica, even at very low concentrations,

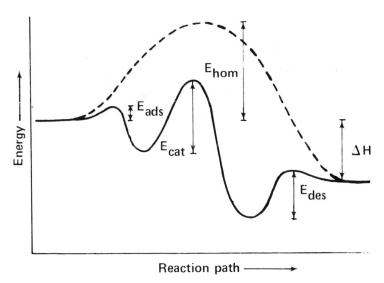

Figure 1.1   An energy diagram for catalytic reactions. (After C.N. Satterfield, *Heterogeneous Catalysis*, © 1980; with permission of McGraw-Hill Book Company, New York.)

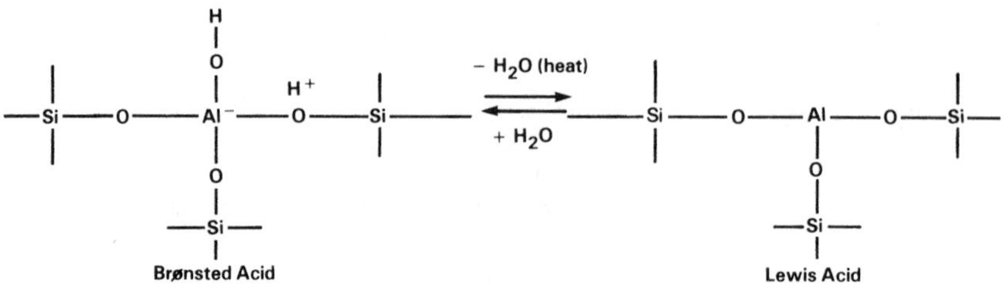

Figure 1.2 Postulated structures of silica-alumina causing Brønsted and Lewis acids. (After C.N. Satterfield, *Heterogeneous Catalysis,* © 1980; with permission of McGraw-Hill Book Company, New York.)

results in the formation of surface sites that catalyze cracking reactions. It is believed that at least some of these sites constitute Brønsted and/or Lewis acids. The source of acidity (Satterfield 1980) may be rationalized in terms of a theory developed largely by Pauling. If an aluminum ion, which is trivalent, is substituted isomorphously for a silicon ion, which is quadrivalent, in a silica lattice comprising silica tetrahedra, the net negative charge must be stabilized by a nearby positive ion such as a proton. This can be produced by the dissociation of water, forming a hydroxyl group on the aluminum atom. The resulting structure, in which the aluminum and the silicon are both tetrahedrally coordinated, is a Brønsted acid. If this is heated, water is driven off and Brønsted acid sites are converted to Lewis acid sites as shown in Figure 1.2. Some metal atoms are now three-coordinated and some four-coordinated. The aluminum atom is electrophilic and can react with hydrocarbons to form an adsorbed carbonium ion, as illustrated below for the two kinds of sites:

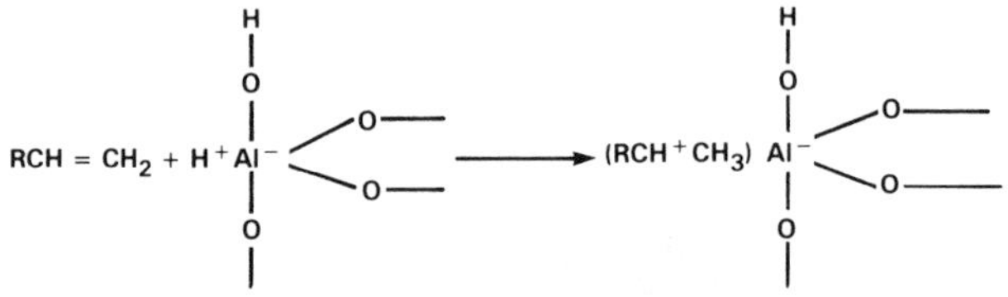

and

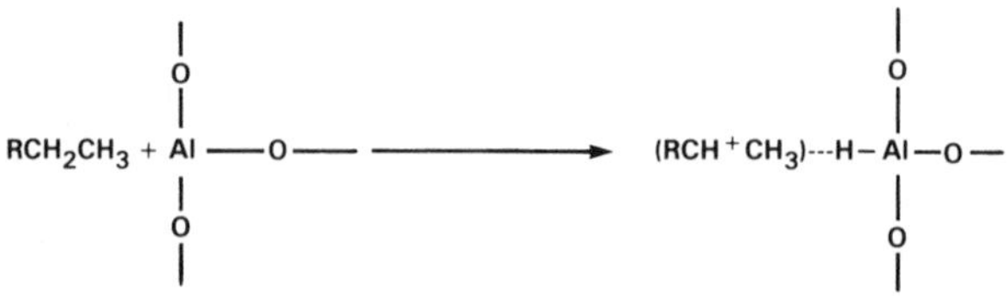

Similar arguments can be advanced to explain the acidity of various other mixed oxides containing metal atoms of differing valence, such as $SiO_2 \cdot MgO$, $SiO_2 \cdot ZrO$,

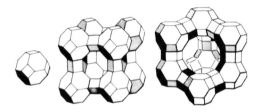

Figure 1.3 The cubo-octahedral unit of faujasite zeolites, structure of A zeolite, and X and Y zeolites (from left to right). (Barrer 1968)

and $Al_2O_3 \cdot MgO$. Mechanistic explanations of catalytic cracking are usually based on the formation of carbonium ion (Voge 1958), which these acidic sites initiate.

The aluminosilicate catalysts considered above are amorphous. In recent years, crystalline aluminosilicates (Bond 1974) have been developed with the general formula $M_\nu(AlO_2)_x(SiO_2)_y \cdot zH_2O$: these are known as zeolites. When M is a mono-positive cation (e.g. sodium or ammonium), ν equals x; for divalent cations, ν is $x/2$, and so on. There are a great many naturally occurring minerals of this kind, but the mordenite and faujasite can now be synthesized. The basic unit of a faujasite is the regular cubo-octahedron or sodalite unit (Fig. 1.3) consisting of 24 tetrahedra of either SiO_4^{4-} or AlO_4^{5-}. Two general types of structure can then arise, as shown in Figure 1.3, depending on the way in which the sodalite units are joined together. When they are joined through their square faces, A zeolite results in which y/x is unity; when joined through their hexagonal faces, the X zeolite (y/x, 1.25) or the Y zeolite (y/x, 1.5 to 3) results. These last two structures have quite large cages joined by smaller openings and therefore a very high internal surface area in the form of pores of fixed geometry. The effective pore diameter is determined by the kind of cation that balances the negative charge on the structure. Cations occupy one of three distinct positions, and their size thus controls the size of the openings. With A zeolite, the effective pore diameter is 0.3 nm when the cation is potassium, 0.4 for sodium, and 0.5 for calcium. Zeolites are usually first made in the sodium form: the sodium ions can then be exchanged quite simply for other cations.

The useful catalytic properties of zeolites hinge on three factors: the regular crystalline structure and uniform pore size that allows only molecules below a certain size to react; the presence of strongly acidic hydroxyl groups that can initiate carbonium-ion reactions; and the presence of very large electrostatic fields in the neighborhood of the cations that can thus induce reactivity in reactant molecules. Catalytic activity therefore depends heavily on the nature of the cation, which also seems able to affect the acidity of the hydroxyl groups. These zeolites are typically used for skeletal rearrangements and cracking of hydrocarbons. Shape-selective catalysis can also be realized using zeolites as reviewed by Chen and Weisz (1967).

Some heterogeneous catalysts of industrial importance are summarized in Table 1.2 (Satterfield 1980) along with the types of reactors commonly used with them.

Table 1.2 Some Heterogeneous Catalysts of Industrial Importance (after C.N. Satterfield, *Heterogeneous Catalysis*, © 1980; with permission of McGraw-Hill Book Company, New York)

Reaction	Catalyst and Reactor Type (*continuous operation unless otherwise noted*)
Dehydrogenation	
C_4H_{10} (butane) $\rightarrow$ butenes and C_4H_6 (butadiene)	$Cr_2O_3 \cdot Al_2O_3$ (fixed bed, cyclic)
Butenes $\rightarrow$ C_4H_6 (butadiene)	Fe_2O_3 promoted with Cr_2O_3 and K_2CO_3, or $Ca_8Ni(PO_4)_6$ (fixed bed, continuous, in presence of steam)
$C_6H_5C_2H_5 \rightarrow C_6H_5CH{=}CH_2$ (ethyl benzene $\rightarrow$ styrene)	Fe_2O_3 promoted with Cr_2O_3 and K_2CO_3 (fixed bed)
CH_4 or other hydrocarbons + $H_2O \rightarrow$ CO + H_2 (steam reforming)	Supported Ni (fixed bed)
$(CH_3)_2CHOH \rightarrow CH_3COCH_3 + H_2$ (isopropanol $\rightarrow$ acetone + hydrogen) $CH_3CH(OH)C_2H_5 \rightarrow CH_3COC_2H_5 + H_2$	ZnO
Hydrogenation	
Of edible fats and oils	Raney Ni, or Ni on a support (slurry reactor, batch or continuous)
Various hydrogenations of fine organic chemicals	Pd on C or other support (slurry reactor, batch or continuous) (other supported metals may also be used)
$C_6H_6 + 3H \rightarrow C_6H_{12}$	Ni or noble metal on support (fixed bed)
$N_2 + 3H_2 \rightarrow 2NH_3$	Fe promoted with Al_2O_3, K_2O, CaO, and MgO (adiabatic fixed beds)
$C_2H_2 \rightarrow C_6H_{12}$ (selective hydrogenation of C_2H_2 impurity in C_2H_4 from thermal-cracking plant)	Pd on Al_2O_3 or sulfided Ni on support (adiabatic fixed bed)
Oxidation	
$SO_2 + \frac{1}{2}O_2 \rightarrow SO_3$	V_2O_5 plus K_2SO_4 on SiO_2 (adiabatic, fixed beds)
$2NH_3 + \frac{5}{2}O_2 \rightarrow 2NO + 3H_2O$	90% Pt–10% Rh wire gauze, oxidizing conditions
$NH_3 + CH_4 + air \rightarrow HCN$ (Andrussow process)	90% Pt–10% Rh wire gauze, under net reduction conditons
$C_{10}H_8$ or $1,2{-}C_6H_4(CH_3)_2 + O_2 \rightarrow C_6H_4(CO)_2O$ (naphthalene or o-xylene + air $\rightarrow$ phthalic anhydride)	Supported V_2O_5 (multitube fixed bed)
$n{-}C_4H_8$ or $C_6H_6 + O_2 \rightarrow C_4H_2O_3$ (butene or benzene + air $\rightarrow$ maleic anhydride)	Supported V_2O_5 (multitube fixed bed)

$C_2H_4 + \frac{1}{2}O_2 \rightarrow (CH_2)_2O$ (ethylene oxide)   Supported Ag

$CH_3OH + O_2 \rightarrow CH_2O + H_2$ and/or H_2O Ag or $Fe_2O_3 \cdot MoO_3$

$C_3H_6 + O_2 \rightarrow CH_2{=}CHCHO$ (acrolein) Cu_2O or multimetallic oxide compositions
and/or $CH_2 = CHCOOH$ (acrylic acid)

$C_3H_6 + NH_3 + \frac{3}{2}O_2 \rightarrow CH_2{=}CHCN + 3H_2O$ Complex metal molybdates or multimetallic oxide compositions (fluid bed)

Complete oxidation of CO, hydrocarbons, in   Pt or Pt $\cdot$ Pd, pellet or monolith support
pollution control, as of auto exhaust

$C_2H_4 + \frac{1}{2}O_2 + CH_3COOH \rightarrow$ Pd on acid-resistant support (vapor phase, multitube fixed bed)

$CH_3COOCH{=}CH_2$ (vinyl acetate)

$C_4H_8 + \frac{1}{2}O_2 \rightarrow C_4H_6 + H_2O$ Promoted ferrite spinels

Acid-catalyzed reactions
Catalytic cracking   Zeolite (molecular sieve) in $SiO_2 \cdot Al_2O_3$ matrix (fluid bed)

Hydrocracking   Metal (e.g., Pd) on zeolite (adiabatic fixed beds)

Isomerization   Metal on acidified Al_2O_3 (fixed bed), zeolites
Catalytic reforming   Pt, Pt $\cdot$ Re, or Pt $\cdot$ Ir on acidified Al_2O_3 (adiabatic, fixed or moving bed)

Polymerization H_3PO_4 on clay (fixed bed)
Hydration, e.g., propylene to isopropyl alcohol   Mineral acid or acid-type ion–exchange resin (fixed bed)

Reactions of synthesis gas
$CO + 2H_2 \rightarrow CH_3OH$ ZnO promoted with Cr_2O_3, or $Cu^1 \cdot ZnO$ promoted with Cr_2O_3 or Al_2O_3 (adiabatic, fixed beds or multitube fixed bed)

$CO + 3H_2 \rightarrow CH_4 + H_2O$ (methanation)   Supported Ni (fixed bed)
$CO + H_2 \rightarrow$ paraffins, etc. (Fischer-Tropsch synthesis)   Fe with promoters (fluid bed)

Other
Oxychlorination (e.g., $C_2H_4 + 2HCl + \frac{1}{2}O_2$ $CuCl_2/Al_2O_3$ with KCl promoter

$\rightarrow C_2H_4Cl_2 + H_2O$)
Hydrodesulfurization   Co $\cdot$ Mo/Al_2O_3 or Ni $\cdot$ —Mo/Al_2O_3 (adiabatic, fixed beds)

$SO_2 + 2H_2S \rightarrow 3S + 2H_2O$ (Claus process) Al_2O_3 (fixed beds)
$H_2O + CO \rightarrow CO_2 + H_2$ (water-gas shift) Fe_3O_4 promoted with Cr_2O_3 (adiabatic fixed bed); for a second, lower temperature stage, Cu $\cdot$ ZnO supported on Al_2O_3 or SiO_2

1-3 CATALYST PREPARATION

Most catalysts are either a finely divided metal in the form of crystallites supported on a carrier such as alumina or silica, or a metal oxide either on a carrier or unsupported. Metal sulfide catalysts are usually prepared first as the oxide and then treated with hydrogen sulfide or another sulfur compound to convert it to the sulfide. For the preparation of supported catalysts, particles of the supporting material are brought into contact with metal solutions in the form of soluble chlorides or nitrates. The resulting mixture is filtered, washed, dried, and then formed into a desired shape. The catalyst is then activated to convert it into its active form by heating to cause calcination or decomposition, followed by reduction if a metallic catalyst is desired. The preparation method is termed the "precipitation method" or the "impregnation method," depending on the manner in which the particles of support are brought into contact with the metal solution.

In the precipitation method, the mixing of two or more solutions or suspensions of material causes precipitation of the catalyst. For instance, a supported nickel catalyst could be prepared from nickel nitrate and a suspension of alumina by precipitation with ammonium hydroxide. Binders, cements, die lubricants, etc., may be added at this or a later stage. The final size and shape of the supported catalyst are determined by the forming process, which may also affect pore size and pore-size distribution. Large pores can be introduced into a catalyst by incorporating fine organic powders into the mixture, which can be burned out later.

The impregnation method is the easiest way of making a catalyst. While fine particles of the supporting material can be used in laboratory preparations, the usual industrial practice is to use the support in its final fabricated form, thus eliminating the filtering and forming steps. Two methods of contacting are used: dry impregnation (impregnation to incipient wetness) and immersion. Dry impregnation is more commonly used industrially since the composition of a batch of solution used for immersion will change as additional supports are impregnated, resulting in nonuniform activity from one pellet to another. In the dry impregnation method, the support is contacted, as by spraying, with a solution of appropriate concentration, corresponding in quantity to the total known pore volume or slightly less. Oxide supports such as alumina and silica are readily wet by aqueous solutions, as are most activated carbons since they have a layer of chemisorbed oxygen on them. Capillary forces ensure that liquid will be sucked up into the entire pore structure. If the support is not readily wetted as in a carbon that is highly graphitized or without chemisorbed oxygen, an organic solvent may be used or the support may be impregnated under vacuum.

The precipitation method generally provides a more uniform catalyst distribution within the pellet than the impregnation method. Some control over pore size and its distribution can be exercised. However, the final structure may be affected if two or more metal compounds are present that precipitate at different rates. The impregnation method is preferred for the preparation of noble metal catalysts since it is economically desirable to spread out the metal in as finely divided a form as possible. The noble metal is usually present at concentrations of 1 wt%

or less of the total. In contrast, the precipitation method is desirable when a high percentage of metal is to be incorporated, up to 20 to 40 wt%, onto the support used for base metals such as copper and nickel. It may be very difficult to obtain such high loadings by impregnation or even by multiple impregnations. While the usual dry impregnation method leads to a supported catalyst with higher catalyst loading toward the surface, use of the immersion method can lead to an essentially uniform deposit of adsorbed catalyst if sufficient time is allowed for the diffusion of reagent species into the interior of the pellet.

The drying process can affect the catalyst distribution within the support. The crystallite size of a supported metal catalyst may also be altered if a considerable portion of the soluble metal is occluded rather than adsorbed. Initially, evaporation occurs at the outer surface, but the liquid evaporated from small pores is replaced by liquid drawn from large pores by capillarity, possibly causing a nonuniform distribution of catalyst. In the precipitation method, the dried catalyst particles are formed into granules, spheres, tablets, and extrudates. In either method, the dried material is calcined to activate the catalyst.

In the calcination process, extraneous material such as binders as well as volatile and unstable anions and cations that have been previously introduced but are not desired in the final catalyst are eliminated. Before calcination, crystallites are usually in the form of metal salts. If a metallic catalyst is the ultimate goal, conversion to the oxide form is frequently sought prior to reduction to the metal. If a mixed oxide catalyst is desired, a substantially elevated firing temperature may be required to cause mixing by diffusion of individual species to form a desired compound or crystalline phase. In any event, the catalyst should be heated to a temperature at least as high as will be encountered in the plant reactor to preserve the integrity of the catalyst under reaction conditions. Otherwise, bound water or carbon dioxide in the catalyst may cause decompositions, leading to breakup and dusting of pellets, which in turn lead to excessive pressure drop and premature reactor shutdown. The choice of temperature, on the other hand, should not be so high that the catalyst forms stoichiometric compounds or a solid solution with the support.

The final form of supported metal catalysts is arrived at by reduction of the oxide at an elevated temperature by contact with flowing hydrogen or hydrogen diluted with nitrogen, the latter for safety reasons. Anderson (1975) has shown from thermodynamic calculations that the reduction of the metal oxide to the metal is highly favored in the region of 570 to 770K for all group VIII elements, whereas the oxide form is highly favored for chromium, vanadium, tantalum, titanium, and manganese. Sometimes reduction is carried out *in situ* in the plant reactor to avoid hazards associated with handling a pyrophoric material.

There are other ways of preparing catalysts. The iron catalyst used for ammonia synthesis is made by fusion of naturally occurring magnetite, Fe_3O_4, with small amounts of potassium carbonate, alumina, and other ingredients. The best known example of a metal catalyst prepared by leaching is Raney nickel, developed by Raney in 1925. The catalyst is prepared from a nickel-aluminum alloy by leaching out much of the aluminum with caustic solution to leave behind a porous nickel catalyst. Typically a 50:50 nickel-aluminum alloy is reacted with a 20% solution of sodium hydroxide. To achieve maximum activity and structural stability some aluminum must be left behind, and some hydrated alumina is also formed and retained. This may act as a physical promoter. Raney nickel is pyrophoric, so it must be handled carefully; it is usually stored under water or an organic solvent. The leaching method can also be applied to other metals alloyed with aluminum to make, e.g., Raney cobalt, Raney iron, etc., and the method may also be used to produce metal catalysts in unusual forms.

It is obvious from the foregoing discussion that many factors are involved in the preparation of a catalyst that undoubtedly affect its final performance. While some efforts are being made to quantify these various effects, catalyst preparation is still an art, and detailed arcane procedures need to be followed to achieve reproducibility and desired properties. The most overriding consideration in preparing a catalyst should be the chemical composition of the surface. The final form of the catalyst should also be stable to heat and to fluctuations in process conditions, and mechanically stable to attrition.

The desired catalyst particle size is determined by the process in which it is to be used. For fluidized-bed or slurry reactors, particles usually range from about 20 to 300 μm in size. In fluidized beds, the lower limit is set by the difficulty of preventing excessive carryover of fines through cyclone separators in the reactors; the upper limit is set by the poorer fluidization characteristics of larger particles. In slurry reactors, powders that are too coarse may be difficult to suspend; those too fine are difficult to remove by filtration. The catalysts used in fixed beds generally range from about 1.5 to 10 mm in diameter and have about the same length. Small pellets (1 ~ 2 mm) may cause excessive pressure drop through the bed; larger ones cause diffusion limitations.

The supports used for the catalyst are usually alumina, silica, or activated carbon. Because stability is needed at high temperatures, the form of alumina is either γ or η-alumina. These are usually the first choice for supports. However, active aluminas can dissolve or become soft and mushy under acidic conditions—conditions under which silica is stable. The relative nonreactivity of silica upon calcination with other catalyst ingredients may also be a significant factor in the choice of supports. Carbon supports are used primarily for noble metals and for reactions in which the strong adsorption of carbon for organic molecules may be an asset. An unusual support form known as a monolith or honeycomb is used in some automobile catalytic converters, where a very low pressure drop is required to minimize power loss from the engine. This is a single block of material containing within it an array of parallel, uniform, straight, nonconnecting channels.

Promoters are often used for supported catalysts. These can be grouped into physical and chemical promoters. A physical promoter is an inert substance that inhibits the sintering of crystallites of the catalyst by being present in the form of very fine particles. An example is the incorporation of a small amount of alumina in the conventional iron catalyst used for the synthesis of ammonia. To be effective, the physical promoter must generally be of considerably smaller particle size than that of the active, catalytic species, it must be well dispersed, and it must not react or form a solid solution with the active catalyst. A chemical promoter causes a chemical effect—it changes the chemical composition of the catalyst.

More details on catalyst preparation can be found in the book by Satterfield (1980), from which the bulk of this section is derived, and the books by Linsen (1970), Anderson (1975), and Moss (1976).

The ultimate goal for catalyst preparation is the capability to tailor a catalyst for any given application. This goal, however, is difficult to attain due to the complexities involved in catalyst preparation. Nevertheless, some progress has been made, which is considered in the following section.

1-4   A DESCRIPTION OF CATALYST IMPREGNATION

The two most important properties of prepared catalysts are the chemical state of the catalyst and its distribution within the support. Consistency is essential even if it does not give complete uniformity since otherwise the behavior of reactions and reactors becomes unpredictable. This is precisely the reason why arcane procedures are followed in preparing catalysts. The preparation step that can change the chemical state of a catalyst is calcination. While all steps involved in the preparation affect the catalyst distribution, impregnation and drying steps are mainly responsible for the final distribution.

The effect of impregnation conditions, and to a certain extent that of drying conditions, on catalyst distribution can be ascertained through an idealized model. To begin, uniform distribution before drying can always be assured by allowing sufficient contact time for the solute equilibration between the impregnating solution and the support surface. While this can easily be accomplished in a laboratory, such an equilibration is not always possible for industrial supported catalysts for which the dry impregnation method is typically used. Since the origin of nonuniformity is the distribution resulting from the impregnation step, attention here is focused on impregnation using the single pore model of Vincent and Merrill (1974).

The impregnation process can be considered to consist of two steps: the initial filling of pores when the impregnating solution is brought into contact with particles of the supporting material and equilibration that follows when the pores are filled. The filling step can be adequately described by a plug-flow model since the ratio of pore length to diameter is much larger than unity. For a cylindrical pore of radius R and length L (Figure 1.4), a solute balance can be written as:

$$\frac{\partial C}{\partial t} + v \frac{\partial C}{\partial z} + \frac{2}{R} \frac{\partial q}{\partial t} = 0 \qquad (1.1)$$

where C = solute concentration of solution in the pore, mol/volume
   v = velocity of penetrating solution
   q = concentration of solute adsorbed on support surface, mol/surface area

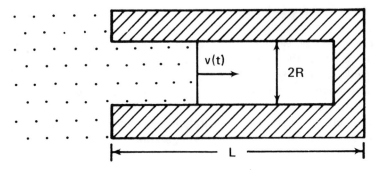

Figure 1.4   Single pore model. (Vincent and Merrill 1974; reprinted with permission from Journal of Catalysis. Copyright by Academic Press.)

The penetration velocity v is not constant and the correlation used by the above authors is:

$$v = \frac{R}{4}\left(\frac{\gamma}{2R\mu}\right)^{1/2} t^{-1/2} \tag{1.2}$$

where γ = surface tension
    μ = viscosity

The physical properties of water may be assumed for the impregnating solution. Typical nonuniform catalyst distributions (Chen and Anderson 1973) shown in Figure 1.5 suggest that there definitely exists a resistance to solute removal from the solution since otherwise the profiles should show stepwise distributions if enough solutes are present. Solute removal from the impregnating solution onto the pore wall can be modeled as the sequential events of mass transfer at the liquid-solid interface followed by adsorption onto the pore wall. In many cases, the adsorption/desorption step is the controlling step. The solute balance for the solid phase in such cases is:

$$\frac{\partial q}{\partial t} = k_a C(1 - \theta) - k_d\theta \tag{1.3}$$

where k_a and k_d are the adsorption and desorption rate constants and q is the solute concentration on the pore wall in mol/surface area. At equilibrium, the coverage θ is given by the Langmuir isotherm:

$$\theta = \frac{q}{q_s} = \frac{kC}{1 + kC} \tag{1.4}$$

where q_s is the saturation concentration in mol/surface area, and $k = k_a/k_d$. The balance equations are rendered dimensionless to give:

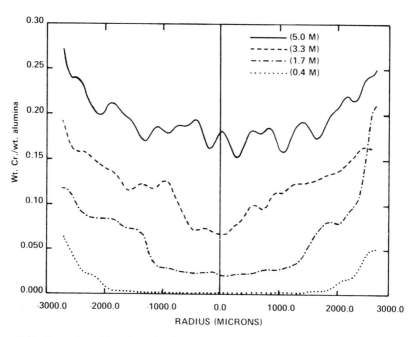

Figure 1.5 Typical profiles of catalyst loading: dry impregnation method. (Chen and Anderson 1973; reprinted with permission from *Industrial and Engineering Chemistry, Product Research and Development.* Copyright by American Chemical Society.)

$$\frac{\partial y}{\partial \tau} + \frac{1}{2\tau^{1/2}} \frac{\partial y}{\partial x} + \eta \frac{\partial \theta}{\partial \tau} = 0 \tag{1.5}$$

$$\frac{\partial \theta}{\partial \tau} = \frac{K}{\eta} [y(1-\theta) - \theta/K_L] \tag{1.6}$$

where $y = C/C_0$, $\tau = t/t_L$, $x = z/L$, $\eta = 2q_s/RC_0$
$K = 2k_a t_L/R$, $K_L = kC_0$, $t_L = 8L^2\mu/R\gamma$
$C_0 =$ concentration of bulk solution
$t_L =$ time required to fill the pore

The concentration as well as the coverage is zero until the front of the penetrating solution reaches a point of interest. Therefore:

$$y(0,\tau) = 1 \qquad y(x,0) = 0 \tag{1.7a}$$

$$\theta(x,0) = 0 \tag{1.7b}$$

The method of characteristics (Forsythe and Wason 1960) can be used in a straightforward manner to solve this system of equations (see Problem 1.10). Typical profiles of coverage at the time pores are just filled ($\tau = 1$) are shown in Figure 1.6 (Vincent and Merrill 1974). The equilibrium parameter K_L determines the

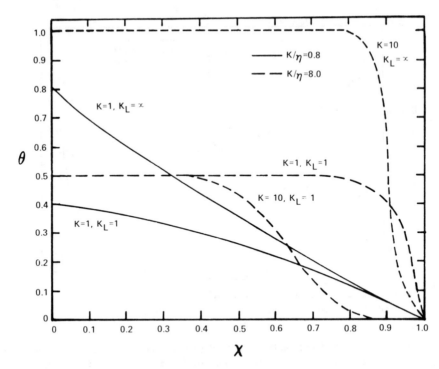

Figure 1.6   Profiles of coverage θ as determined by the single pore model at $\tau = 1$. (Vincent and Merrill 1974; reprinted with permission from Journal of Catalysis. Copyright by Academic Press.)

final steady state level of coverage, and K/η determines the rate of approach to the steady state as evident from Eq. 1.6. It is seen from Figure 1.6 that the steady state coverage is reached at the pore mouth when K/η is relatively large ($K/\eta = .8$). A lower value of η extends the plateau shown in the figure further to the pore center.

Once the pore is filled, the main mechanism for solute movement is that of diffusion. Therefore, the solute equilibration between stagnant solution and the pore wall will be described by:

$$\frac{\partial y}{\partial \tau} - \text{Pe}\,\frac{\partial^2 y}{\partial x^2} + \eta\,\frac{\partial \theta}{\partial \tau} = 0 \qquad (1.8)$$
$$\tau \geq 1$$

$$\frac{\partial \theta}{\partial \tau} = \frac{K}{\eta}(y(1 - \theta) - \theta/K_L) \qquad (1.9)$$

where a Peclet number Pe is given by:

$$\text{Pe} = \frac{Dt_L}{L^2} \qquad (1.10)$$

Here, the diffusivity D is that of solute in solution. Initial and boundary conditions are the profiles of y and θ at $\tau = 1$ obtained from the solutions of Eqs. 1.5 and 1.6. Depending on the impregnation method, one of the boundary conditions takes the following forms:

$$y(0,\tau) = 1 \qquad \text{for the immersion method} \qquad (1.11a)$$

$$\left.\frac{\partial y}{\partial x}\right|_{x=0} = 0 \qquad \text{for the dry impregnation method} \qquad (1.11b)$$

An assumption made on the immersion method has been that the amount of solution used is large enough for the bulk solution concentration to remain constant throughout impregnation. The symmetry around the pore center provides the other boundary condition:

$$\left.\frac{\partial y}{\partial x}\right|_{x=1} = 0 \qquad (1.12)$$

An estimate on the time required for complete equilibration can be obtained by solving:

$$\frac{\partial y}{\partial \tau} - \text{Pe}\,\frac{\partial^2 y}{\partial x^2} = 0 \qquad (1.13)$$

since this corresponds to an equilibration process in which no solute is removed at the pore wall. For the purpose of obtaining the equilibration time, one may envision a pore filled with water being brought into contact with an impregnating solution. This system reaches its steady state (Churchill 1944) exponentially with the factor of $\exp(-\pi^2\text{Pe}\tau)$. When the argument of the exponential is larger than 5, deviation from the steady state is less than 1%. Therefore, the equilibration time may be set as:

$$\tau_{eq} = 5/\pi^2\text{Pe}$$

In order to obtain the final catalyst distribution before calcination, the drying process has to be described with the profiles at the end of impregnation as initial conditions. This drying process is much more complicated than the impregnation process and no satisfactory results are available. For one thing, the single pore model is inadequate. A model pore cannot be treated independently of other pores since the liquid in larger pores is drawn into smaller pores as evaporation proceeds. As the evaporation front moves down a pore, concentrations reach saturation, and diffusion of solute may occur even after complete equilibration. It has been observed (Komiyama et al. 1980), however, that relatively fast drying does not alter significantly the distribution resulting from the equilibration. On the other hand, relatively slow drying results in segregation of the catalyst toward the external surface of the particle.

The parameters defining the catalyst concentration profiles can be obtained from equilibration runs. Let the concentration of solute remaining after complete equilibration and drying be q_m. This quantity, which can be measured, is a combination of adsorbed and occluded solutes, i.e.:

$$\frac{2}{R} q_m = C_0 + \frac{2}{R} q \tag{1.14}$$

Therefore a table of C_0 versus q_m can be obtained from equilibration runs in which the bulk concentration of the impregnating solution is varied for each run. For the equilibration runs, $C = C_0$ and the parameters k and q_s (or K_L and η) can be determined from Eq. 1.4 written in the following form:

$$\frac{1}{q} = \left(\frac{1}{q_s k}\right) \frac{1}{C_0} + \frac{1}{q_s} \tag{1.15}$$

where q is determined from the table of C_0 versus q_m. The coverage on the external surface can be obtained by solving Eq. 1.6 with the condition that y is unity there (see Problem 1.1):

$$\theta_e = \frac{1}{1 + 1/K_L} \left\{ 1 - \exp\left[-\frac{K}{\eta}\left(1 + \frac{1}{K_L}\right)\tau \right] \right\} \tag{1.16}$$

This result is valid for both pore-filling and equilibration periods. The pore radius R is usually obtained by setting the ratio of the pore volume to BET surface area (see Section 1–6) equal to $R/2$.

The single pore model is equivalent to the model for a slab-like particle. In fact, the conservation equation for the liquid phase (Eq. 1.5) can be written in a general form applicable to slab-like, cylindrical, and spherical particles (Lee 1984). The catalyst distribution at the end of the pore-filling period (Z), which is the approximate distribution that results after fast drying, is given by:

$$Z = y + \eta\theta = 1 + \frac{\eta}{1 + 1/K_L} \left\{ 1 - \exp\left[-\frac{K}{\eta}\left(1 + \frac{1}{K_L}\right)(1 - x)^{2(s+1)} \right] \right\} \tag{1.17}$$

where s is 0, 1, and 2 for slab-like, cylindrical, and spherical particles. The catalyst deposited onto the pore wall after drying (Z) consists of occluded (y) and adsorbed ($\eta\theta$) solutes.

1–5 CHARACTERIZATION OF CATALYSTS

It is undoubtedly true that there is no unique way of characterizing a catalyst. There are, however, some basic pieces of information required to determine how effective a catalyst is for a given reaction and how effectively the catalyst is used.

The latter requires knowing the exposed surface area of the catalyst and the former requires knowing the rate of reaction per unit exposed surface area. Since only the exposed catalyst surface is involved in reactions and the metals and metal compounds used as catalysts are usually expensive, the effective utilization of a catalyst requires a good dispersion of the catalyst on a support. For a given catalyst, the exposed surface area is a measure of the number of active sites responsible for the reaction. Therefore, to a first-order approximation, this information can be used to determine the effectiveness of a particular method of catalyst preparation and ultimately to choose the most effective preparation method. The ratio of the number of surface metal atoms to the total number of metal atoms present is termed *dispersion* (percentage exposed). While this term is usually used for metal catalysts, the same definition can be used for catalysts formed of metal compounds. The dispersion of supported platinum catalyst, for instance, can readily exceed 0.5, meaning that more than half the platinum atoms dispersed on the support are surface metal atoms. In contrast, a platinum particle of 1 μm in size yields a dispersion of only about 0.001. *Specific activity* is defined as the number of molecules that react per unit exposed catalyst area per unit time. For metal catalysts, this is frequently referred to as the *turnover number,* or turnover frequency when the basis is on the number of sites rather than the area. The specific activity may be used in place of the turnover number in view of the fact that the number of sites is an elusive quantity to determine, particularly for the catalysts consisting of metal compounds. The specific activity of a catalyst is a basic measure of true catalytic activity: this activity can be used as the basis for choosing a particular catalyst. Here again, knowledge of the exposed surface area is required for the determination of the specific activity.

The catalytic activity inferred from the specific activity is often correlated to certain quantities to gain some understanding of heterogeneous catalytic reactions. As the exposed surface area is required for characterizing the specific activity, methods of determining the surface area will be treated first. This will be followed by correlations developed for characterizing the catalytic activity.

Dispersion and Specific Activity

Chemisorption is the usual method for the measurement of exposed surface area of a metal catalyst. For a catalyst formed of metal compounds, however, the chemisorption method fails, usually yielding a fraction of the total catalyst surface area; only rudimentary methods such as specific poisoning have to be tailored for specific application. In the case of acid catalysts, a measure of the number of acid sites rather than the surface area is determined via titration techniques.

A plot of the specific activity versus the dispersion is often used as a way of characterizing catalytic reactions, as shown in Figure 1.7. The reactions represented by the solid line are termed *structure-insensitive (facile)* reactions. The specific activity of these reactions is independent of the surface area and is thus structure-insensitive. The reactions following the behavior of the dotted lines are

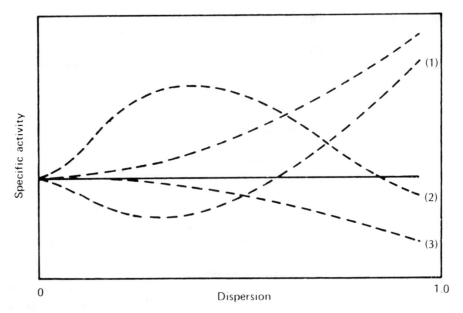

Figure 1.7 Specific activity versus dispersion. (After J.J. Carberry, *Chemical and Catalytic Reaction Engineering*, © 1976; with permission of McGraw-Hill Book Company, New York.)

termed *structure-sensitive* (*demanding*) reactions. The specific activity does depend on the surface area, or more precisely on the size and shape of the crystallites of the catalyst, and therefore is structure-sensitive. A few comments are in order regarding the figure. For one, the rate of most reactions will increase with increasing dispersion, although the possible exceptions are shown in curves (1) through (3), since the actual rate is the specific activity multiplied by surface area. For instance, the rate of structure-insensitive reactions increases linearly with dispersion. For another, the behavior of the specific activity at and close to a dispersion of unity should be viewed with caution. When the dispersion approaches unity, all the metal is present in the form of atoms that are extremely reactive. These metal atoms readily react with gaseous molecules to form, for instance, platinum oxide in the presence of oxygen. The catalytic activity observed in such cases is that of platinum oxide rather than platinum itself. Studies have shown that simple hydrogenation reactions on various metals are usually structure-insensitive but that reactions in which C—C bonds are broken, such as hydrogenolysis and skeletal isomerization, are structure-sensitive. A typical example of a structure-sensitive reaction involving the hydrogenolysis of ethane is given in Table 1.3.

A gas can adsorb on a solid surface either physically (physisorption) or chemically (chemisorption), the latter involving actual bonding at the surface between the gas molecule and the metal as explained in Section 1.2. While the magnitude of the heat of adsorption (defined below) is often used to differentiate physisorption from chemisorption, the most useful way of distinguishing them is the way in which gas molecules cover the surface: only a monolayer of adsorbed gas molecules

Table 1.3 Ethane Hydrogenolysis over Rhodium (Yates and Sinfelt 1967)

State of Dispersion	Crystallite Size ($\overset{\circ}{A}$)	Specific Activity (millimoles ethane converted/hr m² of catalyst)
Very low:		
Bulk Rh	2560	0.79
5% Rh on SiO₂ (sintered)	127	0.41
Intermediate:		
1–10% Rh on SiO₂	12 ~ 41	8 ~ 16
Very high:		
0.1 ~ 0.3% Rh on SiO₂	< 12	4.4

(Reprinted with permission from Journal of Catalysis. Copyright by Academic Press.)

is involved in chemisorption, whereas multilayers are observed in physisorption. Differences between physisorption and chemisorption are summarized in Table 1.4.

The first attempt to quantify adsorption was made by Langmuir based on the assumption of a uniform, energetically homogeneous solid surface. The postulations are that the rate of gas adsorption is proportional to the frequency of gas molecules striking vacant surface sites and that the rate of desorption is proportional to the fraction of surface sites covered by gas molecules. This leads to the expressions:

$$r_{ads} = k_a p(1 - \theta) \tag{1.18}$$

Table 1.4 Physisorption versus Chemisorption (Carberry 1976)[a]

	Physisorption	Chemisorption
Adsorbent	All solids	Some solids
Adsorbate	All gases below critical point	Some chemically reactive solids
Temperature range	Low temperature	Generally high temperature
Heat of adsorption	Low, ~ ΔH liquefaction (< 15 kcal/mol)	High, of the order of chemical reaction (> 15 kcal/mol)
Rate and activation energy	Very rapid, low energy	Nonactivated, low energy; activated, high energy
Coverage	Multilayer	Monolayer and less
Reversibility	Highly reversible	Often irreversible
Use	Surface area, pore size	Active surface area, catalysis, kinetics

[a] P.H. Emmett, Johns Hopkins University (1959).

$$r_{des} = k_d \theta \tag{1.19}$$

where k_a, k_d = adsorption and desorption rate constants, respectively
θ = fraction of surface sites covered by gas molecules, coverage
p = gas partial pressure
$K = k_a/k_d$, adsorption equilibrium constant

At equilibrium, the two rates are equal and therefore the coverage θ can be expressed as:

$$\theta = \frac{Kp}{1 + Kp} \tag{1.20}$$

The temperature dependence of the adsorption equilibrium constant K is expressed as:

$$K = K_0 \exp(Q/R_g T) \tag{1.21}$$

where K_0 is a preexponential factor, and Q is termed the heat of adsorption. According to the energy diagram shown in Figure 1.8, the heat of adsorption is equal to:

$$Q = E_d - E_a \tag{1.22}$$

where E_d and E_a are activation energies for desorption and adsorption, respectively. For gases used for typical chemisorption experiments on transition metals such as H_2, O_2, CO, and CO_2, the activation energy for adsorption is small and the

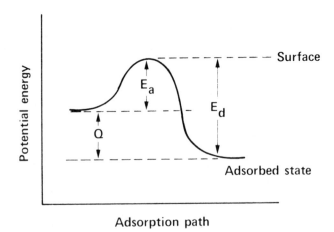

Figure 1.8   Energy diagram for gas adsorption.

activation energy for desorption is often taken as the heat of adsorption. Since adsorption is an exothermic process, the heat of adsorption is positive. Therefore, desorption is favored as the temperature is increased, as evident from Eq. 1.21. Real solid surfaces are not energetically homogeneous as foreseen by Langmuir. He treated such heterogeneous surfaces as a composite of a number of homogeneous surfaces at different energy levels:

$$\theta = \sum_{i=1}^{n} \frac{K_i p}{1 + K_i p} \, n_i \tag{1.23}$$

where K_i's are equilibrium constants for different surfaces, and n_i is the fraction of the total surface occupied by the ith surface. Various models for real surfaces have been proposed since then for the adsorption isotherms and these are summarized in Table 1.5. The BET equation for physisorption will be treated in detail in the section to follow. The difference between the Langmuir model and the other real surface models lies in the dependence of the heat of adsorption on coverage. It has been known for real surfaces that the heat of adsorption decreases with increasing coverage. Assuming a linear dependence of the heat of adsorption on coverage leads to the Tempkin and BLK isotherms. The Freundlich isotherm results when the heat of adsorption is assumed to decrease exponentially with coverage; the heat of adsorption is constant for the Langmuir model.

The fact that chemisorption leads to monolayer coverage is utilized in measuring the exposed surface area of metal catalysts, as was first demonstrated by Spenadel and Boudart (1960) for a platinum catalyst, although the origin of the chemisorption

Table 1.5 Adsorption Isotherms (Thomas and Thomas 1967)

Name	Isotherm Equations	Applications
Langmuir	$\dfrac{v}{v_m} = \theta = \dfrac{kp}{1 + Kp}$	Chemisorption and physisorption
Freundlich	$\theta = Kp^{1/n}$	Chemisorption (real surfaces), physisorption
Tempkin	$\theta = \dfrac{1}{f} \ln \bar{K}p$	Chemisorption (real surfaces), physisorption
Brunauer-Love-Keenan (BLK)	$\theta = \dfrac{RT}{\alpha} \ln \dfrac{1 + a_0 p}{1 + a_0 p \, \exp(-\alpha/RT)}$ $a_0 = A \, \exp(Q_0/RT)$ $Q = Q_0 - \alpha\theta$	Real surfaces
Brunauer-Emmett-Teller (BET)	$\dfrac{p}{v(p_0 - p)} = \dfrac{1}{v_m C} + \dfrac{C-1}{v_m C} \dfrac{p}{p_0}$	Physisorption

f, α, A, C are constants

With permission from *Introduction to the Principles of Heterogeneous Catalysis*. Copyright by Academic Press Inc. (London Ltd.)

work can be traced to Brunauer, Emmett and Teller. A proper choice of gas that chemisorbs on the metal catalyst and the knowledge of stoichiometry of adsorption are essential for the measurement. Gases that chemisorb on various metals are summarized in Table 1.6 along with the relative strength of chemisorption.

In order to arrive at the exposed metal surface area, the volume of the chemisorbed gas is measured, from which the number of adsorbed molecules can be calculated. Given the chemisorption stoichiometry (the number of surface atoms covered for each molecule of gas adsorbed) and the surface area occupied per metal atom, the number of molecules adsorbed can be converted to the exposed surface area of metal. For hydrogen, the stoichiometric number is almost always 2 since the hydrogen molecule usually dissociates upon adsorption and each hydrogen atom is adsorbed on one metal atom. For carbon monoxide, the number is either 1 or 2, depending on whether it adsorbs in a linear form in which it covers one metal atom, or in a bridged form covering two metal atoms. The number of surface atoms per unit area of metal varies slightly with the crystallographic plane, but for all metals it is about 10^{19} atoms per square meter, e.g., 1.5 to 1.6×10^{19} for Fe, Co, and Ni; 1.25 to 1.33×10^{19} for Pt, Pd, Ir, and Rh; and 1.15×10^{19} for Ag (Anderson, 1975). The gas most appropriate for chemisorption is hydrogen, but oxygen and carbon monoxide are also useful. The major problems associated with chemisorption measurements are the preparation of uncontaminated metal surfaces, determination of the presence or absence of adsorption on the support, and determination of the proper stoichiometric number. For instance, the stoichiometric number may be affected by the crystallite size, which may change during the preparation stage of cleaning the metal surface. Clean surfaces can be prepared by reduction with hydrogen at temperatures exceeding about 400°C for several hours, possibly preceded by an oxidation cycle, and then outgassing under high vacuum for several additional hours. The optimum temperature and pressure will vary with the system and must be established experimentally. Studies must also be made with the support by itself to establish its possible contribution to the overall adsorption. High-area supports contribute excessively to hydrogen adsorption at 77 to 90K. Preferred conditions for dispersed platinum are about 273 to 300K and about 0.01 to 0.03 kPa, and for dispersed nickel, about 273 to 300K

Table 1.6 Chemisorption of Gases on Metals (after G.C. Bond 1974)

Group	Metals	O_2 >	C_2H_2 >	C_2H_4 >	CO >	H_2 >	CO_2 >	N_2
A	Ti, Mo, Fe	+	+	+	+	+	+	+
B_1	Ni, Co	+	+	+	+	+	+	−
B_2	Rh, Pd, Pt, Ir	+	+	+	+	+	−	−
B_3	Mn, Cu	+	+	+	+	±	−	−
C	Al, Au	+	+	+	+	−	−	−
D	Li, Na, K	+	+	−	−	−	−	−
E	Mg, Ag, Zn	+	−	−	−	−	−	−

+: strong chemisorption   −: unobservable   ±: weak chemisorption

Table 1.7 Comparison of Methods (Freel 1972)

Chemisorption (H/Pt)	Electron Microscopy	
	Average Crystallite Size (nm)	Pt Surface/Pt Bulk
0.98	1.4	0.73
0.36	3.2	0.32
0.24	4.0	0.25
0.19	8.5	0.13

Reprinted with permission from Journal of Catalysis. Copyright by Academic Press.

and 20 kPa (Satterfield 1980). A recent review by Farrauto (1974) concludes that, with the exceptions of Pt and Ni using H_2 and of Pd using CO, chemisorption results should be used with caution for the determination of the surface area. It should be recognized, however, that a reliable determination of the relative surface area of a metal catalyst can always be made given proper care, even if this measurement does not determine the absolute surface area. A chemisorption measurement should be complemented by the X-ray line broadening method and electron microscopy. An example of a comparison of two of the three methods is given in Table 1.7 in terms of dispersion for a supported platinum catalyst with 2 wt% loading.

The X-ray line broadening method utilizes the Scherrer equation:

$$B = \frac{k\lambda}{r \cos \theta} \tag{1.24}$$

where B = net breadth of an X-ray diffraction line
r = crystallite size
λ = wavelength of X-ray applied
θ = Bragg angle
k = a constant, usually equal to unity

As the equation indicates, the breadth of an X-ray diffraction line is inversely proportional to the crystallite size and therefore the line gets increasingly sharper as the crystallite size increases. The line becomes too diffuse to be useful for sizes smaller than approximately 5 nm, whereas it becomes too sharp for sizes larger than 50 nm. The useful range is 5–50 nm. The method of electron microscopy is treated in some detail in Chapter 6.

A description (Farrauto 1974) of apparatus and procedures used for chemisorption is given below.

Static Vacuum System. Usually made of glass and equipped with appropriate evacuation devices, such as a roughing and diffusion pump, the static vacuum system is the most commonly used apparatus for measuring gas adsorptions. The general principle involves measuring the amount of gas remaining in the manifold system after contact with a sample. By knowing the amount of gas initially present and subtracting from it the amount remaining after equilibration with a sample, the extent of adsorption can be determined. Most often a pressure device, such as a mercury manometer or commercial mechanical device, is used to follow pressure changes caused by gas-sample interactions. Pressure changes in a constant volume system would then be proportional, through the gas laws, to adsorption (or desorption) of a given quantity of gas. It is assumed that the manifold and sample container volumes have been accurately determined.

Prior to gas adsorption, it is common practice to pretreat or condition the catalyst surface. Frequently high temperatures, about 500°C, are employed. Therefore, a sample furnace is an essential part of the apparatus. After pretreatment, evacuation at pretreatment temperature, and cooling to adsorption temperature, it is necessary to determine the dead space volume (the volume in the sample tube which the adsorbate would occupy provided no adsorption occurred). Helium is most often used for this purpose. After evacuation, the adsorbate is added to the manifold and its pressure noted. Subsequently, it is expanded into the sample chamber, and adsorption, if any, commences. The pressure is monitored until no further variation with time is noted. The pressure over the sample can then be increased via a gas burette and readings again taken until equilibrium is established. When there is no longer gas uptake by the sample with increasing pressure, the desirable portion of the isotherm is complete, and the total volume adsorbed, expressed at S.T.P. per gram, can be determined. This procedure must be repeated for the support. The volume adsorbed at any pressure is subtracted from the volume adsorbed on the supported catalyst at the same pressure. Further details can be found in the books by Hayward and Trapnell (1964) and by Anderson (1968).

Continuous Flow Apparatus. A stream of gas composed of adsorbate and inert carrier is passed through a previously outgassed catalyst, usually purged with an inert gas at a high temperature, and the concentration of adsorbate monitored frequently via a thermal conductivity cell. Assuming the latter is used for detection, conditions are adjusted so that no gas will adsorb (accomplished by selecting a suitable temperature or by allowing for sample bypass) and the bridge balanced. When conditions are such that gas will adsorb, the detection system responds to a change in the composition of the gas mixture, and a response occurs proportional to gas adsorbed. One may also adjust the conditions to monitor the desorption of adsorbate from the catalyst. Further details can be found in the article by Eberly (1961).

The major advantage of the method is rapid determinations. However, diffusional limitations and slow significant chemisorption processes may not be readily detected. The adsorption portion of the measurement may also include some physical adsorption. However, this can be minimized by proper temperature and partial pressure adjustments. The apparatus is suited for physical adsorption (that is, BET) as well as chemisorption.

Methods for determining the exposed surface area of oxide catalysts are much more rudimentary than those for metal catalysts and acidic catalysts. Specific poisoning is suggested by Knözinger (1976) for measuring the surface area. Gandhi and Shelef (1973) suggested the use of the Freundlich isotherm for the determination of the surface area of copper oxide using NO as the adsorbate. Recently, a method applicable to any catalyst, including metal compounds, has been developed (Miller and Lee 1984) for the surface area measurement based on selective physisorption.

The quantity typically used in place of the exposed surface area for acidic catalysts is the number of acid sites, or equivalently, the acid amount in mol/g. For the determination of the acid amount (Satterfield 1980), a sample of the solid acid as a powder is suspended in an inert nonaqueous liquid, e.g., benzene, and is titrated with a base, utilizing an indicator. The titrating base must be a stronger base than the indicator, and n-butylamine, $pK \sim +10$, is often used for this purpose. As the base is added it adsorbs on acid sites, the strongest ones first, and ultimately it displaces indicator molecules from the solid. When the indicator has been substantially replaced, the color changes. At this equivalence point $[B]/[BH^+] \approx 1$, where $[B]$ and $[BH^+]$ are the concentrations of the neutral base and its conjugated acid. If the pK_a ($= -\log K_a$) value of the indicator is, say $+3.3$, then the amount of base added is equivalent to the amount of acid sites having the Hammet acidity function $H_0 \leq +3.3$, where:

$$H_0 = pK_a + \log \frac{[B]}{[BH^+]} \qquad (1.25)$$

and where K_a is the equilibrium constant of dissociation of the acid. By amine titration with indicators having different pK_a values, the amount of acid sites having strengths exceeding the various corresponding values of pK_a ($= H_0$) may be determined. This method gives the sum of Brønsted and Lewis acid sites.

The titration method is not suitable for molecular-sieve zeolite catalysts in which the pore size is so small that the indicator molecules cannot penetrate into the interior. Some other disadvantages of the method are that water must be rigorously excluded, since water may react with the surface and alter the acidic character of the solid; the time required for equilibrium may be long, amounting sometimes to days; and measurements are made under conditions much different than those occurring during a reaction.

Another method is to measure the amount of a gaseous base, such as ammonia, pyridine, or quinoline, adsorbed at elevated temperatures under a specified set of conditions. This has the advantage of allowing the study of a catalyst under conditions more nearly similar to those of reaction. Ammonia has been used extensively. Catalysts may be compared in terms of the amount of ammonia adsorbed as a function of the temperature over a range such as 150 to 500°C. A minimum temperature of about 150°C is necessary to eliminate physical adsorption. By infrared spectra it is possible to distinguish between Brønsted and Lewis acid sites. More details on characterizing the acidity of solid catalysts are described in reviews by Formi (1974), and by Bensei and Winquist (1978).

Catalytic Activity and Correlations

The catalytic activity, expressed in terms of the specific activity, is often correlated to the heat of adsorption for metal catalysts. The heat of adsorption is a measure of the stability of the bonding between the adsorbed gas and the surface metal atom. As pointed out in Section 1.2, the stability of this bonding should not be

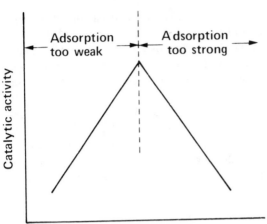

Figure 1.9 Effect of heat of adsorption on catalytic activity: the volcano curve. (After G.C. Bond 1974).

too high or too low for the catalyst to be effective. This effect is known as the *Sabatier's* principle, or as the volcano curve, shown in Figure 1.9. A specific example (Bond 1974) for the hydrogenation of ethylene is shown in Figure 1.10. Here, the rates of hydrogenation relative to rhodium are given in terms of the initial heat of adsorption of hydrogen (ΔH_0). Because of the correlations that exist between the initial heat of adsorption and the periodic group number, or between the initial

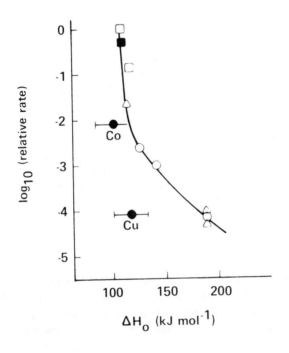

Figure 1.10 An example of the volcano curve for the hydrogenation of ethylene. Open points, evaporated metal films: filled points, silica-supported metals: circles, first row transition metals: squares, second row transition metals: triangles, third row transition metals. (After G.C. Bond 1974).

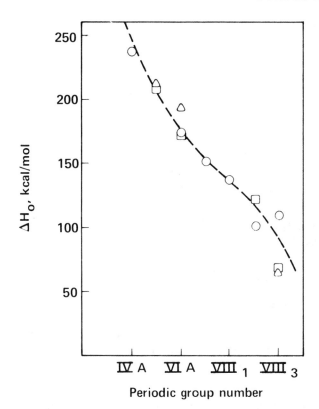

Figure 1.11 Variation of the initial heat of adsorption of oxygen on evaporated films: circles are for the first transition series, squares the second transition series, and triangles the third transition series. (After G.C. Bond 1974.)

Periodic group number

heat of adsorption and the heat of formation of the highest metal oxide per atom of metal, the abscissa can also be the periodic group number or the heat of formation. Some typical examples are given in Figures 1.11 (Bond 1974) and 1.12 (Tanaka and Tamaru 1963).

It is more difficult to find useful correlations between catalytic activity and chemical or physical properties for metal oxides than for metals. One of the problems is the difficulty of obtaining the exposed surface area and thus the specific activity. The chemisorption of gases is more complex on oxides than on metals. To start with, the adsorbed molecule may be attached either to a cation or to an oxide ion. Further, it has long been known that the adsorption of reducing gases such as hydrogen and carbon monoxide is substantially irreversible in the sense that, when desorption is attempted, only water and carbon dioxide, respectively, can be obtained. The process has clearly led to reduction of the surface, and it may be guessed that the adsorbed species resided or were associated at least in part with oxide ions. It is also difficult to find a suitable chemical or physical property to correlate with the catalytic activity. Many of the transition metals exhibit variable valence and in such cases several oxides are possible, each with its distinctive structure, stability, and catalytic ability. Some of the cations at the surface may be in a higher oxidation state than those in the bulk. It is, however, possible to

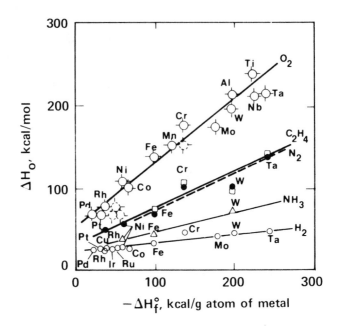

Figure 1.12 Correlations between the heat of adsorption and the heat of formation of the highest metal oxide per atom of metal. (Tanaka and Tamaru 1963; reprinted with permission from Journal of Catalysis. Copyright by Academic Press.)

proceed a little. If the breaking of the bond joining the relevant oxygen species to the surface is the rate-determining step in oxidation reactions, for which oxide catalysts are usually used, an inverse correlation should be found between the catalytic rate and the oxygen-catalyst bond strength. This quantity has been obtained either from the temperature dependence of the oxygen dissociation pressure or from the heat of formation of the oxide per oxygen atom. While reasonable correlations have been found for the oxidation of hydrogen and of many hydrocarbons, they are still only approximate and based on too few examples.

For acidic catalysts, the catalytic activity expressed in terms of rate constant or percent conversion is usually correlated to the acid amount. Typical examples are given in Figures 1.13 (Johnson 1955) and 1.14 (Ward and Hansford 1969). The Brønsted acid groups shown in Figure 1.14 were obtained from infrared peak intensity.

As evident in the preceding discussions on the metal and metal oxide catalysts, the heat of adsorption plays a key role in characterizing a catalyst. The heat of adsorption can also be used to gain an understanding of the nature of catalytic reactions. Three different experimental methods can be used for the determination of heats of adsorption: adsorption experiments (isotherms), the calorimetric method, and temperature-programmed desorption (TPD). In the first method, the adsorption isotherms obtained at different temperatures are used. The values of $d \ln P/dT$ at constant v (volume of gas adsorbed) are calculated as a function of v through

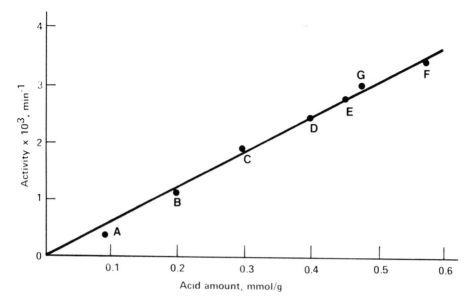

Figure 1.13 Propylene polymerization activity versus acid amount for a series of $SiO_2 \cdot Al_2O_3$ catalysts (Johnson 1955):

Catalyst	A	B	C	D	E	F	G
Al_2O_3 (wt%)	0.12	0.32	1.04	2.05	3.56	10.3	25.1

numerical differentiation of the curve obtained by cross plotting. These values are used in the Clausius-Clapeyron equation:

$$\left(\frac{\partial \ln P}{\partial T}\right)_v = \frac{Q}{R_g T^2}$$

to obtain the (differential) heat of adsorption. The value of Q thus obtained is called the isosteric heat of adsorption. The calorimetric method, in which the heat liberated upon the adsorption of gas on a cooled surface is actually measured, gives the average value of the heat of adsorption over the range of surface coverage studied. Differential heats of adsorption can also be determined calorimetrically by admitting small quantities of vapor at a time or by differentiating integral data. Because heats of adsorption frequently decrease with increasing surface coverage, and not always in the same way, it is convenient to use the heat extrapolated to zero surface coverage, the so-called initial heat of adsorption. As indicated earlier, this initial heat of adsorption is usually used for correlations.

Temperature-programmed desorption (TPD) (Cvetanovic and Amenomiya 1972) yields not only the heat of adsorption but also information on groups of different active sites, which are useful in characterizing reactions. While the flash filament method in which an adsorbate is desorbed from a rapidly heated filament in an ultra-high vacuum environment has been used for a long time, the usefulness of the TPD method lies in the fact that the adsorbed gas is desorbed in a programmed

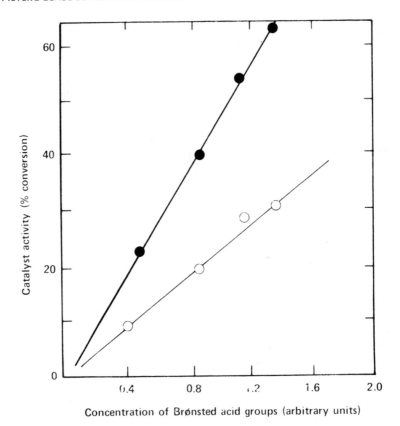

Figure 1.14   Activity for o-xylene isomerization of $SiO_2 \cdot Al_2O_3$ catalysts of various Al_2O_3 contents: dark circles are for samples heated to 500°C and open circles are for those heated to 425°C. (Ward and Hansford 1969; reprinted with permission from Journal of Catalysis. Copyright by Academic Press.)

manner from a conventional catalyst under conditions approaching more closely those prevailing during reaction. The bed of catalyst particles in the sample cell is quite shallow in this method, and in essence quite similar to the arrangement of the differential reactor typically used for kinetic experiments for heterogeneous reactions. The particles are presorbed with a suitable gas that chemisorbs on the solid adsorbent and are then subjected to thermal desorption. The carrier (typically helium), which elutes the desorbed gas, is passed through a conductivity cell for concentration measurements.

In the absence of any appreciable diffusional effects, the mass balance for the shallow bed (differential bed) is that for a continuously-stirred tank reactor:

$$\frac{d}{dt}(\theta v_m V_s) = F(C_{in} - C)$$

$$= - FC$$

(1.26)

where θ = fractional coverage of surface
V_s = volume of solid particles
F = carrier gas flow rate
C = concentration of desorbed gas in the exit carrier stream
v_m = amount of chemisorbed gas when θ is unity per unit volume of solid particles
C_{in} = zero since no adsorbate is fed to the bed at the start of thermal desorption

The heating program is such that the temperature increases linearly with time:

$$T = T_0 + \beta t \tag{1.27}$$

where β is a constant. Combining Eqs. 1.26 and 1.27 yields:

$$C = -\frac{V_s v_m}{F} \frac{d\theta}{dt} = -\frac{V_s v_m}{F} \frac{dT}{dt}\frac{d\theta}{dT}$$

$$= -\left(\frac{V_s v_m \beta}{F}\right)\frac{d\theta}{dT} \tag{1.28}$$

The net rate of desorption is equal to the rate at which the desorbed gas is carried away by the carrier:

$$FC = V_s[v_m k_d \theta - k_a C(1 - \theta)] \tag{1.29}$$

where k_d and k_a are the rate constants for desorption and adsorption, respectively. If the surface is energetically homogeneous, the desorption rate constant is given by:

$$k_d = k_0 \exp(-E_d/R_g T) \tag{1.30}$$

The heating rate β can be set at a value high enough to neglect readsorption. Then, Eqs. 1.28 and 1.29 can be combined to give:

$$\frac{d\theta}{dT} = -\frac{k_d}{\beta}\,\theta = -\frac{\theta k_0}{\beta}\exp(-E_d/R_g T) \tag{1.31}$$

$$C = \frac{V_s v_m k_0}{F}\,\theta \exp(-E_d/R_g T) \tag{1.32}$$

where Eq. 1.30 has been used. Data are obtained in the form of C versus time and T versus time, which yields C versus T. At an extremum on this curve, $dC/dT = 0$, or from Eq. 1.32:

$$\frac{dC}{dT} = 0 = \left(\frac{d\theta}{dT} + \frac{E_d}{R_g T^2}\,\theta\right)\exp(-E_d/R_g T)$$

which yields:

$$\frac{d\theta}{dT} = \frac{-E_d}{R_g T^2} \theta = -\frac{\theta k_0}{\beta} \exp(-E_d/R_g T) \tag{1.33}$$

Equation 1.33 gives:

$$k_0 \exp(-E_d/R_g T_M) = \frac{\beta E_d}{R_g T_M^2} \tag{1.34}$$

where the subscript M denotes the maximum point. Taking the logarithm of both sides of the equation gives:

$$2 \ln T_M - \ln \beta = E_d/R_g T_M + \ln (E_d/k_0 R_g) \tag{1.35}$$

For the determination of E_d and k_0, the heating rate β is varied and the corresponding T_M's are obtained. While E_d and k_0 can be determined from a linear plot of Eq. 1.35, the range of β that can be varied is limited and the resolution is poor. Therefore, a "reasonable" value of k_0 has to be used for the determination of the heat of adsorption E_d. These values are tabulated in the article by Cvetanovic and Amenomiya (1972). When such a value is not available, a plot of Eq. 1.35 gives approximate values of both E_d and k_0. The TPD peaks appearing at different temperatures at a constant heating rate have been used to identify groups of active sites. These results have been used for mechanistic investigation of reactions.

If the peaks do not conform to the theoretical shapes of energetically homogeneous surfaces, for example, if they are appreciably broader, it is assumed that either the desorption process is more complex or a significant diffusional broadening occurs in the carrier gas stream before the desorbed gas reaches the detector. The latter complication can generally be corrected or assessed so that peak shapes should provide information on the complexity of the desorption process. For peak broadening due to heterogeneous surfaces and/or lateral interaction of adsorbed species, a linear dependence of the activation energy on coverage is usually assumed:

$$k_d = k_0 \exp[-(E_{d_0} - \delta\theta)/R_g T] \tag{1.36}$$

An analysis method for this case has been described by Cvetanovic and Amenomiya (1972).

Many instrumental methods have been developed and are still evolving for the characterization of catalyst surfaces. These instruments are capable of revealing information of value on various aspects of catalysis. Electron spectroscopy for chemical analysis (ESCA), which is also known as X-ray photoelectron spectroscopy (XPS), Auger electron spectroscopy (AES), the electron microprobe, electron microscopy, and X-ray diffraction crystallography are better-developed techniques that have been of particular value in working with technical catalysts. AES has been used primarily for analysis of elements and is very sensitive to the top two or three layers of the surface. The excited electrons can be focused down to 5 μm or less for spatial resolution of the AES spectrum. The X-ray excitation involved in ESCA cannot be focused very readily, and the analysis area has to be typically several millimeters across in order to gain sensitivity. However, it provides chemical bonding information, about which little is revealed from AES. The electron microprobe produces an X-ray images that gives the nature and the distribution of a particular element to a high degree of resolution. The electron beam penetrates 200 nm or more, so the composition is averaged through most of a specimen as typically prepared. It is particularly useful for obtaining a profile of the distribution of heavy metals through a catalyst particle by working with a thin slice of the material, or for detecting the build-up of poisons and their distribution through individual pellets. X-ray diffraction may be used to obtain information about the structure and composition of crystalline materials. Details on the methods of surface characterization can be found in the book edited by Kane and Larrabee (1974) and those on the experimental methods for characterizing surface structure in the review by McRae and Hagstrum (1976).

1–6 CHARACTERIZATION OF SUPPORT

As was discussed in Section 1–3, many considerations go into the selection of a suitable support. One of the basic pieces of information is the total surface area. This area can range from 1500 m²/g for highly activated carbon, to 4 m²/g for kieselguhr. The total surface area of the support is usually determined from the method devised by Brunauer-Emmett-Teller (BET method), using N_2 at its boiling point ($-195.8°C$) as the adsorbate. Pore size distribution and average pore diameter are essential in describing the diffusion of gas into pores where most of the catalyst is deposited. These are usually determined from a combination of gas adsorption and mercury penetration methods. The porosity and density of the support are required to relate pellet conditions to reactor volume. Various methods of characterizing the support are treated in this section.

Total Surface Area (BET Method)

Brunauer, Emmett, and Teller (1938) successfully extended the Langmuir adsorption isotherm applicable to monolayer adsorption to multilayer adsorption. While the assumptions made contain some serious deficiencies, the reproducibility of the BET method has made it the standard procedure for determining total surface area. In addition to the assumptions involved in the Langmuir adsorption, the BET method assumes that heats of adsorption for the multilayer adsorption are the same for the second and subsequent layers and are equal to the heat of liquefaction of the adsorbate. Adsorption and desorption rate constants are assumed to be the same for all layers except for the first layer, which is in contact with the solid surface. The main deficiency lies in the assumption of treating all layers except for the first as identical and as liquid-like. As pointed out by Hill (1952), however, it is not too much of an assumption to treat the third and subsequent layers as liquid-like. In multilayer adsorption, different parts of the surface are covered by a different number of layers of adsorbed molecules. Suppose that these layers are rearranged in the manner shown in Figure 1.15 such that s_0, s_1, s_2, . . . , s_i are the surface areas covered by only 0, 1, 2, . . . , i layers of adsorbed molecules. The rate at which the s_1 surface appears is the sum of the rate at which the s_0 surface is covered by molecules due to adsorption and the rate at which the s_2 surface disappears due to desorption from s_2. On the other hand, the rate at which the s_1 surface disappears is the sum of the rate at which the s_1 surface disappears due to desorption to form s_0 surface and the rate at which the s_1 surface is covered by molecules (adsorption on s_1 surface). Therefore, the net change is given by:

$$\frac{ds_1}{dt} = (k_1 s_0 p + k_{-2} s_2) - (k_{-1} s_1 + k_2 p s_1) \tag{1.37}$$

where k_i and k_{-i} are adsorption and desorption rate constants for the ith surface. Likewise, one has:

$$\frac{ds_2}{dt} = (k_2 s_1 p + k_{-3} s_3) - (k_{-2} s_2 + k_3 p s_2) \tag{1.38}$$

$$\frac{ds_3}{dt} = (k_3 s_2 p + k_{-4} s_4) - (k_{-3} s_3 + k_4 p s_3) \tag{1.39}$$

$$\vdots$$

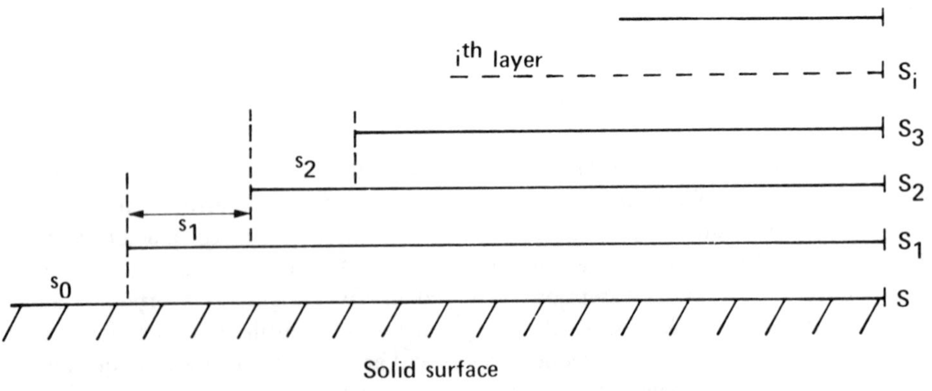

Figure 1.15   Idealized model of physically absorbed gas layers.

Summation gives:

$$\frac{d}{dt} \sum_{i=1}^{\infty} s_i = \frac{dS_1}{dt} = k_1 s_0 p - k_{-1} s_1 \tag{1.40}$$

where S_1 is the total area of the first layer. Proceeding in a similar manner for the total surface area of the ith layer, one gets:

$$\frac{d}{dt} \sum_{j=i}^{\infty} s_j = \frac{dS_i}{dt} = k_i s_{i-1} p - k_{i-1} s_i, \qquad i \geq 1 \tag{1.41}$$

Equations 1.40 and 1.41 can be rewritten at equilibrium as:

$$s_1 = K_1 s_0 p \tag{1.42}$$
$$s_i = K_i p s_{i-1} \tag{1.43}$$

where $K_i = k_i / k_{-i} = K_{10} \exp(+Q_1/RT), \qquad i = 1$
$$= K_0 \exp(+Q_L/RT) = K, \qquad i \geq 2 \tag{1.44}$$

According to the assumptions made, adsorption on the first layer involves a certain heat of adsorption Q_1 different from that for the second and subsequent layers, which is taken to be the same at Q_L. The total solid surface area S ($= S_0$) is given by:

$$S = \sum_{i=0}^{\infty} s_i \tag{1.45}$$

and the total volume of gas adsorbed is given by:

$$v = \frac{v_m}{S} \sum_{i=0}^{\infty} i s_i \tag{1.46}$$

where v_m is the monolayer volume of gas. Combining the two equations gives:

$$\frac{v}{v_m} = \frac{\sum_{i=0}^{\infty} i s_i}{\sum_{i=0}^{\infty} s_i} \tag{1.47}$$

If one lets $x = Kp$, $y = K_1 p$, and $c = y/x$, Eq. 1.47 can be rewritten with the aid of Eqs. 1.42 and 1.43 as:

$$\frac{v}{v_m} = \frac{c s_0 \sum_{i=1}^{\infty} i x^i}{s_0 \left[1 + c \sum_{i=1}^{\infty} x^i \right]} \tag{1.48}$$

Since $\Sigma_{i=1}^{\infty} x^i = x/(1 - x)$ and $\Sigma_{i=1}^{\infty} ix^i = x(d/dx) \Sigma x^i = x(1 - x)^2$, the above equation becomes:

$$\frac{v}{v_m} = \frac{cx}{(1 - x)(1 - x + cx)} \tag{1.49}$$

The free-surface model allows an infinite number of adsorbed layers. Therefore, at the saturation pressure p_0, v must be infinite. According to Eq. 1.49, this is true when $x = 1$. Since $p = p_0$ at $x = 1$, the definition of x ($= Kp$) allows one to write:

$$x = \frac{p}{p_0} \tag{1.50}$$

The usual form of the BET equation is obtained by substituting Eq. 1.50 into 1.49 and rearranging the result:

$$\frac{p}{v(p_0 - p)} = \frac{1}{v_m c} + \frac{c - 1}{v_m c} \frac{p}{p_0} \tag{1.51}$$

A plot of $p/v(p_0 - p)$ versus p/p_0 made from measurements of v versus p gives a straight line with its slope and intercept yielding v_m and c. The usual range of relative pressure (p/p_0) is between 0.05 and 0.3. At higher p/p_0 values, realities of multilayer adsorption and/or pore condensation cause increasing deviation, whereas the amount adsorbed at p/p_0 less than 0.05 is usually too low to give an accurate result. The linear plots made by Brunauer et al. (1938) for the BET equation are shown in Figure 1.16. The partial pressures of nitrogen gas for the p/p_0 range of about 0.05 to 0.30 correspond to the range of 10 to 100 kPa. It should be recognized that the total surface area from the BET method may not be accurate when the surface area is too small ($<$ a few m²) or too large ($>$ 1200 m²/g). While the former can be taken care of by increasing the sample size, the latter is difficult to handle because of condensation that may occur at low values of p/p_0. The continuous flow apparatus described in Section 1–4 can be used for the BET method.

The monolayer volume v_m can be readily converted to the number of molecules adsorbed with a suitable value for the area covered by one adsorbed molecule. If this value is α, the total surface area is given by:

$$S = \left[\frac{v_m N_A}{V} \right] \alpha \tag{1.52}$$

where N_A is the Avogadro's number, and V is the volume per mole of gas at conditions of v_m, which is recorded at standard conditions, giving $V = 22,400$ cm³/gmol. The usual equation used for α (hexagonal close packing) is:

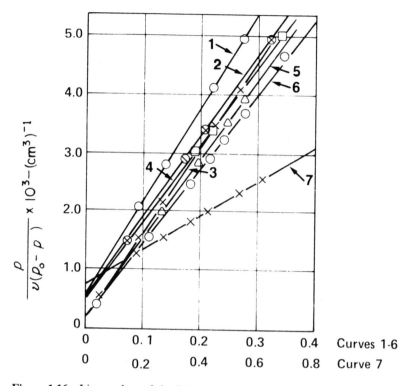

Figure 1.16 Linear plots of the BET equation:

1. CO_2 at $-78°C$	2. A at $-183°C$
3. N_2 at $-183°C$	4. O_2 at $-183°C$
5. CO at $-183°C$	6. N_2 at $-195.8°C$
7. $n\text{-}C_4H_{10}$ at $-0°C$	

(Reprinted from P.H. Emmett (ed.), *Catalysis*, vol. I, Reinhold Publishing Corporation, New York 1954.)

$$\alpha = 1.09 \left[\frac{M}{N_A \rho} \right]^{2/3} \tag{1.53}$$

where M is molecular weight, and ρ is the density of the adsorbed molecules ($\rho = 0.808$ for N_2 at $-195.8°C$). For nitrogen, the projected surface area of the nitrogen molecule has been measured to be 16.2 Å². Therefore, the total surface area when nitrogen is used is given by:

$$S = 4.35 v_m \quad \text{(m}^2\text{/g solid adsorbent)} \tag{1.54}$$

with v_m in cm³/g solid adsorbent.

The physical adsorption isotherm varies somewhat with the nature of the adsorbent. Most physical adsorption isotherms may be grouped into the five types

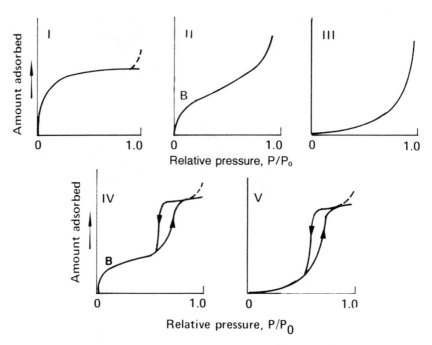

Figure 1.17 Types of physisorption isotherms. (After C.N. Sutterfield, *Heterogeneous Catalysis*, © 1980; with permission of McGraw-Hill Book Company, New York.)

shown in Figure 1.17. In all cases the amount of gas adsorbed gradually increases as its partial pressure is increased, becoming at some point equivalent to a monolayer, but then increasing to a multilayer, that eventually merges into a condensed phase. The adsorption isotherms I, II and IV in Figure 1.17 are suitable for the determination of the total surface area. Nitrogen, krypton, and argon obey these isotherms depending on whether the adsorbent is highly porous or nonporous. Types I and III are commonly observed on porous structures whereas type II is found with nonporous structures. Type I is of the Langmuir type and is observed on porous structures that contain very fine pores. These micropores, it is believed, are completely filled at a relative pressure much less than unity, rather than monolayer adsorption when the asymptotic value is reached. Type IV is also encountered with porous materials. At low relative pressures, the isotherm is similar to type II, but adsorption increases markedly at higher pressures and pore condensation takes place. A hysteresis effect is usually associated with this pore condensation. The Type IV isotherm is often encountered with industrial catalysts. Point *B* in types II and IV occurs at a "knee" and is the stage at which monolayer coverage is complete. Types III and V are rare and are typical of systems where the forces of adsorption are relatively weak. These isotherms are not suitable for the surface area measurement because second and succeeding layers build up before the first is complete.

Pore Size Distribution and Pore Volume

Two different methods may be used for the determination of pore size distribution: physical adsorption of a gas, which is applicable to pores less than about 60 nm in diameter, and mercury porosimetry, applicable to pores larger than about 3.5 nm.

Measurements of the amount of gas desorbed as a function of reduced pressure provide the most commonly used procedure for determining the pore size distribution of fine pores. Desorption measurements at relatively higher values of p/p_0 corresponding to multilayer adsorption are used for the determination; adsorption measurements at relatively low values of p/p_0 (< 0.3) are used for the determination of the total surface area. As the relative pressure is increased, multilayer adsorption occurs and ultimately the adsorbed films are thick enough to bridge the pores. Further uptake of adsorbate will result in capillary condensation. Since the vapor pressure decreases as the capillary size decreases, such condensation will occur first in small pores. Condensation will be complete, as p/p_0 approaches unity, when the entire void region is filled with condensed adsorbate. If the pressure is reduced by a small increment, a small amount of condensed gas will evaporate from the meniscus formed at the ends of the largest pores. Pores that are emptied of condensate in this way will be those in which the vapor pressure of adsorbate is greater than the chosen pressure. The Kelvin equation gives the relationship between vapor pressure and the radius of the concave surface of the meniscus of the liquid:

$$\ln \frac{p}{p_0} = \frac{-2\sigma V_l \cos \theta}{r_c R_g T} \tag{1.55}$$

where V_l = molal volume of liquid adsorbate
$\quad \sigma$ = surface tension of liquid adsorbate
$\quad \theta$ = contact angle between surface and condensate
$\quad r_c$ = radius of curvature
$\quad r$ = physical radius of cylindrical pore

Since some of the adsorbate is adsorbed on the surface, and therefore not present because of capillary condensation, it is assumed that the radius of the meniscus is not the true physical radius r, but rather that the radius has been diminished by the thickness of the adsorbed layer (δ), i.e., $r_c = r - \delta$. The Kelvin equation is then modified to give:

$$r = \frac{-2\sigma V_l}{R_g T \ln (p/p_0)} + \delta \tag{1.56}$$

Studies have shown that values of δ are essentially independent of the chemical nature of the adsorbent for most systems at coverages greater than a monolayer, depending only on the pressure. The value of δ can essentially be calculated by dividing the volume of gas adsorbed by the BET area. In general, Halsey's form (1948) is used, given by:

$$\delta = A \left(\ln \frac{p_0}{p} \right)^{-1/n} \tag{1.57}$$

where A and n depend on the nature of the surface. A form suggested by Wheeler (1955) is:

$$\delta(\mathring{A}) = 7.34 \left(\ln \frac{p_0}{p} \right)^{-1/3} \tag{1.58}$$

Nitrogen has been used almost universally for the determination of pore size distributions. It completely wets the surface so that $\cos \theta = 1$. For nitrogen at its normal boiling point ($-195.8°C$), Eq. 1.56 reduces to:

$$r(\mathring{A}) = 9.52 \left(\ln \frac{p_0}{p} \right)^{-1} + \delta \tag{1.59}$$

For a chosen value of p/p_0, Eqs. 1.58 and 1.59 give the pore radius above which all pores will be empty of capillary condensate. Hence, if the amount of desorption is measured for various p/p_0, the pore volume corresponding to various radii can be evaluated. Differentiation of the curve for cumulative pore volume versus radius gives the distribution of volume as a function of pore radius, as shown in Figure 1.18 (Smith 1981). It is customary to use $dV/d (\ln r)$ for the pore size distribution instead of dV/dr because of the wide range of sizes covered, where V is the cumulative volume.

　　Another method (Wheeler 1955) of calculating the pore size distribution from the same data is to assume a distribution function for the pore size and then determine the parameters in the distribution function from the data. If the total volume of pores occupied by the condensate at the saturation pressure is v_s, and v is the pore volume at an equilibrium pressure p, then $(v_s - v)$ is the volume of unfilled pores, so:

$$v_s - v(r) = \int_r^{\infty} \pi(r - \delta)^2 f(r) dr \tag{1.60}$$

where $f(r)dr$ is the total length of pores of radii between r and $r + dr$. The distribution function $f(r)$ is typically assumed to be either the Maxwellian distribution:

$$f = A_m r \exp \left(-\frac{r}{r_0} \right) \tag{1.61}$$

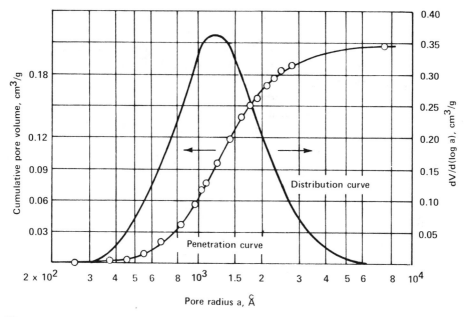

Figure 1.18   Pore-volume distribution in a UO_2 pellet. (After J.M. Smith, *Chemical Engineering Kinetics,* 3rd ed., © 1981; with permission of McGraw-Hill Book Company, New York.)

or the Gaussian distribution:

$$f = A_g \exp \left\{ - \left(\frac{Br_0}{r - r_0} \right)^2 \right\} \tag{1.62}$$

Equations 1.56 and 1.57 are used for r and δ in Eq. 1.60. Values of parameters that best fit the data on v are chosen for the size distribution.

The physical adsorption isotherm in the p/p_0 range for multilayer adsorption usually shows a hysteresis on porous solids, as indicated in Figure 1.17. Although several theories have been put forward to explain the hysteresis, it is not yet totally resolved. However, the theory advanced by Cohan (1944) is helpful in understanding why the desorption isotherm is used for the determination of pore size distributions. He argues that, upon adsorption, pores are not filled vertically but rather radially. On desorption, on the other hand, pores are emptied vertically, for which the Kelvin equation of Eq. 1.55 applies.

The size of the largest pores that can be measured by the desorption method is limited by the marked change of meniscus radius with pressure as the reduced pressure nears unity. Therefore, the method is usually applied to pores of about 60 nm in diameter and less, so surface area in pores larger than 60 nm would be neglected. This method gives satisfactory results for high-area materials such as some aluminas and silica gels in which almost all of the area is in pores of 30 nm or less. The smallest pore sizes that can be determined by this method are about 1.5 to 2nm in diameter. Although measurements corresponding to lower pore sizes can be reported, it becomes increasingly uncertain as to the extent to which the Kelvin equation applies. The derivation assumes that the properties of the condensed phase are the same as those of a bulk liquid, yet the concepts of surface tension or of a hemispherical surface must break down as pore size approaches that of adsorbate molecules.

The mercury penetration method is based on the fact that mercury has a significant surface tension and does not wet most catalytic surfaces. For a cylindrical pore, the force acting against the entrance of mercury to the pores equals $-2\pi r \sigma \cos \theta$. The external pressure applied to overcome this force is $\pi r^2 P$. At equilibrium, the two forces are equal and:

$$r = \frac{-2\sigma \cos \theta}{P} \tag{1.63}$$

For mercury, the contact angle is between 135° and 142° for most surfaces and can be taken as 140°. With σ of 0.48 N/m, the above equation reduces to:

$$r(\text{Å}) = \frac{8.75 \times 10^5}{P(\text{psi})} \tag{1.64}$$

Simple techniques and equipment are satisfactory for evaluating the pore size distribution down to 100Å, but specialized high-pressure apparatus is necessary to go below 100Å (required pressure $> 10,000$ psi). In the experimentation, pressures and corresponding volumes of mercury displaced are measured. The volumes along with Eq. 1.64 then yield the cumulative pore volume as a function of pore radius. The differentiation method discussed previously is used for pore size distribution. A few comparisons between the adsorption and mercury penetration methods, made for the region in which they overlap, show generally good agreement.

Alumina pellets often exhibit a bimodal pore size distribution (bidispersed) as shown in Figure 1.19 (Smith 1981). The void spaces within the particles are commonly termed *micropores,* and the void regions between particles are termed *macropores.* By combining mercury-penetration and nitrogen-desorption measurements, pore-volume information can be obtained over the complete range of radii in a pelletized catalyst containing both macro- and micropores. It is seen in the figure that the micropores are not affected by pelletizing pressure, which seems to suggest that the particles themselves are not crushed significantly during the pelletizing process.

The average pore radius can be found directly from the pore size distribution:

$$\bar{r} = \int_0^\infty rf(r)dr \Big/ \int_0^\infty f(r)dr \tag{1.65}$$

In the absence of information on the pore size distribution, the average radius is obtained from

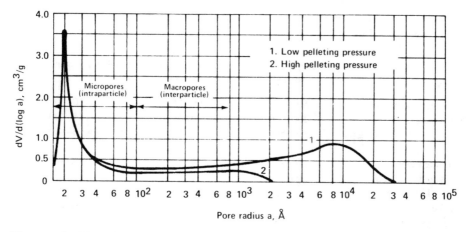

Figure 1.19 Bimodal pore-volume distribution in alumina pellets. (After J.M. Smith *Chemical Engineering Kinetics,* 3rd ed., © 1981; with permission of McGraw-Hill Book Company, New York.)

$$\bar{r} = \frac{2V_g}{S} \qquad (1.66)$$

where V_g and S are the pore volume and BET area per unit weight. Here $\bar{r}$ can be considered as the radius of a cylinder having the same volume/surface ratio as the real pore.

The void volume (pore volume), solid density, and porosity of the catalyst particle can be determined from the *helium-mercury method.* In this method, a container of known volume V is filled with a known weight of pellets or particles W. After evacuation, helium is admitted, and the sum of the volume of the space between the pellets V' and the void volume inside the particles V_g is calculated using the ideal gas law. The true density of the solid is:

$$\rho_T = \frac{W}{V - (V' + V_g)} \qquad (1.67)$$

The helium is then pumped out and the container filled with mercury at atmospheric pressure. Mercury does not penetrate pores at atmospheric pressure. Its volume is that of the space between the particles V'. The porosity or void fraction ϵ_p is the volume of voids per unit volume of particles and is given by:

$$1 - \epsilon_p = \frac{V - (V' + V_g)}{V - V'} \qquad (1.68)$$

The density of the particles is given by:

$$\rho_p = \frac{W}{V - V'} \qquad (1.69)$$

Table 1.8   Surface Area, Pore Volume, and Average Pore Radius for Typical Solid Catalysts (Smith 1981)

Catalyst	Surface Area m²/gr	Pore Volume cm³/gr	Average pore radius Å
Activated carbons	500–1500	0.6–0.8	10–20
Silica gels	200–600	0.4	15–100
$SiO_2 \cdot Al_2O_3$ cracking catalysts	200–500	0.2–0.7	33–150
Activated clays	150–225	0.4–0.52	100
Activated alumina	175	0.39	45
Celite (kieselguhr)	4.2	1.1	11,000
Synthetic ammonia catalysts, Fe	—	0.12	200–1000
Pumice	0.38	—	—
Fused copper	0.23	—	—

In part from A. Wheeler, *Advances in Catalysis*, vol. III; with permission of Academic Press 1950).

These measurements are often made in a mercury porosimeter in conjunction with a measurement of pore-size distribution. Note that the porosity is also obtainable from the density by the expression:

$$\epsilon_p = \frac{\text{void volume}}{\text{total volume}} = \frac{V_g}{1/\rho_p} = \rho_p V_g \tag{1.70}$$

Typical values of ϵ_p are about 0.5, indicating that the particle is about half void space and half solid material. Since overall void fractions in packed beds are about 0.4, a rule of thumb for a fixed-bed catalytic reactor is that about 30% of the volume is pore space, 30% is solid catalyst and support, and 40% is void space between catalyst particles. Typical values of surface area, pore volume, and average pore radius are given in Table 1.8.

Summary

When dealing with reactor design, the catalyst is almost always "given." By now one should have at least a first level of understanding of the nature of catalysis, how the catalyst is prepared, and how the factors involved in the preparation affect the final rate of reaction. One should also have an understanding of characterization techniques, some of which are well established and some still evolving. For any reaction under consideration, well established, basic characterization results should be made available and taken into consideration for reactor design. These measurements would help formulate some fundamental questions regarding reactor performance, answers to which can bring about a significant change in reactor design. It should be clear that the catalyst preparation is still an art. An analytical approach has been given for the catalyst impregnation step to provide a starting point for making catalyst preparation less of an art.

NOTATION

A_g, A_m	constants in distribution functions of Eqs. 1.61 and 1.62
B	net breadth of an X-ray diffraction line; constant in Eq. 1.62
c	constant in Eq. 1.51
C	concentration of impregnating solution; gas concentration
C_0	bulk concentration of impregnating solution
C_w	liquid-side film concentration of impregnating solution
D	diffusivity of solute in solution
E_a	activation energy for adsorption
E_d	activation energy for desorption
$f(r)$	pore size distribution
F	volumetric flow rate
ΔH_0	initial heat of adsorption
k	rate constant in Eq. 1.4; equilibrium constant, k_a/k_d
k_a	adsorption rate constant
k_i	rate constant for i^{th} surface layer in multilayer adsorption
k_d	desorption rate constant
k_m	film mass transfer coefficient (cm/sec)
k_0	preexponential factor in Eq. 1.30
K	k_a/k_d; quantity defined in Eq. 1.6
K_i	K for i^{th} surface layer
K_L	dimensionless rate constant, kC_0
L	particle size; half length of cylindrical pore
M	molecular weight
N_A	Avogadro's number
p	gas partial pressure
P	pressure
p_0	saturation pressure
Pe	Peclet number given by Dt_L/L^2
q	solid phase solute concentration (mol/area)
q_m	measured value of q after equilibration
q_s	saturation value of q
Q	heat of adsorption
r	pore radius; crystallite size
$\bar{r}$	average pore radius
r_c	radius of curvature
R	pore radius in the single pore model
R_g	gas constant
s_i	free surface of ith layer in multilayer adsorption
S	total surface area

t	time
T	temperature
t_L	time required to fill pore
v	velocity of penetrating solution in impregnation; total volume of gas adsorbed
v_m	monolayer volume of gas
V	container volume; gas molal volume; accumulative volume of adsorbed gas
V'	space between particles (macropore volume)
V_g	pore volume
V_l	molal volume of liquid adsorbate
V_s	volume of solid particles in temperature-programmed desorption
W	particle weight
x	dimensionless length, z/L
y	dimensionless concentration, C/C_0
z	space coordinate
Z	catalyst distribution given in Eq. 1.17

Greek Letters

α	area covered by one adsorbed molecule; constant in Table 1.5
β	linear rate of temperature rise
δ	constant in Eq. 1.36; thickness of adsorbed layer
ϵ_p	porosity
η	$2q_s/rC_0$
θ	surface coverage; q/q_s
θ_e	θ at external surface
λ	wavelength of X-ray
μ	viscosity
ρ	density
ρ_p	particle density
ρ_T	true solid density
σ	surface tension
τ	dimensionless time, t/t_L
τ_{eq}	τ required for complete equilibration

PROBLEMS

1.1. The balance equations for the pore filling (Eqs. 1.5 and 1.6) are:

$$\frac{\partial y}{\partial \tau} + \frac{1}{2\tau^{1/2}}\frac{\partial y}{\partial x} + \eta\frac{\partial \theta}{\partial \tau} = 0 \qquad (1.5)$$

$$\frac{\partial \theta}{\partial \tau} = \frac{K}{\eta}[y(1 - \theta) - \theta/K_L] \qquad (1.6)$$

According to the characteristic method, Eq. 1.5 can be written as:

$$\frac{d\tau}{1} = \frac{dx}{1/(2\tau^{1/2})} = \frac{dy}{-\eta(\partial \theta/\partial \tau)} \qquad (1.11.1)$$

On the characteristic lines given by:

$$\tau^{1/2} = x + \tau_e^{1/2} \qquad (1.11.2)$$

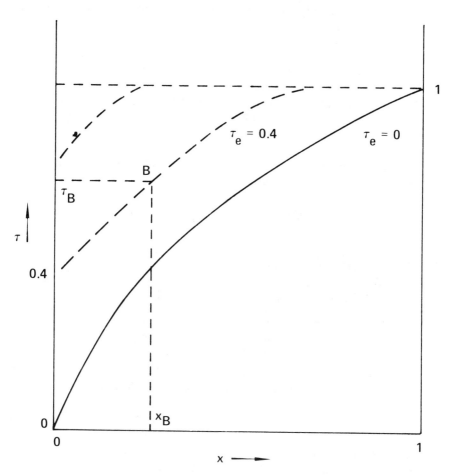

Figure 1.1P   Characteristic lines.

where τ_e is the time at which a plug of interest enters pore mouth, the solution of Eq. 1.5 is obtained from:

$$\frac{dy}{d\tau} = -\eta\frac{d\theta}{d\tau} = -K[y(1-\theta) - \theta/K_L] \qquad (1.11.3)$$

Therefore, simultaneous solution of Eqs. 1.6 and 1.11.3 will yield the profiles y and θ on the lines given by Eq. 1.11.2. These characteristic lines are shown in Figure 1.1P. For any point on the characteristic line, say point B, specification of x (or τ) fixes τ (or x) through Eq. 1.11.2. Therefore, for point B, x of 0.1 automatically gives τ of 0.536.

a)   Give difference versions of Eqs. 1.6, 1.11.2, and 1.11.3 necessary for numerical solution. Give details on how one can solve the differential equations using the difference equations.

b)   Show that

$$y + \eta\theta = \text{constant} = 1 + \eta\theta(0,\tau_e)$$

on the characteristic lines, but that the constant changes with each characteristic line of different τ_e. The expression for $\theta(0,\tau_e)$ is that of Eq. 1.16 with τ replaced by τ_e.

1.2. Melo et al. (1981) give the following experimental data for the impregnation of alumina in $Ni(NO_3)_2$ solution:

Bulk solution concentration (mol/l)	Wt% of Ni on alumina after equilibration
0.08	1.22
0.16	1.66
0.37	2.36
0.58	3.04
0.72	3.25
0.89	3.78

Determine the parameters in the equilibrium relationship of Eq. 1.4. Average pore radius is 50Å, porosity is 0.62, and apparent density of support is 1.38 g/ml.

1.3. Show that Brunauer-Love-Keenan isotherm can be reduced to Langmuir, Freundlich, and Tempkin isotherms under suitable limiting conditions.

1.4. Suppose that a heterogeneous surface can be characterized by a continuous version of the Langmuir isotherm for heterogeneous surfaces (Eq. 1.23) with K_i from K_L to K_U. Obtain an expression for θ by integrating the continuous version of Eq. 1.23. Can you relate the result to any isotherms for heterogeneous surfaces such as the Freundlich and Tempkin isotherms?

1.5. Calculate the temperatures (T_M) at which peaks appear in the TPD method when the heating rate (β) is changed in the increment of 5°C/min from 5°C/min to 20°C/min. This is a typical range of β. Use E_d of 27 kcal/mol and k_0 of $1.6 \times 10^{+15}$ sec^{-1}. Discuss the sensitivity of T_M to a change in k_0.

1.6. Why would the TPD peak broaden for heterogeneous surfaces?

1.7. From the development of the BET equation, obtain an expression for the rate at which the vacant surface area changes with time.

1.8. Given the following nitrogen physisorption data at −195°C ($p_0 = 760$ mm Hg), calculate the BET area.

p (mm Hg)	V cm³/0.6 gr at STP
0	0
10	50
80	80
150	97
200	103
300	124

Now use the Langmuir isotherm for the calculation. Compare the results. Can you tell the p range at which the monolayer formation is complete?

1.9. The desorption data for the above system are given below. Calculate the pore-size distribution by applying the Kelvin equation. Also estimate the average pore size. Here V is the volume of adsorbate still remaining in the pores at the indicated pressure.

p/p_0	V cm³/0.6 gr at STP
0.6	180
0.72	240
0.78	285
0.80	355
0.86	525
0.88	560
0.92	600

1.10. Mercury porosimeter data for a certain pellet (0.8g) are given below. Obtain curves for the cumulative pore volume (cm³/g) and size distribution as a function of pore radius.

Pressure (lb/in²)	Mercury volume penetrating into the pore (cm³)
3310	0.2396
1536	0.2341
921	0.2015
831	0.1888
677	0.1569
612	0.1386
552	0.1196
525	0.1101
498	0.1006
406	0.0654
331	0.0377
256	0.0156
119	0.0000

1.11. Using the helium-mercury method, the following data have been obtained:

Mass of catalyst sample placed in chamber = 99 g
Volume of helium displaced by sample   = 44 cm³
Volume of mercury displaced by sample   = 79 cm³

Calculate the true solid density, porosity, and pore volume.

REFERENCES

Anderson, J.R., *Structure of Metallic Catalysts,* Academic Press, New York (1975).

Anderson, J.R., *Experimental Methods in Catalytic Research,* Chap. 2, Academic Press, New York (1968).

Barrer, R.M., Chem. Ind., *36,* 1206(1968).

Bensei, H.A. and B.H.C. Winquist, Adv. Cat., *27,* 97 (1978).

Bond, G.C., *Heterogeneous Catalysis: Principles and Applications,* p 13, Oxford University Press, London (1974).

Brunauer, S., P.A. Emmett and E. Teller, J. Am. Chem. Soc., *60,* 309 (1938).

Carberry, J.J., *Chemical and Catalytic Reaction Engineering,* McGraw-Hill, New York (1976).

Chen, H.C. and R.B. Anderson, Ind. Eng. Prod. Res. Develop., *12,* 122 (1973).

Chen, N.Y. and P.B. Weisz, Chem. Eng. Progr. Symp. Ser., *67,* 86 (1967).

Churchill, R.V., *Modern Operational Mathematics in Engineering,* p 196, McGraw-Hill, New York (1944).

Cohan, L.H., J. Am. Chem. Soc., *66,* 98 (1944).

Cvetanovic, R.J. and A. Amenomiya, Cat. Rev., *6,* 21 (1972).

Eberly, P.E., Jr., J. Phys. Chem., *65,* 1261 (1961).

Farrauto, R. J., AIChE Symp. Ser., *70,* 9 (1974).

Formi, L., Cat. Rev., *8,* 65 (1974).

Forsythe, G.E. and W.R. Wason, *Finite-Difference Methods for Partial Differential Equations,* Wiley, New York (1960).

Freel, J., J. Catal., *25,* 139 (1972).

Gandhi, H. and M. Shelef, J. Catal., *28,* 1 (1973).

Gates, B.C., J.R. Katzer and G.C.A. Schuit, *Chemistry of Catalytic Processes,* McGraw-Hill, New York (1979).

Halsey, G.D., J. Chem. Phys., *16,* 931 (1948).

Hayward, D.A. and B.M.W. Trapnell, *Chemisorption,* Chap. 2, Butterworth, London (1964).

Hill, T.L., Adv. Cat., *4,* 211 (1952).

Johnson, O., J. Phys. Chem., *59,* 827 (1955).

Kane, P.F. and G.B. Larrabee (eds.), *Characterization of Solid Surfaces,* Plenum, New York (1974).

Komiyama, M., R.P. Merrill and H.F. Harnsberger, J. Catal., *63,* 35 (1980).

Knözinger, H., Adv. Cat. *25,* 184 (1976).

Lee, H.H., Chem. Eng. Sci. *39,* 859 (1984).

Linsen, B.G. (ed.), *Physical and Chemical Aspects of Adsorbents and Catalysts,* Academic Press, New York (1970).

McRae, E.G. and H.D. Hagstrum in N.B. Hannay (ed.), *Treatise on Solid State Chemistry,* p 57, vol. 6A, Plenum (1976).

Melo, F., J. Cervello and E. Hermana, Chem. Eng. Sci., *35,* 2165 (1980).

Miller, D.J. and H.H. Lee, AIChE J., *30,* 185 (1984).

Moss, R.L. in R.B. Anderson and P.T. Dawson (eds.), *Experimental Methods in Catalytic Research,* vol. II, Academic Press, New York (1976).

Satterfield, C.N., *Heterogeneous Catalysis,* McGraw-Hill, New York (1980).

Spenadel, L. and M. Boudart, J. Phys. Chem., *64,* 204 (1960).

Smith, J.M., *Chemical Engineering Kinetics,* 3rd ed., McGraw-Hill, New York (1981).

Tanaka, K. and K. Tamaru, J. Catal., *2,* 366 (1963).

Thomas, J.M. and W.J. Thomas, *Introduction to the Principles of Heterogeneous Catalysis,* Academic Press, New York (1967).

Vincent, R.C. and R.P. Merrill, J. Catal., *35,* 206 (1974).

Voge, H.H., *Catalysis,* vol. VI, p 407, Reinhold, New York (1958).

Ward, J.W. and R.C. Hansford, J. Catal., *13,* 364 (1969).

Wheeler, A., *Catalysis,* vol. II, Reinhold, New York (1955).

Yates, D.J.C. and J.H. Sinfelt, J. Catal., *8,* 348 (1967).

CHAPTER 2

Kinetics of Catalytic Reactions

After a catalyst is selected for the reaction under consideration, the next step is to determine the intrinsic kinetics, which is discussed in Chapter 3. There are many aspects of kinetics that should be investigated before and in conjunction with this determination. This chapter considers some of the aspects that are relevant to reactor design.

The chemical transformation on the catalyst surface takes place in an orderly, sequential manner. A catalytic reaction is initiated by the interaction of reactant(s) with the active sites on the catalyst surface, usually in the form of chemisorption. This causes a chemical transformation on the surface, with the transformed species then leaving the surface. Although the exact nature of the sequence is quite difficult to identify in many cases, this sequence of surface reactions is in general applicable to any catalytic reaction. This framework provides a basis on which rate expressions that typically lead to Langmuir–Hinshelwood kinetic forms can be derived. On the other hand, the kinetics based on a postulated mechanism often do not represent the true picture, and a correlation, usually in a power-law form, may describe observed kinetics equally well. Nevertheless, Langmuir–Hinshelwood kinetics based on some level of mechanistic understanding are much more desirable than brute-force kinetic correlations. The rate expressions resulting from postulated mechanisms can be checked for self-consistency by using some thermodynamic concepts. Consider, first, surface reactions.

2–1 SURFACE REACTIONS

The chemical transformation of reactants on catalyst surfaces take place sequentially in several steps. For the simple example of an isomerization reaction, $A \rightleftarrows B$, which takes place on a surface containing only one type of active site, S, these individual steps are:

$$A + S \rightleftarrows A \cdot S$$
$$A \cdot S \rightleftarrows B \cdot S \qquad (2.1)$$
$$B \cdot S \rightleftarrows B + S$$

The first step in this sequence represents the adsorption of A onto the surface; the second, the surface reaction of adsorbed gas molecule $A \cdot S$ to $B \cdot S$; and the

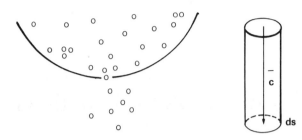

Figure 2.1 Molecular effusion. W.J. Moore, *Physical Chemistry,* 2nd ed., © 1955; with permission of Prentice-Hall, Inc., Englewood Cliffs, NJ.)

third, the desorption of product B from the surface, which frees the active site S for subsequent reactions. As indicated in the sequence for the isomerization reaction, the precursor to any surface reaction is chemisorption, which involves a specific interaction between surface and gas molecule (adsorbate), and thus is a chemical reaction in itself. Although the chemisorption was treated in Chapter 1, it is examined here in some detail for the surface reactions under consideration.

To derive an expression for the rate of adsorption, consider the process of molecular effusion (Moore 1955) (Figure 2.1), which is the effusion of gas molecules from a vessel of gas under pressure through a tiny orifice—an opening so small that the distribution of the velocities of the gas molecules remaining in the vessel is not affected in any way. The number of molecules escaping in unit time is then equal to the number of molecules that, in their random motion, happen to hit the orifice. This number is proportional to the average molecular speed. If an incremental area, *ds,* of the catalyst surface is regarded as an orifice and if there are n molecules per unit volume, the number striking the surface is $n\bar{c}ds$, provided that all the molecules are moving directly perpendicular to the surface at a mean speed of $\bar{c}$. For unit surface area, the number of collisions would be $n\bar{c}$ if all the molecules were moving perpendicular to the surface.

Actually, the number of collisions is only a fraction of $n\bar{c}$ since molecules move randomly. If the direction of the molecules is no longer normal to the wall

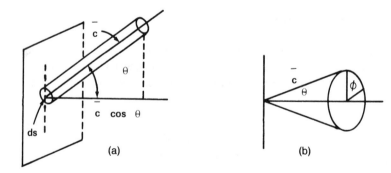

Figure 2.2 Coordinate system for the molecular effusion in Figure 2.1. (W.J. Moore, *Physical Chemistry,* 2nd ed., © 1955; with permission of Prentice-Hall, Inc., Englewood Cliffs, NJ.)

as shown in Figure 2.2(a), for any given direction, the number of molecules hitting ds in unit time will be those contained in a cylinder of base ds and a slant height $\bar{c}$. The volume of this cylinder is $\bar{c} \cos \theta \, ds$, and the number of molecules it contains is $n\bar{c} \cos \theta \, ds$. The question then is how many of these molecules have velocities in the specified direction. Consider the spherical coordinate system of Figure 2.2(b) with its origin at the wall of the vessel. The distance from the origin, $\bar{c}$, defines the magnitude of the velocity, and the angles θ and ϕ represent its direction. Any particular direction from the origin is specified by the differential solid angle dw, which is $d\theta \sin \theta \, d\phi$. This differential solid angle contains molecules having speeds between $\bar{c}$ and $\bar{c} + d\bar{c}$ at positions between θ and $\theta + d\theta$, and ϕ and $\phi + d\phi$. The fraction of the molecules having their velocities within this particular spread of directions is $(\frac{1}{4}\pi)n\bar{c} \cos \theta \sin \theta \, d\theta \, d\phi$. It follows then that the rate at which molecules strike the solid surface is:

$$Z = \int_0^{\pi/2} \int_0^{2\pi} \frac{1}{4\pi} n\bar{c} \cos \theta \sin \theta \, d\phi \, d\theta$$

$$= \frac{1}{4} n\bar{c} = \frac{1}{4} n \left(\frac{8 k_B T}{\pi m} \right)^{1/2}$$

where $m =$ mass of a molecule

$k_B =$ Boltzmann constant

$Z =$ number of collisions per unit time per unit surface area

This rate is the maximum possible rate of adsorption since this is the rate that would result if every molecule striking the surface were adsorbed. In order for adsorption to occur, however, the energy with which a molecule or atom of adsorbate strikes an unoccupied site should be high enough to overcome the energy barrier on the surface. Consider here the simplest model of a surface for the fraction of Z that results in adsorption. This ideal surface is one in which each adsorption site has the same energy of interaction with the adsorbate, which is not affected by the presence or absence of adsorbate molecules on adjoining sites, and in which each site can accommodate only one adsorbate molecule or atom.

In addition to the energy requirement, it is also necessary to know the fraction of vacant sites, since only these sites will be involved in the adsorption. If we designate E as the activation energy required for chemisorption and θ as the fraction of sites occupied, the rate of adsorption is:

$$r_{ad} = \frac{1}{4} n \left(\frac{8 k_B T}{\pi m} \right)^{1/2} e^{-E/k_B T} (1 - \theta)$$

The quantity n in terms of the ideal gas law is given by $p/k_B T$. Therefore, r_{ad} is:

$$r_{ad} = \frac{p}{(2\pi m k_B T)^{1/2}} e^{-E/k_B T} (1 - \theta)$$

Here the sticking probability, σ, may be included as a measure of the deviation of chemisorption rates from this ideal limit:

$$r_{ad} = \frac{\sigma p}{(2\pi \, mk_B T)^{1/2}} \, e^{-E/k_B T} \, (1 - \theta)$$

The preexponential factor, k_a^0, for the adsorption rate constant is thus seen to be $\sigma/(2\pi \, mk_B T)^{1/2}$.

If the vacant fraction $(1 - \theta)$ is expressed as C_v/C_t, where C_v and C_t are the surface concentrations of vacant and total sites, the rate of adsorption can be written as:

$$r_{ad} = k_a p C_v; \qquad k_a = \frac{k_a^0 \, e^{-E/k_B T}}{C_t}$$

An expression for the desorption rate can be written as:

$$r_{de} = k_d^0 \theta \, e^{-E_d/k_B T}$$

where k_d^0 is a desorption rate constant per unit surface area and E_d is the activation energy for desorption. That is, the rate is proportional to the fraction of total sites occupied by the adsorbed species and only these adsorbed species with an energy level high enough to overcome the energy barrier E_d will be desorbed. At equilibrium, $r_{ad} = r_{de}$ and the coverage θ is given by:

$$\theta = \frac{Kp}{1 + Kp} ; \qquad K = \frac{\sigma \exp(Q/k_B T)}{k_d (2\pi \, mk_B T)^{1/2}}, \qquad Q = E_d - E$$

which is the Langmuir isotherm.

The detailed treatment of the adsorption–desorption process as a chemical reaction reveals a few major concepts that are used in developing the kinetics of heterogeneous reactions from a sequence of several surface reactions such as the one in Eq. 2.1. For the purpose of writing the kinetics of each step, each surface reaction can be treated as an elementary step as in homogeneous reactions. The treatment also shows an individual step as a separate entity independent of the other steps, eventually leading to the concept of a rate-limiting (or rate-controlling) step.

The kinetics of the surface reactions of Eq. 2.1 are developed as follows:

$$A + S \underset{k_{-a}}{\overset{k_a}{\rightleftharpoons}} A \cdot S \qquad \text{(adsorption, } K_A = k_a/k_{-a})$$

$$A \cdot S \underset{k_{-s}}{\overset{k_s}{\rightleftharpoons}} B \cdot S \qquad \text{(surface reaction, } K_S = k_s/k_{-s})$$

$$B \cdot S \underset{k_{-d}}{\overset{k_d}{\rightleftharpoons}} B + S \qquad \text{(desorption of product, } K_D = k_d/k_{-d})$$

The net rates for each individual step are:

$$r_a = k_a \left(C_A C_v - \frac{C_{AS}}{K_A} \right) \tag{2.2}$$

$$r_s = k_s \left(C_{AS} - \frac{C_{BS}}{K_S} \right) \tag{2.3}$$

$$r_d = k_d \left(C_{Bs} - \frac{C_B C_v}{K_D} \right) \tag{2.4}$$

where C_v is the surface concentration of vacant sites, and C_{AS} and C_{BS} are the surface concentrations of species $A \cdot S$ and $B \cdot S$, respectively. If the surface reaction is the rate-controlling step, for instance, then:

$$C_{AS} = K_A C_A C_v \tag{2.5}$$

$$C_{BS} = C_B C_v / K_D = K_B C_B C_v; \qquad K_B = 1/K_D \tag{2.6}$$

$$-r_A = r_s \tag{2.7}$$

In addition, a relationship for the active sites is:

$$C_t = C_v + C_{AS} + C_{BS} \tag{2.8}$$

where C_t is the total number of active sites. Using Eqs. 2.5 and 2.6 in 2.8 results in:

$$C_v = \frac{C_t}{1 + K_A C_A + K_B C_B} \tag{2.9}$$

Therefore, the rate of disappearance of A can be written from Eqs. 2.3 and 2.7, using Eq. 2.9, as:

$$-r_A = r_s = \frac{k_s C_t (K_A C_A - K_B C_B / K_s)}{1 + K_A C_A + K_B C_B}$$

$$= \frac{k(C_A - C_B / K)}{1 + K_A C_A + K_B C_B} \tag{2.10}$$

where $k = k_s C_t K_A$ and the thermodynamic equilibrium constant K is given by $K_A K_s / K_B$. The usual practice is to lump unknowns such as C_t into a single rate constant k.

It should be recognized that the net rates for all individual steps are the same and are equal to the rate of disappearance of A:

$$-r_A = r_s = r_a = r_d \qquad (2.11)$$

When one of the steps is rate-controlling, the net rates for the other steps are not zero. Rather, the ratios r_i/k_i are small as compared to r_i/k_i for the rate-controlling step. For the preceding example, in which the surface reaction is the controlling step, r_a/k_a and r_d/k_d are quite small, whereas r_s/k_s is not. Therefore, when none of the steps is rate-controlling, usually two additional relationships obtained from Eq. 2.11 are sufficient to derive the kinetics. For the previous example, $r_s = r_a$ and $r_a = r_d$, and these two conditions can replace Eqs. 2.5 and 2.6. Because very complex kinetics can result from this general treatment, it is often useful to assume that one of the steps is rate-controlling. There are also other reasons, which will be investigated later. For the reactions of Eq. 2.1, the kinetic expression, when Eq. 2.11 is used, becomes:

$$-r_a = \frac{C_t(C_A - C_B/K)}{\left[\left(\dfrac{1}{K_A k_s} + \dfrac{1}{k_a} + \dfrac{1}{K k_d}\right) + \left(\dfrac{1}{K_A k_s} + \dfrac{1+K_s}{K k_d}\right)K_A C_A + \left(\dfrac{1}{K_A k_s} + \dfrac{1+K_s}{K_s k_a}\right)K_B C_B\right]}$$

$$(2.12)$$

When the surface reaction step is rate-controlling, k_a and k_d are much larger than k_s and Eq. 2.12 reduces to:

$$-r_A = \frac{K_A k_s C_t(C_A - C_B/K)}{1 + K_A C_A + C_B C_B}$$

which is Eq. 2.10.

　　For bimolecular reactions, both reactants may react in adsorbed states or one adsorbed reactant may react with an unadsorped molecule. The former is referred to as a Langmuir–Hinshelwood mechanism and the latter as a Rideal or Eley–Rideal mechanism. The kinetics derived from these mechanisms, often with the surface-reaction step as the controlling step, are usually called Langmuir–Hinshelwood (L–H) kinetics, although this approach to their derivation was pioneered by Hougen and Watson (1947). A complex array of L–H kinetics may be derived for various types of reactions, depending on the specific assumptions concerning the rate-controlling step, dissociative or nondissociative adsorption of species, types of sites involved, and so on. Some examples (Butt 1980) are given in Table 2.1. Consider the reaction (5) in Table 2.1, for which the overall reaction is:

$$\frac{1}{2}A_2 + B \rightleftharpoons C + D$$

Table 2.1 Some Examples of Langmuir–Hinshelwood Rate Expressions

Reaction Description	Individual Steps	Equation $(-r_A)$	Constants
(1) Isomerization: $A \rightleftarrows B$	$A + S \rightleftarrows A \cdot S$ $A \cdot S \rightleftarrows B \cdot S$ $B \cdot S \rightleftarrows B + S$	$-r_A = k_{s1} C_{AS} - k_{s2} C_{BS}$ $= \dfrac{k'_{s1} K_A C_A - k'_{s2} K_B C_B}{1 + K_A C_A + K_B C_B}$	$k'_{s1} = k_{s1} C_t$ $k'_{s2} = k_{s2} C_t$
(2) Decomposition: $A \rightleftarrows B + C$	$A + S \rightleftarrows A \cdot S$ $A \cdot S + S \rightleftarrows B \cdot S + C \cdot S$ $B \cdot S \rightleftarrows B + S$ $C \cdot S \rightleftarrows C + S$	$-r_A = k_{s1} C_{AS} C_v - k_{s2} C_{BS} C_{CS}$ $= \dfrac{k'_{s1} K_A C_A - k'_{s2} K_B K_C C_B C_C}{(1 + K_A C_A + K_B C_B + K_C C_C)^2}$	$k'_{s1} = k_{s1} C_t$ $k'_{s2} = k_{s2} C_t$ $C_v = \dfrac{C_t}{(1 + K_A C_A + K_B C_B + K_C C_C)}$
(3) Bimolecular: $A + B \rightleftarrows C + D$	$A + S \rightleftarrows A \cdot S$ $B + S \rightleftarrows B \cdot S$ $A \cdot S + B \cdot S \rightleftarrows C \cdot S + D \cdot S$ $C \cdot S \rightleftarrows C + S$ $D \cdot S \rightleftarrows D + S$	$-r_A = k_{s1} C_{AS} C_{BS} - k_{s2} C_{CS} C_{DS}$ $= \dfrac{k'_{s1} K_A K_B C_A C_B - k'_{s2} K_C K_D C_C C_D}{(1 + K_A C_A + K_B C_B + K_C C_C + K_D C_D)^2}$	$k'_{s1} = k_{s1} C_t$ $k'_{s2} = k_{s2} C_t$
(4) Bimolecular, different sites: $A + B \rightleftarrows C + D$	$A + S_1 \rightleftarrows A \cdot S_1$ $B + S_2 \rightleftarrows B \cdot S_2$ $A \cdot S_1 + B \cdot S_2 \rightleftarrows C \cdot S_1 + D \cdot S_2$ $C \cdot S_1 \rightleftarrows C + S_1$ $D \cdot S_2 \rightleftarrows D + S_2$	$-r_A = k_{s1} (C_{AS})_1 (C_{BS})_2 - k_{s2} (C_{CS})_1 (C_{DS})_2$ $= \dfrac{k'_{s1} (K_A)_1 (K_B)_2 C_A C_B - k_{s2} (K_C)_1 (K_D)_2 C_C C_D}{[1 + (K_A)_1 C_A + (K_C)_1 C_C][1 + (K_B)_2 C_B + (K_D)_2 C_D]}$	$k'_{s1} = k_{s1} C_{t1} C_{t2}$ $k'_{s2} = k_{s2} C_{t1} C_{t2}$
(5) Bimolecular with dissociation of one reactant; $\frac{1}{2} A_2 + B \rightleftarrows C + D$	$A_2 + 2S \rightleftarrows 2A \cdot S$ $B + S \rightleftarrows B \cdot S$ $A \cdot S + B \cdot S \rightleftarrows C \cdot S + D \cdot S$ $C \cdot S \rightleftarrows C + S$ $D \cdot S \rightleftarrows D + S$	$-r_A = k_{s1} C_{AS} C_{BS} - k_{s2} C_{CS} C_{DS}$ $= \dfrac{k'_{s1} (K_A C_A)^{1/2} K_B C_B - k'_{s2} K_C C_C C_D}{[1 + (K_A C_A)^{1/2} + K_B C_B + K_C C_C + K_D C_D]^2}$	$k'_{s1} = k_{s1} C_t$ $k'_{s2} = k_{s2} C_t$
(6) Bimolecular with one reactant not adsorbed; $A(g) + B \rightleftarrows C$	$B + S \rightleftarrows B \cdot S$ $A(g) + B \cdot S \rightleftarrows C \cdot S$ $C \cdot S \rightleftarrows C + S$	$-r_A = k_{s1} C_{BS} C_A - k_{s2} C_{CS}$ $= \dfrac{k'_{s1} K_B C_B (C_A) - k'_{s2} K_C C_C}{1 + K_B C_B + K_C C_C}$	$k'_{s1} = k_{s1} C_t$

From J.B. Butt, *Reaction Kinetics and Reactor Design*. © 1980, p. 151. Reprinted by permission of Prentice-Hall, Inc., Englewood Cliffs, NJ.

One way of writing the individual steps is:

$$A_2 + 2S \underset{k_{-a}}{\overset{k_a}{\rightleftharpoons}} 2A \cdot S \qquad\qquad (k_a, K_A = k_a/k_{-a})$$

$$B + S \rightleftharpoons B \cdot S \qquad\qquad (k_b, K_B = k_b/k_{-b})$$
$$A \cdot S + B \cdot S \rightleftharpoons C \cdot S + D \cdot S \qquad (k_s, K_s = k_s/k_{-s})$$
$$C \cdot S \rightleftharpoons C + S \qquad\qquad (k_c, K_C = k_{-c}/k_c)$$
$$D \cdot S \rightleftharpoons D + S \qquad\qquad (k_d, K_D = k_{-d}/k_d)$$

If the surface reaction is rate-controlling, then:

$$-r_A = r_s = k_s C_{AS} C_{BS} - k_{-s} C_{CS} C_{DS} \qquad\qquad (2.13)$$

For the other steps:

$$K_A C_A C_{v^2} = C_{AS}^2$$
$$K_B C_B C_v = C_{BS}$$
$$C_{CS} = K_C C_C C_v$$
$$C_{DS} = K_D C_D C_v$$

In addition, there is the following site equation:

$$C_t = C_v + C_{AS} + C_{BS} + C_{DS} + C_{CS} \qquad\qquad (2.15)$$

Using Eqs. 2.14 and 2.15 in Eq. 2.13 gives the results shown in Table 2.1.

A generalization regarding the denominator of the L–H kinetics is of interest here. The adsorption terms appearing in the denominator are sometimes called *adsorption inhibition terms*. The denominator essentially represents the fraction of sites that are vacant, resulting from the expression:

$$\frac{C_v}{C_t} = \frac{1}{\left[1 + \sum_i (K_i C_i)^{m_i}\right]^n} \qquad\qquad (2.16)$$

where K_i and C_i are the equilibrium constant and the concentration of species i adsorbed on the surface, respectively. The constant m_i assumes a value of ½ for dissociative adsorption and a value of 1 for molecular adsorption. The constant n can assume a value ranging from 1 to 3 depending on the mechanism and controlling step of surface reactions (Froment and Bischoff 1979). This general expression is used later when catalyst deactivation is considered in Chapter 5.

2.2   MECHANISMS AND KINETIC MODELS

In order that a mechanism be established for a catalytic reaction, it is necessary to show that the reaction proceeds according to a particular sequence of individual steps. This is by no means an easy task. There are only a handful of reactions for which a rather complete understanding exists. Perhaps the most difficult part is that of verifying each step. Adsorption/desorption experiments, isotope-exchange experiments, and spectroscopy involving various instrumental methods are used for this purpose. Nonhomogeneity of a catalyst surface and possible dependence of the mechanism on operating conditions make the verification difficult. As a matter of fact, there can be a number of postulated mechanisms that describe the kinetics of a reaction equally well. Therefore, even though a postulated mechanism describes experimental data satisfactorily, this fact is not sufficient to confirm the postulated mechanism.

Consider now a class of reactions for which we have a fair level of understanding: hydrogenation–dehydrogenation reactions of hydrocarbons on metals. In particular, consider the following reaction:

$$C_2H_4 \rightleftarrows C_2H_6$$

Early work by Turkevich and coworkers (1950) showed that there was extensive H–D exchange between the hydrogens of C_2H_4 and D_2. This exchange indicates that the individual steps are reversible and the following sequence can be postulated (Gates et al. 1979):

$$H_2 + 2S \underset{k_{-a}}{\overset{k_a}{\rightleftarrows}} 2H \cdot S$$

$$H_2C = CH_2 + 2S \underset{k_{-b}}{\overset{k_b}{\rightleftarrows}} \underset{\substack{| \quad | \\ S \quad S}}{H_2C - CH_2}$$

$$H \cdot S + \underset{\substack{| \quad | \\ S \quad S}}{H_2C - CH_2} \underset{k_{-s}}{\overset{k_s}{\rightleftarrows}} \underset{\substack{| \\ S}}{H_2C - CH_3} + 2S$$

$$\underset{S}{\overset{|}{H}} + \underset{\substack{| \\ S}}{H_2C - CH_3} \underset{k_{-d}}{\overset{k_d}{\rightleftarrows}} C_2H_6 + 2S$$

It is seen that hydrogen atoms on the surface add to the adsorbed olefin in two steps, leading first to an adsorbed intermediate. Infrared spectra (Eischens and Pliskin 1958) and interaction studies of $C_2H_4 + D_2$ (Kemball, 1959) provide ample evidence for such a "half-hydrogenated" state.

The oxidation of SO_2 on platinum is believed to occur according to the following sequence:

$$O_2 + 2S \rightleftarrows 2O \cdot S \qquad (K_A)$$

$$SO_2 + O \cdot S \rightarrow SO_3 \cdot S \qquad (k_s, K_B)$$

$$SO_3 \cdot S \rightleftarrows SO_3 + S \qquad (K_c)$$

When the middle step is rate-controlling and irreversible, the kinetics can be represented by:

$$r = \frac{kK_A^{1/2} (O_2)^{1/2} (SO_2)}{1 + K_A^{1/2} (O_2)^{1/2} + K_C(SO_3)}; \qquad k = k_s C_t \qquad (2.17)$$

The following kinetics have also been shown to describe very well the experimental data of the oxidation reaction:

$$r = k_1 (O_2)^{0.25} (SO_2)/(SO_3)^{1/2} - k_2 [(SO_3)^{1/2}/(O_2)]^{0.25} \qquad (2.18)$$

Although it can be shown that the rate expression of Eq. 2.17 can be reduced to Eq. 2.18 with a certain approximation (Carberry 1976), it nevertheless raises an important question: How is a particular form of kinetic expression for the reaction under consideration selected?

In the absence of a mechanistic understanding, any kinetic model, whether in the Langmuir–Hinshelwood or in the power-law form, may be used. The question of choosing a rate form has been treated in detail by Weller (1956) and Boudart (1956), whose arguments are that the rate expressions are nonunique to the kinetic sequence and L–H forms may actually represent situations that are quite different from that model. Given the empirical nature of rate expressions in either case, one might choose the simplest rate equation that will adequately fit the experimental data, recognizing that the validity of the equation is confined to a limited range of operating conditions. On the other hand, it is preferable to have a rate expression based on some level of mechanistic understanding. Therefore, some of the difficulties encountered in obtaining a mechanistic rate expression are examined here, based on fundamental considerations of surface reactions.

When Langmuir–Hinshelwood kinetics are postulated, it is reasonable to expect that the rate constant will show Arrhenius temperature dependence, while the equilibrium constants will decrease with temperature for exothermic reactions. However, any deviation from this usual temperature dependence does not necessarily mean that L–H kinetics are inconsistent. This point has been raised by Sinfelt et al. (1960). Consider a simple irreversible reaction $A \rightarrow B$. If the surface reaction step is rate-controlling, then:

$$r = \frac{kK_a C_A}{1 + K_a C_A} \qquad (2.19)$$

where K_a is the adsorption equilibrium constant. As discussed here, it is expected that k increases with temperature while K_a decreases. Suppose on the other hand that adsorption equilibrium is not established between A and $A \cdot S$. Now, suppose instead, that the process involves:

$$A + S \xrightarrow{k_1} A \cdot S$$

$$A \cdot S \xrightarrow{k_2} B \cdot S$$

$$B \cdot S \xrightarrow{k_3} B + S$$

where each step is irreversible. Therefore,

$$r_1 = k_1 C_A C_v = r_2 = k_s C_{AS} = r_3 = k_3 C_{BS}$$

If C_{AS} is assumed to be small, the site balance gives $C_v = C_t / (1 + (k_1/k_3)C_A)$, and the rate expression is:

$$r = \frac{k_1 C_t C_A}{1 + (k_1/k_3)C_A} = \frac{(k_3 C_t)(k_1/k_3)C_A}{1 + (k_1/k_3)C_A}$$

$$= \frac{k_3 C_t K C_A}{1 + K C_A}$$

(2.20)

It can readily be seen that K_a in Eq. 2.19 decreases with temperature for exothermic reactions but that K in Eq. 2.20 can decrease or increase with temperature depending on the relative magnitudes of activation energies for k_1 and k_3. The point here is that the equilibrium constants may decrease, increase, or in some cases, go through a maximum as the temperature changes.

The treatment so far has been based on ideal surfaces, which are rare to nonexistent. Although nonideal surface models, such as those discussed in Chapter 1, where the heat of adsorption depends on coverage, can be used for the derivation of kinetics, it is often sufficient to base the kinetics on ideal surface theory. One reason for this is the insensitivity of the L–H form to the precise sort of the kinetic scheme. The insensitivity can readily be seen from the preceding example, whereby two completely different sequences give similar forms for the kinetics as in Eqs. 2.19 and 2.20. Another reason is the fact that plausible assumptions regarding the nature of nonuniform surfaces often give kinetic expressions resembling those for uniform surfaces. For instance, Boudart and coworkers (1967) showed for a relatively simple example that the rate expression for an ideal surface differed only in the exponent of the denominator of the L–H form when compared to that for a nonideal surface in which heats of chemisorption varied linearly with coverage.

2-3   INTERPRETATION OF L-H KINETICS

When the L–H form of kinetics is postulated and the necessary constants are determined from experimental data, there is always the question of the accuracy of these constants, and by inference the accuracy of the L–H form. A very useful method, based on the compensation effect, has been advanced by Boudart et al (1967). This effect, first noted by Constable (1925), refers to a linear relationship between the logarithm of the preexponential factor of the rate equation and the activation energy that is often noted in the kinetics of catalytic reactions on a series of related catalysts. The compensation effect has also been reported for physical adsorption equilibria by Everett (1950) in terms of a linear relationship between the entropy and enthalpy changes for adsorption:

$$(\Delta S_a)^0 = m(\Delta H_a)^0 + b \tag{2.21}$$

where $(\Delta S_a)^0$ is the entropy change referred to a standard state of 1 atm at the temperature of adsorption and $(\Delta H_a)^0$ is the corresponding enthalpy change. From experimental data on the physical adsorption of a large number of gases at less than monolayer coverage on charcoal, Everett obtained the following correlation:

$$(\Delta S_a)^0 = 0.0014 \, (\Delta H_a)^0 - 12.2 \tag{2.22}$$

where $(\Delta S_a)^0$ is in cal/gmol °K and $(\Delta H_a)^0$ in cal/gmol. Following these ideas, Boudart et al. (1967) calculated $(\Delta S_a)^0$ and $(\Delta H_a)^0$ for a number of chemisorbed systems, using adsorption constant data reported from L–H interpretations in the literature, which were deemed to be reasonable results. $(\Delta S_a)^0$ was determined from an experimentally obtained K value, the thermodynamic relationship:

$$K = e^{(\Delta S_a)^0 / R} \, e^{-(\Delta H_a)^0 / RT} \tag{2.23}$$

and values of $-(\Delta H_a)^0$, which may be determined for the particular chemisorption involved. Their results are shown in Figure 2.3 (Butt 1980). It appears that there is no linear relationship. Nevertheless, the plot shows that the Everett correlation is a limiting case for the $(\Delta S_a)^0$ versus $(\Delta H_a)^0$ relationship since almost all the data fall below the line. The result of Figure 2.3 thus provides some guide as to reasonable values of $(\Delta S_a)^0$ in chemisorbed-reacting systems and can be used as a further test of the significance of the adsorption constant that appears in L–H kinetics. These results form the basis for "less rigorous" rules that they established:

$$10 < (-\Delta S_a)^0 < 12.2 - 0.0014 \, (\Delta H_a)^0 \tag{2.24}$$

The nature and the state of adsorbed molecules provide "more rigorous rules" concerning reasonable values of $(\Delta S_a)^0$. First, $(\Delta S_a)^0$ should be negative. This follows from the fact that the chemisorbed layer is a more ordered state than the gas phase. Second, $(\Delta S_a)^0$ should be smaller than the standard state entropy of the gas phase, $(S_g)^0$, since the entropy change for adsorption cannot be larger than the entropy of the nonadsorbed state. These facts can be stated as the following rule:

$$0 < (-\Delta S_a)^0 < (S_g)^0 \tag{2.25}$$

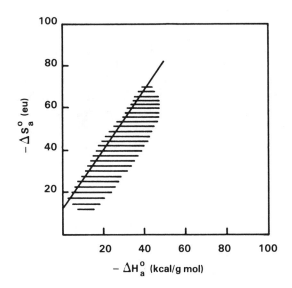

Figure 2.3 Relationship between $(\Delta S_a)^0$ and $(\Delta H_a)^0$ for most chemisorption data. The line is given by Eq. 2.22. (From J.B. Butt, *Reaction Kinetics and Reactor Design*, © 1980, p. 169. Reprinted by permission of Prentice-Hall, Inc., Englewood Cliffs, NJ.)

These two rules (Eqs. 2.24 and 2.25) can be used to check the accuracy of the value of $(\Delta S_a)^0$ determined from kinetic interpretation.

There are also other ways of checking the consistency of L–H kinetics. As Yang and Hougen (1950) first proposed, various rate-controlling steps can be discriminated on the basis of the total pressure dependence of the initial rate. Experimental kinetic data can also be used in a variety of ways (Kittrell and Mezaki 1967) to check the proposed mechanism and controlling step. Some of the rate constants associated with individual steps can be checked by independent experiments or compared with theoretical values. For example, adsorption constants determined by correlation of kinetic data can be compared with those constants obtained directly from adsorption experiments (Kabel and Johanson 1962). The magnitude of rate constant can also be compared with the theoretical value obtained from transition state theory (Sinfelt et al. 1960). Spectroscopic studies can lead to information on individual steps.

For many real reactions, a mechanistically rigorous L–H formulation can become quite complex due to the involvement of several surface intermediates in the elementary steps. The resulting kinetics can be so complex as to lose much of their utility. Many such cases can be usefully treated based on the two-step kinetic model proposed by Boudart (1972). Two simplifying assumptions for the model are that: (1) one step is the rate-determining step, and (2) one surface intermediate is dominant and thus all the other intermediates are present in relatively insignificant amounts. Under the model, it is not necessary to assume details of the overall mechanism. For a given reaction, the formulation depends on the assumptions made as to the rate-determining step and the most abundant surface intermediate.

Summary

This chapter has shown that the general framework of surface reactions leads to Langmuir–Hinshelwood kinetics. Although there are pitfalls in applying the framework to specific cases, some internal checks can be made for self-consistency. In the absence of mechanistic understanding, any rate expression, whether in the Langmuir–Hinshelwood or power-law form, may be used for the purpose of reactor design, with a clear comprehension of the limitations of each. However, the Langmuir–Hinshelwood kinetics should be preferred in the presence of some understanding of the mechanism.

NOTATION

b	constant in Eq. 2.21
C_A, C_B	concentrations of species A and B
C_{AS}, C_{BS}	concentrations of adsorbed species
C_v	surface concentration of vacant sites
C_t	surface concentration of total active sites
$\bar{c}$	average molecular velocity
E	activation energy; activation energy for adsorption
E_d	activation energy for desorption
$(\Delta H_a)^0$	enthalpy change for adsorption at 1 atm and at the temperature of adsorption
k	rate constant
k_a	adsorption rate constant
k_B	Boltzmann constant
k_d	desorption rate constant
k_s	surface reaction rate constant
k_a^0	preexponential factor for k_a
k_d^0	preexponential factor for k_d
K	equilibrium constant
K_i	equilibrium constants
K_a	equilibrium constant in Eq. 2.19
K_A	equilibrium constant for adsorption step
K_s	equilibrium constant for surface reaction step
K_D	equilibrium constant for desorption step
m	mass of a molecule; constant in Eq. 2.21
m_i	constant in Eq. 2.16
n	number of molecules per unit volume; constant in Eq. 2.16
p	partial pressure of adsorbate
Q	heat of adsorption
r_a	net rate of adsorption
r_d	net rate of desorption
r_s	net rate of surface reaction

r_{ad}	rate of adsorption
r_{de}	rate of desorption
R	gas constant
$(-\Delta S_a)^0$	entropy change for adsorption at 1 atm and at the temperature of adsorption
$(S_g)^0$	standard state entropy of gas phase
T	temperature
Z	number of collisions per unit time per unit area

Greek Letters

σ	sticking probability
θ	fractional coverage; angle shown in Figure 2.2
ϕ	angle shown in Figure 2.2

PROBLEMS

2.1. Calculate the rate at which oxygen molecules strike an object exposed to air at 25°C.

2.2. If the object in Problem 2.1 is activated carbon kept in a nitrogen blanket before being exposed to air, what is the initial rate of adsorption? Assume a sticking probability of 0.1 and an activation energy of 1 kcal/mol. If the total surface area of the activated carbon is 2000 m², how long would it take to reach a coverage of 0.95? Assume that the heat of adsorption is high enough to neglect desorption at 25°C.

2.3. The coverage determined by a static apparatus with oxygen is 0.98 at 25°C. On raising the temperature by 100°C at the same partial pressure of oxygen, the coverage changes to 0.1. What is the heat of adsorption?

2.4. Derive the rate expression of Eq. 2.12 and show that it reduces to Eq. 2.10 when the surface reaction step is rate-controlling.

2.5. Show for reaction (5) in Table 2.1 that the fraction of vacant sites is given by:

$$\frac{C_v}{C_t} = \frac{1}{[1 + (K_A C_A)^{1/2} + K_B C_B + K_c C_c + K_D C_D]^2}$$

Compare this with Eq. 2.16.

2.6. Derive the rate expression of Eq. 2.17 and show that it can be reduced to Eq. 2.18.

2.7. If the desorption step in the elementary steps shown for the rate expression of Eq. 2.20 is reversible, how would it change the conclusion regarding the behavior of the equilibrium constant?

2.8. Furusaki (1973) studied the kinetics of the following Deacon process for the production of chlorine.

$$HCl + \frac{1}{4}O_2 \rightleftarrows \frac{1}{2}Cl_2 + \frac{1}{2}H_2O$$

The rate of disappearance of HCl could be correlated to the following expression:

$$r = \frac{k[C_{HCl}C_{O_2}^{1/4} - (1/K)C_{Cl_2}^{1/2}C_{H_2O}^{1/2}]}{[1 + K_1 C_{HCl} + K_2 C_{Cl_2}]^2}$$

Postulate a series of elementary steps that will yield this rate expression.

REFERENCES

Boudart, M., AIChE J., *2*, 62 (1956).

Boudart, M., AIChE J., *18*, 465 (1972).

Boudart, M., D.E. Mears and M.A. Vannice, Ind. Chim. Belge, *32*, 281 (1967).

Butt, J.B., *Reaction Kinetics and Reactor Design*, Prentice-Hall, Englewood Cliffs (1980).

Carberry, J.J., *Chemical and Catalytic Reaction Engineering*, McGraw-Hill, New York (1976).

Constable, F.H., Proc. Roy. Soc. (London), *A108*, 355 (1925).

Eischens, R.P. and W.A. Pliskin, Adv. Cat., *9*, 1 (1958).

Everett, D.H., Trans. Faraday Soc., *46*, 942 (1950).

Froment, G.F. and K.B. Bischoff, *Chemical Reactor Analysis and Design*, Wiley, New York (1979).

Furusaki, S., AIChE J., *19*, 1009 (1973).

Gates, B.C., J.R. Katzer and G.C.A. Schuit, *Chemistry of Catalytic Processes*, McGraw-Hill, New York (1979).

Hougen, O.A. and K.M. Watson, *Chemical Process Principles*, Vol. 3, Wiley, New York (1947).

Kable, R.L. and L.N. Johanson, AIChE J., *8*, 621 (1962).

Kemball, C., Adv. Cat., *10*, 223 (1959).

Kittrell, J.R. and R.A. Mezaki, AIChE J., *13*, 389 (1967).

Moore, W.J., *Physical Chemistry*, 2nd ed., Prentice-Hall, Englewood Cliffs (1955).

Sinflet, J.H., H. Hurwitz and R.A. Shulman, J. Phys. Chem., *64*, 1559 (1960).

Turkevich, J.F. et al., Discuss. Faraday Soc., *8*, 352 (1950).

Weller, S.W., AIChE J., *2*, 59 (1956).

Yang, K.H. and O.A. Hougen, Chem. Eng. Prog., *46*, 146 (1950).

CHAPTER 3

Experimental Reactors and Intrinsic Kinetics

The appropriate first step in designing a heterogeneous reactor is the determination of the intrinsic kinetics for the given catalyst. Unlike homogeneous reactions, the actual rate at which the heterogeneous catalytic reaction takes place, which is termed *global rate,* is not necessarily the same as its intrinsic rate, which is the unique rate for a given catalyst. For example, if the support pores are sufficiently small such that the gaseous reactant molecules cannot move freely into the support interior, the actual rate that we observe is not at its intrinsic rate but rather the rate at which the molecules diffuse into the support pores. If experimentally correlated rate expressions were to be used for the observed rate for reactor design, one would need many such rate expressions to cover all design alternatives since the rate expressions would be different for different situations. On the other hand, only one rate expression for the global rate is needed if it is given in terms of intrinsic kinetics and the conditions affecting the intrinsic rate, thus providing the predictive capability.

There are many types of experimental reactors that can be used for the generation of intrinsic kinetic data. Certain conditions have to be met, however, to assure that the kinetics are truly intrinsic. These conditions are given in this chapter. The reduction of kinetic data to a rate expression is a trial and error process. This process of determining kinetic parameters is treated in some detail. For complex reactions with unknown reaction paths, however, the usual methods may fail. Methods of synthesizing a consistent kinetic structure for such complex reactions are also treated in this chapter.

3–1 EXPERIMENTAL REACTORS AND TRANSPORT CRITERIA

There are many types of laboratory reactors that can be used to determine intrinsic kinetics, provided that proper steps are taken to eliminate various factors obscuring the intrinsic kinetics. These factors include: resistances to mass and heat transport and catalyst deactivation, which is time-dependent.

The laboratory reactors typically used to obtain kinetic data are shown in Figure 3.1. The first two in the figure are fixed-bed reactors: depending on the

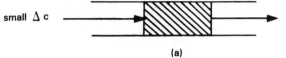

small Δc

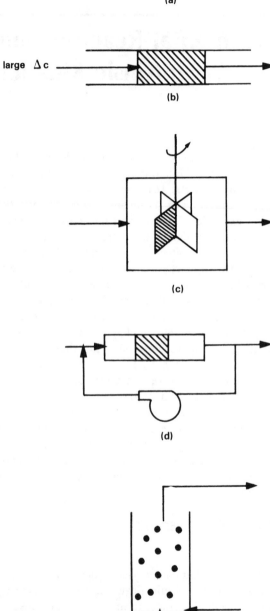

(a)

large Δc

(b)

(c)

(d)

(e)

Figure   3.1 Experimental reactors: (a) differential reactor (b) integral reactor (c) CSTR (d) recycle reactor (e) transport reactor.

concentration difference across the packing (shaded area), a fixed-bed reactor is called differential (small ΔC) or integral (large ΔC). The continuously-stirred-tank catalytic reactor first used by Carberry (1964) is equipped with a stirrer with blades that are packed with catalyst pellets and essentially operates as a CSTR. The recycle reactor in Figure 3.1 is a fixed-bed with a recycle loop (Berty 1974). The transport reactor, in which the catalyst particles are continuously fed from the bottom and removed from the top by the fluid, is useful for reactions affected by rapid catalyst deactivation.

Various types of reactors, including those shown in Figure 3.1, have been evaluated by Weekman (1974) and Berty (1979) with respect to the ease of construction, sampling, isothermality, and contact time. These evaluations show that the usual differential and integral reactors are poor devices compared with the other types. The recycle reactor is perhaps the best overall. The transport reactor is quite useful for rapidly decaying catalytic reactions since steady state conditions can be achieved despite catalyst deactivation.

Given an experimental reactor, catalyst pellet size and fluid velocity are usually chosen to eliminate the disguise of intrinsic kinetics by mass and heat transport resistances. To a certain extent, the fluid concentration and temperature can also be used for that purpose. Small pellet sizes and high fluid velocities tend to lower the concentration and temperature differences caused by the resistances.

The extent to which intrinsic kinetics are disguised by transport resistances can be quantified in a precise manner as will be shown in Chapter 4. The relationships for these effects can be used to arrive at various transport criteria that indicate the importance of temperature and concentration gradients. These criteria are summarized here in Table 3.1 so that they can be used for the specification of reaction conditions leading to intrinsic kinetic data. Details are given in Chapter 4.

All but the transport and integral reactors shown in Figure 3.1 yield kinetic data directly in terms of rates. For the differential and integral reactors, a mass balance yields:

$$F_{A0} \frac{dx_A}{dV} = r_A \tag{3.1}$$

where r_A = rate of reaction per unit catalyst volume
$\quad V$ = volume of catalyst pellets (excluding interparticle voids)
$\quad F_{A0}$ = molar flow rate of species A
$\quad x_A$ = conversion of species A
$\quad N_A$ = moles A

For the differential reactor in which the concentration change across the packing is small, Eq. 3.1 can be rewritten as:

$$r_A = \frac{F_{A0} \Delta x_A}{V} \tag{3.2}$$

Table 3.1 Transport Criteria for Negligible Transport Effects

a. *When the rate expression is not known:*

1. For the absence of concentration gradients in an isothermal pellet (Weisz and Prater 1954):

$$\frac{R_G L^2}{C_b D_e} < 1 \quad \text{(first order reaction)}$$

In general, 0.1 in place of 1 may be used for arbitrary kinetics.

2. For the absence of temperature gradients (within 5% of isothermal rate) (Anderson 1963):

$$\frac{|\Delta H| R_G L^2 \epsilon}{k_e T_b} < 1$$

3. For the absence of temperature gradients across the gas-solid interface (within 5% of the rate of reaction in the absence of the gradients) (Mears 1971):

$$\frac{(-\Delta H) L R_G \epsilon}{h T_b} < 0.15$$

b. *When the rate expression is known:*

1. For the absence of concentration gradients in an isothermal pellet (Petersen 1965; Bischoff 1967):

$$\frac{R_G L^2 f(C_b)}{2 \int_0^{C_b} D_e f(C) dC} < 1; \quad r_c = k f(C)$$

R_G	=	observed rate; C_b = bulk concentration; T_b = bulk temperature
L	=	characteristic length (half-thickness for slab, $R/2$ for cylinder, $R/3$ for sphere)
R	=	radius; D_e = effective diffusivity; k_e = effective thermal conductivity
$(-\Delta H)$	=	heat of reaction; $\epsilon = E_a/R_g T_b$ (E_a = activation energy, R_g = gas constant)
h	=	film heat transfer coefficient; r_c = intrinsic rate of reaction

The rate thus calculated is for the concentration averaged over the catalyst packing, typically the arithmetic average. The data obtained from an integral reactor are analyzed using Eq. 3.1 based on integral analysis (Levenspiel 1972). For the CSTR shown in Figure 3.1, a mass balance yields:

$$r_A = \frac{C_i - C_o}{(V/Q)} \tag{3.3}$$

where C_i = inlet concentration of species A
C_o = outlet concentration of species A
Q = volumetric flow rate

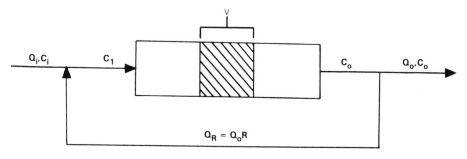

Figure 3.2 Recycle reactor.

It is noted that V here is not the volume of the vessel but rather that of the catalyst. The recycle reactor is redrawn in Figure 3.2. A mass balance around the entire system gives:

$$Q_i C_i - Q_o C_o - \int_v r_A dV = 0$$

Suppose that the reactor has a small volume and operates with a large recycle ratio so that the change in concentration between the inlet and outlet of the reactor is small. As shall soon be shown, the rate is nearly the same in all parts of the packed reactor. Then the above equation can be reduced to:

$$r_A = \frac{Q_i C_i - Q_o C_o}{V} \tag{3.4}$$

Usually $Q_i = Q_o$ and this expression reduces to Eq. 3.3. In order to show that $C_o \approx C_i$ for a large recycle ratio, a mass balance is written around the reactor:

$$C_1 - C_o = \left[\frac{V}{Q_o(R + 1)} \right] r_A \tag{3.5}$$

It is seen that for small V and large R, C_1 is nearly equal to C_o, indicating that the concentration is nearly the same in all parts of the reactor. Equation 3.5 can be used to determine the error involved in treating the recycle reactor as a CSTR. For the transport reactor in Figure 3.1, a mass balance yields:

$$\frac{F_{A0}}{V_v} \frac{dx_A}{dz} = (1 - \epsilon_B) r_A; \qquad 0 \le z \le 1 \tag{3.6}$$

where V_v is the vessel volume and ϵ_B is the bed voidage. This factor is needed for the mass balance since the rate of reaction r_A is based on catalyst volume.

It is seen that measurements of concentration and flow rate lead to kinetic data in terms of rates for differential, CSTR, and recycle reactors. On the other hand, integral kinetic data are obtained in integral and transport reactors. Note that the concentration corresponding to a given rate is either the outlet concentration (CSTR, integral reactor) or an average concentration (differential reactor).

3–2 DETERMINATION OF INTRINSIC KINETICS

Determination of rate expressions from experimental data, even when free from transport effects and deactivation, is at best a trial and error process. Unless there are some experimental evidences regarding the nature of the surface reactions, it may be prudent to correlate the data as a power function of species concentrations:

$$r_A = kC_A^a C_B^b C_D^d \qquad (3.7)$$

The rate constant and the exponents (orders of reaction) can be determined in the usual way using, for instance, initial rate data or the excessive concentration method (Levenspiel 1972). From the power law thus determined, one may postulate a L-H rate expression that in turn can be verified kinetically up to a certain point. As discussed in Chapter 2, it may suffice just to use the power law rate expression in some cases.

When L-H kinetics are postulated, initial rate data are often used not only to determine the kinetics but also to find the rate-controlling step. For the dual site decomposition reaction $A \rightleftarrows B + C$ given in Table 2.1, for instance, one has:

$$r_A = \frac{k_a(C_A - C_B C_C/K)}{1 + K_A C_B C_C/K + K_B C_B + K_C C_C} : \quad \text{adsorption controlling} \qquad (3.8)$$

$$r_A = \frac{k_s K_A(C_A - C_B C_C/K)}{(1 + K_A C_A + K_B C_B + K_C C_C)^2} : \quad \text{surface-reaction controlling} \quad (3.9)$$

$$r_A = \frac{k_d K(C_A/C_C - C_B/K)}{1 + K_A C_A + K K_B C_A/C_C + K_C C_C} : \quad \text{desorption of } B \text{ controlling} \ (3.10)$$

where $K = K_A K_S/K_B K_C$

If there is only A in the feed with no products ($C_B = C_C = 0$), the initial rate expressions can be rewritten directly from Eqs. 3.8 through 3.10 with the aid of the ideal gas law:

$$r_{A_0} = \frac{k_a P}{RT} : \qquad \text{adsorption controlling}$$

$$r_{A_0} = \frac{k_s K_A P/RT}{(1 + K_A P/RT)^2} : \qquad \text{surface-reaction controlling}$$

$$r_{A_0} = \frac{k_d}{K_B} : \qquad \text{desorption of } B \text{ controlling}$$

where P is total pressure. The behavior of these initial rates is shown in Figure 3.3. If initial rates are determined from the experiments in which the feed contains only A, the plot of the initial rates should reveal which step is rate-controlling (Yang and Hougen 1950) in accordance with the behavior shown in Figure 3.3.

While the example just considered is clear-cut, it is often quite difficult to distinguish subtle differences in the shapes of the initial rate data because of the limitations on the range of total pressure that can be covered and the precision of the measurements. Nonetheless, the initial rate experiments are useful to eliminate obviously inappropriate rate forms even when the experimental results do not give any definitive answer.

As suggested earlier, the experimental data from transport and integral reactors cannot be reduced to rates except by numerical differentiation. Therefore, concentration measurements are used as such for data analysis. This requires assuming a rate form and then integrating it for the purpose of estimating those parameters that best fit the experimental data. Needless to say, use of these integral reactors leads to a much more complicated analysis than would be the case with the other reactors that yield rates directly. When an experimental fixed-bed is used in the integral mode (integral reactor), usually the catalyst volume or feed rate of reactant is changed in each run to effect a change in the outlet concentration.

Whether the kinetic data can be reduced to rates or not, the determination of kinetics eventually involves model discrimination in which rival kinetic models are tested against experimental results and further experimental tests are designed to select a model that best describes the kinetics. Readers are referred to the work by Box and Hill (1967) and papers by Hosten and Froment (1976), Dumez et al. (1977), Kittrell (1970), and Buzzi-Ferraris and Forzatti (1983) for model discrimination techniques based on statistical arguments. This section concludes with an interesting example due to Sinfelt et al. (1960).

Sinfelt et al. (1960) obtained initial rate data using a differential reactor for the dehydrogenation of methylcyclohexane (M) over a 0.3% $Pt \cdot Al_2O_3$ catalyst in the presence of H_2 to reduce coking. The reaction is:

$$M \xrightarrow[Pt]{H_2} T$$

The rate data are given in Table 3.2. If the data are correlated in a power law form, one has:

$$r = kM^a(H_2)^b$$

It is readily seen from the data in Table 3.2 that the rate is independent of H_2 pressure. Therefore, the rate expression reduces to:

$$r = k'M^a$$

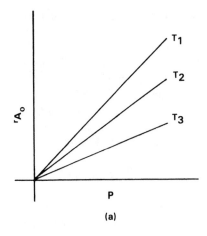

(a)

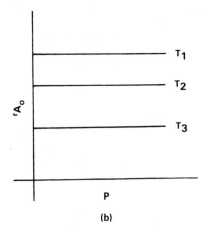

(b)

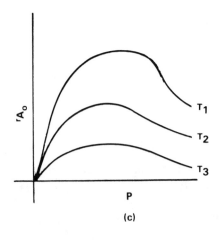

(c)

Figure 3.3   Patterns of initial rates: (a) adsorption controlling (b) desorption controlling (c) surface-reaction controlling.

Table 3.2 Rate Data for Dehydrogenation of Methylcyclohexane (Sinfelt et al. 1960)

Run	T (°C)	M (atm)	H_2 (atm)	r, mol toluene/(hr·gcat)
A a	315	0.36	1.1	1.2
b		0.36	3.0	1.2
c		0.07	1.4	0.86
d		0.24	1.4	1.1
e		0.72	1.4	1.3
B a	344	0.36	1.1	3.0
b		0.36	3.1	3.2
c		0.08	1.4	2.0
d		0.24	1.4	3.4
e		0.68	1.4	3.4
C a	372	0.36	1.1	7.6
b		0.36	4.1	8.0
c		0.26	4.1	12.4
d		0.22	4.1	13.1

The order of reaction, a, determined from the data at three different temperatures gave three different values, suggesting a L-H rate expression of the form:

$$r = \frac{kM}{(1 + KM)^n}$$

This rate expression for $n = 1$ can be rewritten as:

$$\frac{M}{r} = \frac{K}{k} M + \frac{1}{k}$$

The linear plots of this expression yielded straight lines at three temperatures giving the values of K/k and k in Table 3.3. This suggests the following sequence of individual steps:

$$M + S \rightleftarrows M \cdot S \ (K_A)$$

$$M \cdot S \xrightarrow{k_s} T$$

This sequence in which the surface-reaction is irreversible and rate-controlling yields the following rate expression:

$$r = \frac{C_t k_s (M) K_A}{1 + K_A(M)} = \frac{k_s' M K_p}{1 + K_p M} \tag{3.11}$$

where $k_s' = C_t k_s$

$\quad\quad K_p = K_A/RT$

The rate constants appearing in Eq. 3.11 are listed in the last two columns of Table 3.3. As expected from the normal behavior of the rate constant and adsorption equilbrium constant, the Arrhenius plots of k_s' and K_p gave straight lines with k_s' increasing and K_p decreasing with temperature. The activation energy for k_s' was calculated to be 30 kcal/mol. The apparent enthalpy for the equilibrium constant was $(-\Delta H) = 19$ kcal/mol. This value, which can be taken as the activation energy for adsorption, is rather high for the energy barrier for adsorption. The criteria of Boudart and coworkers (1967) discussed in Chapter 2 were also satisfied.

While all would seem to be in accord with the assumed mechanism, they nevertheless found that the presence of aromatics, which are known to chemisorb strongly and preferentially on the catalyst, did not affect the rate much. This suggested that the methylcyclohexane is not in a state of chemisorbed equilibrium. This evidence, along with a rather large value of activation energy for adsorption, prompted them to propose the following nonequilibrium sequence:

$$M + S \xrightarrow{k_1} M \cdot S$$

$$M \cdot S \xrightarrow{k_2} T \cdot S$$

$$T \cdot S \xrightarrow{k_3} T + S$$

in which each step is irreversible. As shown in Chapter 2, this sequence yields:

$$r = \frac{k_3'(k_1/k_3)M}{1 + (k_1/k_3)M}$$

Note that the concentration C_{MS} has been assumed to be small in deference to the observation that the aromatics affect the rate only modestly. With this formulation, the activation energy of $E_3 = 30$ kcal/mol is that of desorption of the strongly bound toluene. From the rate constant ratio (k_1/k_3), it is seen that the activation energy for the chemisorption of methylcyclohexane is 11 kcal/mol. They went on to show that the ratio of preexponential factors determined from experimental data compared reasonably well with a value calculated from transition state theory.

Table 3.3   Rate Constants for the Reaction in Table 3.2 (Sinfelt et al. 1960)

T (°C)	K/k	k	$K(= K_p)$	$k_s'(= k/K_p)$
315	70.67	0.3	21.2	0.0014
344	22.73	0.44	10.0	0.044
372	7.46	0.67	5.0	0.134

This example illustrates the extent to which a surface reaction can be examined based on kinetic measurements. It is seen that even when all reasonable checks are satisfied, there still are possibilities that the reaction may proceed in a way different from the sequence believed to be correct from kinetic information.

3-3 COMPLEX REACTIONS: SYNTHESIS OF KINETIC STRUCTURE

Determination of intrinsic kinetics for complex reactions is relatively simple if the kinetic structure and all species involved in the structure are known. As indicated earlier, the methods due to Box and Hill (1967) and Hosten and Froment (1976) may be used for the determination of kinetic parameters. There is a class of reactions, however, for which the kinetic structure (reaction paths) is not necessarily known. This situation arises when the reactant is a composite of many species as in catalytic cracking of gas oil, which consists of hundreds of species. A simpler and yet similar situation can arise even when only pure species are involved as in chlorination of benzene, in which there are so many dichlorides and trichlorides formed that it is impractical to follow each reaction path. In such a case, the usual practice is to lump a number of species into one group and then treat this group (lump) as a pure species for the purpose of kinetic description (Aris and Gavalas 1966; Hutchinson and Luss 1970; Liu and Lapidus 1973; Luss and Golikeri 1975). This lumping may be done in terms of molecular weight, chemical structure, and so on (Nace et al. 1971), but usually the concentration measurements are available for those lumps that are the final products of the process. The main problem for these complex reactions is to find all reaction paths connecting all the lumps, known and unknown, in the kinetic network in such a way that the kinetic structure is consistent. A kinetic structure is said to be consistent if it describes the behavior of all lumps in the structure and is invariant with respect to the changes in reactant concentration. The construction of a consistent kinetic structure is very difficult. One factor in our favor, however, is that irreversible first-order kinetics are often sufficient to describe the kinetic network if the lumps are chosen properly. A good example is the reaction network for the catalytic cracking of gas oil (Jacob et al. 1976).

Consider a reaction network consisting of irreversible first-order reactions. The basic unit for the most complex case is shown in Figure 3.4. Suppose that I is the lumped reactant from which all products originate, and B and C are the product lumps that are the end products of the process. Suppose further that the problem is to find a consistent kinetic structure. The lump A_1 is an unknown lump added to the structure for consistency as shall be seen. Given these three lumps (I, B, and C), the task is then to find the reaction paths connecting these lumps and the rate constants in such a way that the kinetic structure is consistent. The first step is to find out which lump does not have any disappearance (as opposed to formation) reaction paths. Such a lump, which is lump C in Figure 3.4, should be at the end of the kinetic structure as shown since otherwise it

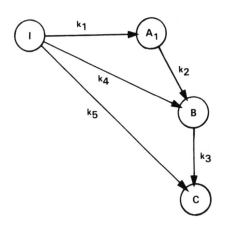

Figure 3.4   Example of a kinetic network.

should have at least one disappearance path. This terminating lump can readily be identified since the concentration of this lump will increase continuously if one lets the reaction proceed for a long period of time. The process of identifying the terminating lump should reveal that B is not a terminating lump. In order to start constructing a consistent kinetic structure, it is necessary to determine whether there are direct paths connecting I to B and C, and B to C. If there is a direct path between I and B, and I and C, the initial concentration gradient (Lee 1978) should be positive:

$$\frac{dC_i}{dt}\bigg|_{t=0} > 0 \qquad i = B, C \tag{3.12}$$

Further, the concentration versus time curve at around time zero should show an exponential increase when a step change in the concentration of I is made. The CSTR reactor in Figure 3.1, for instance, can be used in a batch mode for the purpose of the identification. A similar result can be obtained for the direct

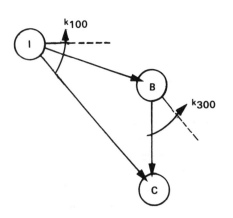

Figure 3.5   First stage of synthesis.

path between B and C when we use only lump B for the reaction in the absence of I. These results lead to the structure shown in Figure 3.5. By now, the values of k_{100} and k_{300}, which are the sums of the rate constants for all disappearance paths involving I and B, respectively, are known since simple linear plots of ln $C_i (i = I, C)$ versus time will yield these values. The next step is to find out whether there are any unknown lumps that need to be identified to ensure consistency. In order to proceed further, a transfer function needs to be introduced.

Suppose that there are n kinetic lumps that need to be added to the structure between I and B, as shown in Figure 3.6. The transfer function (Caughanowr and Koppel 1965) relating a change in the concentration of I to the corresponding change in B in terms of a Laplace transformation variable s is:

$$G_B(s) = \frac{\bar{B}(s)}{\bar{I}(s)} = \frac{k_4}{(s + k_{100})(s + k_{300})}$$

$$+ \frac{k_1}{(s + k_{100})(s + k_{300})}\left(\frac{k_{u\,11}}{s + k_{u\,11}}\right)$$

$$+ \frac{k_1}{(s + k_{100})(s + k_{300})}\sum_{j=1}^{2}\left(\frac{k_{ukj}}{s + k_{ukj}}\right)$$

$$+ \frac{k_1}{(s + k_{100})(s + k_{300})}\sum_{j=1}^{n}\left(\frac{k_{uj}}{s + k_{uj}}\right) \qquad (3.13)$$

where the overbar denotes the deviation variable. When a pulse change in I $[\bar{I}(s) = \bar{I}_0]$ is introduced, the deviation due to this change can be measured for $\bar{B}(t)$. Traditionally, the amplitude ratio is obtained by substituting jw for s. Here,

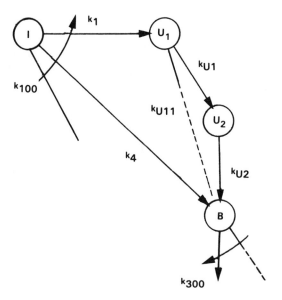

Figure 3.6 Identification of kinetic lumps.

however, a transform ratio (Akella and Lee 1981) (TR) is used in place of the amplitude ratio, which is as accurate but much simpler to evaluate than the amplitude ratio. The transform ratio applicable to any real number is defined by:

$$TR \equiv \frac{\overline{B}(s)}{\overline{I}(s)} \qquad s = \text{real} \qquad (3.14)$$

For the purpose of obtaining the transform ratio, the value of $\overline{B}_1(s)$ for any given real number s can be obtained from:

$$\overline{B}(s) = \int_0^\infty \overline{B}(t) \exp(-st) dt \qquad (3.15)$$

which follows from the definition of the Laplace transformation. If the transform ratio is evaluated at a sufficiently high value of s, Eqs. 3.13 and 3.14 yield:

$$TR(s_1) \approx \frac{k_4}{s_1^2} \qquad (3.16)$$

where s_1 is the real number chosen and $TR(s_1)$ is:

$$TR(s_1) = \frac{1}{\overline{I}_0} \int_0^\infty \overline{B}(t) \exp(-s_1 t) dt \qquad (3.17)$$

Equations 3.16 and 3.17 yield the rate constant for the direct path between I and B. To find any additional paths, $TR1(s_1)$ is defined as:

$$TR1(s_1) \equiv TR(s_1) - \frac{k_4}{(s_1 + k_{100})(s_1 + k_{300})}$$

$$\approx \frac{k_1 k_{u\,11}}{s_1^3} \qquad (3.18)$$

The last part of Eq. 3.18 follows from Eq. 3.13 for a large value of s. The value of $TR1(s_1)$ can readily be calculated since k_4, k_{100}, and k_{300} are now known. Therefore, the value of $k_1 k_{u\,11}$ can be determined from Eq. 3.18. If there is only one kinetic lump, Eq. 3.13 can be written as:

$$G_B(s) = \frac{k_4}{(s + k_{100})(s + k_{300})} + \frac{k_1}{(s + k_{100})(s + k_{300})} \left(\frac{k_{u\,11}}{s + k_{u\,11}}\right) \qquad (3.19)$$

If we evaluate $TR1(s)$ at two large values of s (s_1 and s_2) and take the ratio, it follows from the first part of Eq. 3.18 and Eq. 3.19 that:

$$\frac{TR1(s_1)}{TR1(s_2)} \approx \left(\frac{s_2}{s_1}\right)\left[\frac{(s_2 + k_{100})(s_2 + k_{300})}{(s_1 + k_{100})(s_1 + k_{300})}\right] \tag{3.20}$$

Here the quantity in the square brackets is included for better accuracy since k_{100} and k_{300} are known. The left hand side of Eq. 3.20 should match the right hand side if there is only one kinetic lump between I and B. In general, if there are n kinetic lumps (Akella and Lee 1981), one has:

$$\frac{TR1(s_1)}{TR1(s_2)} = \left(\frac{s_2}{s_1}\right)^n\left[\frac{s_2 + k_{100})(s_2 + k_{300})}{(s_1 + k_{100})(s_1 + k_{300})}\right] \tag{3.21}$$

It follows then that if the right hand side of Eq. 3.21 is evaluated for various n at different values of s and if a value of n is found that matches the left hand side, this n is the number of kinetic lumps that should be added to the kinetic structure. For the structure shown in Figure 3.4, one should find that $n = 1$, leading to the structure in Figure 3.7. The same procedures can be followed to show that there are no kinetic lumps between B and C. The only question remaining is whether there is a direct reaction path between A_1 and C. The transfer function relating I to C in terms of deviation variables is:

$$G_C = \frac{\overline{C}(s)}{\overline{I}(s)} = \frac{k_1 k_2 k_3}{s(s + k_{100})(s + k_2)(s + k_3)} + \frac{k_1 k_w}{s(s + k_{100})(s + k_w)}$$
$$+ \frac{k_3 k_4}{s(s + k_{100})(s + k_3)} + \frac{k_5}{s(s + k_{100})} \tag{3.22}$$

where k_{300} has been set to k_3 in view of the fact that there are no kinetic lumps between B and C. Using the transform ratio again at a large value of s (say s_1):

$$TR(s_1) = \frac{k_5}{s_1^2} \tag{3.23}$$

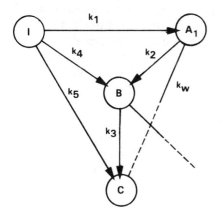

Figure 3.7   Final stage of synthesis.

Table 3.4   Determination of the Number of Kinetic Lumps in Figure 3.6

$k_1 = 0.4$, $k_2 = 0.2$, $k_{u\,1} = 0.1$, $k_{u\,2} = 1.5$, $k_{100} = 0.64$, $k_{300} = 0.15 h^{-1}$

Using the first part of Eq. 3.18

At $s = 10$	$TR1(s)_R = 0.4780 \times 10^{-5}$
$s = 5$	$TR1(s)_R = 0.6231 \times 10^{-4}$
$s = 20$	$TR1(s)_R = 0.3300 \times 10^{-6}$

Using Eq. 3.21 with $s_1 = 10$ and $s_2 = 5$

For $n = 1$	$TR1(5)_1 = 0.3555 \times 10^{-4}$
$n = 2$	$TR1(5)_2 = 0.7110 \times 10^{-4}$
$n = 3$	$TR1(5)_3 = 1.4220 \times 10^{-4}$

Using Eq. 3.21 with $s_1 = 10$ and $s_2 = 20$

For $n = 1$	$TR1(20)_1 = 0.6200 \times 10^{-6}$
$n = 2$	$TR1(20)_2 = 0.3100 \times 10^{-6}$
$n = 3$	$TR1(20)_3 = 0.1550 \times 10^{-6}$

---

which yields the value of k_5. This in turn gives the value of k_1 ($k_{100} = k_1 + k_4 + k_5$) and also that of k_2 ($= k_{u\,11}$) since $k_1 k_{u\,11}$ is known from Eq. 3.18. Now that all rate constants are known with the exception of k_w, the values of $G_c(s)$ at various values of s can be compared with the right hand side of Eq. 3.22 excluding the term with k_w to conclude that there is no direct path ($k_w \approx 0$) between A_1 and C, thus completing the structure.

Consider an example for the determination of the number of kinetic lumps n in Figure 3.6. If one assigns the values of the rate constants as given in Table 3.4, the deviation variable $\bar{B}(t)$ due to a change in I can be calculated, which is used as the process response in this example. Then, $TR(s)$ can be calculated from Eq. 3.17. With this value of $TR(s)$, $TR1(s)$ can also be calculated for a given value of s from the first part of Eq. 3.18. This is given in Table 3.4 for $s = 10$, $s = 5$, and $s = 20$. On the other hand, $TR1(s)$ can also be calculated from Eq. 3.21 for an assumed value of n. These values are given in Table 3.4 for two pairs of s_1 and s_2. It is quite clear that there is a match between the value of $TR1$ obtained from Eq. 3.18 and that from Eq. 3.21 for $n = 2$ (underlined in Table 3.4) at two values of s_2, indicating that there are two kinetic lumps between I and B as shown in Figure 3.6.

The synthesis method given here has been shown (Akella and Lee 1981) to be insensitive to the noise in experimental data and quite effective even when there are as many as eight lumps for which there could conceivably be 7^8 possible kinetic structures. For those complex reaction systems involving pure species with known reaction paths, the method provides an efficient way of identifying rate constants accurately.

Summary

Many types of laboratory reactors are available for the determination of intrinsic kinetics. For the generation of intrinsic kinetic data, the transport criteria given in Table 3.1 can be used to specify experimental conditions, in particular with respect to catalyst particle size and fluid velocity, although it may be necessary to modify these conditions in the course of the experiment. Once the intrinsic kinetic data are generated, the data are reduced to a rate expression, which is a trial and error process at best. The data should be initially correlated in a power-law rate expression. It is desirable to transform this correlation into a rate expression consistent with available understanding of the surface reactions. For complex reactions, this is often impractical or impossible. In such a case, model discrimination methods based on statistical arguments can be used for the correlation. For complex reactions involving many species with unknown reaction paths, the synthesis method given in this chapter can be used.

NOTATION

$\bar{B}$	deviation of concentration of species B from the reference concentration of B when a pulse of species I is introduced
C_i	concentration of species i; inlet concentration
C_o	outlet concentration
$\bar{C}$	deviation of concentration of species C from the reference concentration when a pulse of species B is introduced
F_{A0}	inlet molar flow rate of species A
G_B, G_C	transfer function relating $\bar{I}$ to $\bar{B}$ and $\bar{B}$ to $\bar{C}$
$\bar{I}$	a pulse of species I
$\bar{I}_0$	magnitude of $\bar{I}$
k	rate constant
k_a	rate constant for the adsorption step
k_d	rate constant for the desorption step
k_s	rate constant for the surface reaction step
k_{100}	sum of all rate constants for the disappearance paths of I
k_{300}	sum of all rate constants for the disappearance paths of B
K	equilibrium constant; overall equilibrium constant
K_i	equilibrium constant for species i
K_p	equilibrium constant in terms of pressure
M	concentration of methylcyclohexane
N_A	moles A
n	number of unknown lumps; order of reaction
P	total pressure
Q	volumetric flow rate
Q_i	inlet value of Q

Q_o	outlet value of Q
r_A	rate of reaction for species A
R	gas constant; radius; recycle ratio
s	Laplace transformation variable
t	time
T	temperature
TR	transform ratio defined by Eq. 3.14
$TR1$	transform ratio defined in Eq. 3.18
x_A	conversion of species A
V	catalyst volume
V_v	reactor volume
w	frequency
z	normalized axial coordinate
ϵ_B	void fraction of bed

PROBLEMS

3.1. The intrinsic kinetic data obtained from an integral reactor operated at 4 atm and 488°K are given below for the catalytic reaction, $A \rightarrow 2B + D$. The bulk volume of catalyst is 20 cm³ and the feed rate is 30 l/hr. Obtain a power-law rate expression that best fits the data.

Run number	1	2	3	4
% inerts in the feed	0	20%	40%	60%
$(C_A)_{out}$, gmol/l	0.079	0.065	0.05	0.0344

Note that the volume change due to the reaction has to be taken into consideration.

3.2. The following data are available from a differential reactor for the reaction, $A \rightarrow 3B + D$. The bulk volume of catalyst is 10 cm³ and the feed rate is 20 l/hr. Obtain a power-law rate expression that best fits the data.

Run	C_{Ain} (mol/l)	C_{Aout} (mol/l)
1	0.08	0.07
2	0.06	0.055
3	0.04	0.038

3.3. For a recycle reactor, V/Q_o is 2 ($s \cdot$pellet vol./fluid vol.). If the feed of pure reactant is at 1 atm and 350°K and $r_A = 10^{-6}$ mol/($s \cdot$cm³ pellet vol.), what should the recycle ratio be for the recycle reactor to approach 99% CSTR behavior? Use Eq. 3.5. [$R = 4.8$]

3.4.   If we use the recycle reactor in Prob. 3.3, what is the change in the concentration of the reactant due to the reaction?

3.5.   At relatively low temperatures, the ethylene oxidation reactions over a silver catalyst are:

$$C_2H_4 + \frac{1}{2}O_2 \rightarrow C_2H_4O \qquad (\text{rate} = r_1)$$

$$C_2H_4 + 3O_2 \rightarrow 2CO_2 + 2H_2O \qquad (\text{rate} = r_2)$$

Show that the selectivity S defined by $r_1/(r_1 + r_2)$ can be expressed as

$$S = y/x$$

where x is the conversion of ethylene and y is given by:

$$y = x - \frac{(\%CO_2)_{out}}{2(\%C_2H_4)_{in}}$$

First, derive the y expression. When a recycle reactor is used in excess oxygen with a recycle ratio of 20, the outlet stream contains 6.65 mole% C_2H_4 and 1.32% CO_2. If the feed contains 8.16% C_2H_4, what are the rates r_1 and r_2? Ethylene is fed at the rate of 5.549×10^{-8} mol/($s \cdot g$ catalyst). To what extent does the recycle reactor approach CSTR behavior at a recycle ratio of 20?

3.6.   For the dual site decomposition ($A \rightleftharpoons B + C$) given in Table 2.1, the following initial rate data are available:

Initial Rate [mol/($s \cdot cm^3$ pellet vol.)]	Total Pressure (atm)
5.36×10^{-5}	1
9.60×10^{-5}	2
23.40×10^{-5}	10

If the reaction was carried out at $400°K$ using pure A, what can you say about the rate-controlling step? Determine as many rate and equilibrium constants as you can. How would you design your experiments if you want to determine the rest of the rate and equilibrium constants?

3.7.   Use Eq. 3.11 and the rate and equilibrium constants given in Table 3.3 to show that the criteria due to Boudart and coworkers (1967) discussed in Chapter 2 are satisfied.

3.8.   Use the transform ratios TR and $TR1$ to show that there are two unknown lumps (n) in Figure 3.6. For this purpose, use $k_1 = 5$, $k_2 = 1$, $k_{u1} = 1.5$, $k_{u2} = 0.75$, $k_{100} = 6.5$, $k_{300} = 0.01$. Note that this is a stiff system, the ratio k_{100}/k_{300} being 650. Use the following for the real values of $TR1$ at $s = 50$, 100, and 200:

$$TR1(s) = \frac{5.625}{(s + 6.5)(s + 1.5)(s + 0.75)(s + 0.01)}$$

Then, use Eq. 3.21 for various values of n to show that $n = 2$.

3.9. Use the transform ratios to show how the first order rate constants in the following reaction network can be determined. Give specific steps for the determination.

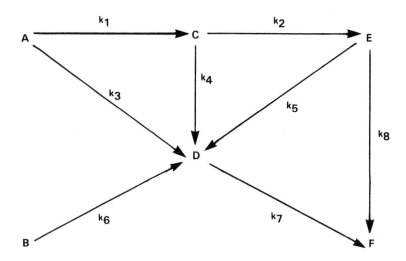

REFERENCES

Akella, L.M. and H.H. Lee, Chem. Eng. J., *22*, 25 (1981).

Anderson, J.B., Chem. Eng. Sci., *18*, 147 (1963).

Aris, R. and G.R. Gavalas, Royal Soc. London Phil. Trans., *260*, 351 (1966).

Berty, J.M., Chem. Eng. Prog., *70*, 78 (1974).

Berty, J.M., Cat. Rev., *20*, 75 (1979).

Bischoff, K.B., Chem. Eng. Sci., *22*, 525 (1967).

Box, G.E.P. and W.J. Hill, Technometrics, *9*, 57 (1967).

Buzzi-Ferraris, G. and P. Forzatti, Chem. Eng. Sci., *38*, 225 (1983).

Carberry, J.J., Ind. Eng. Chem., *56*, 39 (1964).

Caughanowr, D.R. and L.B. Koppel, *Process Systems Analysis and Control,* McGraw-Hill, New York (1965).

Dumez, F., J. Hosten and G.F. Froment, Ind. Eng. Chem. Fund., *16*, 298 (1977).

Hosten, L.H. and G.F. Froment, Proc. 4th Int. Symp. Chem. React. Eng., Heidelberg (1976).

Hutchinson, P. and D. Luss, Chem. Eng. J., *1*, 129 (1970).

Jacob, S.M., B. Gross, S.E. Voltz and V.W. Weekman, Jr., AIChE J., *22*, 701 (1976).

Kittrell, J.R., Adv. Chem. Eng., *8*, 97 (1970).

Lee, H.H., AIChE J., *24*, 116 (1978).

Levenspiel, O., *Chemical Reaction Engineering,* 2nd ed., Wiley, New York (1972).

Liu, Y.A. and L. Lapidus, AIChE J., *19*, 467 (1973).

Luss, D. and S.V. Golikeri, AIChE J., *21*, 865 (1975).

Mears, D.E., J. Catalysis, *20*, 127 (1971); Ind. Eng. Chem. Proc. Des. Dev., *10*, 541 (1971).

Nace, D.M., S.E. Voltz and V.W. Weekman, Jr., Ind. Eng. Chem. Proc. Des. Develop., *10*, 530 (1971).

Petersen, E.E., Chem. Eng. Sci., *20*, 587 (1965).

Sinfelt, J.H., H. Hurwitz and R.A. Shulman, J. Phys. Chem., *64*, 1559 (1960).

Weekman, V.W., Jr., AIChE J., *20*, 833 (1974).

Weisz, P.B. and C.D. Prater, Adv. Cat., *6*, 143 (1954).

Yang, K.H. and O.A. Hougen, Chem. Eng. Prog., *46*, 146 (1950).

PART II

Catalyst Pellets and Particles

CHAPTER 4

Reaction/Heat and Mass Transport

4-1 INTRODUCTION

Catalytic reactions take place on the exposed surface of a catalyst. Consequently, a higher surface area available for the reaction yields a higher rate of reaction. It is therefore necessary to disperse an expensive catalyst on a support of small volume and high surface area. Use of such a supported catalyst in the form of a pellet is not without its adverse effects. Reactants have to diffuse through the pores of the support for the reaction to take place, and therefore, the actual rate can be limited by the rate at which the diffusing reactants reach the catalyst. This actual rate can be determined in terms of intrinsic kinetics and pertinent physical parameters of the diffusional rate process.

Throughout this book, the rate of reaction is based on the apparent volume of catalyst pellets. The intrinsic rate of reaction is represented by:

$$r_c = k_{sp}g(S_r)f(C) \tag{4.1}$$

where the specific rate constant k_{sp} is the rate constant per unit active catalyst surface area, $g(S_r)$ is the functional dependence of the rate constant on active surface area S_r per unit volume, and $f(C)$ is the functional dependence of the intrinsic rate on concentration C. For a given fresh catalyst with uniform activity, $g(S_r)$ is constant everywhere and the intrinsic rate can be written as:

$$r_c = kf(C) \tag{4.2}$$

where $k = k_{sp}g(S_r)$. Here, the units of k for a first-order reaction, for instance, are (volume)/(time-pellet volume).

The factor that relates the intrinsic rate to the global rate R_G is termed the *effectiveness factor*, defined as:

$$\eta = \frac{\text{global rate}}{\text{intrinsic rate}} = \frac{R_G}{r_c} \tag{4.3}$$

95

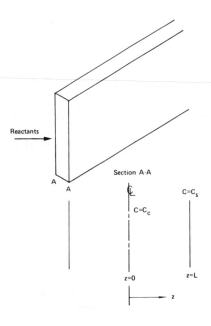

Figure 4.1 Coordinate system for a thin slab-like pellet.

By definition, the global rate is simply the intrinsic rate multiplied by the effectiveness factor. In order to obtain an expression for the effectiveness factor, conservation equations for the diffusion and reaction taking place in a pellet need to be solved. Take as an example the simple case of a first-order, isothermal reaction occurring in a thin slab-like pellet as shown in Figure 4.1. If only bulk diffusive flux is considered, with constant effective diffusivity D_e, a steady state mass balance in one dimension gives:

$$D_e \frac{d^2C}{dz^2} = r_c = kC \qquad (4.4a)$$

with
$$dC/dz = 0 \qquad \text{at } z = 0$$
$$C = C_s \qquad \text{at } z = L$$

where z is the pellet coordinate with the pellet center at the origin, and C_s is the concentration at the pellet surface. The first boundary condition results from the symmetry. The definition of the effectiveness factor gives:

$$\eta = \frac{(1/L)\int_0^L kC dz}{kC_s} = \frac{D_e(dC/dz)_L}{LkC_s}$$

The global rate is the measured rate, which is the rate averaged over the pellet length L or the flux at the pellet surface divided by L. If dimensionless variables are used, the equation can be rewritten as:

$$\frac{d^2\overline{C}}{dy^2} = \phi^2\overline{C}; \qquad \phi^2 = \frac{L^2 k}{D_e}, \qquad \overline{C} = C/C_s, \qquad y = z/L \qquad (4.4b)$$

with
$$\begin{aligned} d\overline{C}/dy &= 0 \qquad \text{at } y = 0 \\ \overline{C} &= 1 \qquad \text{at } y = 1 \end{aligned}$$

$$\eta = \int_0^1 \overline{C}\, dy = \frac{1}{\phi^2}\left(\frac{d\overline{C}}{dy}\right)_{y=1}$$

The dimensionless quantity ϕ is called the *Thiele modulus* after Thiele (1939) who originally studied the effect of diffusion on reaction rate. The significance of this modulus is that one single quantity represents the relative importance of diffusion and reaction taking place in the pellet. When integrated for η, solution of Eq. 4.4 gives:

$$\eta = \frac{\tanh \phi}{\phi} \qquad (4.5)$$

When the Thiele modulus is large, say $\phi \geq 3$, $\tanh \phi$ approaches unity. In this diffusion-limited region, the effectiveness factor is approximated by:

$$\eta_\infty = \frac{1}{\phi} \qquad \phi \geq 3 \qquad (4.6)$$

Thus, the intrinsic rate is reduced by a factor of $1/\phi$ when diffusion limitations are severe. Properties of the effectiveness factor emerge from Eqs. 4.5 and 4.6: the effectiveness factor approaches unity when ϕ is small (diffusion-free reaction) and the product of η and ϕ approaches unity when ϕ is large (diffusion-limited reaction). It is clear from the definition of the Thiele modulus that a higher intrinsic rate of reaction, a larger pellet, or a lower effective diffusivity results in a higher Thiele modulus and thus a relatively lower global rate.

A few assumptions have been made regarding the diffusive flux in treating this simple case. These assumptions are probed in this chapter. It is to be shown that the effectiveness factors for different geometric shapes of pellets converge to the same η–ϕ relationship, when a characteristic pellet dimension is used. A general diffusion-reaction problem is then treated. The diffusion-reaction taking place in a pellet is coupled to the external (interphase) mass and heat transfer through the pellet-bulk fluid interface to show that the major transport resistances lie in the internal diffusion process and the external heat transfer process. It is then possible to explain why the internal diffusion plays a pivotal role in the overall effects of transport processes on the reaction rate. This is followed by the development of a generalized internal effectiveness factor for arbitrary intrinsic kinetics. The result leads to the concept of a general purpose reactor point effectiveness, which relates the intrinsic rate to the global rate at any point in a reactor. This is followed by the development of the reactor point effectiveness for a pellet with a certain activity profile. The chapter ends by considering the influence of pellet design on transport effects.

4-2 FLUXES

The usual treatment of the diffusion-reaction problem leads to a balance equation such as Eq. 4.4. Conditions under which such a treatment is valid are considered in this section. The effective diffusivity D_e is defined by the following relationship between flux and concentration gradient:

$$N_1 = -D_{e_1} \frac{dC_1}{dz} \tag{4.7}$$

where N_1 is the flux of species 1 based on the pellet surface area, consisting of both pores and solids.

Consider the flow and diffusion of n species in one direction. The dusty gas model (Mason and Evans 1969) as utilized by Feng and Stewart (1973) is treated with some modifications. The flux of species i consists of:

$$N_i = \underset{\substack{\text{viscous}\\\text{flow}}}{N_i^{(v)}} + \underset{\substack{\text{gaseous}\\\text{diffusion}}}{N_i^{(g)}} + \underset{\substack{\text{surface}\\\text{diffusion}}}{N_i^{(s)}} \qquad i = 1, \ldots, n \tag{4.8}$$

The viscous flow contribution is:

$$N_i^{(v)} = -C_i \frac{B_0}{\mu} \frac{dP}{dz} \tag{4.9}$$

where B_0 is the Darcy permeability characteristic of the pore geometry, μ is viscosity, and P is total pressure. The gaseous diffusion term $N_i^{(g)}$ is obtained by solving the extended Stefan-Maxwell equation (Bird et al. 1960) which includes Knudsen diffusion:

$$\frac{dC_i}{dz} = \frac{-N_i^{(g)}}{D_{e,ik}^{(r)}} + \sum_{j=1}^{n} \frac{N_j^{(g)} x_i - N_i^{(g)} x_j}{D_{e,ij}} \tag{4.10}$$

where $D_{e,ik}^{(r)}$ is an "effective Knudsen diffusivity" for species i, which is taken as the limiting value for free-molecule flow in a tube and corrected for pellet characteristics, such as voidage (ϵ) and tortuosity (κ) (see Chapter 14 for details):

$$D_{e,ik}^{(r)} = 9.7 \times 10^3 \, r \left(\frac{T}{M_i}\right)^{1/2} f_e \left(\frac{\epsilon}{\kappa}\right) \tag{4.11a}$$

where $f_e(\epsilon/\kappa)$ is the correction factor for the pellet characteristics, r is the average pore radius, T is temperature, and M_i is the molecular weight of species i. The effective molecular diffusivity $D_{e,ij}$ is given by:

$$D_{e,ij} = D_{m,ij}\, f_e\left(\frac{\epsilon}{\kappa}\right) \qquad (4.11b)$$

where $D_{m,ij}$ is the molecular diffusivity given by the Chapman-Enskog formula (see Chapter 14). The surface diffusion term is:

$$N_i^{(s)} = -D_{e,is}^{(r)}\frac{dC_i}{dz} \qquad (4.12)$$

with the flux and diffusivity based on the pellet cross section. The assumption here is that surface coverage is low, and therefore, there is no interaction between adsorbed species. The solution for the fluxes in one direction is:

$$\mathbf{N} = -\mathbf{C}\frac{B_0}{\mu}\frac{dP}{dz} - \mathbf{F}(r)^{-1}\frac{d\mathbf{C}}{dz} - \mathbf{D}_{e,s}\frac{d\mathbf{C}}{dz} \qquad (4.13)$$

$$(\mathbf{F})_{ij} = \frac{\delta_{ij}}{D_{e,ik}^{(r)}} + \sum_{l=1}^{n}\frac{x_l(\delta_{ij}-\delta_{il})}{D_{e,il}} \qquad (4.14)$$

where $\mathbf{D}_{e,s}$ is a diagonal matrix of the effective surface diffusivities, the boldface for vector notation is for all species i, and δ is a Kronecker delta function. An order of magnitude comparison is useful to estimate the relative magnitude of each contribution to the flux: the Darcy permeability is a few orders of magnitude smaller than the effective surface diffusivity, which in turn is approximately three orders of magnitude smaller than the effective molecular diffusivity $D_{e,ij}$. The effective Knudsen diffusivity is approximately an order of magnitude smaller than the effective molecular diffusivity.

For a binary system, Eq. 4.13 reduces to:

$$N_1 = \left[\frac{-D_{e,12}}{1+(m^{1/2}-1)x_1 + D_{e,12}/D_{e,1k}}\right]\frac{dC_1}{dz} - D_{e,1s}\frac{dC_1}{dz}$$

$$- \left[\frac{B_0}{\mu}C_1 + \frac{m^{1/2}D_{e,1k}/R_gT}{1+(m^{1/2}-1)x_1 + D_{e,12}/D_{e,1k}}\right]\frac{dP}{dz} \qquad (4.15)$$

where $m = M_1/M_2$. The corresponding expression for N_2 can be obtained by simple exchange of subscripts. Equation 4.15 reduces to Eq. 4.7 when the pellet is isobaric and surface diffusion is negligible. Then the effective diffusivity in Eq. 4.7 is equal to:

$$\frac{1}{D_{e,1}} = \frac{1+(m^{1/2}-1)x_1}{D_{e,12}} + \frac{1}{D_{e,1k}} \qquad (4.16)$$

This effective diffusivity can be considered constant if the mole fraction of species 1 remains relatively constant or when an average value is used. For a binary system, therefore, the usual treatment of the diffusion-reaction problem is valid for an isobaric pellet and negligible surface diffusion with D_e given by Eq. 4.16. For a multicomponent system, Eqs. 4.13 and 4.14 show that the usual treatment of diffusion-reaction is also valid under the conditions of an isobaric pellet and negligible surface diffusion with the use of average compositions. The pressure effect is usually negligible unless the support is of large pore material (micron size pores) or there exists a very large pressure drop (Gunn and King 1969; DiNapoli et al. 1975).

One particular form of the dusty gas model has been utilized by Hite and Jackson (1977) to arrive at a condition sufficient for a pellet to be considered isobaric. Their form for the relationship between flux and pressure gradient is:

$$\sum_i \frac{N_i}{D_{e,ik}} = \frac{-1}{R_g T}\left(1 + \frac{B_0 P}{\mu}\sum_i \frac{x_i}{D_{e,ik}}\right)\frac{dP}{dz} \tag{4.17}$$

It follows from this relationship that the isobaric condition is guaranteed if:

$$\sum \frac{N_i}{D_{e,ik}} = 0$$

which can be rewritten with the aid of Eq. 4.11 as:

$$\sum_i N_i (M_i)^{1/2} = 0 \tag{4.18}$$

This is known as Graham's relationship for isobaric diffusion. This relationship, together with the flux expression (Eq. 4.17) yields the following relationship often used for a binary system when the isobaric condition prevails:

$$\frac{D_{e,1}}{D_{e,2}} = \left(\frac{M_2}{M_1}\right)^{1/2} \tag{4.19}$$

Since Eq. 4.18 must be satisfied everywhere within the pellet, we may take the divergence of the relationship written in vector form to obtain:

$$\sum_i \text{div } N_i (M_i)^{1/2} = 0$$

If we let ν_i be the stoichiometric coefficient for species i, a mass balance on this species gives:

$$-\text{div } N_i = \nu_i r_c$$

It follows from the above relationship that the isobaric condition is:

$$\sum_i \nu_i (M_i)^{1/2} = 0 \qquad (4.20)$$

This result shows that the generally used relationship of $\Sigma_i \nu_i = 0$ is not necessarily correct for the isobaric condition. For binary systems, however, the condition of $\Sigma_i \nu_i = 0$ requires that $M_1 = M_2$ and therefore the condition is equivalent to that of Eq. 4.20.

With the preliminaries on the fluxes and effective diffusivities established, the diffusion-reaction problem for a single reaction can be stated as follows:

$$D_e \nabla^2 C = r_c$$

with $\qquad\qquad C = C_s$ on pellet surface

and $$\eta = \frac{D_e \iint_{\text{surface}} (\partial C / \partial \mathbf{n})\, ds}{r_c (\text{volume})}$$

where $\mathbf{n}$ is an outward normal length coordinate. Here, the subscripts have been omitted since the balance equation is written for the key species of interest. Pellets of practical interest are usually symmetrical. Furthermore, one dimension of these pellets is usually much shorter than the other dimensions such that one-dimensional mass balance can be used.* For these pellets, the above formulation reduces to Eq. 4.4a. Unless otherwise noted, the pellet is treated as isothermal and surface diffusion is neglected in what follows. Details on the experimental and theoretical aspects of pore diffusion and effective diffusivities are given in Chapter 14.

4–3 INTERNAL EFFECTIVENESS FACTOR AND PELLET SHAPE

The effectiveness factor was defined as the ratio of global rate to intrinsic rate. If both rates are expressed in terms of pellet surface conditions such that the effectiveness factor represents the effects of internal (intraphase) transport resistances, it is termed the internal effectiveness factor as opposed to the external effectiveness factor, which represents the effects of transport resistances external to the pellet surface. The internal effectiveness factor for various pellet shapes will now be derived.

The steady state, one-dimensional mass balance for slab-like, cylindrical and spherical pellets can be written as:

---

* For transverse diffusion effect, see Bischoff, K. B., AIChE J., 5, 135 (1966).

$$\frac{1}{z^m}\frac{d}{dz}\left(z^m D_e \frac{dC}{dz}\right) = r_c; \qquad m = \begin{cases} 0 & \text{slab,} \\ 1 & \text{cylinder,} \\ 2 & \text{sphere} \end{cases} \tag{4.21}$$

The boundary conditions are the same as in Eq. 4.4a. Normalization of the pellet coordinate with respect to the radius for both cylindrical and spherical pellets gives:

$$\frac{1}{y^m}\frac{d}{dy}\left(y^m \frac{d\overline{C}}{dy}\right) = \phi_m \overline{C}^n \tag{4.22}$$

$$\phi_m = \frac{L(kC_s^{n-1}/D_e)^{1/2}}{R(kC_s^{n-1}/D_e)^{1/2}} \qquad \begin{array}{l} m = 0 \text{ (slab)} \\ m = 1, \text{ (cylinder)}, 2 \text{ (sphere)} \end{array} \tag{4.23}$$

where an nth-order reaction and constant diffusivity has been assumed. Analytical solution of Eq. 4.22 for a first-order reaction yield the results given in Table 4.1. This table uses the following definition of the effectiveness factor:

$$\eta_m = \frac{m+1}{(L_m)^{m+1}}\frac{\int_0^{L_m} z^m r_c \, dz}{r_c(C_s)} = \frac{m+1}{\phi_m^2}\left(\frac{d\overline{C}}{dy}\right)_{y=1} \tag{4.23}$$

$$L_m = \begin{array}{ll} L & m = 0 \\ R & m = 1, 2 \end{array}$$

In Table 4.1, I_0 and I_1 are modified Bessel functions of zeroth and first–order. A few observations can be made from Table 4.1. The product of the Thiele modulus ϕ_m and the effectiveness factor η_m approaches unity, two, and three for slab-like, cylindrical, and spherical pellets, respectively, as ϕ_m approaches infinity. On the other hand, the characteristic length L defined as the ratio of apparent pellet

Table 4.1 Internal Effectiveness Factor for Pellets of Various Shapes, First-Order Reaction (Aris, *Elementary Chemical Reactor Analysis,* Prentice-Hall 1969)

Pellet geometry	η_m	$\eta_m\vert_{\phi_m \to \infty}$	$L = \dfrac{\text{volume}}{\text{external surface area}}$ $= V_p/S_x$	$\eta_{\phi(L) \to \infty}$
slab	$\dfrac{\tanh \phi_0}{\phi_0}$	$\dfrac{1}{\phi_0}$	L	$\dfrac{1}{\phi}$
cylinder	$\dfrac{2\,I_1(\phi_1)}{\phi_1\,I_0(\phi_1)}$	$\dfrac{2}{\phi_1}$	$\dfrac{R}{2}$	$\dfrac{1}{\phi}$
sphere	$\dfrac{3}{\phi_2}\left(\dfrac{1}{\tanh \phi_2} - \dfrac{1}{\phi_2}\right)$	$\dfrac{3}{\phi_2}$	$\dfrac{R}{3}$	$\dfrac{1}{\phi}$

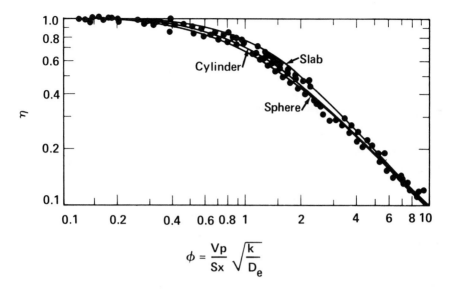

$$\phi = \frac{V_p}{S_x} \sqrt{\frac{k}{D_e}}$$

Figure 4.2   Isothermal effectiveness factor for first-order reactions. The dots represent values for pellets of various shapes. (Rester and Aris 1969. Reprinted with permission from *Chemical Engineering Science.* Copyright by Pergamon Press, Inc.)

volume (V_p) to external pellet surface area (S_x) varies inversely with the same factor. Therefore, the product of the Thiele modulus and the effectiveness factor will approach unity for large values of the Thiele modulus if the characteristic dimension L is used in the definition of the Thiele modulus. This is shown in the last column of Table 4.1. It has been shown* that the use of the characteristic length in the definition of the Thiele modulus brings together all curves of $\eta -\phi$ corresponding to different pellet shapes. This is illustrated in Figure 4.2 (Rester and Aris 1969) for a first-order reaction. The agreement is very good at small and large values of the normalized Thiele modulus. The deviation of the exact value of η from the asymptotes $\eta = 1$ and $\eta\phi = 1$ depends on the specific kinetic expression and the value of ϕ. However, the normalized curve of $\eta - \phi$ for the first-order reaction gives a reasonable approximation in many cases for all values of ϕ, and especially for small and large values of ϕ. In view of these facts, the slab-like geometry will be used in what follows since the half-thickness L is the characteristic length for this shape. This considerably simplifies the mathematics. Shape normalization based on the characteristic length $L(= V_p/S_x)$ enables one to relate the effectiveness factor for one shape to that for any other shape.

Given the internal effectiveness factor, the global rate can be expressed as:

$$R_G = \eta k(T_s)f(C_s)$$

---

* For the general treatment of arbitrary shapes involving a first-order reaction by Murray, see the book by Aris, *The Mathematical Theory of Diffusion and Reaction in Permeable Catalysts,* Vol. I, Oxford University Press, London (1975).

For a rate constant obeying the Arrhenius relationship, differentiation gives:

$$\frac{d \ln R_G}{d(1/T_s)} = \frac{d \ln k}{d(1/T_s)} + \frac{d \ln \eta}{d(1/T_s)}$$

or

$$(E_a)_{obs} = E_a - R_g \frac{d \ln \eta}{d \ln \phi} \frac{d \ln \phi}{d(1/T_s)} \qquad (4.24)$$

where E_a is the intrinsic activation energy, $(E_a)_{obs}$ is the observed activation energy and R_g is the gas constant. For small values of ϕ, $(d \ln \eta/d \ln \phi)$ approaches zero, and the observed and intrinsic activation energies are the same. On the other hand, this derivative approaches (-1) for large values of ϕ and Eq. 4.24 gives:

$$(E_a)_{obs} = \frac{E_a}{2} \qquad (4.25)$$

where the temperature dependence of the effective diffusivity has been neglected. Thus the observed activation energy in the asymptotic region of strong diffusion

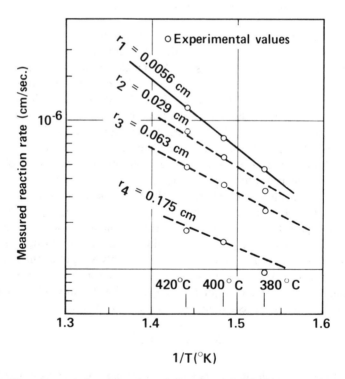

Figure 4.3 Experimental demonstration of the effect of diffusion on apparent activation energy. (Weisz and Prater 1954. Reprinted with permission from *Advances in Catalysis.* Copyright by Academic Press.)

effects is one half the intrinsic value. Experimental evidence of the above result is provided by the study of Weisz and Prater (1954) shown in Figure 4.3. They obtained these results for the cracking of cumene on a silica-alumina catalyst in which the change in Thiele modulus was effected by varying the pellet size. It is seen that the slope of the ln R_G versus $1/T$ plot for severely diffusion-limited pellets of 0.175 cm radius is approximately one-half that of the 0.0056 cm pellets. Following similar procedures as in Eq. 4.24, it can be shown for an nth-order reaction that:

$$(n)_{obs} = n + \left(\frac{n-1}{2}\right)\frac{d \ln \eta}{d \ln \phi} \tag{4.26}$$

It follows from Eq. 4.26 that the observed order of reaction $(n)_{obs}$ is $(n+1)/2$ in the asymptotic region of strong diffusion effects.

A comparison between theoretical and experimental effectiveness factors has been made by Dente and Pasquou (1965) using the methanol oxidation reaction over an Fe-molybdenum oxide catalyst in a differential reactor. Results are shown in Figure 4.4. With excess O_2 the intrinsic rate is one-half order in methanol. The experimental effectiveness factors in Figure 4.4 were calculated from:

$$\eta = \frac{R_G}{r_c}$$

using the experimental differential rates while the Thiele moduli were calculated from the intrinsic kinetics. The solid line is the theoretical curve for one-half

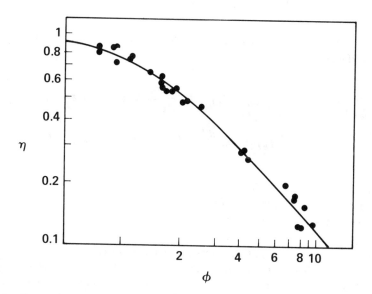

Figure 4.4 Comparison between theoretical and experimental effectiveness factors. (Dente and Pasquou 1965)

order kinetics. Good agreement is seen in the figure. Under these experimental conditions, the external mass transport resistance was negligible. The pellet was assumed to be isothermal at the temperature dictated by the external heat transport process. The validity of these assumptions will now be examined.

4–4 INTERNAL AND EXTERNAL TRANSPORT PROCESSES

Only isothermal pellets and internal effectiveness factors have been treated so far. Limitations on internal heat transfer, external mass transfer, and external heat transfer can all affect the reaction rate. Consider a pellet placed in a fluid medium. Assuming constant physical properties, one-dimensional mass and heat balances yield for a slab-like pellet:

$$D_e \frac{d^2 C}{dz^2} = r_c = k(T)f(C) \tag{4.27}$$

$$-\lambda_e \frac{d^2 T}{dz^2} = (-\Delta H)k(T)f(C) \tag{4.28}$$

$$\begin{array}{ll} C = C_s, \; T = T_s & \text{at } z = L \\ dC/dz = dT/dz = 0 & \text{at } z = 0 \end{array} \tag{4.29}$$

where λ_e = effective thermal conductivity
$(-\Delta H)$ = heat of reaction
$\quad s$ = subscript denoting pellet surface

The concentration and temperature are also affected by the resistances at the fluid-pellet interface:

$$k_g(C_b - C_s) = D_e \frac{dC}{dz} = R_G L \qquad \text{at } z = L \tag{4.30}$$

$$h(T_s - T_b) = -\lambda_e \frac{dT}{dz} = (-\Delta H)R_G L \qquad \text{at } z = L \tag{4.31}$$

where k_g = film mass transfer coefficient
$\quad h$ = film heat transfer coefficient
$\quad b$ = subscript denoting bulk-fluid condition

These relationships state that the mass and energy transferred across the interface are equal to the fluxes at the pellet surface, which in turn are respectively equal to the global rate of reaction per unit area and the rate at which the heat is generated by the reaction. These relationships can be used if desired, to eliminate the usually unknown surface quantities C_s and T_s. In dimensionless form, Eqs. 4.27 through 4.31 become:

$$\frac{d^2c}{dy^2} = \phi^2 F(c) \exp[-\epsilon(1/t - 1)] \tag{4.32}$$

$$\frac{d^2t}{dy^2} = -\beta\phi^2 F(c) \exp[-\epsilon(1/t - 1)] \tag{4.33}$$

$$c = c_s, \, t = t_s \quad \text{at } y = 1$$

$$\frac{dc}{dy} = \frac{dt}{dy} = 0 \quad \text{at } y = 0 \tag{4.34}$$

$$1 - c_s = \frac{1}{(Bi)_m} \frac{dc}{dy} = (\eta Da) \quad \text{at } y = 1 \tag{4.35}$$

$$t_s - 1 = -\frac{1}{(Bi)_h} \frac{dt}{dy} = \beta_x(\eta Da) \quad \text{at } y = 1 \tag{4.36}$$

where $c = C/C_b$, $t = T/T_b$, $y = z/L$, and $F(c) = f(C)/f(C_b)$.

The Biot numbers for mass and heat, which are a measure of the ratio of internal to external resistance, are given by:

$$(Bi)_m = \frac{k_g L}{D_e} \tag{4.37}$$

$$(Bi)_h = \frac{hL}{\lambda_e} \tag{4.38}$$

The Damkoehler number Da, which is a measure of the ratio of chemical reaction velocity to external mass transport velocity, is an observable quantity when multiplied by the effectiveness factor η:

$$(\eta Da) = \frac{LR_G}{k_g C_b} \tag{4.39}$$

The dimensionless quantities β (sometimes called Prater number) and β_x relate the concentration differences to the temperature differences within the pellet and external to the pellet, respectively, as will be seen later.

$$\beta = \frac{(-\Delta H)D_e C_b}{\lambda_e T_b} \tag{4.40}$$

$$\beta_x = \frac{(-\Delta H)k_g C_b}{hT_b} \tag{4.41}$$

It follows from Eqs. 4.37, 4.38, 4.40, and 4.41 that:

$$\frac{\beta_x}{\beta} = \frac{(\text{Bi})_m}{(\text{Bi})_h} \equiv B_r \qquad (4.42)$$

The Arrhenius number ϵ and the Thiele modulus are given by:

$$\epsilon = \frac{E_a}{R_g T_b} \qquad (4.43)$$

$$\phi^2 = \frac{L^2 k_b f(C_b)}{D_e C_b} \qquad (4.44)$$

where k_b is the rate constant evaluated at T_b. The overall effectiveness factor expressed in terms of bulk-fluid conditions is obtained by:

$$\eta_0 = \frac{D_e (dC/dz)_{z=L}}{L r_c (C_b, T_b)} \qquad (4.45)$$

The above can be rewritten to give:

$$\eta_0 = \left(\frac{D_e (dC/dz)_{z=L}}{L r_c (C_s, T_s)}\right) \left(\frac{r_c (C_s, T_s)}{r_c (C_b, T_b)}\right) \qquad (4.46)$$

$$\equiv \eta_i \eta_x$$

where by definition the first quantity is the internal effectiveness factor η_i and the second is the external effectiveness factor η_x. Although the symbol η was used previously for the internal effectiveness factor, the subscripts are now used to differentiate internal and external effectiveness factor. Nonisothermal effectiveness factors are considered next before proceeding to the determination of the relative importance of the various transport processes.

In order to obtain the nonisothermal internal effectiveness factor, Eqs. 4.27 through 4.29, or equivalently Eqs. 4.32 through 4.34, have to be solved numerically for use in the definition of the internal effectiveness factor given in Eq. 4.46. No analytical solutions exist even for a first-order reaction. Shown in Figure 4.5 are the numerical results obtained by Weisz and Hicks (1962) for an irreversible first-order reaction in a spherical pellet. It is seen that the effectiveness factor for endothermic reactions ($\beta < 0$) is always smaller than that for the isothermal case ($\beta = 0$) at the same ϕ value, whereas the effectiveness factor for exothermic reactions ($\beta > 0$) is always larger. Furthermore, the effectiveness factor for exothermic reactions exceeds unity in certain ranges of ϕ values. This occurs when the increase in reaction rate due to temperature rise within the pellet exceeds the decrease in the rate due to concentration gradients. Figure 4.5 also reveals that the effectiveness factor for exothermic reactions is not unique in certain ranges

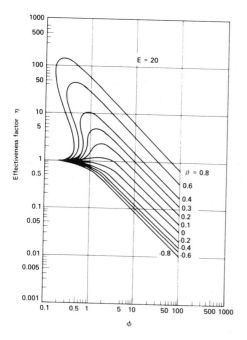

Figure 4.5 Nonisothermal effectiveness factors for irreversible first-order reactions in spherical pellets. (Weisz and Hicks 1962. Reprinted with permission from *Chemical Engineering Science.* Copyright by Pergamon Press, Inc.)

of ϕ values, i.e., steady state multiplicity exists. It should be pointed out (Luss 1977), however, that the multiplicity due to the internal temperature gradient usually occurs at unrealistic values of the parameters and is mainly of academic interest. When a single reaction is involved, the external and not the internal gradient is usually responsible for any steady state multiplicity.

When the catalyst is not supported, as in the case of metal wire or foil, only the external transport processes affect the reaction rate. In such a case, Eqs. 4.35 and 4.36 with η replaced by η_x can be used in the definition of η_x given in Eq. 4.46:

$$\eta_x = \frac{k_s f(C_s)}{k_b f(C_b)} = \frac{f(C_s)}{f(C_b)} \exp[-\epsilon(1/t_s - 1)] \tag{4.47}$$

Using Eqs. 4.35 and 4.36 in 4.47 yields the following implicit relationship:

$$\eta_x = \frac{f[C_b(1 - \eta_x \mathrm{Da})]}{f(C_b)} \exp\left[-\epsilon\left(\frac{1}{1 + \beta_x \eta_x \mathrm{Da}} - 1\right)\right] \tag{4.48}$$

where

$$\eta_x \mathrm{Da} = \frac{L R_G(C_b, T_b)}{k_g C_b} \tag{4.49}$$

For a first-order reaction, this reduces to:

$$\eta_x = (1 - \eta_x \mathrm{Da}) \exp\left[-\epsilon\left(\frac{1}{1 + \beta_x \eta_x \mathrm{Da}} - 1\right)\right] \tag{4.50}$$

Shown in Figure 4.6 is this relationship obtained by Carberry and Kulkarini (1973). Similar behavior as in Figure 4.5 can be observed. However, multiple steady states do not exist. It is obvious that the external effectiveness factor by itself is mainly of academic interest except for those unusual cases where the catalyst has only an external surface.

It is clear from the foregoing discussion that the general problem of diffusion-reaction for the overall effectiveness factor is quite involved. Fortunately, however, the physical and chemical processes at work under realistic conditions favor isothermal pellets and negligible external mass transfer resistances. A more detailed examination of this is in order. Combining Eqs. 4.32 and 4.33 results in:

$$\frac{d^2t}{dy^2} = -\beta\frac{d^2c}{dy^2} \tag{4.51}$$

Integration of this equation with respect to y once using Eqs. 4.34 through 4.36 gives:

$$t_s - 1 = \beta B_r(1 - c_s) \tag{4.52}$$

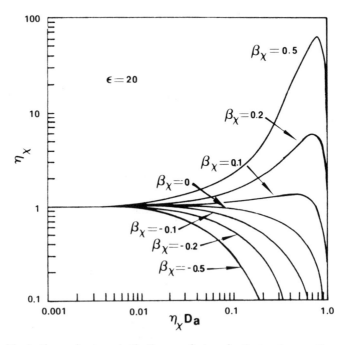

Figure 4.6 Nonisothermal external effectiveness factors for first-order reactions. (Carberry and Kulkarini 1973. Reprinted with permission from Journal of Catalysis. Copyright by Academic Press.)

Integrating the same equation first from zero to y and then from zero to unity results in:

$$t_c - t_s = \beta(c_s - c_c) \qquad (4.53)$$

which is the Prater relationship. Here, the subscript c denotes the pellet center. Dividing Eq. 4.53 by 4.52 gives:

$$\frac{\Delta T_i}{\Delta T_x} = \frac{1}{B_r}\frac{\Delta C_i}{\Delta C_x} \qquad (4.54)$$

where $\Delta T_i = T_c - T_s$, $\Delta C_i = C_s - C_c$, $\Delta T_x = T_s - T_b$, and $\Delta C_x = C_b - C_s$. The maximum value of ΔT_i is that corresponding to $C_c = 0$. It follows from Eqs. 4.35 and 4.53 that:

$$(\Delta T_i)_{max} = \beta T_b(1 - \eta\mathrm{Da}) \qquad (4.55)$$

Also, from Eqs. 4.36 and 4.42:

$$\Delta T_x = T_b\beta_x\eta\mathrm{Da} = T_b\beta B_r\eta\mathrm{Da} \qquad (4.56)$$

Combining Eqs. 4.55 and 4.56 gives, after some manipulations:

$$\frac{\Delta T_x}{\Delta T_x + (\Delta T_i)_{max}} = \frac{\Delta T_x}{(\Delta T_0)_{max}} = \frac{B_r\eta\mathrm{Da}}{1 + \eta\mathrm{Da}(B_r - 1)} \qquad (4.57)$$

The temperature ratios are shown in Figure 4.7 as a function of $\eta\mathrm{Da}(\leq 1)$ with B_r as the parameter (Carberry 1975). It is seen that the overall temperature differ-

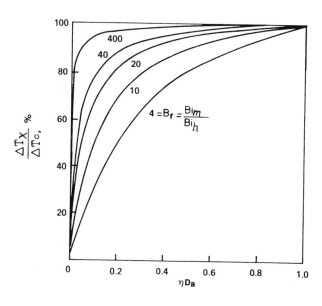

Figure 4.7 Ratio of external to total temperature difference for catalytic reactions. (Carberry 1975. Reprinted with permission from *Industrial and Engineering Chemistry Fundamentals.* Copyright by American Chemical Society.)

ence ΔT_0 is almost entirely due to the external temperature difference when the Biot number ratio B_r is large. The various dimensionless groups and typical ranges of parameters are summarized in Table 4.2. Here, k_g and h for the Biot numbers are obtained from j factor correlations (see Chapter 14 for details):

$$j_D = \frac{\epsilon_B k_g}{v}\, Sc^{2/3} = j_H = \frac{h\epsilon_B}{\rho v C_p}\, Pr^{2/3} = K \frac{1}{Re^\alpha} \qquad (4.58)$$

where v is superficial velocity, and K and α are correlation constants. It follows from the above relationship and the Biot number definitions that:

$$B_r = \frac{\lambda_e}{D_e \rho C_p Le^{2/3}}$$

which can be used to arrive at the typical range of B_r in Table 4.2.

The ranges of parameters given in Table 4.2 show that the Biot number ratio is large (>10) for gas-solid systems. This suggests that the pellet can be treated as isothermal (Butt et al 1977). In fact, the ratio is in the range of hundreds under typical reaction conditions. Kehoe and Butt (1972) measured temperature profiles of a supported Ni pellet during the hydrogenation of benzene in a single-pellet reactor. Their measurements, shown in Figure 4.8, confirm the theoretical

Table 4.2 Ranges of Parameters and Dimensionless Quantities

a. *Ranges of parameters*

Quantity	Gas	Liquid	Porous Solid	Gas–Solid	Liquid–Solid
D_e(cm²/sec)	1–0.1	10^{-5}–10^{-6}	10^{-1}–10^{-3}	—	—
λ_e(cal/cm sec °C)	10^{-4}–10^{-5}	10^{-2}–10^{-4}	10^{-3}–10^{-4}	—	—
ρC_p(cal/cm³ °C)	10^{-2}–10^{-4}	10^{-1}–2	1–0.4	—	—
ϵ	—	—	—	5–30	5–30
β	—	—	—	0.001–0.3	0.001–0.1
β_x	—	—	—	0.01–2.0	0.001–0.05
$B_r = (Bi)_m/(Bi)_h$	—	—	—	1.0–10^4	10^{-4}–10^{-1}

b. *Dimensionless groups*

$(\eta Da) = LR_G/k_g C_b$

$(Bi)_m = k_g L/D_e$

$(Bi)_h = hL/\lambda_e$

$\beta = (-\Delta H)D_e C_b/\lambda_e T_b$

$\beta_x = (-\Delta H)k_g C_b/hT_b$

$B_r = \beta_x/\beta = (Bi)_m/(Bi)_h$

$\epsilon = E_a/R_g T_b$

$Da = Lr_c/k_g C_b$

$\eta\phi^2 = R_G L^2/D_e C_b$

$Sc = \mu/\rho D_m$

$Pr = C_p \mu/\lambda$

$Le = Sc/Pr$

$Re = d_p v/v$

After J.J. Carberry, *Chemical and Catalytic Reaction Engineering,* © 1976; with permission of McGraw-Hill Book Company, New York.

results. It is seen that the pellet is indeed isothermal when the Biot number ratio is large, whereas there can exist an internal temperature gradient when the ratio is relatively small.

The fact that the Biot number of mass (a measure of the ratio of internal diffusion to external mass transfer resistance) is much larger than unity, implies that the major resistance lies in the internal diffusion process. A simple analysis can be made to assess the relative importance of internal and external mass transport processes now that the pellet can be considered isothermal. For an isothermal pellet, Eq. 4.32 can be written as:

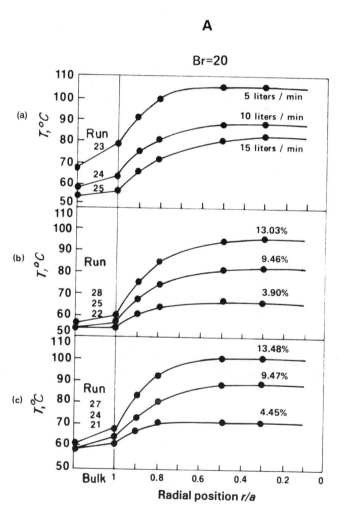

Figure 4.8a Measured temperature profiles in pellets for $(Bi)_m/(Bi)_h = 20$ with a feed temperature of 52°C: (a) as a function of flow rate at about 10% C_6H_6 (b) as a function of feed composition at 15 l/min (c) at 10 l/min. (Kehoe and Butt 1972)

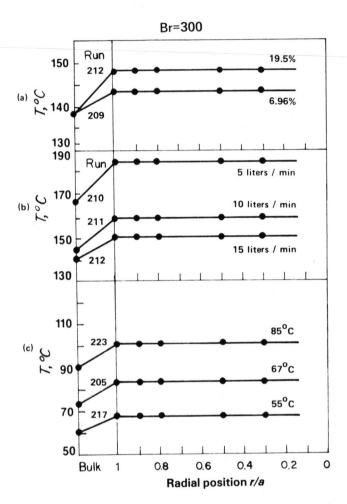

Figure 4.8b Measured profiles in pellets for $(Bi)_m/(Bi)_h = 300$: (a) as a function of feed composition (b) as a function of flow rate (c) as a function of feed temperature. (Kehoe and Butt 1972)

$$\frac{d^2\bar{c}}{dy^2} = \phi_s^2 F(\bar{c})$$

(4.59)

$$\phi_s^2 = \frac{k_s L^2 f(C_s)}{D_e C_s}, \qquad F(\bar{c}) = \frac{f(C)}{f(C_s)}, \qquad \bar{c}_A = \frac{C}{C_s}$$

where k_s is the rate constant at T_s. Integrating this equation with the boundary conditions of Eqs. 4.34 and 4.35 gives:

$$1 - c_s = \frac{\phi_s^2}{(Bi)_m} \int_0^1 F(\bar{c}) \, dy \qquad (4.60)$$

Since the internal effectiveness factor is given by:

$$\eta_i = \int_0^1 F(\bar{c}) \, dy \qquad (4.61)$$

Equation 4.60 can be rewritten as:

$$1 - c_s = \frac{\phi_s^2 \eta_i}{(Bi)_m} \qquad (4.62)$$

Therefore:

$$\frac{C_b - C_s}{C_b} = \frac{\phi_s^2 \eta_i}{(Bi)_m} \qquad (4.63)$$

This result reveals that the external concentration difference is strongly dependent on the internal diffusion process: when the internal diffusion resistance is negligible, $\phi_s^2 \eta_i$ approaches zero and so does the concentration difference, whereas this difference increases with increasing ϕ_s and approaches $\phi_s/(Bi)_m$ in the asymptotic region of strong diffusion effects ($\eta_i \phi_s \rightarrow 1$). As indicated earlier, the value of $(Bi)_m$ is typically in the range of hundreds. Therefore, the external concentration difference is negligible unless the Thiele modulus assumes a value close to that of the Biot number for mass.

It has been shown that under realistic reaction conditions, the major transport resistances are external heat transport and internal diffusion. The pellet can be treated as isothermal and the external mass transfer resistance can be neglected. While these conclusions are valid for gas–solid systems, they are not valid for liquid–solid systems as evident from the range of the Biot number ratio (10^{-4}–10^{-1}) given in Table 4.2 for the liquid–solid systems.

4–5 WHY IS DIFFUSION IMPORTANT?

One may wonder at this point why so much attention is being paid to the internal diffusion. If indeed the diffusional process reduces the rate of reaction, wouldn't it be better to eliminate the diffusion resistance altogether? This question was addressed at the beginning of this chapter. A systematic approach is taken here to probe the matter further. Consider the Thiele modulus for a first-order reaction defined in Eq. 4.4b:

$$\phi = L \left(\frac{k}{D_e} \right)^{1/2} = L \left(\frac{k_{sp} g(S_r)}{D_e} \right)^{1/2} \qquad (4.64)$$

where Eq. 4.1 has been used for k. The Thiele modulus is a composite of three factors: the characteristic pellet size L, a measure of the catalyst dispersion $g(S_r)$ for a given catalyst loading, and the type of support with a certain surface area and pore characteristics that determine the value of D_e. For a given support, the pellet size and the method of impregnation that yields a certain dispersion $[g(S_r)]$ determine the modulus.

It is always possible to reduce the pellet size and thus to reduce the diffusional resistance. The extent to which this is practical is limited by the pressure drop in the reactor that a tight packing of small pellets can cause. This brings up an important question: Is it desirable to have a higher dispersion that will result in a higher diffusional resistance? The answer is rather obvious if it is recognized that a higher dispersion, despite the higher diffusional resistance it causes, will result in a higher apparent rate of reaction or a higher rate constant. This can be seen by examining the apparent rate constant $(k)_a$ for a first-order reaction:

$$(k)_a = k\eta = k \tanh \phi/\phi$$

$$= (k)^{1/2} \frac{(D_e)^{1/2}}{L} \tanh [L(k/D_e)^{1/2}] \tag{4.65}$$

It is clear from the examination of this equation that the apparent rate constant $(k)_a$ increases with increasing values of k or with increasing dispersion. Therefore, it is desirable for a given support and pellet size to have as high a dispersion as practical despite the diffusion resistance that highly dispersed catalyst brings about. It shall soon be shown that this desire has to be moderated for various reasons. If a reaction is highly exothermic, too high a dispersion or too high a rate of reaction may cause a runaway situation. It can also cause faster sintering of catalyst crystallites, resulting in a higher rate of deactivation.

Consider the choice of a support. For a given pellet size, a support with a higher surface area gives a higher rate constant, since the rate constant is proportional to the surface area of the support for the same catalyst loading per unit pellet volume. On the other hand, the support with a higher surface area has smaller pores, which tend to cause a reduction in effective diffusivity. Therefore, detailed knowledge of the support is necessary to make definitive conclusions regarding the choice of support and its effect on the apparent rate constant. If the average pore radius of a support is used as a rough measure of surface area, the rate constant can be considered to be inversely proportional to the square of the average pore radius r since the surface area is inversely proportional to r. Therefore, the apparent rate constant will increase proportionally to $(1/r)$ according to Eq. 4.65 if $g(S_r)$ is simply S_r, which is the active surface area per unit volume. Now consider the effect of r on the effective diffusivity. For the simple case of equimolar counterdiffusion, Eq. 4.16 reduces to:

$$\frac{1}{D_e} = \frac{1}{D_{e,12}} + \frac{1}{D_{e,k}} \tag{4.66}$$

where $D_{e,12} = D_m f_e(\epsilon/\kappa)$
$$D_{e,k} = 9.7 \times 10^3 \, r \, (T/M)^{1/2} \, f_e(\epsilon/\kappa)$$

Here, the subscript 1 in Eq. 4.16 was dropped for the counterdiffusion. The correction factor for pore characteristics, $f_e(\epsilon/\kappa)$ is, for instance, simply ϵ^2 if the random pore model (see Chapter 14) is used. In general, the voidage ϵ decreases with decreasing r. Therefore, the effective diffusivity will increase with decreasing r, the exact manner in which r affects the diffusivity being dependent on the factor $f_e(\epsilon/\kappa)$. It is seen then that depending on the relative effects of r on k and D_e, a support with a higher surface area may or may not result in a higher apparent rate constant. This analysis shows that the choice of the support with a higher surface area, which causes a higher diffusional resistance, is not necessarily undesirable for a higher apparent rate constant.

Admittedly, many constraints have been placed on these analyses to better focus on the point being made: diffusional resistance is not necessarily undesirable as long as the net effect of the factor causing the resistance is in the direction of a higher apparent rate of reaction. Otherwise, the diffusional resistance should be minimized.

The importance of internal diffusion can also be appreciated from a different point of view: the fact that the internal diffusion plays a pivotal role in internal and external transport processes. For negligible concentration gradient in the pellet, Eq. 4.57 still holds. However, the value of ηDa will be larger than that for diffusion-limited case for the same intrinsic rate since η is larger and therefore the pellet will be more isothermal as Figure 4.7 reveals. Further, a relatively large Biot number for mass under realistic conditions still ensures negligible external mass transfer resistance. It is seen then that in the absence of diffusional resistance, the pellet tends to be more isothermal and the only major resistance is likely to be external heat transfer.

4–6 GENERALIZED INTERNAL EFFECTIVENESS FACTOR

This section develops generalized internal effectiveness factors (Aris 1965; Bischoff 1965; Petersen 1965), generalized in the sense that they are applicable to arbitrary kinetics. For a slab-like, isothermal pellet, Eq. 4.21 reduces to:

$$\frac{d}{dz}\left(D_e \frac{dC}{dz}\right) = r_c(C) \tag{4.67}$$

where the effective diffusivity can depend on concentration. If P is defined as dC/dz, a change of variables gives:

$$\frac{d}{dz} = \left(\frac{dC}{dz}\right)\left(\frac{d}{dC}\right) = P\left(\frac{d}{dC}\right) \tag{4.68}$$

Then, Eq. 4.67 can be written as:

$$\frac{d}{dC}(D_eP)^2 = 2r_cD_e \tag{4.69}$$

Integrating this equation with respect to C from the pellet center to any point in the pellet yields:

$$(D_eP)^2 = 2\int_{C_c}^{C} D_e(\alpha)r_c(\alpha)d\alpha \tag{4.70}$$

where the boundary condition at the center ($P = 0$) has been used and C_c is the concentration at the center. Solving this equation for P, one gets:

$$\frac{dC}{dz} = \frac{2^{1/2}}{D_e}\left[\int_{C_c}^{C} D_e(\alpha)r_c(\alpha)d\alpha\right]^{1/2} \tag{4.71}$$

The internal effectiveness factor is obtained from Eqs. 4.46 and 4.71:

$$\eta_i = \frac{2^{1/2}}{Lr_c(C_s)}\left[\int_{C_c}^{C_s} D_e(\alpha)r_c(\alpha)d\alpha\right]^{1/2} \tag{4.72}$$

A property of the effectiveness factor is that the product of the Thiele modulus and effectiveness factor approaches unity as the value of the Thiele modulus approaches infinity. In this asymptotic region of strong diffusion effects, the pellet center concentration approaches zero. Therefore, a generalized Thiele modulus can be defined as:

$$\phi_G \equiv Lr_c(C_s)\left[2\int_{0}^{C_s} D_e(\alpha)r_c(\alpha)d\alpha\right]^{-1/2} \tag{4.73}$$

The pellet center concentration C_c is still an unknown quantity. It can be obtained by integrating Eq. 4.71:

$$L = \frac{1}{2^{1/2}}\int_{C_c}^{C_s} D_e(\beta)\left[\int_{C_c}^{\beta} D_e(\alpha)r_c(\alpha)d\alpha\right]^{-1/2}d\beta \tag{4.74}$$

which is an implicit expression for C_c. Consider a first-order reaction for which $r_c = kC$. Using Eq. 4.73 for the Thiele modulus with constant diffusivity gives:

$$\phi_G = LkC_s\left[2D_e\int_{0}^{C_s} ka\,d\alpha\right]^{-1/2}$$

$$= L\left(\frac{k}{D_e}\right)^{1/2}$$

This result is identical to the Thiele modulus derived earlier. For the unknown C_c, Eq. 4.74 can be used to arrive at:

$$L(k/D_e)^{1/2} = \phi_G = \int_{C_c}^{C_s} \frac{dC}{(C^2 - C_c^2)^{1/2}}$$

which can be solved for C_c. Following similar procedures for an nth-order reaction gives:

$$\phi_G = L \left(\frac{n+1}{2} \frac{kC_s^{n-1}}{D_e} \right)^{1/2}$$

This generalized modulus is different from the traditional modulus ϕ by a factor of $[(n+1)/2]^{1/2}$. A plot of generalized effectiveness factor versus generalized modulus is shown in Figure 4.9 (Bischoff 1965). It is seen that the product $\eta_i \phi_G$ approaches unity when $\phi_G \geqslant 3$, i.e.:

$$\eta_i \phi_G = 1 \qquad \text{when } \phi_G \geqslant 3 \tag{4.75}$$

This fact suggests that the reaction may be considered to be in the asymptotic region of strong diffusion effects whenever $\phi_G \geqslant 3$, which in turn means that C_c in Eq. 4.72 can be set to zero for all practical purposes when $\phi_G \geqslant 3$. A comparison between theoretical and experimental effectiveness factors has been made by Brown and Bennett (1970) using the approach of generalized effectiveness factors. Their results for the synthesis of methanol from carbon monoxide and hydrogen are shown in Figure 4.10. Good agreement is shown, especially in the region of strong diffusion effects.

The usefulness of the generalized modulus lies in the fact that it is applicable to arbitrary kinetics. Further, the product $\eta_i \phi_G$ approaches unity for all forms

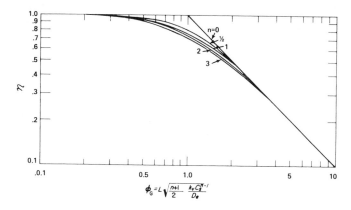

Figure 4.9 Generalized effectiveness factor versus generalized modulus for nth-order reactions. (Bischoff 1965)

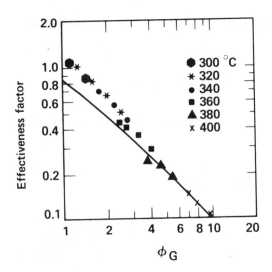

Figure 4.10 Comparison between theoretical and experimental generalized effectiveness factors. (Brown and Bennet 1970)

Legend within figure:
● 300 °C
* 320
• 340
■ 360
▲ 380
x 400

of rate expression in the asymptotic region of strong diffusion effect ($\phi_G \geq 3$). However, it is not without its limitations. It is obvious from Eq. 4.73 that ϕ_G is indeterminate when the order of reaction is negative. If there are multiple solutions (Satterfield et al. 1967) to steady state mass balances, Eq. 4.73 also fails in the sense that it gives one single ϕ_G for the reaction. It should be recognized on the other hand that multiplicity of solutions usually occurs for unrealistic values of the physical parameters, as pointed out earlier, even when an internal thermal gradient exists. In general, the approach of the generalized modulus is not applicable to those reactions that exhibit empirical or approximate negative-order behavior.

4–7   REACTOR POINT EFFECTIVENESS

If all the transport resistances affecting the reaction rate can be combined into a single factor Λ, the global rate at any point in the reactor can be written as:

$$R_G(x) = \Lambda(x)r_c(x) \tag{4.76}$$

where x is the reactor coordinate. This factor Λ is termed the reactor point effectiveness (Lee, 1981b) since it represents the effectiveness of a catalytic reaction at any point in the reactor. If this reactor point effectiveness is expressed in terms of bulk-fluid conditions, Eq. 4.76 can be used directly in the overall conservation equations for the reactor without having to solve the pellet conservation equations. The reactor point effectiveness is defined by:

$$\Lambda = \frac{R_G(C_b, T_b)}{r_c(C_b, T_b)} \tag{4.77}$$

It has been shown that, under realistic reaction conditions, the pellet can be considered isothermal and the external mass transfer resistance can be neglected. Under these conditions, the reactor point effectiveness can be written for the arbitrary kinetics of Eq. 4.2 as:

$$\Lambda = \frac{k_s f(C_s)}{k_b f(C_b)} \eta_i = \frac{k_s}{k_b} \eta_i \qquad (4.78)$$

where

$$\eta_i = \frac{1}{Lr_c(k_s, C_b)} \left[2D_e \int_{C_c}^{C_b} r_c(\alpha) d\alpha \right]^{1/2} \qquad (4.72)$$

Here, C_s has been set to C_b since the external mass transfer resistance is negligible. The ratio of rate constants can be written as:

$$\frac{k_s}{k_b} = \exp[-\epsilon(1/(T_s/T_b) - 1)] \qquad (4.79)$$

The heat transferred across the external film is equal to the heat generated by the reaction:

$$h(T_s/T_b - 1) = L(-\Delta H)R_G/T_b = L(-\Delta H)\Lambda k_b f(C_b)/T_b \qquad (4.80)$$

The ratio of temperatures T_s/T_b is quite close to unity and therefore a Taylor series expansion of the exponential in Eq. 4.79 can be truncated and still gives an adequate representation. Taking only the first-order term in the expansion, the rate constant ratio can be written as:

$$\left(\frac{k_s}{k_b}\right)^{1/2} = 1 + \epsilon(T_s/T_b - 1)/2$$

This ratio can be better approximated by introducing a factor of 1.2 for small values of the argument of the exponential:

$$\left(\frac{k_s}{k_b}\right)^{1/2} = 1 + 1.2\,\epsilon(T_s/T_b - 1)/2 \qquad (4.81)$$

Use of Eq. 4.80 in 4.81 gives:

$$\left(\frac{k_s}{k_b}\right)^{1/2} = 1 + \frac{1.2\,\epsilon}{2}\left[\frac{L(-\Delta H)\Lambda k_b f(C_b)}{hT_b}\right] \qquad (4.82)$$

This equation together with Eq. 4.72 can be used in Eq. 4.78 to arrive at the reactor point effectiveness expressed in terms of bulk-fluid quantities:

$$\Lambda = \frac{[2D_e I(C_b)/k_b]^{1/2}/Lf(C_b)}{1 - \dfrac{1.2\epsilon(-\Delta H)[2D_e k_b I(C_b)]^{1/2}}{2hT_b}}$$

(4.83)

where

$$I(C_b) = \int_{C_c}^{C_b} f(C)\,dC$$

(4.84)

The reactor point effectiveness of Eq. 4.83 is in a form suitable for direct inclusion in reactor conservation equations. However, the pellet center concentration C_c has to be specified. As discussed earlier, C_c can be set to zero when $\phi_G \geqslant 3$. For ϕ_G less than 3, C_c can be calculated in principle using Eq. 4.74, but this involves trial and error procedures, which are cumbersome. Instead, an approximate solution for C_c can be based on the fact that the value of the integral I in Eq. 4.84 is rather insensitive to errors in the estimated value of C_c. If we let ΔC_c be the error in C_c, the corresponding maximum relative error in the integral I for a second-order reaction, for instance, is $|3 \, \Delta C_c C_c^2 / C_b^3|$. A 30% error in C_c will result in a maximum error of less than 10% for $\phi = 1$.

Consider the estimation of the center concentration, C_c. As indicated earlier, the center concentration can be set to zero for $\phi_G \geqslant 3$. On the other hand, the reaction can be considered to be free from diffusional effects for all practical purposes when ϕ_G is less than 0.3. Therefore, estimation is necessary only for the intermediate range of ϕ_G, i.e., $0.3 < \phi_G < 3$. As the results in Figure 4.9 show, the maximum error in using the effectiveness factor of a first-order reaction for a third-order reaction is less than 10% when the generalized modulus is used. This suggests that the center concentration corresponding to a first-order reaction may be used for arbitrary kinetics:

$$C_c/C_b = \frac{1}{\cosh \phi_G}$$

(4.85)

where ϕ_G is given by Eq. 4.73. Note that the generalized modulus can be determined without the knowledge of the center concentration. Although a more accurate method is available (Lee and Ruckenstein 1983) for the estimation of the center concentration, the center concentration given by Eq. 4.85 is sufficiently accurate for the purpose of obtaining the global rate as the results in Section 4.9 for multiple reactions will show.

The linearization given by Eq. 4.81 needs further explanation in light of the fact that there can be as many as three solutions (steady states) to Eq. 4.80. Equation 4.80 can be rewritten as follows:

$$Q_c \equiv h(T_s - T_b) = L(-\Delta H) \, \Lambda \, k_b f(C_b) \equiv Q_h$$

(4.80a)

The left hand side designated as Q_c represents the rate of cooling by the bulk fluid; the right hand side designated as Q_h represents the rate of heat generation due to reaction. For steady state operation, Q_c should be equal to Q_h. However, there can be no solution as shown in Figure 4.11 for $(Q_c)_1$ if the operating conditions and particularly the fluid velocity are chosen in such a way that the heat transfer coefficient h is too small. When there is no solution, the heating rate is larger than the cooling rate and there will be accumulation of heat, leading to a runaway. In such cases, the reactor point effectiveness given by Eq. 4.83 would become negative, indicating that the selected operating conditions would lead to a runaway. The heat transfer coefficient h is strongly dependent on the fluid velocity, its value increasing with increasing velocity. Therefore, heat transfer between the bulk fluid and the pellet places a constraint on the fluid velocity that can be used. The velocity must be high enough to ensure there is at least one solution to Eq. 4.80a for steady state operation. With further increases in velocity, two steady states, and then three steady states will arise. The latter case is shown in Figure 4.11.

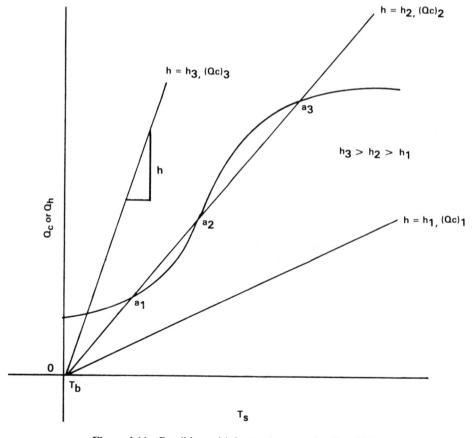

Figure 4.11   Possible multiple steady states for Eq. 4.80a.

The steady state a_2 is "unstable" in the sense that a slight change in operation either way will move the steady state condition to either a_1 or a_3. The steady state a_3 is possible only when $T_s/T_s \gg 1$. Since T_s/T_b is close to unity under typical operating conditions, the steady state corresponds to the state a_1. The steady state corresponding to Eq. 4.81 is that of a_1, i.e., the steady state at the lowest T_s, when there are multiple solutions to Eq. 4.80.

The utility of the reactor point effectiveness can be best illustrated through an example. Consider an adiabatic plug-flow reactor in which a catalytic reaction takes place. Suppose that ϕ_G is greater than 3 and that the intrinsic rate expression is:

$$r_c = kf(C) = k \left(\frac{C}{1 + KC} \right)$$

Assuming constant velocity and physical properties, the steady state mass and heat balances are:

$$\frac{dC_b}{d\bar{x}} = -\tau R_G$$

$$\frac{dT_b}{d\bar{x}} = \left(\frac{-\Delta H}{\rho c_p} \right) \tau R_G$$

where $\bar{x}$ = normalized reactor coordinate, x/Z
$\quad \tau = Z(1 - \epsilon_B)/v$
$\quad Z$ = reactor length
$\quad \epsilon_B$ = void fraction of packed bed
$\quad v$ = superficial fluid velocity

For reactor design purposes, these equations have to be solved along with the pellet conservation equations 4.27 through 4.29 and the relationships at the pellet-bulk fluid interface (Eqs. 4.30 and 4.31). With the use of reactor point effectiveness, however, solution of these equations can be avoided. The global rate is:

$$R_G(C_b, T_b) = \Lambda r_c(C_b, T_b)$$

The integral I for Λ is:

$$I = \int_0^{C_b} f(C)dc = \int_0^{C_b} \frac{C}{1 + KC} dC = \frac{1}{K} \left[C_b - \frac{1}{K} \ln(1 + KC_b) \right]$$

Using this in Eq. 4.83 and then in Eq. 4.77 for the global rate R_G, gives:

$$R_G = \cfrac{(2D_e k_b/K)^{1/2} \left[C_b - \dfrac{1}{K} \ln (1 + KC_b) \right]^{1/2} \Big/ L}{1 - \dfrac{1.2\epsilon(-\Delta H)}{hT_b} (2D_e k_b/K)^{1/2} \left[C_b - \dfrac{1}{K} \ln (1 + KC_b) \right]^{1/2}}$$

This global rate can be used directly in the reactor conservation equations, leading to the specification of reactor size in a straightforward manner since now the global rate is expressed in terms of the bulk fluid quantities C_b and T_b. Otherwise, the pellet conservation equations would have to be solved first. Details on this approach of reactor point effectiveness and comparison with the traditional approach are given in Chapter 10.

It is appropriate at this time to point out that any rate expression for a single reaction can always be reduced to the rate expression of Eq. 4.2, which contains only the concentration of the main reactant, C. Consider the following reaction:

$$A + bB \rightarrow gG + wW \tag{4.86}$$

where A is the main reactant, and the lower case letters are the stoichiometric coefficients. Let the rate expression for the main reactant be:

$$r_c = kf_c (C, C_B, C_G, C_W) \tag{4.87}$$

where C_B, C_G, and C_W are the concentrations of species B, G, and W, respectively. The steady state conservation equations for species A and G, for instance, are:

$$D_e \frac{d^2 C}{dz^2} = r_c \tag{4.88}$$

$$(D_e)_G \frac{d^2 C_G}{dz^2} = -g r_c \tag{4.89}$$

where D_e and $(D_e)_G$ are the effective diffusivities for the main reactant and species G, respectively. Combining these two equations yields:

$$D_e \frac{d^2 C}{dz^2} + \frac{(D_e)_G}{g} \frac{d^2 C_G}{dz^2} = 0 \tag{4.90}$$

Integrating with respect to z from the center to z, one has:

$$D_e \frac{dC}{dz} + \frac{(D_e)_G}{g} \frac{dC_G}{dz} = 0 \tag{4.91}$$

where the boundary condition at the center has been used. Integrating once again, but this time from z to the pellet surface gives:

$$C_G = (C_G)_s - g\frac{D_e}{(D_e)_G}(C - C_s) \tag{4.92}$$

where $(C_G)_s$ and C_s are the concentrations at the surface for species G and the main reactant, respectively. For any number of species involved in the reaction, in general, the concentration of species i in the pellet can be related to that of the main reactant as follows:

$$C_i = (C_s)_i + \nu_i\frac{D_e}{(D_e)_i}(C - C_s) \tag{4.93}$$

where ν_i is the ratio of the stoichiometric coefficient of species i to that of the main reactant, with the convention that it is negative for products and positive for reactants. Since the concentrations of all species can be expressed in terms of the main reactant concentration C as given by Eq. 4.93, the substitution of Eq. 4.93 into Eq. 4.87, for instance, reduces the rate expression to that given by Eq. 4.2.

4-8   NONUNIFORM ACTIVITY DISTRIBUTIONS

So far in this chapter, the catalyst activity has been assumed uniform throughout a pellet. Nonuniform activity can result from poor impregnation, deliberate partial impregnation, or from deactivation by poisoning species. In such cases, the intrinsic kinetics can be represented by:

$$\begin{aligned} r_c(C,z) &= \bar{k}G(z)f(C) \\ &= G(z)r(C); \qquad r(C) = \bar{k}f(C) \end{aligned} \tag{4.94}$$

where $\bar{k}$ is a normalizing factor such that $G(z) \leqslant 1$. Here, $G(z)$ is an arbitrary but presumably monotonic function for the activity distribution. The steady-state pellet mass balance written for a slab-like pellet in terms of $P(= dC/dz)$ for constant effective diffusivity is:

$$D_e\frac{dP}{dz} = G(z)r(C) \tag{4.95}$$

The boundary conditions are the usual ones. Both sides of this equation are multiplied by P and integrated from 0^+ to z to give:

$$\frac{D_e}{2}P^2\Big]_{0^+}^{z} = \int_{0^+}^{z} G(\zeta)r(C)P\,d\zeta \tag{4.96}$$

The right hand side can be integrated by parts to give:

$$\frac{D_e}{2} P^2 \Big]_{0^+}^z = G(z) \int_0^C r(\alpha) d\alpha \Big]_{0^+}^z$$

$$- \int_{0^+}^z \left[\int_0^C r(\alpha) d\alpha \right] G' dz; \qquad G' = dG/dz \qquad (4.97)$$

where the following identity has been used:

$$\frac{d}{dz} \int_0^C r(\alpha) d\alpha = r(C)P \qquad (4.98)$$

Repeated integration by parts of the last term in Eq. 4.97 yields the following integro-differential series (Lee 1981a), evaluated at $z = L$ and valid in the asymptotic region of strong diffusion effects;

$$\frac{D_e}{2} \bar{P}_\infty^2 = \bar{G} \int_0^{C_b} r(C) dC - \frac{\bar{G}'}{\bar{P}_\infty} \int_0^{C_b} \int_0^c r(\alpha) d\alpha dC$$

$$+ \sum_{i=1}^\infty \frac{(-1)^{i+1}}{\bar{P}_\infty} \left[\frac{d(\bar{J}_{i-1})}{dz} \right] \int_0^{C_b} \int_0^c \int_0^\beta \cdots \int_0^\zeta r d\zeta \cdots d\beta d\alpha dC \qquad (4.99)$$

$$\bar{J}_i = \frac{1}{\bar{P}_\infty} \frac{d\bar{J}_{i-1}}{dz} \qquad \text{with } \bar{J}_0 = \bar{G}'/\bar{P}_\infty \qquad (4.100)$$

where the number of times $r(C)$ is integrated is equal to $(i + 2)$. Here, the overbar denotes evaluation at $z = L$, and the subscript ∞ denotes the region of strong diffusion effects. In accordance with the previous conclusions on pellet isothermality and negligible external mass transfer resistance, which still hold, the pellet surface concentration has been set to C_b. The integro-differential series in Eq. 4.99 has been shown (Lee 1981a) to be a converging series. The flux at the surface can be adequately represented by the first term:

$$\bar{P}_\infty = \left[\frac{2\bar{G}}{D_e} \int_0^{C_b} r(C) dC \right]^{1/2} \qquad (4.101)$$

provided the ratio R given below is much less than unity:

$$R \equiv \left| \frac{\bar{G}' D_e^{1/2}}{\bar{G}(2\bar{G})^{1/2}} \right| \left\{ \int_0^{C_b} \int_0^c r(\alpha) d\alpha dC \Big/ \left[\int_0^{C_b} r(C) dC \right]^{3/2} \right\} \qquad (4.102)$$

This ratio is essentially the maximum fractional error involved in using the approximate relationship of Eq. 4.101. For an nth-order reaction, this ratio reduces to:

$$R = \left|\frac{\overline{G}'L}{\overline{G}^{3/2}}\right| \left(\frac{1}{\phi}\right) \left[\frac{(n+1)^{1/2}}{2^{1/2}(n+2)}\right] \tag{4.103}$$

In the asymptotic region of strong diffusion effects, therefore, this ratio is much less than unity unless $\overline{G}$ is small. When the activity monotonically decreases toward the center of the pellet, this ratio R is usually much less than unity. When the activity increases toward the centers, however, the value of R should be checked before using Eq. 4.101: for $\overline{G} \geq 0.5$ it has been shown (Lee 1981a) for various forms of the activity distribution that R is much less than unity.

The generalized internal effectiveness factor in the asymptotic region of strong diffusion effects is obtained from:

$$\eta_i = \frac{D_e \overline{P}_\infty / L}{r(C_b) \int_0^L G(z)dz/L} \tag{4.104}$$

Since $\eta_i \phi_G$ should approach unity in this asymptotic region, the generalized modulus can be defined as:

$$(\phi_G)_n = \frac{1}{\eta_i} = \frac{r(C_b) \int_0^L G(z)dz}{D_e \overline{P}_\infty} \tag{4.105}$$

In an analogy with the case of a uniform activity distribution, a reaction may be considered to be in the region of strong diffusion effects when $(\phi_G)_n$ is greater than 3. The analytical results obtained by Wang and Varma (1980) for a first-order reaction confirms this limit of 3 for the asymptotic region.

For all practical purposes, then, the generalized internal effectiveness factor is:

$$\eta_i = \frac{\left[2D_e\overline{G} \int_0^{C_b} r(C, T_s)dC\right]^{1/2}}{r(C_b, T_s) \int_0^L G(z)dz}; \qquad \begin{aligned} r(C) &= \overline{k}f(C) \\ r_c &= G(z)r(C) \end{aligned} \tag{4.106}$$

when $(\phi_G)_n$ is greater than 3. This relationship follows from Eqs. 4.101 and 4.104. When the ratio R given by Eq. 4.102 is not much less than unity, more terms of the integro-differential series in Eq. 4.99 can be retained for the calculation of $\overline{P}_\infty$, which then can be used for η_i (Eq. 4.104). If one more term is used, for instance, Eq. 4.99 becomes a cubic equation in $\overline{P}_\infty$ and a numerical search for $\overline{P}_\infty$ can start with the value of $\overline{P}_\infty$ obtained from Eq. 4.101. In general, this procedure will be required for a pellet with activity increasing toward the center for $\overline{G} <$ 0.5. It has been shown for a first-order reaction (Wang and Varma 1980) that the use of the characteristic length L for shape normalization is also valid for pellets with nonuniform activity when the surface activity is not zero. It should be noted however that for pellets of geometric shape other than slab, the line integral appearing in the denominator of Eqs. 4.104 and 4.106 must be changed to a volume integral.

Suppose that a slab-like pellet of initially uniform activity is deactivated such that the activity profile is given by $k[0.2(1 - z/L)^2 + 0.8]$ when the surface activity is reduced to 80% of its original activity. Suppose further that the intrinsic rate r_c for the fresh catalyst is kC^2. Then, in the asymptotic region of strong diffusion effects, one has:

$$\bar{k} = k$$
$$G(z) = 0.2(1 - z/L)^2 + 0.8 \qquad (\bar{G} = 0.8)$$
$$r(C) = kC^2$$

Therefore, the internal effectiveness for this case is:

$$\eta_i = \frac{(0.533 kD_e C_b^3)^{1/2}}{0.867 L k C_b^2}$$

which is the generalized effectiveness factor for the second-order reaction multiplied by $(L\bar{G}^{1/2}/\int_0^L Gdz)$. Note that the activity of the deactivated pellet was used here as the intrinsic activity for the purpose of illustration. The activity of fresh catalyst should be used as the intrinsic activity for the combined effects of deactivation and diffusion as will be seen in Chapter 5.

For a pellet of nonuniform activity, the reactor point effectiveness is:

$$\Lambda = \frac{R_G(C_b, T_b)}{\bar{k}_b \left[\frac{1}{L}\int_0^L G(z)dz\right] f(C_b)} \qquad (4.107)$$

Following the same procedures as for a pellet of uniform activity, one can arrive at the following expression for Λ:

$$\Lambda = \frac{[2D_e\bar{G}I(C_b)/\bar{k}_b]^{1/2} / \left[f(C_b)\int_0^L G(z)dz\right]}{1 - \frac{1.2\epsilon(-\Delta H)[2D_e\bar{k}_b\bar{G}I(C_b)]^{1/2}}{2hT_b}}; \qquad (\phi_G)_n \gg 1 \quad (4.108)$$

with $I(C_b) = \int_0^{C_b} f(C)dC$

where Eq. 4.106 has been used in 4.107. The restrictions placed here are that the maximum fractional error given by Eq. 4.102 is small and that ϕ_G is much larger than unity.

4-9 MULTIPLE REACTIONS

So far, only a single reaction has been considered. While the reactor point effectiveness cannot be expressed explicitly for a reversible reaction, the internal effectiveness factor can readily be obtained analytically using the generalized modulus (see Problem 4.23). For complex multiple reactions, however, it is not possible to obtain analytical expressions for the global rates and one has to solve the conservation equations numerically. The numerical solution of nonlinear, coupled diffusion equations with split boundary conditions is by no means trivial and often presents convergence difficulties. In this section, the same approach is taken as was used for the reactor point effectiveness. This enables the global rates to be obtained in a straightforward manner and the diffusion equations to be solved as an initial value problem (Akella 1983).

Consider the following reaction network:

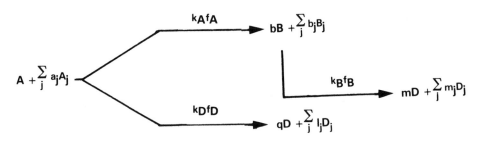

where the lower case letters denote the ratios of stoichiometric coefficients, and the capital letters denote the species. This is the basic unit of any complex reaction network. Since the reaction stoichiometry allows one to express the concentrations of all species in terms of the concentrations of species A and B and the diffusivity ratios, as shown in Section 4-7 for a simple case, it is sufficient to consider only the concentrations of A and B for rate expressions. As was done in Section 4-6 (Eq. 4.69), the steady state mass balance equations are written on concentration coordinates:

$$\frac{D_A}{2}\frac{dP_A^2}{dA} = k_A f_A(A,B) + k_D f_D(A,B); \qquad P_A = dA/dz \qquad (4.109)$$

$$\frac{D_B}{2}\frac{dP_B^2}{dB} = -bk_A f_A(A,B) + k_B f_B(A,B); \qquad P_B = dB/dz \qquad (4.110)$$

$$\frac{D_D}{2}\frac{dP_D^2}{dD} = -qk_D f_D(A,B) - \frac{m}{b} k_B f_B(A,B); \qquad P_D = dD/dz \qquad (4.111)$$

where the D_i ($i = A$, B, and D) are the effective diffusivities of species i and are assumed constant, and k_i are the rate constants. The usual boundary conditions apply.

Consider the relationship between the concentrations of A and B. If the relationship is known, Eqs. 4.109 and 4.110 can be integrated to yield:

$$P_A(z) = \left\{ \frac{2}{D_A} \int_{A(0)}^{A(z)} [k_A f_A(A,B) + k_D f_D(A,B)] dA \right\}^{1/2} \tag{4.112}$$

$$P_B(z) = \left\{ \frac{2}{D_B} \int_{B(0)}^{B(z)} [-bk_A f_A(A,B) + k_B f_B(A,B)] dB \right\}^{1/2} \tag{4.113}$$

Since $P_A = dA/dz$, and $P_B = dB/dz$, Eqs. 4.112 and 4.113 can be integrated and then combined to give:

$$\int_{A(z_i)}^{A(z_{i+1})} \frac{dA}{\left\{ \frac{2}{D_A} \int_{A(0)}^{\zeta} (k_A f_A + k_D f_D) d\zeta \right\}^{1/2}}$$

$$= \int_{B(z_i)}^{B(z_{i+1})} \frac{dB}{\left\{ \frac{2}{D_B} \int_{B(0)}^{\zeta} (-bk_A f_A + k_B f_B) d\zeta \right\}^{1/2}} \tag{4.114}$$

where the integration is from any arbitrary position z_i to a position z_{i+1} such that $\Delta z = |z_i - z_{i+1}|$ is small. If the small changes in A and B corresponding to the small change in z (Δz) are denoted as ΔA and ΔB, then the trapezoidal rule of integration can be used to expand Eq. 4.114 as:

$$\frac{\Delta A}{2} \{[P_A(z_{i+1})]^{-1} + [P_A(z_i)]^{-1}\} = \frac{\Delta B}{2} \{[P_B(z_{i+1})]^{-1} + [P_B(z_i)]^{-1}\} \tag{4.115}$$

where

$$P_A(z_{i+1}) = \left[\frac{2}{D_A} \int_{A(0)}^{A(z_{i+1})} (k_A f_A + k_D f_D) dA \right]^{1/2} \tag{4.116}$$

$$P_A(z_i) = \left[\frac{2}{D_A} \int_{A(0)}^{A(z_i)} (k_A f_A + k_D f_D) dA \right]^{1/2} \tag{4.117}$$

as given by Eq. 4.112. Similar expressions can be written for $P_B(z_{i+1})$ and $P_B(z_i)$. Now, Eq. 4.116 can be rewritten as:

$$P_A^2(z_{i+1}) = \frac{2}{D_A} \int_{A(0)}^{A(z_i)} (k_A f_A + k_D f_D) dA + \frac{2}{D_A} \int_{A(z_i)}^{A(z_{i+1})} (k_A f_A + k_D f_D) dA$$

$$= P_A^2(z_i) + \epsilon_A(z_{i+1}) \tag{4.118}$$

where

$$\epsilon_A(z_{i+1}) = \frac{2}{D_A} \int_{A(z_i)}^{A(z_{i+1})} (k_A f_A + k_D f_D) dA \tag{4.119}$$

Similarly for $P_B^2(z_{i+1})$:

$$P_B^2(z_{i+1}) = P_B^2(z_i) + \epsilon_B(z_{i+1}) \tag{4.120}$$

where

$$\epsilon_B(z_{i+1}) = \frac{2}{D_B} \int_{B(z_i)}^{B(z_{i+1})} (-bk_A f_A + k_B f_B) dB \tag{4.121}$$

Substituting Eqs. 4.118 and 4.120 into Eq. 4.115 yields:

$$\frac{\epsilon_B}{\epsilon_A} = \left(\frac{P_B}{P_A}\right)^2 \tag{4.122}$$

Here, the following relationship has been used:

$$\frac{P_B}{P_A} = \frac{dB}{dA} = \frac{\Delta B}{\Delta A} \tag{4.123}$$

It follows from the definitions of ϵ_A and ϵ_B and Eqs. 4.122 and 4.123 that:

$$\left(\frac{\Delta B}{\Delta A}\right)_{i+1}^2 = \left(\frac{D_A}{D_B}\right) \left[\frac{\int_{B(z_i)}^{B(z_{i+1})} (-bk_A f_A + k_B f_B) dB}{\int_{A(z_i)}^{A(z_{i+1})} (k_A f_A + k_D f_D) dA} \right] \tag{4.124}$$

If the trapezoid rule is used again for the integration, one obtains:

$$\left(\frac{\Delta B}{\Delta A}\right)_{i+1} = \left(\frac{D_A}{D_B}\right) \left[\frac{(-bk_A f_A + k_B f_B)_i + (-bk_A f_A + k_B f_B)_{i+1}}{(k_A f_A + k_D f_D)_i + (k_A f_A + k_D f_D)_{i+1}} \right] \tag{4.125}$$

This is the relationship that can be used to generate the concentration profile of B as a function of A, thus providing the relationship between A and B needed for the integration of Eqs. 4.112 and 4.113. Since the position i is arbitrary, one can start at the pellet surface where $A_i = A_b$ and $B_i = B_b$, A_b and B_b being the bulk concentrations. In order to calculate the concentrations A and B at a distance Δz away from the pellet surface, one can choose a small value for ΔA_1. This immediately gives $A_1 (= A_b - \Delta A_1)$ for $i = 0$. For the calculation of ΔB_1 corresponding to the chosen value of ΔA_1, Eq. 4.125 can be written as:

$$\frac{\Delta B_1}{\Delta A_1} = \left(\frac{D_A}{D_B}\right)\left[\frac{\begin{matrix}-bk_Af_A(A_b,B_b) + k_Bf_B(A_b,B_b) - bk_Af_A(A_b - \Delta A_1,B_b \\ + \Delta B_1) + k_Bf_B(A_b - \Delta A_1,B_b + \Delta B_1)\end{matrix}}{\begin{matrix}k_Af_A(A_b,B_b) + k_Df_D(A_b,B_b) + k_Af_A(A_b - \Delta A_1,B_b \\ + \Delta B_1) + k_Df_D(A_b - \Delta A_1,B_b + \Delta B_1)\end{matrix}}\right] \quad (4.126)$$

since, for instance, $(k_Af_A)_{i+1} = k_Af_A(A_1,B_1) = k_Af_A(A_b - \Delta A_1,B_b + \Delta B_1)$ for $i = 0$. This result uses the fact that the reactant concentration decreases toward the pellet center, whereas the product concentration increases. Equation 4.126 contains only one unknown, ΔB_1, which is thus determined. Now that all information at $i = 0$ is available, one can increment i successively and repeat the whole procedure until the pellet center is reached, thereby obtaining the profile of B as a function of A. The step size ΔA_i may be made different for each step or constant for all steps.

The numerical method based on Eq. 4.125 does not involve any two point boundary value calculations: only straightforward marching is involved. Furthermore, the integrated values obtained in the process of solving Eq. 4.125 can be used directly in Eqs. 4.112 and 4.113 for the global rates:

$$(R_G)_A = \frac{D_A}{L} P_A(L) = \frac{1}{L}\left[2D_A \int_{A_c}^{A_b} (k_Af_A + k_Df_D)dA\right]^{1/2} \quad (4.127)$$

$$(R_G)_B = \frac{D_B}{L} P_B(L) = \frac{1}{L}\left[2D_B \int_{B_c}^{B_b} (-bk_Af_A + k_Bf_B)dB\right]^{1/2} \quad (4.128)$$

where A_c and B_c are the pellet center concentrations. A similar expression for $(R_G)_D$ can also be written. In order to evaluate the integrals, however, the center concentrations have to be known.

For an estimate of the center concentration of A, one can define the effectiveness factor for the reactions involving the consumption of A in accordance with the usual practice:

$$\eta_G = \frac{D_A P_A(L)}{L r_A} \quad (4.129)$$

It is sufficient to estimate only A_c since all the other center concentrations can be found from it. In anticipation of the development to follow, the subscript G was used to denote the generalized effectiveness factor. As was done in Section 4–6, the generalized modulus ϕ_G is defined by:

$$\phi_G = \frac{1}{\eta_G}\bigg|_{\phi_G \gg 1} = \frac{1}{\eta_G}\bigg|_{A_c=0,B_c=B_c} \quad (4.130)$$

This follows from the fact that A_c approaches zero for large ϕ_G but B_c does not. The definition of η_G, when used in Eq. 4.130 yields:

$$\phi_G = \frac{L[k_A f_A(A_b, B_b) + k_D f_D(A_b, B_b)]}{\left[2D_A \int_0^{A_b} (k_A f_A + k_D f_D) dA\right]^{1/2}} \tag{4.131}$$

where $r_A = k_A f_A + k_D f_D$. Since the relationship between A and B is now known through Eq 4.125, the generalized modulus ϕ_G can be evaluated readily, using the value of B corresponding to $A_c = 0$ as B_c. As was done in Section 4–7, the center concentration is approximated by:

$$\frac{A_c}{A_b} = \frac{1}{\cosh \phi_G} \tag{4.132}$$

where ϕ_G is given by Eq. 4.131.

It is seen that the calculation of the global rates involves only the solution of Eq. 4.125, from which A_c, B_c, $P_A(L)$, and $P_B(L)$ can be directly calculated with the aid of Eqs. 4.131 and 4.132. The original split boundary value problem has been transformed into an initial value problem.

Consider the following reaction network for an illustration of the numerical method:

$$k_A f_A = k_A A/(1 + K_1 A + K_2 B)$$
$$k_B f_B = k_B A^2/(1 + K_1 A + K_2 B)$$
$$k_D f_D = k_D AB/(1 + K_1 A + K_2 B)$$

The rate expressions can be substituted into Eq. 4.125 for the calculation of B as a function of A. For instance, the numerator terms in Eq. 4.125 become:

$$(-bk_A f_A + k_B f_B)_i = \frac{-k_A A_i + k_B A_i^2}{1 + K_1 A_i + K_2 B_i}$$

$$(-bk_A f_A + k_B f_B)_{i+1} = \frac{-k_A(A_i - \Delta A_{i+1}) + k_B(A_i - \Delta A_{i+1})^2}{1 + K_1(A_i - \Delta A_{i+1}) + K_2(B_i + \Delta B_{i+1})}$$

Here again, one must remember that the concentration of species A decreases toward the pellet center ($A_i - \Delta A_{i+1}$), but that of species B increases ($B_i + \Delta B_{i+1}$). Starting at the pellet surface ($i = 0$) at which A and B are at bulk-fluid concentrations, i.e., $A_0 = A_b$ and $B_0 = B_b$, Eq. 4.125 can be solved for ΔB_{i+1} at each step i for a chosen value of ΔA_i. This forward marching yields not only the corresponding values of A and B, but also those of $k_A f_A$, $k_B f_B$, and $k_D f_D$. The forward marching can continue by incrementing the index i until $A = 0$ is reached. The values of $k_A f_A$, $k_D f_D$, and A thus generated are then used for the evaluation of ϕ_G using Eq. 4.131. Since the pellet center concentration of A, A_k, is given by $A_k/A_b = 1/\cosh \phi_G$, the calculated value of ϕ_G can be used to locate the index k, yielding the center concentrations of A and B. Now that all quantities

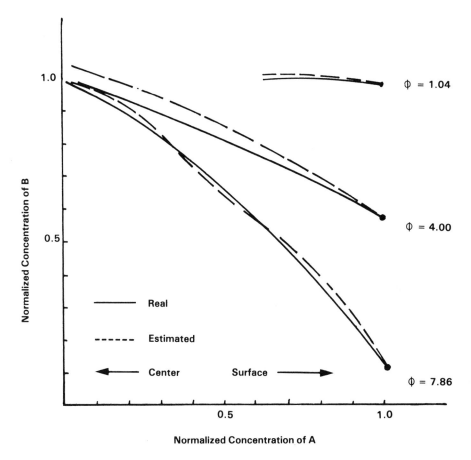

Figure 4.12   Comparison between correct and approximate solutions for concentration profiles.

necessary to evaluate the integrals in Eqs. 4.127 and 4.128 are known, the global rates can be calculated from these equations.

Results obtained from the example reaction network are summarized in Table 4.3 for various sets of rate constants and surface concentrations (see Problem 4.24), which were selected to give ϕ_G ranging from 0.1 to 8. The concentration profiles for some selected cases are shown in Figure 4.12. The example system turned out to be very sensitive at low concentrations of A and the usual trial and error method failed to give any reliable results. For the purpose of obtaining the real concentration profiles, therefore, the system equations were solved by assuming pellet center concentrations and treating the resulting surface concentrations as boundary conditions. The solutions thus obtained are listed as "real" solutions in Table 4.3. It is seen that the maximum error in global rates is about 5% at $\phi_G = 4$ and the error is substantially less for other ϕ_G values.

Table 4.3   Comparison of Global Rates for Various ϕ_G Values

ϕ_G	$(R_G)_A$ (mol/cm³·s)		$(R_G)_B$ (mol/cm³·s)		% Error In	
	Real	Calculated*	Real	Calculated	$(R_G)_A$	$(R_G)_B$
7.86	$1.907\,(10^{-3})$	$1.907\,(10^{-3})$	$3.06\,(10^{-4})$	$3.08\,(10^{-4})$	0.00	0.70
4.00	$2.09\,(10^{-4})$	$2.09\,(10^{-4})$	$5.82\,(10^{-5})$	$6.13\,(10^{-5})$	0.05	5.27
1.04	$5.19\,(10^{-6})$	$5.24\,(10^{-6})$	$1.02\,(10^{-6})$	$1.03\,(10^{-6})$	0.83	1.47
0.33	$4.44\,(10^{-5})$	$4.46\,(10^{-5})$	$8.56\,(10^{-8})$	$8.62\,(10^{-8})$	0.61	0.68
0.07	$1.86\,(10^{-9})$	$1.89\,(10^{-9})$	$3.58\,(10^{-10})$	$3.60\,(10^{-10})$	0.65	0.67

* Calculated based on Eq. 4.125.

The numerical method presented for the basic reaction unit can be extended in a straightforward manner to any complex reaction network (Akella 1983). Consider the general reaction network shown in Figure 4.13. If there are n species $(A_1, A_2, A_3, \ldots , A_n)$, the rate of consumption of a species A_j $(j = 1, 2, \ldots , n)$ is given by:

$$R_{Aj} = (r_{j,1} - v_{j,1}r_{1,j}) + (r_{j,2} - v_{j,2}r_{2,j}) + \cdots + (r_{j,n} - v_{j,n}r_{n,j}) \quad (4.133)$$

where $v_{i,j}$ is the appropriate stoichiometric coefficient, and where a general rate expression for the reaction path between A_i and A_j is given by:

$$r_{i,j} = k_{i,j}(T)\,f_{i,j}\,(A_1, A_2, \ldots , A_n) \quad (4.134)$$

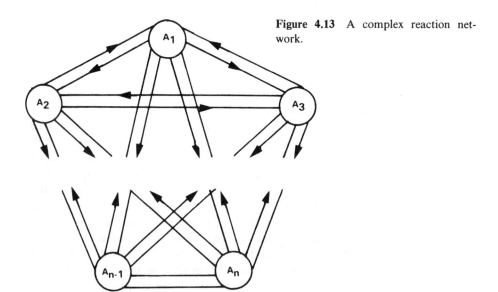

Figure 4.13   A complex reaction network.

The steady state mass balance for species A_j can be written as:

$$\frac{D_{A_j}}{2}\frac{dP_{A_j}^2}{dA_j} = R_{A_j} \qquad (j = 1, 2, \ldots, n): \qquad P_{A_j} = \frac{dA_j}{dz} \qquad (4.135)$$

where D_{A_j} is the effective diffusivity of the species A_j. If one follows the same procedures as for the basic unit considered earlier, the results corresponding to Eqs. 4.125, 4.127, and 4.131 are:

$$\frac{(\Delta A_j)_{i+1}}{(\Delta A_1)_{i+1}} = \left(\frac{D_{A_1}}{D_{A_j}}\right)\left\{\frac{R_{A_j}(A_1, A_2, \ldots, A_n)|_i + R_{A_j}(A_1, A_2, \ldots, A_n)|_{i+1}}{R_{A_1}(A_1, A_2, \ldots, A_n)|_i + R_{A_1}(A_1, A_2, \ldots, A_n)|_{i+1}}\right\} \qquad (4.136)$$

$$(R_G)_{A_j} = \frac{D_{A_j}}{L}P_{A_j}(L) = \frac{1}{L}\left\{2D_{A_j}\int_{A_j(0)}^{A_j(L)} R_{A_j}dA_j\right\}^{1/2} \qquad (4.137)$$

$$\phi_G = \frac{LR_{A_j}[A_1(L), A_2(L), \ldots, A_n(L)]}{\left\{2D_{A_1}\int_0^{A_1(L)} R_{A_1}dA_1\right\}^{1/2}} \qquad (4.138)$$

where $(j = 1, 2, \ldots, n)$

The same procedures as for the basic reaction unit can be followed to calculate the global rates. However, the solution of Eq. 4.136 involves simultaneous solution of n nonlinear algebraic equations, which is by no means trivial. One way of circumventing this problem is to use the rectangular rule of integration instead of the trapezoidal rule in Eq. 4.136, i.e.:

$$\frac{(\Delta A_j)_{i+1}}{(\Delta A_1)_{i+1}} = \left(\frac{D_{A_1}}{D_{A_j}}\right)\left\{\frac{R_{A_j}(A_1, A_2, \ldots, A_n)|_i}{R_{A_1}(A_1, A_2, \ldots, A_n)|_i}\right\} \qquad (4.139)$$

This approximation allows one to calculate the only unknown $(\Delta A_j)_{i+1}$ directly, but at the expense of accuracy. The rectangular rule leads to an integration error that is first-order in magnitude with respect to the step size. In order to avoid such errors, one can use the $(\Delta A_j)_{i+1}$ value calculated from Eq. 4.139 only as a predictor and then use these predictor values to obtain corrected values. For this purpose, let $(\Delta A_j)_{i+1}^*$ be the predictor value calculated from Eq. 4.139. When this is used in Eq. 4.136 for the corrected value,

$$\frac{(\Delta A_j)_{i+1}}{(\Delta A_1)_{i+1}} = \left(\frac{D_{A_1}}{D_{A_j}}\right)\left\{\frac{R_{A_j}(A_1, A_2, \ldots, A_n)|_i + R_{A_j}[(A_1)_{i+1}^*, (A_2)_{i+1}^*, \ldots, (A_n)_{i+1}^*]}{R_{A_1}(A_1, A_2, \ldots, A_n)|_i + R_{A_1}[(A_1)_{i+1}^*, (A_2)_{i+1}^*, \ldots, (A_n)_{i+1}^*]}\right\}$$

Calculations involving this equation and Eq. 4.139 are straightforward since no trial and error is involved. Further, the accuracy should be much better than that possible with only Eq. 4.139. In fact, the results obtained by the above predictor-corrector method for the example problem in Table 4.3 have been found to be identical with those obtained by solving Eq. 4.136 by an iterative bisection method.

A major subset of the general reaction network shown in Figure 4.13 is the one in which all the reactions involved are irreversible and the rate of reaction for a given species depends only on the concentration of the species in question and those from which the species in question is formed. The steady state mass balances can then be written as:

$$\frac{D_{A_1}}{2} \frac{dP_{A_1}^2}{dA_1} = R_{A_1}(A_1) \tag{4.140}$$

$$\frac{D_{A_2}}{2} \frac{dP_{A_2}^2}{dA_2} = R_{A_2}(A_1, A_2) \tag{4.141}$$

$$\frac{D_{A_3}}{2} \frac{dP_{A_3}^2}{dA_3} = R_{A_3}(A_1, A_2, A_3) \tag{4.142}$$

$$\vdots \qquad \qquad \vdots$$

$$\frac{D_{A_n}}{2} \frac{dP_{A_n}^2}{dA_n} = R_{A_n}(A_1, A_2, \ldots, A_n) \tag{4.143}$$

It is obvious from the balance equations that they can be solved two at a time. The results obtained for the basic unit can be used to solve Eqs. 4.140 and 4.141, which in turn can be used to solve the entire set.

This section has shown that the global rates for multiple reactions can be obtained in a straightforward manner. In order to use the global rates in reactor conservation equations, however, the pellet surface temperature corresponding to the bulk-fluid temperature has to be known. This involves satisfying the heat balance around the film between pellet and bulk-fluid (Eq. 4.31) with $(-\Delta H)R_G L$ replaced by $\Sigma_j (-\Delta H_j)(R_{G_j})L$ for the calculation of the pellet surface temperature T_s. This aspect will be explored in detail in Chapter 10.

4–10   TRANSPORT CRITERIA AND PELLET DESIGN

The developments so far are based on the premise that intrinsic rate expressions are available. As discussed in Chapter 3, these rate expressions are determined from the kinetic data obtained under experimental conditions that ensure negligible transport effects. Various criteria of negligible transport effects have been developed to provide guidance on the experimental conditions and these were summarized in Chapter 3 without derivation. Consider these criteria in light of the understanding gained in this chapter. In view of the fact that external transport effects usually can be neglected in the absence of internal diffusion resistance, it is necessary that the internal diffusion resistance be negligible. For a first-order reaction, the internal effectiveness factor η_i approaches unity when $\eta_i \phi^2$ ($= \phi \tanh \phi$) is much less than unity. On the other hand, the definition of the internal effectiveness factor gives:

$$\eta_i = \frac{R_G}{r_c} = \frac{R_G}{k_s C_s} \qquad (4.144)$$

Since $\phi^2 = L^2 k_s / D_e$ for the first-order reaction, it follows from Eq. 4.144 that the product $\eta_i \phi^2$ is $R_G L^2/(D_e C_s)$. Therefore, a criterion of negligible diffusion effects for a first-order reaction can be stated as follows:

$$\eta_i \phi^2 = \frac{R_G L^2}{D_e C_s} \ll 1 \qquad (4.145)$$

This is known as the Weisz-Prater (1954) criterion. The usefulness of the criterion lies in the fact that all quantities appearing in the right hand side of Eq. 4.145 are measurable. While this relationship has been derived for a first-order reaction, it is nevertheless applicable to many reactions for which the kinetics are not known. Mears (1971), for instance, suggested the use of $1/n$ in place of 1 in Eq. 4.145 for an nth-order reaction. A conservative number, say 0.1, may be used in the criterion for reactions with unknown kinetics. It should be recognized, however, that the criterion does not necessarily work in all cases. This is particularly so when the reaction is strongly inhibited by one of the products (Froment and Bischoff 1979).

In experiments to obtain intrinsic kinetic data, the pellet size is often chosen small enough to ensure that diffusion is negligible. This involves carrying out a series of isothermal experiments in which the pellet size starts at a very small value and is successively increased. If the initial rate decreases in successive experiments above a certain pellet size, say L_f, then the experiments to obtain intrinsic kinetics are carried out with pellets of size smaller than L_f. A more efficient method is an experiment in which the reaction temperature is increased progressively in a planned manner. As discussed in Section 4–3, the slope of a plot of the logarithm of measured rates (ln R_G) versus $1/T$ changes from $(-E_a/R_g)$ to $(-E_a/2R_g)$ as the reaction regime changes from a diffusion-free to a diffusion-limited regime with the increase in temperature. In the transition regime, the plot is curved, as shown in Figure 4.14 (see also Problem 5.16). Therefore, a reaction can be considered free from diffusional effects, in a certain temperature range and for the pellet size used in the experiment, if a plot of ln R_G versus $1/T$ gives a straight line for that temperature range and then a curved line at higher temperatures, the straight line representing the intrinsic kinetics.

A criterion on the importance of external heat transport can be arrived at by combining Eqs. 4.80 and 4.81:

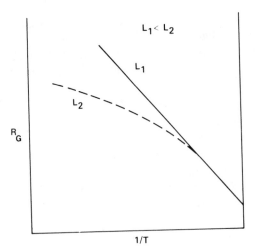

Figure 4.14 A test for the effect of diffusion.

$$\frac{1.2\epsilon L(-\Delta H)R_G}{hT_b} \ (= k_s/k_b - 1) < 0.05 \qquad (4.146)$$

where the deviation in the value of the rate constant due to the external temperature difference is restricted to less than 5%. This criterion is almost identical to that developed by Mears (1971).

Much more rigorous tests of the importance of diffusional effects can be made if the intrinsic kinetics are known on the basis of the generalized internal effectiveness factor and reactor point effectiveness. Using the rate expression of Eq. 4.2 in Eq. 4.73 for ϕ_G^2, one gets:

$$\phi_G^2 = \frac{L^2 k_s f^2(C_s)}{2 \displaystyle\int_0^{C_s} D_e f(C) \, dC} \qquad (4.147)$$

Applying the Weisz-Prater condition that $\eta_i \phi_G^2$ be much less than unity, the same procedures used in obtaining Eq. 4.145 give the following criterion of negligible diffusion effect (Petersen 1965; Bischoff 1967):

$$\frac{R_G L^2 f(C_s)}{2 \displaystyle\int_0^{C_s} D_e f(C) \, dC} \ (= \eta_i \phi_G^2) \ll 1 \qquad (4.148)$$

where $R_G/[k_s f(C_s)]$ has been used for η. The reactor point effectiveness can be used as a criterion of negligible transport effects by insisting that it be greater than some value, say 0.95:

$$\Lambda \sim \frac{(2D_e)^{1/2}}{Lr_c(C_b)}\left[\int_{C_c}^{C_b} f(C)\, dC\right]^{1/2} + \frac{1.2\epsilon(-\Delta H)2D_e}{LhT_br_c(C_b)}\left[\int_{C_c}^{C_b} f(C)\, dC\right] > 0.95 \quad (4.149)$$

Here, the inverse of the denominator of Eq. 4.83 has been approximated by the first-order term in a Taylor series.

Pellet design is concerned with the selection of pellet size, average pore size, pore size distribution, and the specification of the activity distribution function $g(S_r)$ in Eq. 4.1. Pellet design cannot be considered separately from reactor design. Nevertheless, certain pertinent conclusions can be made. It is almost always true, for instance, that the characteristic pellet size be chosen as small as the pressure drop and temperature rise in the reactor will allow. A support has to be chosen for the desired pore size distribution. There is rarely any control over the pore size distribution. If selectivity is severely affected by diffusion, a support of low surface area should be used. A support of higher surface area can give a higher rate of reaction per unit pellet volume but also gives a higher diffusion resistance, which reduces selectivity.

The catalyst surface area per unit volume, $g(S_s)$, can be made to vary with pellet coordinate by choosing an appropriate impregnation method. Hence, this function represents not only the level of dispersion but also activity distribution. A partially impregnated (or equivalently hollow) pellet is a typical example of a pellet with a certain activity distribution. The motivation for making such a pellet becomes obvious if it is recognized that the reactant concentration becomes almost zero at some point in the pellet when the reaction is diffusion-limited. The fraction of the volume of the pellet for which the concentration is zero is not utilized at all. If this fraction is made hollow or inert, then the observed rate on a per pellet basis should be the same as the fully impregnated pellet. Let us examine this further. Suppose that a pellet is hollow or partially impregnated for a distance L_i from the center. Consider a diffusion-limited, first-order reaction. The internal effectiveness factor for this hollow pellet is:

$$(\eta_i)_h = \frac{1}{\phi_h}; \qquad \phi_h = (L - L_i)\left(\frac{k_s}{D_e}\right)^{1/2}$$

For the fully impregnated pellet, it is:

$$(\eta_i)_f = \frac{1}{\phi_f}; \qquad \phi_f = L\left(\frac{k_s}{D_e}\right)^{1/2}$$

Therefore, the internal effectiveness factor increases as a result of making the pellet hollow by the following factor:

$$\frac{(\eta_i)_h}{(\eta_i)_f} = \frac{\phi_f}{\phi_h} = \frac{L}{(L - L_i)} > 1$$

While the internal effectiveness factor increases, the reaction rate per pellet remains the same since:

$$\frac{R_{Gf}V_f}{R_{Gh}V_h} = \frac{k_s C_b L (\eta_i)_f}{k_s C_b (L - L_i)(\eta_i)_h} = 1$$

where V_f and V_h are the pellet volume for the fully impregnated pellet and that for the hollow pellet, respectively. It is seen that the use of hollow pellets does not increase the amount of reactant converted but does result in a saving of catalyst in the ratio of $(L - L_i)/L$. For the same dispersion and catalyst loading, then, it is best, in the absence of attrition and poisoning, to have all catalyst deposited in the thinnest possible exterior layer of the pellet at least for the simple reactions now under consideration. The rationale behind this is that the highest possible rate constant can be attained with this arrangement. The observed rate constant is proportional to the square root of the intrinsic rate constant and this intrinsic rate constant increases with increasing surface area per unit volume. However, this approach to pellet design is not practical even in the absence of chemical and physical deactivation since most of the catalyst will be lost due to attrition in the reactor. In practice, therefore, a compromise has to be made that is suitable for the application of interest. The optimal catalyst distribution for relatively complex reactions can depend on several factors (Morbidelli et al. 1982). While the activity profile can be tailored to suit a certain application, uniformity of activity is often an overriding concern in practice because the repeatability of catalyst performance is one of the overriding factors in plant operation.

As is apparent from the j-factor correlations of Eq. 4.58, external transport effects can be minimized by a proper choice of fluid velocity. While the transfer coefficients are also affected by the pellet size, the fluid velocity is often the choice for the manipulation. As discussed earlier, the external mass transfer resistance should not pose a problem at throughputs commonly encountered in typical operation.

Summary

This chapter has shown that internal and external transport processes can significantly alter rates of reaction. Since the actual conversion of reactants occurs at the global rate as affected by the transport processes, pellet conservation equations have to be solved first to relate the intrinsic rate to the global rate for reactor design. The fact that the major resistances are usually internal diffusion and external heat transfer allows one to express the global rate in terms of intrinsic kinetics and pertinent transport properties. This result, in turn allows one to deal directly with the reactor conservation equations for reactor design without having to solve the pellet conservation equations. For multiple reactions, however, the pellet conservation equations have to be solved as part of the reactor design. The numerical method given in this chapter allows this calculation in a straightforward manner.

NOTATION

$A_j(L)$	pellet surface concentration of species A_j
$A_j(0)$	pellet center concentration of species A_j
B_0	Darcy's permeability
$(\text{Bi})_h$	Biot number for heat, defined in Table 4.2
$(\text{Bi})_m$	Biot number for mass, defined in Table 4.2
B_r	$(\text{Bi})_h/(\text{Bi})_m$
c	C/C_b
c_0	solution of Eq. 4.85 when $F(c) = c$
C	fluid concentration
C_b	bulk-fluid concentration
C_i	concentration for species i
C_c	pellet center concentration
$\overline{C}$	C/C_s
C_p	specific heat capacity
$\mathbf{C}$	concentration vector for all species
ΔC_i	$C_s - C_c$
ΔC_x	$C_b - C_s$
D_e	effective diffusivity
$D_{e,1}$	D_e for species 1
$D_{e,ij}$	effective molecular diffusivity of species i in species j
$D_{e,ik}$	effective Knudsen diffusivity for species i
$D_{e,k}$	effective Knudsen diffusivity for species of interest in binary system
$D_{e,is}$	effective surface diffusivity for species i
D_m	molecular diffusivity
D_{A_j}	effective diffusivity of species A_j
E_a	activation energy
$f(C)$	concentration dependence of intrinsic kinetics $[r_c = kf(C)]$
f_e	correction factor given in Eq. 4.11
$F(c)$	$f(C)/f(C_b)$
$g(S_r)$	catalyst surface area per unit pellet volume
G	catalyst activity distribution function ($G \leq 1$)
G'	dG/dz
$\overline{G}, \overline{G}'$	G and G' evaluated at $z = L$
h	film heat transfer coefficient
$(-\Delta H)$	heat of reaction
I	integral defined by Eq. 4.84
j_D	j factor for mass given in Eq. 4.58
j_H	j factor for heat given in Eq. 4.58
$\overline{J}$	quantity defined in Eq. 4.100
k	rate constant
$\overline{k}$	rate constant used for normalization of k $[k = \overline{k}G(z)]$
k_g	film mass transfer coefficient
k_{sp}	rate constant per unit catalyst surface area

K	equilibrium constant; a constant in j-factor correlation
L	characteristic length of pellet ($= V_p/S_x$), position for pellet surface as measured from pellet center for slab
Le	Lewis number defined in Table 4.2
L_i	distance from pellet center up to which the pellet is inert or hollow
L_m	L for various pellet shapes
m	M_1/M_2
M_i	molecular weight for species i
n	order of reaction
N_i	flux for species i
$\mathbf{N}_i$	directional flux vector for species i
P	dC/dz; total pressure
P_{A_j}	dA_j/dz
$\bar{P}$	P evaluated at $z = L$
$\bar{P}_\infty$	$\bar{P}$ in the region of strong diffusional effect
Pr	Prandtl number defined in Table 4.2
r	rate of reaction defined in Eq. 4.94; average pore radius
r_c	intrinsic rate of reaction given in Eq. 4.1
r_{ij}	intrinsic rate of reaction for the reaction path from species A_i to species A_j
R_{A_j}	intrinsic rate of consumption of a species A_j
R	radius; ratio defined by Eq. 4.102
R_G	global rate
R_g	gas constant
Re	Reynolds number defined in Table 4.2
Sc	Schmidt number defined in Table 4.2
S_r	active catalyst surface area
S_x	external pellet surface area
T	temperature
T_b	bulk-fluid temperature
ΔT_i	$T_c - T_s$
ΔT_x	$T_s - T_b$
t	T/T_b
v	superficial fluid velocity
V_p	pellet volume
x_i	mole fraction of species i
y	z/L
z	pellet coordinate with pellet center as the orgin
Z	reactor length

Greek Letters

| α | dummy variable for integration; a constant in j factor correlation |
| β | quantity defined in Table 4.2; dummy variable for integration |

β_x	quantity defined in Table 4.2
ϵ	pellet porosity; Arrhenius number defined in Table 4.2
ϵ_A	quantity defined by Eq. 4.119
ϵ_B	bed porosity; quantity defined by Eq. 4.121
κ	tortuosity
ζ	dummy variable for integration
η	effectiveness factor
η_i	internal effectiveness factor defined in Eq. 4.46
η_m	η for various pellet shapes
η_o	overall effectiveness factor defined in Eq. 4.46
η_x	external effectiveness factor defined in Eq. 4.46
η_∞	η in the asymptotic region of strong diffusion effect
ηDa	quantity defined in Table 4.2
λ_e	effective thermal conductivity
Λ	reactor point effectiveness defined by Eq. 4.77
μ	fluid viscosity
ρ	fluid density
ν_i	stoichiometric coefficient for species i
ϕ	Thiele modulus defined by Eq. 4.44
ϕ_0, ϕ_1, ϕ_2	Thiele modulus for slab-like, cylindrical, and spherical pellets
ϕ_G	generalized modulus defined by Eq. 4.73
$(\phi_G)_n$	generalized modulus for nonuniform pellet, Eq. 4.105
ϕ_s	Thiele modulus defined in Eq. 4.59

Subscripts

b	at bulk-fluid conditions
c	at pellet center, at $z = 0$
f	fully impregnated pellet
h	hollow pellet
i	index for concentration grid
j	species
L	at pellet surface, at $z = L$
m	pellet shape ($m = 0$, slab; $m = 1$, cylinder; $m = 2$, sphere)
s	at pellet surface

PROBLEMS

4.1. Obtain the internal effectiveness factor for a first-order reaction taking place in a spherical pellet. Use the following transformation of variables: $v = Cr$ where r is the radial coordinate.

4.2. Suppose that a pellet is isobaric. Does this mean that the convective flux can be neglected? What is the rationale behind using Eq. 4.4a in most treatments of reaction-diffusion problems?

4.3. Show that Eq. 4.15 follows from Eq. 4.13 for a binary system.

4.4. Extract as much information as you can on the intrinsic kinetics and diffusional effect from the results shown in Figure 4.3.

4.5. The observed rate using a 2 cm cubic pellet is 10^{-8} mol/(cm^3 pellet sec). If the intrinsic rate is 10^{-7}, what is the half-thickness of a slab that would yield the same observed rate? Suppose further that conversion is independent of initial concentration. What is the intrinsic rate constant? The effective diffusivity is 0.01 cm^2/sec.

4.6. Calculate the surface concentration and the concentration and temperature at the pellet center for the following conditions:

$$R_G = 10^{-5} \text{ mol/(cm}^3 \text{ pellet sec)} \qquad \phi = 5$$
$$C_b = 10^{-3} \text{ mol/cm}^3 \qquad\qquad \epsilon = 0.5$$
$$k_g = 1 \text{ cm/sec} \qquad\qquad\qquad L = 2 \text{ cm}$$
$$T_b = 110°C \qquad\qquad\qquad\quad T_s = 120°C$$
$$B_r = 60$$

Suppose that $E_a/R_g T_b = 10$. What is the effectiveness factor if the external temperature difference is taken into account?

4.7. Derive the effectiveness factor for a first-order reaction when the catalyst is deposited only on the external surface of the support.

4.8. When the CO content is quite low, the oxidation of CO over Pt can be represented by $r_c = k/(CO)$. Discuss this behavior and obtain the internal effectiveness factor.

4.9. Show that for a slab-like pellet, the effectiveness factor for a first-order reaction is given by:

$$\eta = \frac{\tanh \phi/\phi}{1 + \phi \tanh \phi/(\text{Bi})_m}$$

Assume that the pellet is isothermal and that the external heat transfer resistance is negligible. What can you conclude about the relative importance of external mass transfer? What if the internal diffusion resistance is very high?

4.10. Liu (1969) gave the following approximate relationship for the nonisothermal, internal effectiveness factor:

$$\eta_i = \frac{\tanh \phi}{\phi} \exp(\epsilon\beta/5)$$

Find an expression for the locus of the maximum values of η_i corresponding to different values of $\epsilon\beta$. Note that both ϕ and $\epsilon\beta$ change with temperature.

4.11. For uniform catalyst pellets with a dispersion of 0.2, the Thiele modulus is 5. If the same pellets with a dispersion of 0.8 are used under identical reaction conditions, what is the change in the observed rate?

4.12. Draw conclusions on the effect of internal diffusion on the internal and external temperature and concentration gradients from the following information for a first-order reaction:

$$\beta = 0.01, \quad B_r = 100, \quad T_b = 400°K$$
$$\phi = 5, \quad D_e = 0.01 \text{ cm}^2/\text{sec}$$
$$\epsilon = 20, \quad C_b = 0.6 \times 10^{-5} \text{ mol/cm}^3$$
$$(Bi)_m = 200$$

The values of ϕ and $(Bi)_m$ are for L of 2 cm. Calculate the gradients (ΔT's and ΔC's) corresponding to $L = 0.1$ cm, 1 cm, and 10 cm. Is it true that all gradients increase as ϕ increases?

4.13. Extract as much numerical information as you can from the results shown in Figure 4.8 on the values of dimensionless groups given in Table 4.2 and on the values of concentration differences (ΔC's).

4.14. Show that the reactor point effectiveness for a first-order reaction reduces to:

$$\Lambda = \frac{\tanh \phi}{\phi[1 - (\eta Da)_r]}$$

$$(\eta Da)_r = \frac{1.2\epsilon(-\Delta H)(D_e k_b C_b)^{1/2}}{2hT_b}\left[1 - \frac{1}{\cosh^2\phi}\right]^{1/2}$$

$$\phi = L\left(\frac{k_b}{D_e}\right)^{1/2}$$

4.15. Consider a reaction, $A + B \rightarrow 2D$. The intrinsic rate expression is:

$$r_c = \frac{kC_B}{(1 + K_1 C_D + K_2 C_B)}$$

Rewrite the rate expression in the form of

$$r_c = kf(C_B)$$

Obtain the generalized modulus for this reaction. Assume equal D_e's.

4.16. Petersen (1965) used the data of Austin and Walker (1963) for the reaction, $C + CO_2 \rightarrow 2CO$. Use the result of Problem 4.15 and the following data to check the importance of diffusion limitation, first using the Weisz and Prater criterion and then the criterion based on the generalized modulus.

$$(C_{CO_2})_s = 1.22 \times 10^{-5} \text{ mol/cm}^3, \quad K_1 = 4.15 \times 10^9 \text{ cm}^3/\text{mol}$$
$$K_2 = 3.38 \times 10^5 \text{ cm}^3/\text{mol}, \quad D_e = 0.1 \text{ cm}^2/\text{sec}$$
$$L = 0.7 \text{ cm}, \quad R_G = 4.67 \times 10^{-9} \text{ mol/cm}^3 \text{ sec}$$

What conclusions can you make regarding the suitability of the criteria for this reaction that is inhibited by product? Assume that $(C_{CO_2})_{inlet} = 1.22 \times 10^{-5}$ mol/cm^3.

4.17. The following has been proven by Corbett and Luss (1974): For two identical pellets in which a first-order irreversible reaction occurs, which have the same volume averaged rate constant but different rate constant density functions, the effectiveness factor corresponding to the pellet with the larger surface activity exceeds that of the second pellet if the difference between the rate constant density functions of two spherical pellets changes monotonically with the radial distance. Is the statement correct for slab-like pellets in the asymptotic region of strong diffusion effects? Use the generalized effectiveness factor for nonuniform activity.

4.18. In the region of strong diffusion effects, approximate as well as exact solutions for the effectiveness factor can be obtained by various methods for a first-order reaction taking place in a pellet of nonuniform activity. Obtain the effectiveness factor when the activity distribution function $G(z)$ is given by $(z/L)^\alpha$, using the generalized modulus approach, the method of exact solution in the form of Bessel functions and the method of Liouville-Green approximation (WKB approximation) (Yortsos and Tsotsis 1982). Discuss and compare the results.

4.19. Calculate the effectiveness factors as a function of $\bar{k}$ for a pellet of nonuniform activity for which $r_c = \bar{k} G C^n$. Use the conditions given below.

$$G(z) = 0.1 + 0.9 (1 - z/L)^3$$
$$n = 3$$
$$C_b = 4 \times 10^{-5} \text{ mol/cm}^3$$
$$L = 1 \text{ cm}, \quad D_e = 0.05 \text{ cm}^2/\text{sec}$$

Use Eq. 4.101 first. Add one and then two more correction terms for $\bar{P}_\infty$ (Eq. 4.99) to compare these results with those obtained from Eq. 4.101. Redo the problem for $G = 0.271 + (0.9 - z/L)^3$.

4.20. Does an increase in the Thiele modulus result in an increase, decrease, or no change in the internal and external concentration and temperature differences? Answer the question when the increase is due to an increase in pellet length, a decrease in effective diffusivity, or an increase in temperature.

4.21. Some supports have a bimodal pore-size distribution consisting of macropores and micropores. Show for a first-order reaction that the isothermal, internal effectiveness factor (Mingle and Smith, 1961) is given by:

$$\eta = \eta_\mu \eta_M \begin{cases} \eta_\mu = \tanh \phi_\mu / \phi_\mu & \phi_\mu = L_\mu (k/D_\mu)^{1/2} \\ \eta_M = \tanh \phi_M / \phi_M & \phi_M = L(k\eta_\mu/D_e)^{1/2} \end{cases}$$

where the subscript μ denotes micropores. Explain why there usually is no micropore diffusion limitation unless there is also a macropore limitation.

4.22. Consider the following parallel reaction network:

$$A \overset{\displaystyle B}{\underset{\displaystyle C}{\diagdown}}$$

$$r_A = -k_1 f_1(A) - k_2 f_2(A)$$
$$r_B = k_2 f_2(A)$$

Show that the concentration gradients at the pellet surface are given by:

$$\frac{dA}{dz}\Big|_{z=L} = \left\{ \frac{2}{D_A} \left[k_{1s} \int_0^{A_s} f_1(A) \, dA + k_{2s} \int_0^{A_s} f_2(A) \, dA \right] \right\}$$

$$\frac{dB}{dz}\Big|_{z=L} = -\frac{1}{D_B} \left\{ \int_0^{A_s} \left[\frac{\sqrt{D_A/2} \; k_{2s} f_2(A)}{\left(k_{1s} \int_0^A f_1(\alpha) \, d\alpha + k_{2s} \int_0^A f_2(\alpha) \, d\alpha \right)} \right] dA \right\}$$

4.23. The rate expression for the following reversible reaction is:

$$A + 2B \underset{k_2 f_2}{\overset{k_1 f_1}{\rightleftharpoons}} D$$

where $k_1 f_1 = k_1 A B^2$
$k_2 f_2 = k_2 D$

Obtain the internal effectiveness factor for species A. Use the generalized modulus approach.

4.24. For the example network in Section 4–9, the following parameters and concentrations were used:

ϕ_G	k_A	k_B	k_D	k_1	k_2	$A(L)$	$B(L)$
7.86	12.0	$6(10^4)$	$3(10^5)$	$1(10^3)$	$2(10^3)$	$5.274(10^{-4})$	$3.812(10^{-6})$
4.00	6.0	$3(10^4)$	$1.5(10^5)$	$1(10^3)$	$2(10^3)$	$1.042(10^{-4})$	$1.779(10^{-5})$

$$D_A = 0.05 \text{ cm}^2/\text{s}, \quad D_B = 0.08 \text{ cm}^2/\text{s}, \quad L = 1 \text{ cm}$$

Calculate the global rates, $(R_G)_A$ and $(R_G)_B$, using the numerical method given in Section 4–9.

4.25. Derive Eq. 4.122 and Eq. 4.139.

REFERENCES

Akella, L.M., Ph.D. Thesis, Dept. of Chem. Eng., University of Florida (1983).
Aris, R., Ind. Eng. Chem. Fund., 4, 227 (1965).

Aris, R., *Elementary Chemical Reactor Analysis*, Prentice-Hall, Englewood Cliffs, N.J. (1969).

Aris, R., *Mathematical Theory of Diffusion and Reaction in Permeable Catalysts*, Vol. I, Oxford University Press, London (1975).

Austin, L.G. and P.L. Walker, AIChE J., *9*, 303 (1963).

Bird, R.B., W.E. Stewart and E.N. Lightfoot, *Transport Phenomena*, Wiley, New York (1960).

Bischoff, K.B., AIChE J., *11*, 351 (1965).

Bischoff, K.B., Chem. Eng. Sci., *22*, 525 (1967).

Brown, C.E. and C.O. Bennett, AIChE J., *16*, 817 (1970).

Butt, J.B., D.M. Downing and J.W. Lee, Ind. Eng. Chem. Fund., *16*, 270 (1977).

Carberry, J.J. and A.A. Kulkarini, J. Catalysis, *31*, 41 (1973).

Carberry, J. J., Ind. Eng. Chem. Fund., *14*, 129 (1975).

Carberry, J.J., *Chemical and Catalytic Reaction Engineering*, p 230, McGraw-Hill, New York (1976).

Corbett, W.E. and D. Luss, Chem. Eng. Sci., *29*, 1473 (1974).

Dente, M. and I. Pasquou, Chim. Ind. (Milan), *47*, 359 (1965).

DiNapoli, N.M., R.J.J. Williams and R.E. Cunningham, Lat. Am. J. Chem. Eng. Appl. Chem., *5*, 101 (1975).

Feng, C.F. and W.E. Stewart, Ind. Eng. Chem. Fund., *12*, 143 (1973).

Froment, G.F. and K.B. Bischoff, *Chemical Reactor Analysis and Design*, p 194, Wiley, New York (1979).

Gunn, R.D. and C.J. King, AIChE J., *15*, 507 (1969).

Hite, R.H. and R. Jackson, Chem. Eng. Sci., *32*, 703 (1977).

Kehoe, J.P.G. and J.B. Butt, AIChE J., *18*, 347 (1972).

Lee, H.H., Chem. Eng. Sci., *36*, 1921 (1981a).

Lee, H.H., AIChE J., *27*, 558 (1981b).

Lee, H.H. and E. Ruckenstein, Cat. Rev., *25*, 475 (1983).

Liu, S.L., AIChE J., *15*, 337 (1969).

Luss, D., in *Chemical Reactor Theory*, eds. Amundson and Lapidus, Chap. 4, Prentice-Hall, Englewood Cliffs, N.J. (1977).

Mason, E.A. and R.B. Evans, J. Chem. Ed., *46*, 358 (1969).

Mears, D.E., Ind. Eng. Chem. Proc. Des. Dev., *10*, 541 (1971).

Mingle, J.O. and J.M. Smith, AIChE J., *7*, 243 (1961).

Morbidelli, M., A. Servida and A. Varma, Ind. Eng. Chem. Fund., *21*, 278 (1982).

Petersen, E.E., Chem. Eng. Sci., *20*, 587 (1965).

Rester, S. and R. Aris, Chem. Eng. Sci., *24*, 793 (1969).

Satterfield, C.N., G.W. Roberts and J. Hartman, Ind. Eng. Chem. Fund., *6*, 80 (1967).

Thiele, E.W., Ind. Eng. Chem. *31*, 916 (1939).

Wang, J.B. and A. Varma, Chem. Eng. Sci., *35*, 613 (1980).

Weisz, P.B. and C.D. Prater, Adv. Cat., *6*, 144 (1954).

Weisz, P.B. and J.S. Hicks, Chem. Eng. Sci., *17*, 265 (1962).

Yortsos, Y.C. and T.T. Tsotsis, Chem. Eng. Sci., *37*, 237 (1982).

CHAPTER 5

Chemical Deactivation

5-1 INTRODUCTION

A catalyst is not consumed during a reaction and thus can be used indefinitely unless it loses its activity through deactivation. Unfortunately, however, all catalysts are to a certain extent susceptible to deactivation, the cause of which can be chemical or physical in nature. Physical deactivation is treated in the next chapter. Chemical deactivation in which the loss of catalytic activity is caused by strong chemisorption of some impurities, normally contained in the reacting mixture, is termed *poisoning*. Sulfur, phosphorus, and nitrogen compounds are typical poisons. Chemical deactivation is termed *fouling* when the loss of activity is caused by reactant or product degradation on the catalyst surface. Coke formation is the most important example. Chemical deactivation is also termed *independent* or *dependent* depending on whether the species causing deactivation is involved in the main reaction (dependent) or not (independent). While reactions on the catalyst surface are the main source of deactivation, noncatalytic reactions such as thermal cracking, where coke simply deposits on the surface, can also cause deactivation.

The present level of understanding of chemical deactivation, although not complete, does allow a quantitative description of deactivation for purposes of reactor design and optimization. The unique feature of deactivation as related to reactor design is the time dependence of catalytic activity: the activity decreases with time on stream. In fact, studies on deactivation have been centered around the time dependence of activity so that it can be incorporated into reactor studies. The earliest work by Voorhies (1945), for instance, treated the correlation of coke concentration with time on stream. Various models result depending on how the time dependence is represented. Another important factor for reactor design is the interaction between diffusional and deactivational effects. Limiting cases of this interaction as well as more general cases can be treated with an understanding of noncatalytic gas-solid reactions. The first half of this chapter, therefore, deals with mechanisms of deactivation, noncatalytic gas-solid reactions, and deactivation-diffusion models.

As the catalyst deactivates, the active surface area decreases. If deactivation is assumed nonselective, the effective rate of reaction is simply the intrinsic rate multiplied by the fraction of surface area still remaining active. This effective rate can then be used in pellet conservation equations to arrive at the global rate, provided that the deactivation kinetics are known. It is then possible to combine the effects of diffusion and deactivation into a reactor point effectiveness, which gives the ratio of global rate to intrinsic rate in terms of bulk fluid quantities. These subjects are treated in the second half of this chapter.

5–2   NATURE OF DEACTIVATION

In Chapter 1, catalysts were grouped into three types: metals, acid oxides, and semiconducting oxides. The nature of chemical deactivation will be examined in accordance with this division.

The common poisons for metal catalysts can be classified into two groups (Maxted 1951): (a) molecules containing elements of the periodic groups Vb and VIb, namely: N, P, As, and Sb from group Vb, and O, S, Se, and Te from group VIb, including the free elements with the exception of nitrogen, and (b) compounds of catalytically toxic metals and metallic ions: Cu^+, Cu^{2+}, Ag^+, Au^+, Zn^{2+}, Cd^{2+}, Hg^+, Hg^{2+}, In^{3+}, Tl^+, Sn^{2+}, Pb^{2+}, Bi^{3+}, Mn^{2+}, Fe^{2+}, Co^{2+}, Ni^{2+}. A poison that deactivates metal catalysts does so by being more strongly adsorbed than the reactant. It is not surprising then that the poison should have unshared electron pairs or empty valency orbitals available for forming stable surface compounds with the catalyst. This fact is exemplified in Table 5.1 (Maxted 1951) in which toxic and nontoxic compounds of group Vb and VIb elements are shown with their electronic configurations. The argument of unshared electron pairs can be used to explain why a substance poisons a catalyst for one reaction but not another. For example, arsenic compounds are general poisons, via arsine formation, for platinum in hydrogenation reactions:

$$\left[\text{H} \; : \; \overset{..}{\text{As}} \; : \; \text{H} \atop \overset{.}{\text{H}} \right]$$

However, they have no effect on catalytic activity for the decomposition of hydrogen peroxide, presumably by remaining in a "shielded" or saturated form such as the arsenate under strong oxidizing conditions

$$\left[\text{O} \; : \; \overset{\overset{\displaystyle O}{..}}{\underset{\overset{..}{\text{O}}}{\text{As}}} \; : \; \text{O} \right]^{3-}$$

The catalytically toxic metallic ions listed above show the existence of some connection between the toxicity of a metallic ion and the structure of its d band: all have occupied d orbitals. It is presumed that these metallic compounds or ions are involved via the d-shell in the formation of chemisorbed complexes resembling intermetallic species. While most of these results pertain to observations made in hydrogenations with Pd, Pt, or Ni, the arguments presented can serve as a basis for examining the nature of deactivation for other types of reactions.

Acidic oxides, notably silica-alumina, suffer from both poisoning and fouling.

Table 5.1 Influence of Electronic Configuration on Toxicity

Toxic Types	Nontoxic Types (shielded structure)

$$H : \overset{..}{\underset{..}{S}} : H \qquad\qquad H : \overset{H}{\underset{..}{P}} : H$$

Hydrogen sulfide   Phosphine

$$\left[O : \overset{O}{\underset{O}{P}} : O \right]^{3-}$$

Phosphate ion

$$\left[O : \overset{O}{\underset{..}{S}} : O \right]^{2-}$$

Sulfite ion
(also selenite and tellurite)

$$\left[O : \overset{O}{\underset{O}{S}} : O \right]^{2-}$$

Sulfate ion
(also selenate and tellurate)

$$(R)C : \overset{..}{\underset{..}{S}} : H$$

Organic thiol

$$(R)C : \overset{O}{\underset{O}{S}} : OH$$

Sulfonic acid

$$(R)C : \overset{..}{\underset{..}{S}} : C(R')$$

Organic sulfide

$$(R)C : \overset{O}{\underset{O}{S}} : C(R')$$

Sulfone

Pyridine

Pyridinium ion

(Maxted 1951. Reprinted with permission from *Advances in Catalysis.* Copyright by Academic Press.)

While fouling is not confined to acid oxides, these catalysts suffer the most from fouling. Organic base, particularly nitrogen compounds, are effective poisons for these catalysts. In view of the fact that these catalysts owe their activity to their surface acid sites, it is expected that basic compounds would poison the catalysts and that their effectiveness as poisons would be related to basicity. In fact, the basic nitrogen compounds have been found in poisoning tests (Mills et al. 1950) to be more poisonous in the order of quinaldine > quinoline > pyrrole > piperidine > decylamine > aniline, which is in the order of increasing basicity with the exception of piperidine, the most basic compound. It has been found, however, that a significant amount of piperidine cracked in the tests while others did not. It can be postulated, therefore, that the mechanisms of poisoning for these catalysts is chemisorption of the poison on incompletely coordinated aluminum or silicon ions, which results in a reduction of acid sites, of acid strength, or both.

The redox mechanism for catalysis by semiconducting oxides suggests that the poison would be a species that chemisorbs strongly on lattice oxygen that is presumably in the form of the O^{2-} ion. Sulfur and sulfur compounds are typical poisons.

While there are other modes of fouling, such as metals deposition on catalysts in desulfurization of residual oils, emphasis here will be mainly with coking because of its importance and the relatively abundant literature available on the subject. The coke typically encountered in fouling is a carbonaceous material ranging from pure carbon to a hydrogen-deficient carbon matrix of high molecular weight. Carbon deposition in steam reforming, for instance, occurs through decomposition of carbon monoxide, or its reduction by hydrogen, and by dehydrogenation of hydrocarbons. On most catalysts used in hydrocarbons (Butt 1972) where coking is observed, highly unsaturated species of high molecular weight are adsorbed preferentially, and in particular, polyring aromatics have been associated with coke formation. After adsorption, these aromatics undergo further condensation and hydrogen elimination on the surface to form hydrogen-deficient coke. The amount of coke thus formed can be well in excess of 50% in terms of weight percent on the catalyst and this can lead to pore blocking. When pore blocking occurs, the reactants cannot get into the interior of pores.

Although the kinetics of coke formation have been relatively thoroughly investigated, the chemical and physical nature of coke formation has not. In particular, it is difficult to comprehend why pore blocking should occur if coking takes place on the catalyst surface. The point here is that once the whole surface is eventually covered by coke, no further coking should occur, and that one layer of coke cannot be thick enough to cause pore blocking. Some light on this aspect has been shed by the work of Baker et al. (1972; 1980). They studied the metal catalyzed decomposition of acetylene using an *in situ* electron microscope. They observed that flocculent amorphous carbon formed initially around metal particles. On raising the temperature, the metal particles became mobile and carbon filaments were observed to grow beneath them, eventually lifting the particles from the support. The filaments grew with one metal particle at the end of each filament. The growth of the filaments stopped when the bare front of the metal particle was encapsulated by deposited carbon. The mechanism of this growth was attributed to the diffusion through the metal particle of carbon deposited on the bare front, which provided carbon for the filament growth. The supports used for the metal catalysts were silica and graphite, on which the mobility of metal particles is relatively high (see Chapter 6). On alumina, however, the mobility is lower and thus the same mechanism may not be operative. While there are other questions to be resolved, the observations of Baker and coworkers do provide a basis for further studies on coking and pore blocking. Models for catalyst deactivation by site coverage and pore blockage have been proposed by Beekman and Froment (1980; 1982).

A series of coking experiments on silica-alumina catalysts by Levinter et al. (1967) demonstrated that pore blocking can occur on coke deposition to various extents depending strongly on catalyst properties and reaction conditions. Under extreme conditions, for instance, the coking can be localized at the pore mouth, thereby completely eliminating accessible surface at very low total levels of coke. Under more moderate conditions, near-theoretical limiting amounts would be required before surface access is denied. Coking, therefore, can substantially alter physical and transport properties of catalysts as well as their activity. Deactivation by coking and its effect on the transport properties is treated in Section 5–8.

5–3 DEACTIVATION KINETICS

The earliest effort to correlate the time dependence of activity was that of Voorhies (1945). He found:

$$C_c = At^n \tag{5.1}$$

where C_c is the concentration of carbon on the catalyst (which may be related to the activity), A and n are the correlation constants, and t is time on stream. This correlation fits a very large amount of catalyst deactivation data, not only on coking, and is encountered throughout the literature. Based on various correlations in the literature, the following power law form has been proposed by Szepe and Levenspiel (1970):

$$\frac{da}{dt} = -ka^m \tag{5.2}$$

where a is activity. This has been extended (Khang and Levenspiel 1973) to:

$$r_p = k_d(C^{m_1}, N^{m_2})a^d \tag{5.3}$$

where the rate of deactivation is expressed in terms of powers of reactant concentration C, concentration of poisoning species N, and activity.

Since deactivation affects the rate of the main reaction through the activity, the deactivation kinetics are invariably tied to main reaction kinetics. When the effect of deactivation can be factored out in such a way that kinetic dependencies, which are time independent, and activity dependencies, which are not, are separable, the rate expressions are termed separable (Szepe and Levenspiel 1970). Separable kinetics can be expressed as:

$$r = r_1(C, T)r_2(a) \tag{5.4}$$

$$r_p = r_3(C, T)r_4(a) \tag{5.5}$$

where r_1 and r_3 are the usual rate expressions without deactivation, and r_2 and r_4 are functions of activity. These kinetics in many, but not all, cases represent real catalytic deactivation mechanisms. One apparent advantage of this approach, even when empiricisms are involved, over the pure activity correlations such as Eq. 5.1 is that the time dependence of activity is not tacked on only after the fact. The latter approach gives the following deactivation kinetics, when written in the separable form:

$$r_p = r_3(C,T)a(t) \tag{5.6}$$

When the mechanism of deactivation is the chemisorption of poisoning species, the rate at which active sites disappear due to poisoning is equal to the rate at which the poisoning species chemisorbs on these sites. Therefore, the rate of deactivation is given by:

$$r_p = k_p s \, N \tag{5.7}$$

assuming that the desorption rate is negligible. Here, s is the fraction of vacant active sites per unit volume of catalyst pellet, and k_p is the adsorption constant. If it is assumed that the surface is energetically uniform, the rate of the main reaction, when subjected to deactivation, is simply:

$$r = ksf(C) \tag{5.8}$$

where k is the rate constant for fresh catalyst. A balance on the total number of sites gives:

$$C_t = C_p + C_v \left[1 + \sum_i (K_i C_i)^{m_i}\right]^n \tag{5.9}$$

where C_t = total number of sites/pellet volume
C_p = number of sites poisoned/pellet volume
C_i = reactant and product concentrations excluding poisoning species involved in elementary steps
K_i = equilibrium constants
n = constant ranging from 1 to 3 depending on the mechanism and controlling step in the main reaction

The value of m_i is ½ or 1 depending on the mode of adsorption (Chapter 2): ½ for dissociative adsorption and 1 for molecular adsorption. Rearrangement of Eq. 5.9 gives:

$$s = \frac{C_v}{C_t} = \frac{1 - \gamma}{\left[1 + \sum_i (K_i C_i)^{m_i}\right]^n} \tag{5.10}$$

since by definition, $C_t - C_p = C_t(1 - \gamma)$. Here, γ is the fraction of catalyst deactivated. Combining Eqs. 5.8 and 5.10 yields:

$$r_p = \frac{k_p N(1-\gamma)}{\left[1 + \sum\limits_i (K_i C_i)^{m_i}\right]^n} \equiv \frac{k_p N(1-\gamma)}{[1 + G(C,K_i)]^n} \tag{5.11}$$

The term $\sum_i K_i C_i$ can always be rewritten in terms of the concentration of key species C and equilibrium constants. Take, as an example, a simple reaction with independent poisoning:

$$A + S \rightleftharpoons A \cdot S \quad (K_A)$$
$$A \cdot S \rightleftharpoons B \cdot S \quad (k_s, \text{ controlling step})$$
$$B \cdot S \rightleftharpoons B + S \quad (K_B)$$

which are elementary steps for the reaction $A \rightleftharpoons B$, and:

$$N + S \rightarrow N \cdot S \quad (k_p)$$

where S represents the active site(s). These steps lead to the following rate expressions:

$$r_A = \frac{k_s C_t K_A (C_A - C_B/K)(1-\gamma)}{1 + K_A C_A + K_B C_B}; \qquad K = \frac{K_A K_s}{K_B}$$
$$r_p = \frac{k_p N(1-\gamma)}{1 + K_A C_A + K_B C_B}$$

Therefore, the quantity G is:

$$G(C,K_i) = K_A C_A + K_B C_B = C_A (K_A - K_B) + K_B C_{As}$$

which is expressed solely in terms of the key species A and equilibrium constants. Here, C_{As} is the concentration of species A at the pellet surface. The deactivation kinetics of Eq. 5.11 apply to both dependent and independent poisoning.

The chemical nature and mechanisms of fouling are not as well understood as that of poisoning. Nevertheless, deactivation kinetics can still be treated in the general framework of chemisorption. Fouling is usually a dependent deactivation process, and this dependent fouling can be represented by the following models (Butt 1972):

$$\text{Parallel:} \quad \begin{array}{l} A + S \rightarrow B + S \\ A + S \rightarrow A \cdot S \end{array}$$

$$\text{Series:} \quad \begin{array}{l} A + S \rightarrow B + S \\ B + S \rightarrow B \cdot S \end{array}$$

where A is the main reactant and B is an intermediate product. Following similar procedures as in poisoning, one can arrive at the deactivation kinetics:

$$\text{Parallel:} \quad r_p = \frac{k_p C_A (1 - \gamma)}{\left[1 + \sum_i (K_i C_i)^{m_i}\right]^n} \tag{5.12}$$

$$\text{Series:} \quad r_p = \frac{k_p C_B (1 - \gamma)}{\left[1 + \sum_i (K_i C_i)^{m_i}\right]^n} \tag{5.13}$$

These simple models can be expanded in their forms to represent various situations such as fouling of polyfunctional catalysts, reversible chemisorptions, etc.

An approach by Froment and Bischoff (1979) for coking is to write the deactivation kinetics as:

$$r_p = (r_p)_0 \phi_c \tag{5.14}$$

where $(r_p)_0$ is the initial coking rate and the deactivation function ϕ_c is correlated to coke concentration by:

$$\phi_c = \exp(-\alpha C_c) \tag{5.15}$$

where α is a constant.

The various forms of deactivation kinetics show dependence not only on the mechanism of deactivation, but on the composition of reactants and products. Consequently, deactivation would not take place uniformly throughout a pellet whenever there exist concentration gradients within the pellet. In order to handle deactivation in the presence of diffusion limitations, one needs an understanding of noncatalytic gas-solid reactions.

5-4   NONCATALYTIC GAS-SOLID REACTIONS

Regeneration of coke-laden catalyst pellets by combustion in an oxygen-bearing gas stream, calcination, coal gasification, and oxidation of solids in blast furnaces, involve gas-solid reactions in which a component of the solid reactant phase is consumed. A transient exists because of this consumption of solid reactant. A major concern here, as related to catalyst deactivation, is the manner in which the solid reactant is consumed since there is a close parallel between this and the manner in which the catalyst is deactivated.

In general, gas-solid reactions can be represented by:

$$A(\text{gas}) + bB(\text{solid}) \rightarrow pP(\text{gas}) + mM(\text{solid}) \tag{5.16}$$

where lower case letters are stoichiometric coefficients normalized with respect to the coefficient of the species of interest, A. A one dimensional mass balance for the species A can be written as:

$$\epsilon_p \frac{\partial C}{\partial t} = D \frac{\partial^2 C}{\partial x^2} - r_A \qquad (5.17)$$

where ϵ_p = particle porosity
$\quad D$ = effective diffusivity
$\quad r_A$ = rate of reaction of A on a volume basis
$\quad C$ = concentration of species A

It is assumed that ϵ_p and D are constant and that the particle is a slab. Other geometries can be handled in a similar manner. The boundary conditions are the usual ones:

$$\frac{\partial C}{\partial x}\bigg|_{center} = 0; \qquad D \frac{\partial C}{\partial x}\bigg|_{surface} = k_g(C_b - C_s) \qquad (5.18)$$

The last boundary condition can be replaced by $C|_{surface} = C_s$. The particle may be assumed isothermal without loss of generality. The rate at which the solid reactant disappears is obtained through stoichiometry:

$$r_A = \frac{1}{b} r_B$$

$$\text{or } \frac{dN_A}{dt} = \frac{1}{b} \frac{dN_B}{dt} \qquad (5.19)$$

Two rate processes are at work here: the chemical reaction and the transportation of gaseous reactant to the solid reactant surface at which the reaction takes place. The relative magnitudes of these two rates determine the manner in which the solid reactant is consumed. Two limiting cases are of interest: one in which the potential rate of transport is much higher than the rate of reaction, which leads to uniform consumption of the solid reactant throughout the particle, and the other in which the potential rate of transport is much lower than the rate of reaction, which leads to formation of a distinguishable reaction front separating the outer product zone from the inner reactant zone (Figure 5.1). In the former, the reaction is completely free from diffusion limitations of the gaseous reactant; in the latter it is diffusion-limited. Concentration profiles within the particle are shown in Figure 5.1. It is seen from Figure 5.1(a) that the concentrations are uniform within the particle when the gas-solid reaction is diffusion-free. The mass balance of Eq. 5.17 reduces to:

$$\epsilon_p \frac{\partial C}{\partial t} = -r_A \qquad (5.20)$$

In the other limiting case shown in Figure 5.1(b), the gaseous concentration at the reaction boundary approaches zero because of a very high consumption rate

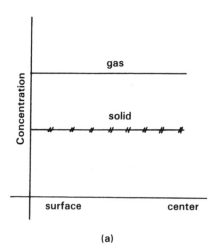

(a)

Figure 5.1 Concentration profiles for various cases: (a) no diffusion limitation (b) diffusion limitation (shell-progressive reaction) (c) intermediate between the two extremes.

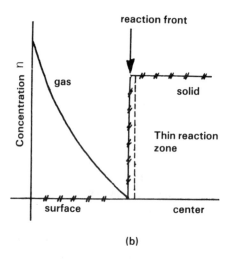

(b)

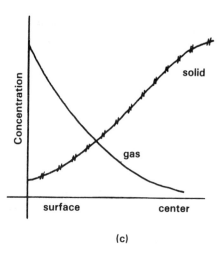

(c)

of the reactant at the reaction front. This moves inward with time, leaving behind a completely reacted zone (product zone). No reaction takes place in the product zone and the balance equation reduces to:

$$\epsilon_p \frac{\partial C}{\partial t} = D \frac{\partial^2 C}{\partial x^2} \tag{5.21}$$

Equation 5.20 applies at the reaction front where a very thin band of reaction zone exists.

When the potential rate of transport of gaseous reactants is much lower than the rate of chemical reaction, a *shell-progressive model* (SPM) results since the completely reacted outer shell progressively moves inward with time. For gas-solid reactions, this automatically results in a pseudo-steady-state situation since the speed at which the reaction front moves is much slower than the speed of transport of gaseous reactants. For this steady-state, Eq. 5.21 further reduces to:

$$D\frac{\partial^2 C}{\partial x^2} = 0 \tag{5.22}$$

with

$$C = C_b \qquad x = L \tag{5.22a}$$
$$C = 0 \qquad x = x_d$$

Referring to the coordinate system shown in Figure 5.2, Eq. 5.19 gives:

$$\frac{dN_A}{dt} = \frac{1}{b}\frac{dN_B}{dt} = \frac{1}{b}\frac{d}{dt}\left(\frac{Ax_d\rho_B}{M_B}\right) \tag{5.23}$$

where ρ_B and M_B are the density and molecular weight of the solid reactant B, and A is the cross-sectional area. On the other hand, one has at $x = x_d$:

$$\frac{dN_A}{dt} = -AD\frac{\partial C}{\partial x}\bigg|_{x_d} \tag{5.24}$$

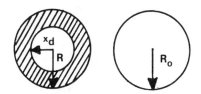

Figure 5.2   Shell-progressive model for non-catalytic gas-solid reactions.

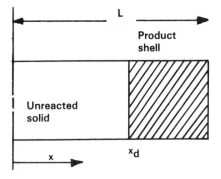

Combining these equations gives:

$$\frac{\rho_B}{bM_B}\frac{dx_d}{dt} = -D\frac{\partial C}{\partial x}\Big|_{x_d} \tag{5.25}$$

Substituting into the right hand side of the above equation the solution of Eq. 5.22 yields:

$$\frac{\rho_B}{bM_B}\frac{dx_d}{dt} = -D\left(\frac{C_b}{L - x_d}\right) \tag{5.26}$$

The solution of this equation gives the following relationship:

$$y = \left(\frac{2bDC_bt}{\rho'_B L^2}\right)^{1/2}; \qquad \rho'_B = \frac{\rho_B}{M_B} \tag{5.27}$$

since for the slab geometry the conversion y is equal to $(1 - x_d/L)$.

The fact that the SPM for gas-solid reactions automatically results in a pseudo-steady-state SPM can be verified by comparing the solution of Eq. 5.21 with that of Eq. 5.22, as shown by Bischoff (1963). The error in using the solution of Eq. 5.22 in place of that of Eq. 5.21 is of the order of $C_b/6b\rho_B$ for slab geometry. For gas-solid reactions, C_b is of the order of $10^{-3}\rho_B$. Therefore, the error is quite small. On the other hand, the pseudo-steady-state assumption is not good for liquid-solid reactions since C_b then is of the order of ρ_B.

The change of particle size due to reaction is often of interest for noncatalytic gas-solid reactions. The SPM for the spherical particle shown in Figure 5.2 is called shrinking core model (Wen 1968). Referring to the figure, the moles of B reacted for the particle of initial radius R_0 are:

$$\Delta N_B = \rho_B \frac{(4/3)\pi(R_0^3 - x_d^3)}{M_B} \tag{5.28}$$

assuming the reaction of Eq. 5.16. Similarly for the solid product M:

$$\Delta N_M = \rho_M \frac{(4/3)\pi(R^3 - x_d^3)}{M_M} \tag{5.29}$$

Combining Eqs. 5.28 and 5.29 through the stoichiometry, $\Delta N_B/b = \Delta N_M/m$, yields:

$$(R/R_0)^3 = z + (1 - z)\delta^3 \tag{5.30}$$

where
$$\delta = \frac{x_d}{R_0}$$

$$z = \frac{\rho_B M_M b}{\rho_M M_B m}$$

The parameter z determines whether the particle grows or shrinks as the reaction proceeds: expansion occurs for $z > 1$, constant size for $z = 1$, and shrinkage for $z < 1$.

With the SPM assumptions used in the shrinking core model, the major resistances are diffusion through the product zone and external mass transfer, the case of chemical reaction controlling being inconsistent with the SPM. The time required for a given level of conversion when one of the resistances is controlling can be obtained in the usual manner. Since these resistances are in series, the time when both resistances are important is obtained by simple addition, i.e., $t = t_D + t_m$, where t_D and t_m are the times obtained when the diffusion and the external mass transfer, respectively, are controlling. Take, as an example, the case of diffusion controlling. Equation 5.22 for spherical geometry becomes:

$$\frac{d^2C}{dx^2} + \frac{2}{x}\frac{dC}{dx} = 0$$

where x is radial coordinate. The solution with the boundary conditions of Eq. 5.22a is:

$$\frac{C}{C_b} = \frac{1 - x_d/x}{1 - x_d/R}$$

The equation equivalent to Eq. 5.26 for the spherical geometry is:

$$\frac{\rho_B 4\pi x_d^2}{bM_B}\frac{dx_d}{dt} = \frac{-4\pi C_b D}{1/x_d - 1/R}$$

Using Eq. 5.30 for R in the above equation and integrating, one gets:

$$t_D = \begin{cases} \dfrac{\rho_B R_0^2}{2bDC_b M_B}\left\{\dfrac{[z + (1-z)\delta^3]^{2/3}}{1-z} + (1-\delta^2) - \dfrac{1}{1-z}\right\} \\ \dfrac{\rho_B R_0^2}{bDC_b M_B}\left\{\dfrac{1-\delta^2}{2} - \dfrac{1-\delta^3}{3}\right\}; \qquad z = 1 \end{cases} \qquad (5.31)$$

which is the time required for the conversion $(1 - \delta^3)$ when diffusion through the product zone is controlling.

5-5   DIFFUSION-DEACTIVATION

When reactions are affected by both deactivation and diffusion, one-dimensional, steady state pellet conservation equations are:

$$D \frac{d^2C}{dx^2} = k(1 - \gamma)f[C] \tag{5.32}$$

$$\lambda \frac{d^2T}{dx^2} = -(-\Delta H)k(1 - \gamma)f[C] \tag{5.33}$$

$$D_p \frac{d^2N}{dx^2} = r_p[k_p(1 - \gamma), N, C] \tag{5.34}$$

The usual boundary conditions apply. Here, the kinetics of the main reaction are expressed as $r = k(1 - \gamma)f[C]$ where k is the rate constant for fresh catalyst; $k(1 - \gamma)$ is the effective rate constant for the catalyst being deactivated. It has been assumed for the energy balance that the heat of reaction for the main reaction is much greater than that for the deactivation reaction. Following the same procedures as in the case of diffusion-limited reactions (Chapter 4), it can be readily shown that the pellet is practically isothermal under typical reaction conditions. Therefore, only mass balance equations will be dealt with here. It shall soon be seen, however, that this assumption of isothermality is not good in the limiting case of shell-progressive deactivation. The fraction of catalyst deactivated, γ, depends not only on time but also on the spatial position within the pellet. In order to quantify this dependence, Q is defined as the maximum uptake of poisoning species by the catalyst in moles of poisoning species per unit apparent volume of catalyst pellet. Then:

$$\frac{d}{dt}(Q\gamma) = r_p \tag{5.35}$$

This relationship states that the uptake rate of poisoning species by the catalyst is equal to the rate at which the poisoning takes place. The pseudo steady state assumption has been made with respect to concentrations. Validity of this assumption comes from the fact that the rate of poisoning is much slower than the rate at which the concentrations reach steady state. If this were not true, the catalyst would lose its usefulness because it would have to be regenerated too frequently. For instance, the useful life of catalyst pellets is of the order of months, whereas the time required to reach steady state is of the order of $L/D^{1/2}$.

For reactor design, it is important to know how the solutions of Eqs. 5.32, 5.34, and 5.35 affect the intrinsic rate of reaction. Wheeler (1955) treated this deactivation-diffusion problem for two limiting cases: uniform and pore-mouth (shell-progressive) poisoning. As described in the previous section for noncatalytic gas-solid reactions, the poison will deposit preferentially on the pore-mouth initially and grow progressively inward with time, if the rate of poisoning is rapid relative to intraparticle transport. This shell-progressive poisoning is termed pore-mouth poisoning. Uniform poisoning, in which the active surface is poisoned uniformly throughout the pellet, should occur if the rate of poisoning is low relative to the potential transport rate. The effect of diffusion on deactivation was treated concisely by Wheeler for the two limiting cases in terms of an activity factor defined as:

$$A = \frac{\text{rate of reaction for deactivated pellet}}{\text{rate of reaction for fresh pellet}} \tag{5.36}$$

If the pellet is uniformly deactivated, γ is the same everywhere and A, for diffusion-free main reactions, is simply that given by:

$$A = 1 - \gamma \tag{5.37}$$

For diffusion-limited reactions, $A = (1 - \gamma)^{1/2}$ (see Problem 5.3).

The devastating effect of diffusion on the residual activity left after poisoning for the pore-mouth poisoned pellet, as compared to uniform poisoning, is amply revealed in Figure 5.4 (Wheeler 1955), where the activity factor is given in terms of γ with the Thiele modulus as parameter. This is due to the presence of the completely deactivated outer shell, as shown in Figure 5.3 for a slab-like pellet. Referring to the figure, we set the rate of the main reaction within the fresh inner core equal to the transport rate through the deactivated outer shell at steady state:

$$(\text{area})D_e \left.\frac{dC}{dx}\right|_{x=x_d} = (\text{area})D_e \left(\frac{C_s - C_d}{l}\right) = L(\text{area})R_G$$

$$= L(\text{area})(1 - \gamma)k_s \eta_{in} C_d$$

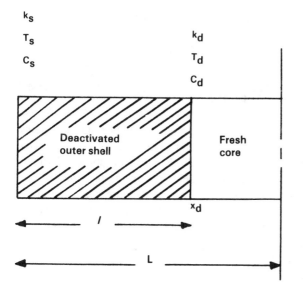

k_s
T_s
C_s

k_d
T_d
C_d

Deactivated outer shell

Fresh core

x_d

l

L

Figure 5.3 Shell-progressive poisoning, slab-like pellet.

This reduces to:

$$C_s - C_d = \left(\frac{L^2 k_s}{D_e}\right) \gamma (1 - \gamma) \eta_{in} C_d \qquad (5.38)$$

Here, the effectiveness factor for the fresh inner core based on the conditions at $x = x_d$ is given by:

$$\eta_{in} = \frac{\tanh \phi_{in}}{\phi_{in}} ; \qquad \phi_{in} = (L - l)(k_s/D_e)^{1/2}$$

and γ is simply l/L. While a pore-mouth poisoned pellet is not, in general, isothermal, it has been assumed isothermal in accordance with the original work of Wheeler. The definition of the activity factor gives:

$$A = \frac{\eta_{in}(1 - \gamma)C_d}{\eta_f C_s} \qquad (5.39)$$

since the global rate for the deactivated pellet is $k_s(1 - \gamma)\eta_{in} C_c$ on the basis of whole pellet volume. The effectiveness factor for the fresh pellet is:

$$\eta_f = \frac{\tanh \phi}{\phi} ; \qquad \phi = L(k_s/D_e)^{1/2}$$

Substitution of Eq. 5.38 into 5.39 gives:

$$A = \left[\frac{\eta_{in}(1 - \gamma)}{\eta_f}\right]\left[\frac{1}{1 + \phi^2 \gamma (1 - \gamma) \eta_{in}}\right] \qquad (5.40)$$

which reduces to:

$$A = \frac{1}{1 + \phi\gamma} \qquad (5.41)$$

when ϕ is large since then $\eta_f \phi \rightarrow 1$ and $\phi_{in}/\phi = (1 - \gamma)$. For slab geometry, this result is the same as that obtained by Wheeler. The activity factor derived originally by Wheeler for spherical geometry (see Problem 5.4) is shown in Figure 5.4. The approximation of deactivation-diffusion behavior by the pore-mouth poisoning model has been found in many cases to be quite adequate when the initial rate of deactivation is relatively fast, although one should be judicious in applying the approximation (Petersen 1982). One would expect the pore-mouth poisoning model to be valid for independent deactivation when the main reaction is quite diffusion-limited since the rate constant for the independent deactivation reaction is inherently higher than that of the main reaction, and, therefore, the deactivation

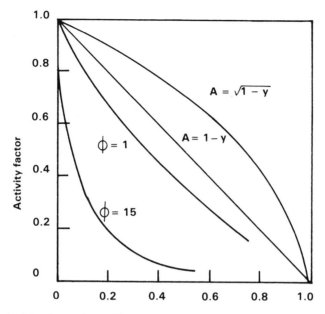

Figure 5.4 Activity factor for uniform and shell-progressive poisoning. (Wheeler 1955)

reaction is likely to be more diffusion-limited than the main reaction. This, however, is not necessarily true for dependent deactivation as shall be seen shortly.

Masamune and Smith (1966) solved the conservation equations of Eq. 5.32, 5.34, and 5.35 for various deactivation cases in which both the main and deactivation reactions are of first-order. They found that a catalyst with the lowest intraparticle diffusion resistance gave the maximum activity for any time on stream when a series dependent deactivation is involved; for a parallel deactivation, a catalyst with an intermediate diffusion resistance is less easily deactivated and gives a higher conversion to desirable product, particularly at long on-stream times. Their simulation results for the overall effectiveness factor representing the combined effects of deactivation and diffusion are shown in Figure 5.5 for independent poisoning. With the assumption of negligible external transport resistances, the overall effectiveness factor can be plotted, as shown, in terms of time with the Thiele moduli for the main and deactivation reactions as parameters. As expected, the effectiveness factor decreases with time.

Carberry and Gorring (1966) treated the deactivation-diffusion problem as a noncatalytic gas-solid reaction to arrive at the time dependence of the fraction of catalyst deactivated. Following the same procedures as in arriving at Eq. 5.27, one can get:

$$\gamma = -\alpha + (\alpha^2 + \beta^2)^{1/2} \tag{5.42}$$

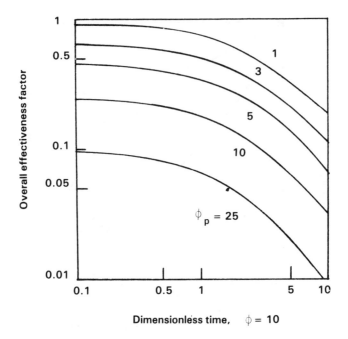

Figure 5.5 Overall effectiveness factor for independent poisoning and diffusion-affected first-order reaction. (Masamune and Smith 1966)

where $\alpha = \dfrac{1}{(Bi)_m}$; $(Bi)_m = \dfrac{k_g L}{D_e}$

$$\beta^2 = \dfrac{2 D_e N_b t}{b \rho_B' L^2}$$

(5.43)

They found that this result for slab geometry would apply equally well to spherical geometry up to γ of 0.3 if L is defined as the ratio of volume to surface area ($L = R/3$ for sphere). This result gives a square root dependence of γ on time.

In order to obtain the global rates for reactor design, Eqs. 5.32, 5.34, and 5.35 have to be solved numerically, given the kinetics of the main and deactivation reactions. In the asymptotic region of strong diffusion effects, this deactivation-diffusion problem for the global rates can be simplified considerably as will be shown in the next section.

5–6   REACTOR POINT EFFECTIVENESS AND GLOBAL RATES

Chapter 4 has demonstrated that the global rate, which appears in reactor conservation equations, can be obtained by simply multiplying the intrinsic rate by the reactor point effectiveness. In this section, the reactor point effectivenesses is obtained for three cases: general, uniform, and pore-mouth deactivation. As in Chapter 4, realistic assumptions are made of negligible external mass transfer resistance and isothermal pellets. In the case of pore-mouth deactivation, however, the nonisothermal nature of the pellet is taken into consideration.

General

For a pellet of nonuniform activity, it has been shown in Chapter 4 that the internal effectiveness factor in the asymptotic region of strong diffusion effects can be accurately approximated by:

$$\eta_i = \frac{\left[2\bar{G}D_e \int_0^{C_b} r_c(C)dC\right]^{1/2}}{L\bar{r}_c} \tag{4.104a}$$

where the overbar denotes the quantity evaluated at surface conditions, and G is the activity distribution function normalized with respect to the surface activity of a fresh pellet. Note that the effectiveness factor here is based on the activity of fresh catalyst and therefore the denominator of Eq. 4.104a is simply $L\bar{r}_c$ as opposed to that given in Chapter 4 (Eq. 4.104). A restriction placed on the relationship was that the following conditon be met:

$$\left|\frac{\bar{G}'D_e^{1/2}}{\bar{G}(2\bar{G})^{1/2}}\right| \left\{ \frac{\int_0^{C_b}\int_0^C r_c(\alpha)d\alpha dC}{\left[\int_0^{C_b} r_c(C)dC\right]^{3/2}} \right\} \ll 1 \tag{4.102}$$

where $G' = dG/dx$. Since deactivation results in a nonuniform activity and this activity distribution is $(1 - \gamma)$ for the deactivated catalyst, these results can be used as such with G equated to $(1 - \gamma)$. The left hand side of Eq. 4.102 is essentially the maximum fractional error involved in using Eq. 4.102. This condition is usually satisfied unless $\bar{G} \ll \bar{G}'$ since the second term in the left hand side is smaller than unity and D_e is of the order of 10^{-2} cm²/sec. For the deactivation being considered, which is intermediate between uniform and pore-mouth deactivation, the slope of the activity function $(1 - \gamma)$ at the surface is not very steep. Further, the surface activity is not likely to be less than 0.4 if the catalyst is regenerated before its average activity becomes less than 70% of its original activity. Therefore, the condition of Eq. 4.102 is likely to be satisfied in many cases. For instance, this condition for an nth order reaction becomes:

$$\left|\frac{\bar{G}'L}{\bar{G}^{3/2}}\right| \left(\frac{1}{\phi}\right)\left(\frac{(n+1)^{1/2}}{2^{1/2}(n+2)}\right) \ll 1$$

where ϕ is the usual Thiele modulus. In what follows, it is assumed that this condition is satisfied.

The reactor point effectivness has been defined as:

$$\Lambda = \frac{\text{observed rate}}{\text{intrinsic rate for fresh catalyst at bulk fluid conditions}}$$

which can be expressed as:

$$\Lambda = \frac{R_G}{r[C_b, k_b]} = \frac{k_s}{k_b}\, \eta_i \tag{4.78}$$

under the assumption of negligible external mass transfer resistance. Since the rate constant ratio in the above relationship has been shown (Chapter 4) to be well approximated by:

$$\left(\frac{k_s}{k_b}\right)^{1/2} = 1 + 1.2\epsilon \left[\frac{L(-\Delta H)\Delta k_b f(C_b)}{2hT_b}\right] \tag{4.82}$$

this can be used along with Eq. 4.104a in Eq. 4.81 above to give:

$$\Lambda = \frac{1}{k_b f(C_b)}\left\{\frac{J^{1/2}/L}{1 - [1.2\epsilon L(-\Delta H)/2hT_b]J^{1/2}}\right\}; \qquad R_G = \Lambda k_b f(C_b) \tag{5.44}$$

where

$$J = 2k_b(1 - \bar{\gamma})\, D_e \int_0^{C_b} f[C]\,dC \tag{5.45}$$

Here, $\bar{G}$ has been replaced by $(1 - \bar{\gamma})$. For the global rate of the deactivation reaction, Eq. 5.34 is rewritten for independent deactivation:

$$D_p \frac{d^2 N}{dx^2} = r_p = \frac{k_p N(1 - \gamma)}{[1 + G(C)]^n} \tag{5.46}$$

where Eq. 5.11 has been used for r_p. Since the inhibition term $G(C)$ is quite a bit smaller than unity, an arithmetic average can be used to approximate the denominator:

$$H(C_b) = \frac{[1 + G(C_b)]^n + [1 + G(C = 0)]^n}{2} \tag{5.47}$$

Then procedures identical to those used in obtaining Eq. 4.5 lead to the following internal effectiveness factor for the deactivation reaction:

$$(\eta_i)_p = \frac{1}{L}\left[\frac{D_p H(C_b)}{(k_p)_s(1-\bar{\gamma})}\right]^{1/2}$$ (5.48)

The rate constant ratio for the deactivation reaction equivalent to Eq. 4.86 is:

$$\left[\frac{(k_p)_s}{(k_p)_b}\right]^{1/2} = 1 + 1.2\epsilon_p\left[\frac{L(-\Delta H)R_G}{2hT_b}\right]$$ (5.49)

where ϵ_p is the Arrhenius number for the deactivation reaction. An assumption used here is that the heat liberated by reaction is mainly due to the main reaction. Use of Eqs. 5.48 and 5.49 in the definition of the reactor point effectiveness for the deactivation reaction gives:

$$(R_G)_p = \frac{[(k_p)_b(1-\bar{\gamma})D_pH(C_b)]^{1/2}}{L[1+G(C_b)]^n}\left\{1+\frac{1.2\epsilon_p L(-\Delta H)R_G}{2hT_b}\right\}$$ (5.50)

It is seen that the global rates for the main and deactivation reactions contain only the surface activity $(1-\bar{\gamma})$, and not the activity profile. Therefore, Eq. 5.35 is evaluated at the pellet surface for independent deactivation:

$$\frac{d}{dt}(Q\bar{\gamma}) = \frac{(k_p)_s N_b(1-\bar{\gamma})}{[1+G(C_b)]^n}$$ (5.51)

Global rates are now expressed in terms of bulk fluid quantities and the surface activity required for the rates is given by Eq. 5.51. These are summarized in Table 5.3. Unlike the cases of uniform and shell-progressive deactivation, the results obtained for the general case are restricted to diffusion-limited reactions. They are also subject to the limitation of Eq. 4.102.

Uniform

One would expect the rate constant of an independent deactivation reaction to be inherently larger than that of the main reaction. One would therefore expect that the independent deactivation reaction is also diffusion-limited, meaning nonuniform deactivation, if the main reaction is. Nonetheless, uniform deactivation can take place even when the main reaction is diffusion-limited. In such a case, $G(C)$ appearing in the deactivation kinetics would be much smaller than unity so that the kinetics of independent deactivation can be reduced to:

$$r_p = k_p N(1-\gamma)$$ (5.52)

since, otherwise, uniform deactivation cannot take place. If $G(C)$ were not much smaller than unity, γ would depend on position within the pellet because of the dependence of $G(C)$ on the position, which contradicts the assumption of uniform deactivation.

Since the fraction of catalyst deactivated is uniform throughout the pellet, the reactor point effectiveness is simply that for a diffusion-limited reaction (Eq. 4.87) with k replaced by the effective rate constant $k(1 - \gamma)$:

$$\Lambda = \frac{\left[\frac{2D_eI(C_b)}{k_b(1-\gamma)}\right]^{1/2} \Big/ Lf(C_b)}{1 - \frac{1.2\epsilon(-\Delta H)[2D_ek_b(1-\gamma)I(C_b)]^{1/2}}{2hT_b}} \quad (5.53)$$

where

$$I(C_b) = \int_{C_c}^{C_b} f(C)dC$$

The time dependence of γ is given by:

$$\frac{d}{dt}(Q\gamma) = r_p = (k_p)_sN_b(1-\gamma) \quad (5.54)$$

Pore-Mouth (Shell-Progressive)

The global rate of the main reaction is obtained from:

$$D_e\frac{dC}{dx}\bigg|_{x=r_c} = D_e\left(\frac{C_b - C_d}{l}\right) = LR_G$$

or $\quad R_G = \dfrac{D_e(C_b - C_d)}{L^2\gamma}; \quad \Lambda = \dfrac{D_e(C_b - C_d)}{k_bf(C_b)L^2\gamma} \quad (5.55)$

The concentration at the boundary between fresh inner core and the completely deactivated outer shell (C_d) is obtained from Eq. 5.38 written for arbitrary kinetics, $r_c = kf(C)$:

$$C_b - C_d = \left(\frac{L^2k_d}{D_e}\right)\gamma(1-\gamma)\eta_{in}f(C_d) \quad (5.56)$$

For the nonisothermal pellet being considered, k_s in Eq. 5.38 has been replaced by k_d, i.e., the rate constant evaluated at the conditions at the boundary. With the assumption of negligible external mass transfer resistance, C_s has also been replaced by the bulk fluid concentration C_b. The internal effectiveness factor for the fresh inner core is rewritten as:

$$\eta_{in} = \frac{1}{(L-l)k_{df}(C_d)}\left[2\int_{C_c}^{C_d} D_e k_{df}(\alpha)d\alpha\right]^{1/2} \tag{5.57}$$

Combining Eqs. 5.56 and 5.57 gives:

$$C_b - C_d = \frac{L\gamma}{D_e}\left[2D_e k_d \int_{C_c}^{C_d} f(\alpha)d\alpha\right]^{1/2} \tag{5.58}$$

It has been shown both experimentally and theoretically (Butt et al. 1977; Lee et al. 1978) for a pore-mouth poisoned pellet that the main intraparticle temperature gradient lies in the deactivated outer shell and that the fresh inner core is isothermal. This is the reason why the fresh inner core was treated as isothermal in writing Eq. 5.57. The intra- and interparticle temperature gradients (Lee 1980a) are related by:

$$\Delta T_d = (Bi)_h \gamma \Delta T_e \tag{5.59}$$

$$(\Delta T_d + \Delta T_e) = [1 + (Bi)_h \gamma]\,\Delta T_e \tag{5.60}$$

where $\Delta T_d[= (T_d - T_s)]$ is the temperature difference in the deactivated outer shell, and $\Delta T_e[= (T_s - T_b)]$ is the external temperature difference. These two relationships can be combined and rearranged to give:

$$1 - T_b/T_d = \frac{[1 + (Bi)_h\gamma](T_s/T_b - 1)}{1 + [1 + (Bi)_h\gamma](T_s/T_b - 1)} \tag{5.61}$$

Setting the heat transferred across the external film equal to the heat liberated by the reaction gives:

$$T_s/T_b - 1 = \frac{(-\Delta H)D_e(C_b - C_d)}{hT_bL\gamma} \tag{5.62}$$

In order to express k_d appearing in Eq. 5.58 in terms of bulk quantities and C_d, one writes:

$$\left(\frac{k_d}{k_b}\right)^{1/2} = \exp\left\{\frac{1}{2R_gT_b}\left(1 - \frac{T_b}{T_d}\right)\right\} \tag{5.63}$$

Using Eqs. 5.61 and 5.62 in 5.63 and then in 5.58 gives:

$$C_b - C_d = \left(\frac{L\gamma}{D_e}\right)\left[2D_e k_b \int_{C_c}^{C_d} f(\alpha)d\alpha\right]^{1/2}$$

$$\exp\left\{\left(\frac{E_a}{2R_gT_b}\right)\left[\frac{(-\Delta H)D_e[1 + (Bi)_h\gamma](C_b - C_d)}{hT_bL\gamma + [1 + (Bi)_h\gamma](C_b - C_d)(-\Delta H)D_e}\right]\right\} \tag{5.64}$$

This is an implicit expression for C_d in terms of bulk-fluid quantities. The global rate can readily be calculated from Eq. 5.55 and C_d obtained from 5.64, given the kinetics and bulk-fluid conditions.

As was seen in the section on noncatalytic gas-solid reactions, there must be a very thin deactivation reaction zone at the forefront of the moving boundary separating the deactivated outer zone from the still active inner zone in order for pore-mouth deactivation to take place. In this small band, the rate of deactivation or the rate of chemisorption of poisoning species must be much greater than the rate of chemisorption of the other species. Therefore, the rate of deactivation is practically independent of the fraction of vacant sites in this small band and the deactivation kinetics of Eq. 5.7 can be approximated by:

$$r_p = k_p N \tag{5.65}$$

It follows then that:

$$\frac{d}{dt}(Q\gamma) = (k_p)_d N_d \tag{5.66}$$

Setting the rate of diffusion of poisoning species through the deactivated zone equal to the rate of deactivation yields:

$$N_b = [1 + (\phi_p)_d^2 \gamma] N_d; \qquad (\phi_p)_d^2 = \frac{L^2(k_p)_d}{D_p} \tag{5.67}$$

Combining Eqs. 5.66 and 5.67 gives:

$$\frac{d\gamma}{dt} = \frac{(k_p)_d}{Q} \left[\frac{N_b}{1 + (\phi_p)_d^2 \gamma} \right] \tag{5.68}$$

Therefore, the global rate of reaction as affected by diffusion and pore-mouth deactivation is described by Eqs. 5.55, 5.64, and 5.68. These results are also summarized in Table 5.3. When the Thiele modulus for the deactivation reaction $(\phi_p)_d$

Table 5.2   Parameters: Shell-Progressive Poisoning in the Hydrogenation of Benzene on Ni/Kieselguhr by Thiophene (Lee et al. 1978)

$D_e = D_p = 0.052$ cm²/s (counter diffusion) or 0.03 cm²/s (random pore)
$L = 0.65$ cm
$(k_p)_d = 28$ s⁻¹
$Q = 4 \times 10^{-4}$ mol/cm³
$C_s = 5.7 \times 10^{-6}$ mol/cm³
$k_d = 0.15$ s⁻¹
$N_s = 3.6 \times 10^{-7}$ mol/cm³ $\Big\}$ for run 24
$k_d = 0.10$ s⁻¹
$N_s = 1.35 \times 10^{-7}$ mol/cm³ $\Big\}$ for run 14

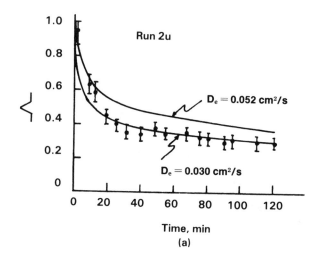

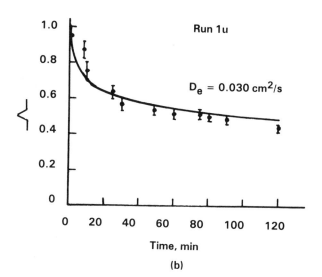

is very large, which is often the case for pore-mouth deactivation, the solution of Eq. 5.68 reduces to:

$$\gamma = \left[\frac{2D_p}{QL^2}\int_0^t N_b \, dt\right]^{1/2} \tag{5.69}$$

A comparison (Lee and Butt 1982) between theoretical and experimental results is shown in Figure 5.6 for a single pellet undergoing pore-mouth poisoning. Parameter values for the experiments, which were determined by independent measurements, are given in Table 5.2. For the first-order main reaction being considered, the hydrogenation of benzene, Eqs. 5.55, 5.58, and 5.68 reduce to:

$$\Lambda = \frac{D_e}{L^2 k_b \gamma}\left(\frac{C_b - C_d}{C_b}\right) \tag{5.70}$$

$$C_d = \frac{C_b}{1 + L\gamma(k_d/D_e)^{1/2}} \tag{5.71}$$

$$\gamma = -\frac{1}{(\phi_p)_d^2} + \left[\frac{1}{(\phi_p)_d^4} + \frac{2D_p N_b t}{QL^2}\right]^{1/2} \approx \left[\frac{2D_p N_b t}{QL^2}\right]^{1/2} \tag{5.72}$$

The concentration of poisoning species was kept constant during the experiment, which permits direct integration of $\int_0^t N_b dt$ to $N_b t$. It is seen that the agreement between theory and experiment is satisfactory. It is also seen that the effective diffusivity is a sensitive parameter in calculating the reactor point effectiveness. The upper line in Figure 5.6(a) is for the diffusivity determined from the counter-diffusion method* ($D_e = 0.052$ cm²/sec) and the lower line is for the diffusivity obtained from the random-pore theory· ($D_e = 0.03$ cm²/sec).

The result obtained so far for independent deactivation can be extended to dependent deactivation. In contrast to independent deactivation, the rate constant of a dependent deactivation must be much smaller than that for the main reaction since otherwise the rate of the deactivation reaction could become greater than that of the main reaction. Such a catalyst could not serve any useful purpose in the form of pellets. In fact, this is precisely the reason why catalytic cracking of hydrocarbons, in which the coking rate is high, is carried out in a fluidized bed with continuous regeneration and recycling of deactivated silica-alumina catalyst in the form of very small particles. Note that the same species involved in the main reaction is also involved in the deactivation reaction as can be seen from the dependent deactivation models given previously:

Parallel:
$$A + S \xrightarrow{k} B + S$$
$$A + S \xrightarrow{k_p} A \cdot S$$

Series:
$$A + S \xrightarrow{k} B + S$$
$$B + S \xrightarrow{k_p} B \cdot S$$

If the rate constant of the deactivation reactions k_p is much smaller than that of the main reaction k, as would be the case for dependent deactivation when catalyst pellets are used, the rate of reaction for the species of interest (A) can be approximated in both cases by:

$$r_c = kf(C)$$

---

* See Chapter 14, Section 1.

where C is the concentration of species A. In both cases, then, the reactor point effectiveness and global rate are given by Eqs. 5.44 and 5.45. The surface activity $(1 - \bar{\gamma})$ follows directly from Eqs. (5.12) and (5.13):

$$\text{Parallel:} \quad \frac{d}{dt}(Q\bar{\gamma}) = \frac{(k_p)_s (C_A)_b (1 - \bar{\gamma})}{\left[1 + \sum_i (K_i C_i)^{m_i}\right]^n_{\text{bulk}}} \tag{5.73}$$

$$\text{Series:} \quad \frac{d}{dt}(Q\bar{\gamma}) = \frac{(k_p)_s (C_B)_b (1 - \bar{\gamma})}{\left[1 + \sum_i (K_i C_i)^{m_i}\right]^n_{\text{bulk}}} \tag{5.74}$$

Reaction stoichiometry can be used in these equations to express $(C_B)_b$ in terms of $(C_A)_b$.

As discussed earlier, catalyst particles instead of pellets are used when the rate of deactivation is high. The typical size of the catalyst particles used in fluidized-beds ranges from 20 to 300 μm, and therefore, diffusional effects can be neglected. In the absence of diffusional effects, the particles are isothermal and the external mass transfer resistance is negligible (Chapter 4). The only transport resistance is external heat transfer. For these particles then, the global rate is simply:

$$R_G = k_s (1 - \gamma) f(C_b) \tag{5.75}$$

where γ depends only on time. Expressing k_s in terms of bulk quantities with the aid of Eq. 4.82, one gets:

$$R_G = \frac{k_b (1 - \gamma) f(C_b)}{1 - \dfrac{1.2\epsilon L(-\Delta H)}{2hT_b} k_b (1 - \gamma) f(C_b)} \tag{5.76}$$

The space-independent γ is obtained from:

$$\frac{d}{dt}(Q\gamma) = r_p \tag{5.77}$$

It is of interest to note that the catalytic activity $(1 - \gamma)$ decays exponentially with time according to Eq. 5.77 if the bulk-fluid concentrations are constant. In well-mixed fluidized-beds, concentrations can be considered to be uniform everywhere, approaching CSTR behavior. This exponential decay of catalytic activity has been found (Weekman 1970) satisfactory for the simulation of a fluidized-bed catalytic cracker.

Table 5.3   Global Rates and Reactor Point Effectiveness

$r_c = kf(C)$
$\Lambda = R_G/k_b f(C_b)$: $k_b = k(T_b)$ for fresh catalyst

---

a. Independent deactivation: diffusion-affected main reactions

---

General (restricted to diffusion-limited reactions such that $C_c = 0$)

$$R_G = \frac{(J/L)^{1/2}}{1 - \dfrac{1.2\epsilon L(-\Delta H)}{2hT_b} J^{1/2}}$$

$$J = 2k_b(1 - \bar{\gamma})D_e \int_0^{C_b} f(C)dC$$

$$(R_G)_p = \frac{[(k_p)_b(1 - \bar{\gamma})D_p H(C_b)]^{1/2}}{L[1 + G(C_b)]^n} \left\{ 1 + \frac{1.2\epsilon_p L(-\Delta H)}{2hT_b} R_G \right\}$$

$$H(C_b) = \frac{[1 + G(C_b)]^n + [1 + G(C = 0)]^n}{2}$$

$$\frac{d}{dt}(Q\bar{\gamma}) = \frac{(k_p)_s N_b(1 - \bar{\gamma})}{[1 + G(C_b)]^n}$$

Uniform

$$R_G = \frac{[2D_e k_b(1 - \gamma)I]^{1/2}/L}{1 - \dfrac{1.2\epsilon(-\Delta H)[2D_e k_b(1 - \gamma)I]^{1/2}}{2hT_b}}$$

$$I = \int_{C_c}^{C_b} f(C)dC$$

$$\frac{d}{dt}(Q\gamma) = (k_p)_s N_b(1 - \gamma)$$

Pore-mouth

$$R_G = \frac{D_e(C_b - C_d)}{L^2\gamma}$$

$$C_b - C_d = \left(\frac{L\gamma}{D_e}\right)\left[2D_e k_b \int_{C_c}^{C_d} f(C)dC\right]^{1/2} \exp(Z)$$

$$Z = \left(\frac{E_a}{2R_g T_b}\right)\left[\frac{(-\Delta H)D_e[1 + (Bi)_h\gamma](C_b - C_d)}{hT_b L\gamma + [1 + (Bi)_h\gamma](C_b - C_d)(-\Delta H)D_e}\right]$$

$$\frac{d}{dt}(Q\gamma) = (k_p)_d\left[\frac{N_b}{1 + (\phi_p)_d^2}\right]$$

b. Dependent deactivation: general, diffusion-limited: $k_p \ll k$

$$R_G = \frac{J^{1/2}/L}{1 - \dfrac{1.2\epsilon L(-\Delta H)}{2hT_b} J^{1/2}}$$

$$J = 2k_b(1-\bar{\gamma})D_e \int_0^{C_b} f(C)dC$$

Parallel: $\quad \dfrac{d}{dt}(Q\bar{\gamma}) = \dfrac{(k_p)_s(C_A)_b(1-\bar{\gamma})}{\left[1 + \sum\limits_i (K_iC_i)^{m_i}\right]^n_{\text{bulk}}}$

Series: $\quad \dfrac{d}{dt}(Q\bar{\gamma}) = \dfrac{(k_p)_s(C_B)_b(1-\bar{\gamma})}{\left[1 + \sum\limits_i (K_iC_i)^{m_i}\right]^n_{\text{bulk}}}$

c. Small catalyst particles: diffusion-free

$$R_G = \frac{k_b(1-\gamma)f(C_b)}{1 - \dfrac{1.2\epsilon L(-\Delta H)}{2hT_b} k_b(1-\gamma)f(C_b)}$$

$$\frac{d}{dt}(Q\gamma) = \bar{r}_p$$

where $\bar{r}_p = \dfrac{(k_p)_s N_b(1-\bar{\gamma})}{[1 + G(C_b)]^n}$ $\qquad$: independent

$\qquad\quad = \dfrac{(k_p)_s(C_A)_b(1-\bar{\gamma})}{\left[1 + \sum\limits_i (K_iC_i)^{m_i}\right]^n_{\text{bulk}}}$ $\qquad$: parallel

$\qquad\quad = \dfrac{(k_p)_s(C_B)_b(1-\bar{\gamma})}{\left[1 + \sum\limits_i (K_iC_i)^{m_i}\right]^n_{\text{bulk}}}$ $\qquad$: series

$$\epsilon = E_a/R_gT_b, \qquad \epsilon_p = (E_a)_p/R_gT_b$$

$$C_c/C_b = \begin{cases} 0 \text{ for } \phi_G \geqslant 3 \\ \dfrac{1}{\cosh \phi_G} \text{ for } 0.3 < \phi_G < 3 \end{cases} \qquad \text{(A)}$$

$$\phi_G = Lk_b(1-\gamma)f(C_b) \Big/ \left[2\int_0^{C_b} D_e k_b(1-\gamma)f(C)dC\right]^{1/2}$$

Replace C_b by C_d in Eq. (A) for the pore-mouth poisoned case.
A reaction can be considered to be diffusion-free if $\phi_G < 0.3$.

For the purpose of using the results obtained so far, a reaction is considered to be diffusion-limited if the generalized modulus defined by:

$$\phi_G = \frac{Lk_b(1 - \bar{\gamma})f(C_b)}{\left[2\int_0^{C_b} D_e k_b(1 - \bar{\gamma})f(C)dC\right]^{1/2}} \qquad (5.78)$$

is greater than 3. The global rates and reactor point effectivenesses are summarized in Table 5.3. These results are used in Chapter 10 for reactor design.

5-7   CATALYST REGENERATION

Deactivated catalyst can be regenerated by simple heating if poisoning is due to reversible chemisorption. If the poisoning species is irreversibly chemisorbed, however, it has to be removed by oxidation or some other chemical means. It is not unusual, when deactivation is a severe problem, to have a guard reactor in which the impurities in the feed are captured before they enter the primary reactor.

The subject most studied is the regeneration of coked catalyst because coking is important in many reactions such as cracking, reforming, hydrodesulfurization, and various dehydrogenations. The buildup of coke may become significant in less than a minute, as in catalytic cracking, or only over a period of months, as in some catalytic reforming processes. The coke is removed by combustion in an oyxgen-bearing gas stream. The combustion of coke is a complicated phenomenon because of the composition of coke, which can vary from carbon (formed from relatively high temperature reactions) to a high molecular weight, hydrogen-deficient carbon matrix (relatively low temperature), and because of the manner in which the coke is deposited within the catalyst pellet. The diffusion characteristics of the main and deactivation reactions as well as the mode of the deactivation reaction (parallel versus series) determine the manner in which the coke is deposited. In diffusion-limited reactions, for instance, coke is deposited preferentially at the pore-mouth in parallel coking whereas it is deposited at the core in series coking. Adding complication to these is sintering (Chapter 6) that can take place during regeneration, resulting in reduced surface area.

The fact that coke consists of a spectrum of compounds is a complicating factor in determining the intrinsic kinetics of coke combustion, in particular the order of reaction with respect to coke concentration. It is well established, however, that the kinetics can be represented by (Weisz and Goodwin 1966):

$$r_c = -kp_{O_2}C_c \qquad (5.79)$$

for combustion of monolayer coke deposits when they are formed on inert supports or acidic cracking catalysts not containing transition metals. Here, P_{O_2} is the partial pressure of oxygen, and C_c is the coke concentration in moles of coke per unit volume of pellet. The activation energy is 37.6 kcal/mol with the preexponential factor of 1.9×10^8 sec^{-1} atm^{-1}. The stipulation on coke coverage has to do with the accessibility of coke to oxygen. If each atom of carbon in the coke occupies the same area, a monoatomic layer of carbon on an oxide-type catalyst or support would comprise about 5 wt% carbon for each 100 m^2/g of surface area: an atom of carbon in graphite occupies about 4Å^2. Therefore, not all carbon will be accessible to oxygen if this wt% is higher than 5 for each 100 m^2/g of surface area. In fact, it has been found that, at higher coke contents, the rate for the first 50 percent or so of the reaction is less than proportional to the amount of carbon present, indicating that some of the carbon atoms are initially inaccessible, but in the latter part of the reaction the rate again becomes first-order with respect to total carbon present. While the combustion of hydrogen contained in coke is also involved in coke combustion, this complication may be neglected for purposes of regeneration analysis and design, since the combustion of carbon is the slow reaction.

If the useful life of the catalyst is of the order of a minute or less, very fine particles instead of pellets are used, as practiced in fluidized-beds. In such cases, the coke distribution in the pellet is uniform and the global rate of combustion is essentially equal to the intrinsic rate. If the useful life of the catalyst is of the order of months, the rate of coking would be much smaller than that of the main reaction. Nonetheless, there should exist a nonuniform coke distribution within the pellet if the main reaction is diffusion-limited. A complication involved in treating the combustion of nonuniformly deposited coke is the possible variation of combustion kinetics with coke content.

An instructive example of coke burning is provided by Weisz and Goodwin (1966). As one would expect from the discussion in Section 5.4, the mode of coke removal will depend on the combustion temperature. The removal will be uniform at low temperatures, intermediate between uniform and shell-progressive at intermediate temperatures, and strictly shell-progressive at high temperatures. This is illustrated in Figure 5.7. Silica-alumina pellets (beads) can be made transparent by immersing them in a high-refractive index liquid such as carbon tetrachloride. With such immersion the distribution of deposited coke becomes apparent. Photographs of partially regenerated coke catalyst beads obtained by the original authors are represented in Figure 5.7 (Weisz and Goodwin 1966). In the region in which the SPM mode of coke burning prevails ($T > 625°C$), they have found that the fraction of coke remaining, δ, can be represented by:

$$\frac{1}{2}(1 - \delta^{2/3}) - \frac{1}{3}(1 - \delta) = \frac{bDC_bM_B}{\rho_B R_0^2} t$$

which is Eq. 5.31 for $z = 1$ (no change in particle size). Here, the subscript B denotes coke, and C_b denotes the bulk concentration of oxygen. A thorough analysis with respect to the effects of bead size, diffusivity, and coke content confirmed the soundness of the above result.

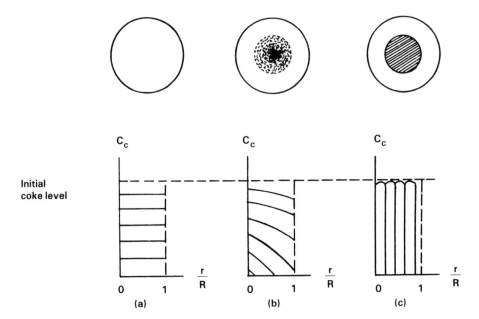

Figure 5.7 Coke burning as affected by oxidation temperature and the progression of coke concentration: (a) low temperature, (b) intermediate temperature, (c) high temperature. (Weisz and Goodwin 1963. Reprinted with permission from Journal of Catalysis. Copyright by Academic Press.)

5–8 DEACTIVATION BY COKING

Because of some unique features, coking is perhaps the most challenging problem of chemical deactivation. The activity of coked catalyst does not decrease linearly with coke content, which is typically expressed as weight percent on catalyst pellet, even for structure-insensitive reactions. This is opposed to poisoning in which the activity decreases linearly due to the monolayer nature of the adsorbed poisons. Coking can also cause severe diffusion limitation as the pores get filled with coke. These unique features are a direct result of coke growth, which was discussed earlier. In addition to the uncertainty regarding the chemical nature of coke and coke precursors, there are also complications due to the specificity of coking with respect to catalyst and reaction conditions. The mechanism of coke growth on a nickel catalyst, for instance, is not necessarily the same for different reactions catalyzed by the same catalyst. In spite of these difficulties, one can still deal with coking provided certain relationships are available. One of the required relationships is that between catalytic activity remaining after coking and coke content, since the residual activity can then be determined from the coke content. The other needed relationship is the manner in which the coke content lowers the effective diffusivity. In this section, we provide these relationships based on a multi-layer kinetic model that accounts for coke growth.

Consider coking taking place on an energetically uniform catalyst surface. Suppose that a coke precursor, P, can form coke upon deposition onto sites whether the sites are vacant or occupied by coke. This multilayer coking can be represented by the following irreversible elementary surface reactions:

$$P + S \xrightarrow{k_m} P \cdot S\downarrow \qquad : \text{monolayer coking}$$

$$\left.\begin{array}{l} P + P \cdot S \xrightarrow{k_p} (2P \cdot S)\downarrow \\[6pt] P + (2P \cdot S) \xrightarrow{k_p} (3P \cdot S)\downarrow \\[4pt] \qquad\qquad \cdot \\[2pt] \qquad\qquad \cdot \\[2pt] \qquad\qquad \cdot \\[4pt] P + (nP \cdot S) \xrightarrow{k_p} [(n+1)P] \cdot S\downarrow \end{array}\right\} \quad \text{multilayer coking}$$

where it is assumed that the rate constant is the same for all multilayer coking reactions. The first step is monolayer coking; one "molecule" of coke precursor adsorbs onto a vacant site, S, to form monolayer coke. Multilayer coke forms, as shown, when the precursor deposits onto mono- or multilayer coke. The net rate of formation of n-layer ($n = 1, \ldots, N$) coke, $R_{np \cdot s}$, can be written based on the elementary steps shown above. Thus, we can write, in general,

$$\frac{R_{(n-1)p \cdot s}}{R_{np \cdot s}} = \frac{dC_{(n-1)p \cdot s}}{dC_{np \cdot s}} = \frac{C_{(n-2)p \cdot s} - C_{(n-1)p \cdot s}}{C_{(n-1)p \cdot s} - C_{np \cdot s}}, \quad n \geqslant 3 \qquad (5.80)$$

where $C_{np \cdot s}$ ($n = 3, 4, \ldots, N$) is the surface concentration of the sites occupied by n-layer coke and not the concentration of n-layer coke. A solution to Eq. 5.80 is

$$\frac{C_{np \cdot s}}{C_{(n-1)p \cdot s}} = f \qquad (5.81)$$

where f is a distribution factor relating the surface concentration of $(n-1)$ layer coke to that of n-layer coke. This distribution factor is constant over short time intervals, but does change over long intervals. The ratio of the net rates of formation for the mono- and di-layer coke is:

$$\frac{R_{p \cdot s}}{R_{2p \cdot s}} = \frac{dC_{p \cdot s}}{dC_{2p \cdot s}} = \frac{K_p C_v - C_{p \cdot s}}{C_{p \cdot s} - C_{2p \cdot s}}, \quad K_p = k_m/k_p \qquad (5.82)$$

which also follows from the elementary steps. Here, C_v is the surface concentration of vacant sites. If we assume that Eq. 5.81 holds for mono- and di-layer coke such that $fC_{p \cdot s} = C_{2p \cdot s}$, it follows from Eq. 5.82 that:

$$\frac{1}{f} = K_p \frac{C_v}{C_{p \cdot s}} \qquad (5.83)$$

Consider site balances for the determination of the distribution factor, f. A balance for the total sites is:

$$C_t = C_v + \sum_{n=1}^{N} C_{np \cdot s} + \sum_{i} C_{i \cdot s} \qquad (5.84)$$

where C_t is the number of total sites per unit area and $C_{i \cdot s}$ is the surface concentration of surface intermediate, i, involved in the main reaction and the reaction leading to the formation of the coke precursor. If we restrict attention to coking in which both the main and coking reactions occur on the same sites,

$$C_t \gamma = \sum_{n=1}^{N} C_{np \cdot s} \qquad (5.85)$$

where γ is the fraction of active sites deactivated. Note that $C_{np \cdot s}$ is the number of sites occupied by n-layer coke per unit area. Thus, from Eqs. 5.84 and 5.85:

$$C_t(1 - \gamma) = C_v + \sum_{i} C_{i \cdot s} \qquad (5.86)$$

As in Eqs. 5.9 and 5.10, we express $\sum C_{i \cdot s}$ in terms of equilibrium constants and concentrations to rewrite Eq. 5.86 as follows:

$$\frac{C_v}{C_t} = \frac{C_t(1 - \gamma)}{(1 + G)^n} \qquad (5.87)$$

Since the distribution factor, f, is given by Eq. 5.81, it can be used in Eq. 5.85 to give:

$$C_t \gamma = C_{p \cdot s} \left(1 + \sum_{n=1}^{N} f^n \right) = C_{p \cdot s} \left(\frac{1 - f^N}{1 - f} \right) \qquad (5.88)$$

provided f is less than unity. Eqs. 5.87 and 5.88 can be used in Eq. 5.83 to arrive at the following expression for the distribution factor for N approaching infinity:

$$f = \frac{1}{1 + \bar{K}_p(1 - \gamma)/\gamma}, \qquad (N \to \infty) \qquad (5.89)$$

where $\bar{K}_p$ is given by:

$$\bar{K}_p = \frac{K_p}{(1+G)^n} \tag{5.90}$$

It is seen from Eq. 5.89 that $f < 1$ for $\gamma < 1$.

Now that the distribution is known, the relationship between coke content and the fraction of catalyst deactivated can readily be obtained. Let q be the monolayer coke weight per site per unit pellet weight. Then

$$C_c = q \sum_{n=1}^{N} n C_{np \cdot s} = q C_{p \cdot s} \left\{ \frac{f(1-f^N) - Nf^N(1-f)}{(1-f)^2} + \frac{1-f^N}{1-f} \right\} \tag{5.91}$$

where C_c is the coke content in weight per unit pellet weight. The following relationship results when Eq. 5.88 is used in Eq. 5.91:

$$\frac{C_c}{Q} = \gamma \left[\frac{1}{1-f} - \frac{Nf^N}{1-f^N} \right] \tag{5.92}$$

where Q $(= C_t q)$ is the weight of coke when all sites are covered by monolayer coke. Let N approach infinity and use Eq. 5.89 in 5.92, to obtain

$$\frac{C_c}{Q} = \gamma \left[1 + \frac{\gamma}{\bar{K}_p(1-\gamma)} \right] \tag{5.93}$$

The coke content due to multilayer coke or coke growth is contained in the second term in the bracket of Eq. 5.93 since $C_c/Q = \gamma$ when only monolayer coke is involved. For strictly monolayer coking, $\bar{K}_p$ approaches infinity, k_p approaches zero, and Eq. 5.93 reduces to that for monolayer coking. For structure-insensitive reactions, the fractional catalytic activity remaining after deactivation, A, is equal to $(1 - \gamma)$. This can be used in Eq. 5.93 to obtain the following relationship:

$$\frac{C_c}{1-A} = Q + \left(\frac{Q}{\bar{K}_p}\right)\left(\frac{1-A}{A}\right) \tag{5.94}$$

The coke content and the residual activity can be determined experimentally. Therefore, a linear plot of Eq. 5.94 yields the monolayer coke content, Q, and the rate constant ratio $\bar{K}_p$.

The coking kinetics for the rate at which the active sites disappear follow directly from the multilayer model. This rate, R_p, is simply the rate of monolayer formation. Thus, we have:

$$\frac{d}{dt}(C_t \gamma) = R_p = k_m C_p C_v \tag{5.95}$$

With the aid of Eq. 5.87, this can be rewritten as:

$$\frac{d\gamma}{dt} = \frac{k_m C_p (1 - \gamma)}{(1 + G)^n} \tag{5.96}$$

As illustrated earlier for poisoning, expressions for G and C_p can be derived for the rate of change of γ with time if the elementary steps leading to the formation of the coke precursor and the reaction products are given. Thus, Eqs. 5.93 and 5.96 form the basis for the intrinsic coking kinetics. It will become clear why Eq. 5.93 is necessary when the influence of coking on effective diffusivity is considered.

We have so far treated only the case of an infinite layer of coke ($N \rightarrow \infty$). The infinite layer model should be sufficient in most cases since the distribution factor for infinite layers is essentially the same as that for finite layers when γ is relatively small. Consider, as an example, the coking data of Dumez and Froment (1976), which were obtained under diffusion-free conditions. Since Eq. 5.93 or 5.94 is applicable when γ is relatively small, a plot of $C_c/(1 - A)$ versus $(1 - A)/A$ should still yield a straight line for the initial stages of coking, i.e. for small values of γ. Such a plot for the data of Dumez and Froment is shown in Figure 5.8. It is seen that a deviation from linearity occurs at around $\gamma = 0.55$. The values of Q and $\bar{K}_p$, determined from the linear portion of the plot, are also applicable to the finite layer case since the distribution factors for both finite and infinite cases are essentially the same for small values of γ. Since the values of Q and $\bar{K}_p$ are determined from the linear portion of the plot ($Q = 0.015$, $\bar{K}_p = 0.96$), the value of N is the only unknown in the finite layer case, which can be determined from Eq. 5.92 and the distribution factor given by

$$\frac{1 - f}{f(1 - f^N)} = \frac{\bar{K}_p (1 - \gamma)}{\gamma} = \frac{\bar{K}_p A}{1 - A} \tag{5.97}$$

This expression follows from Eqs. 5.83, 5.87, and 5.88. The value of N, determined from Eqs. 5.92 and 5.97 for the data of Dumez and Froment using the stepwise procedures detailed in Klingman and Lee (1984), is 8.4. The comparison between theoretical and experimental results is shown in Figure 5.9 for both infinite and finite cases. It is seen that the infinite layer case gives satisfactory results down to 30% residual activity, indicating that the infinite layer model should be sufficient for all practical purposes since a catalyst usually is regenerated before it loses so much of its activity.

The multilayer model is a physical model and therefore the parameters contained have physical meanings. The monolayer coke content, Q, determined from a linear plot of Eq. 5.94 can be compared with estimates obtained by independent means to give credence to the model. For instance, it is known that a monoatomic layer of carbon on an oxide-type catalyst or support would comprise about 5 wt

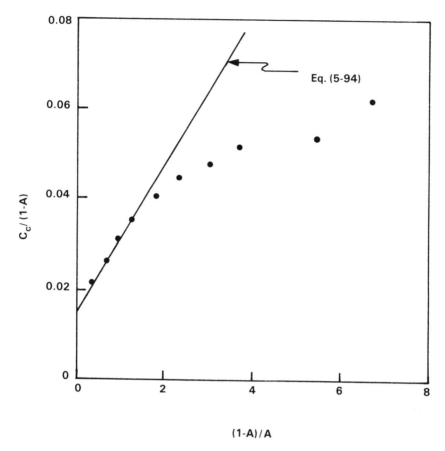

Figure 5.8 Determination of parameters in Eq. 5.94.

% carbon for each 100 m²/g of surface area (Butt 1972). For coking experiments carried out under constant concentration conditions, $\bar{K}_p$ is simply a ratio of two rate constants. Thus, the values of $\bar{K}_p$ determined at different temperatures should yield a straight-line Arrhenius plot. This Arrhenius temperature dependence is expected to hold only when the nature of the coke precursor does not change with temperature. The chemical nature of coke precursors, which have been lumped as a single compound, may change with temperature. It is a good indication that the coke precursor changes over the temperature range of interest if the value of Q depends on temperature and the $\bar{K}_p$ values do not obey the Arrhenius relationship. In such a case, Eq. 5.94 would be valid for segments of the temperature range with a different coke precursor (different Q) for each segment.

Consider now the effect of coking on effective diffusivity. As a result of coke accumulation on the catalyst surface, the effective diffusivity will usually

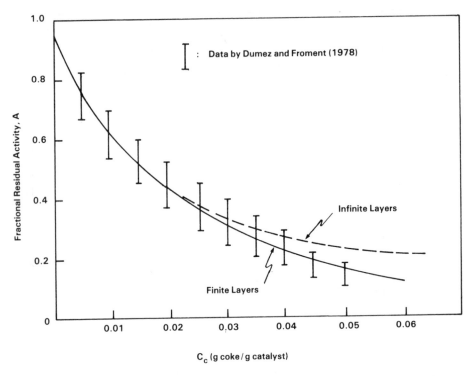

Figure 5.9 Comparison of catalytic activity by multilayer technique to experimental data.

decrease. The coke occupies pore volume that would otherwise be available for the transport of reacting species. The decrease in the pore volume due to coking is equal to the volume of the deposited coke, as shown by Haldemann and Botty (1959). Thus, we may set the volume occupied by coke deposits equal to the reduction in pore volume:

$$V_p(\epsilon_o - \epsilon) = \frac{C_c}{\rho_c} \tag{5.98}$$

where ϵ_o and ϵ are the catalyst pellet porosities for fresh and coked catalyst, respectively, V_p is the pellet volume per unit weight, C_c is the coke content in weight per unit pellet weight, and ρ_c is the coke density. Eq. 5.98 can be rewritten as:

$$\frac{\epsilon}{\epsilon_o} = 1 - \alpha C_c, \ \alpha = \frac{1}{V_p \rho_c} \equiv \frac{1}{(C_c)_s} \tag{5.99}$$

where $(C_c)_s$ is the saturation coke content corresponding to the amount per unit pellet weight required to fill the initial pore volume.

The change in the pellet porosity due to coking can be related to the corresponding change in the diffusivity through the models already available for effective diffusivity. In a simplified version of Feng and Stewart's model (1973) (see Chapter 14), the ratio of the diffusivity for the coked catalyst to that for a fresh catalyst is:

$$\frac{D_e}{D_{eo}} = \frac{\epsilon/\kappa}{\epsilon_0/\kappa_0} \tag{5.100}$$

where κ is the tortuosity factor and the subscript o is for a fresh, uncoked catalyst pellet. The tortuosity factor is a correction for the zigzag nature of the diffusion path along the pores, although constrictions can also play a role. Neglecting any constriction effects, Eq. 5.100 can be written as:

$$D_e(x) = D_{eo} [1 - \alpha C_c(x)] \tag{5.101}$$

where Eq. 5.99 has been used. Eq. 5.101 has been made dependent on the pellet coordinate x since Eq. 5.99 and Eq. 5.100 are valid at any point in the pellet. If we use the random pore model of Wakao and Smith (1962),* on the other hand,

$$\frac{D_e}{D_{eo}} = \left(\frac{\epsilon}{\epsilon_o}\right)^2 \tag{5.102}$$

In this case, the local effective diffusivity $D_e(x)$ is given by:

$$D_e(x) = D_{eo} [1 - \alpha C_c(x)]^2 \tag{5.103}$$

which follows from Eqs. 5.99 and 5.102. In general, the local effective diffusivity can then be written as:

$$D_e(x) = D_{eo} [1 - \alpha C_c(x)]^\beta \tag{5.104}$$

where β is unity for the simple version of the Feng and Stewart model (1973) and is two for the random pore model. If we include the tortuosity factor in the Feng and Stewart model as a power of (ϵ/ϵ_o), the value of β will be something other than unity or two. The relationship between coke content and the fraction of catalyst deactivated is applicable to any point in the pellet. Thus, combining Eq. 5.93 with Eq. 5.104, gives:

$$D_e(x) = D_{eo} \left\{1 - \alpha Q \gamma(x) \left[1 + \frac{\gamma(x)}{K_p(1 - \gamma)}\right]\right\}^\beta \tag{5.105}$$

which is the expression for the local effective diffusivity, $D_e(x)$.

---

* See Chap. 14.

The relationships required for the analysis of a reaction affected by coking can be summarized as:

$$\frac{d}{dx}\left(D_e(x)\frac{dC}{dx}\right) = k(1-\gamma)f(C) \tag{5.106}$$

$$\frac{d\gamma}{dt} = \frac{k_m C_p(1-\gamma)}{(1+G)^n} \tag{5.96}$$

$$R_G = \frac{1}{L}\left[D_e\frac{dC}{dx}\right]_{surface} \tag{5.107}$$

where $D_e(x)$ is given by Eq. 5.105 and the expressions for C_p and G follow from the surface elementary steps of the reaction. It is seen that only the fraction of the catalyst deactivated, which is the quantity relevant to catalytic activity rather than coke content, appears in the conservation equations. Thus, the complex problem of dealing with a reaction affected by coking can be treated in a systematic and consistent way based on the relationships summarized above.

Summary

Chemical deactivation by poisoning is well understood and permits a logical representation of the deactivation kinetics in terms of chemisorption. While our understanding of fouling is not as good, it still permits the representation of the kinetics in the general framework of parallel and series chemisorption. Coking is very difficult to deal with because of the specificity of coking with respect to catalyst and reaction conditions. Nevertheless, the framework of the multilayer kinetic model permits a logical and systematic treatment of the problem. The time dependence of catalytic activity due to deactivation must be taken into account for reactor design. The global rates and reactor point effectiveness developed here permit heterogeneous fixed- and fluidized-beds, affected by deactivation and diffusion, to be treated as homogeneous reactors. The single pellet reactor of Hegedus and Petersen (1974), though not treated here, is useful in determining the intrinsic kinetics of deactivation. Two specialized books on catalyst deactivation are available (Delmon and Froment, eds., 1980; Butt and Petersen, in press).

NOTATION

a	activity of catalyst
A	activity factor defined by Eq. 5.36; area; a constant
b, m, p	stoichiometric coefficient ratios

$(Bi)_m$	Biot number for mass
$(Bi)_h$	Biot number for heat
C	concentration of key species
C_c	pellet center concentration; coke concentration (mole/pellet volume); coke content in weight coke per weight catalyst. coke content in wt coke/wt cat. pellet
$(C_c)_s$	quantity defined in Eq. 5.99
C_d	concentration at the boundary between fresh inner core and completely deactivated outer shell
C_p	concentration of poisoning species or coke precursor
C_v	surface concentration of vacant sites
C_t	total number of sites per unit surface area
$C_{np \cdot s}$	surface concentration of the sites occupied by n-layer coke
d, n	exponents
D_e	effective diffusivity
D_p	effective diffusivity of poisoning species
Da	Damkohler number
E_a	activation energy
f	distribution factor in Eq. 5.81
$f(C)$	functional dependence of rate of reaction on concentration
G	activity distribution normalized with respect to surface activity of fresh catalyst; $G' = dG/dx$
$G(C, K_i)$	inhibition term in Eq. 5.11
h	film heat transfer coefficient
$H(C_b)$	quantity defined by Eq. 5.47
k	rate constant for fresh catalyst, main reaction
k_m	rate constant for the formation of monolayer coke
k_p	rate constant for deactivation reaction; rate constant for the formation of multilayer coke
K_p	k_m/k_p
$\overline{K_p}$	defined in Eq. 5.90
l	length of the portion of pellet completely deactivated
L	characteristic length of pellet; half-thickness of slab
M	molecular weight
n	number of coke layers
N	concentration of poisoning species; number of moles; number of maximum coke layers
P	coke precursor
Q	poison capacity of catalyst (moles poisoning species/pellet volume); monolayer coke content
r, r_c	rate of reaction, main reaction
r_p	rate of deactivation reaction
R	radius at any given time
R_o	original radius

R_g	gas constant
R_G	global rate of main reaction
$(R_G)_p$	global rate of deactivation reaction
$R_{np \cdot s}$	net rate of formation of n-layer coke
s	fraction of vacant active sites
S	active site
t	time
T	temperature
T_d	temperature at x_d
V_p	pellet volume
x	pellet coordinate as measured from center
x_d	x at which fresh inner core is separated from completely deactivated outer shell
y	conversion
z	quantity defined in Eq. 5.30

Greek Letters

α	constant, quantity defined in Eq. 5.43; quantity defined in Eq. 5.99
β	quantity defined in Eq. 5.43
γ	fraction of catalyst deactivated
δ	quantity defined in Eq. 5.30
ϵ	Arrhenius number, $E_a/R_g T_b$; pellet porosity
ϵ_p	pellet voidage; $(E_a)_p/R_g T_b$
Λ	reactor point effectiveness
κ	tortuosity factor
η	effectiveness factor
η_i	internal effectiveness factor
$(\eta_i)_p$	internal effectiveness factor for poisoning reaction, Eq. 5.48
η_{in}	$\tanh \phi_{in}/\phi_{in}$; effectiveness factor given by Eq. 5.57
η_f	$\tanh \phi/\phi$
ρ	density
ρ'	molar density
ρ_c	coke density
ϕ	usual Thiele modulus
ϕ_G	generalized Thiele modulus
ϕ_{in}	$(L - l)(k_s/D_e)^{1/2}$
$(\phi_p)_d$	Thiele modulus defined in Eq. 5.67

Subscripts

A	species A
b	at bulk conditions

B	species B
c	pellet center
d	at the boundary between fresh inner core and completely deactivated outer shell
M	species M
0	fresh catalyst
p	poisoning reaction; pellet
s	at pellet surface conditions

Superscript

$-$ evaluated at surface conditions

PROBLEMS

5.1. For the following mechanism, show that the kinetics of the main and deactivation reactions can be written in the form of Eqs. 5.4 and 5.5:

$$A + S \rightleftharpoons A \cdot S \qquad (K_A)$$
$$A \cdot S \rightleftharpoons B \cdot S \qquad (K_S)$$
$$B \cdot S \rightarrow B + S \qquad (k_d, \text{ controlling step})$$
$$N + S \rightarrow N \cdot S \qquad (k_p, \text{ poisoning})$$

5.2. For noncatalytic gas-solid reactions, write at least four quantities which are identical to the quantity given below:

$$\frac{dN_B}{dt} = b\frac{dN_A}{dt}$$

Assume the SPM mode of reaction for:

$$A(g) + bB(s) \rightarrow cC(g) + dD(s)$$

5.3. Show for general kinetics, $r_c = kf(C)$, that the activity factor A for uniform poisoning is given by:

$$A = (1 - \gamma)^{1/2}$$

when the main reaction is diffusion-limited. Use the following global rate for a deactivated pellet:

$$R_G = \eta_d k(1 - \gamma)f(C_b)$$

$$\eta_d = \frac{[2D_e k(1 - \gamma)]^{1/2}}{Lk(1 - \gamma)f(C_b)} \left[\int_{C_c}^{C_b} f(C)dC \right]^{1/2}$$

First show that:

$$A = \frac{\eta_d}{\eta_f}(1 - \gamma)$$

where η_f is the effectiveness factor for fresh pellet.

5.4. Equation 5.41 applies to slab geometry. Obtain the activity factor for spherical geometry.

5.5 Fill the marked corners with the types of catalyst pellet given below and give reasons for the choices. The shaded part of the pellets is inert. Refer to Becker and Wei (1977). For optimal catalyst distribution, see Varghese and Wolf (1980) and Dadyburjor (1982).

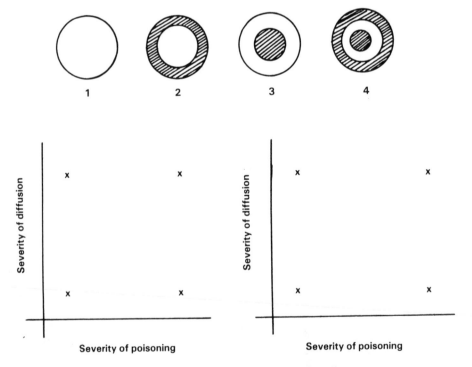

Figure 5.5P   Various types of catalyst pellets.

5.6. In the catalytic cracking of gas oil, deactivation is caused by two types of coking: fast and slow. One possible model is:

$$A \underset{\searrow\ B_p}{\overset{\rightarrow B}{}} \quad \begin{array}{l} \text{carbon (slow coking)} \\ \text{(fast coking)} \end{array}$$

The mechanism of deactivation is chemisorption for fast coking and simple deposition for slow coking. Assume that all reactions are of pseudo-first-order. Assume also uniform deactivation. Derive an expression for the time progression of the fraction of the catalyst deactivated.

5.7. It has been shown in Chapter 4 that a reduction in the amount of catalyst used can be realized for diffusion-limited reactions when hollow pellets or pellets with an inert core are used. Determine whether the same is true when the reactions are also affected by deactivation (Lee 1980).

5.8. Write global rates for the main and deactivation reactions in the asymptotic region of strong diffusion effects for the following cases: (1) uniform, (2) pore-mouth, (3) general. Repeat the same when heat effects are negligible. Assume independent poisoning.

5.9. Show for an nth-order reaction represented by:

$$r_c = GkC^n$$

that the ratio of Eq. 4.102 is:

$$R = \left| \frac{L\bar{G}'}{\bar{G}^{3/2}} \right| \left(\frac{1}{\phi} \right) \left(\frac{(n+1)^{3/2}}{2^{1/2}(n+1)(n+2)} \right)$$

For surface activities of 0.3, 0.5, and 0.7 ($\bar{\gamma}$ of 0.7, 0.5, and 0.3, respectively), determine the value of R for $n = 1$. To obtain γ profiles, first find the times at which $\bar{\gamma}$ becomes equal to 0.3, 0.5, and 0.7. Use the results of Problem 5.8 (part 3) for G. Assume that the Thiele modulus for the deactivation reaction is 6 when γ is zero, and that $N_b/Q = 3.6 \times 10^{-7}/4 \times 10^{-4}$ and $D_p/L = 10^{-2}$cm/sec.

5.10. Suppose that a pellet is deactivated according to the shell-progressive model. Show that the activity factor can be determined from the measurements of external temperature differences:

$$A = \frac{\Delta T_e}{(\Delta T_e)_f}$$

where $(\Delta T_e)_f$ is the external temperature difference for a fresh pellet. A comparison between A calculated from temperature differences and experimental results is given in the article by Lee (1980a).

5.11. A single pellet (Kehoe and Butt 1972) is often used to study various aspects of a heterogeneous reaction. For a cylindrical pellet in which a first-order reaction takes place and where deactivation occurs according to the shell-progressive model, show that:

$$\frac{C_d}{C_b} = 1 - (\eta Da)_d \left(1 + (Bi)_m \ln \frac{1}{\gamma}\right); \qquad \gamma = x_d/R$$

where $(\eta Da)_d = LR_G/k_g C_b$ for a deactivated pellet. Show also that the activity factor A in the asymptotic region of strong diffusion effect is given by:

$$A = \gamma \left[\frac{k_d}{(k_b)_f}\right]^{1/2} \left\{1 - (\eta Da)_d \left(1 + (Bi)_m \ln \frac{1}{\gamma}\right)\right\}$$

$$= \frac{\Delta T_e}{(\Delta T_e)_f}$$

where $(k_b)_f$ is the rate constant for fresh catalyst at the bulk temperature. It is seen that the fraction of catalyst deactivated γ can be calculated iteratively from temperature measurements and the observed rate of reaction for a single pellet. Note that T_d for k_d can be obtained from Eq. 5.60.

5.12. Derive Eqs. 5.72 and 5.76.

5.13. Show for very fine catalyst particles that:

$$1 - \gamma = \exp(-\alpha t)$$

where α is a constant if the bulk reactant concentration is also constant.

5.14. Bartholomew et al. (1979) studied H_2S poisoning of a Ni bimetallic catalyst on alumina for methanation activity. Poisoning tests were conducted at 1

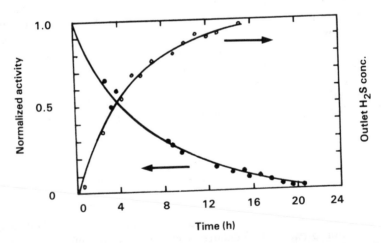

Figure 5.14P H_2S poisoning of a Ni bimetallic catalyst. (Bartholomew et al. 1979. Reprinted with permission from Journal of Catalysis. Copyright by Academic Press.)

atm, 525°K, and a space velocity of 30,000 hr^{-1} using 1 cm^3 of powdered catalyst particles in an isothermal glass reactor. The feed consisted of 95% N_2, 4% H_2, 1% CO, and 10 ppm H_2S. The normalized activity and outlet H_2S concentrations that they obtained are shown in Figure 5.14P. Would first-order deactivation kinetics fit the experimental results? Give reasons for your conclusion. Note that the relative activity is equal to $(1 - \gamma)$ and that:

$$\frac{d}{dt}(Q\gamma) = r_p$$

5.15. A study of coke formation in cumene cracking over silica-alumina has been made by Plank and Nace (1955). They found that the coke is due to an impurity in cumene feed. They proposed the following for the inhibition of the reaction by the impurity:

$$C + S \underset{k_2}{\overset{k_1}{\rightleftharpoons}} C \cdot S \overset{k_3}{\longrightarrow} C_6H_6 + C_3H_6 + S$$

$$N + S \underset{k_5}{\overset{k_4}{\rightleftharpoons}} N \cdot S$$

where S is acidic site. They assumed that the number of vacant sites is much smaller than that of occupied sites such that

$$S_t = \text{total sites} = (N \cdot S) + (C \cdot S)$$

The rate in the absence of the impurity then is:

$$R_0 = k_3 S_t$$

and the rate in the presence of the impurity is:

$$R_p = k_3 (C \cdot S)$$

Show that

$$\frac{R_0}{R_0 - R_p} = 1 + \frac{C_0(1 - f)}{K P_0}$$

where f = conversion of cumene
$K = (k_4 k_5)/[k_1/(k_2 + k_3)]$
C_0 = cumene concentration in feed
P_0 = inhibitor concentration in feed

Their experimental results are shown in Figure 5.15P.

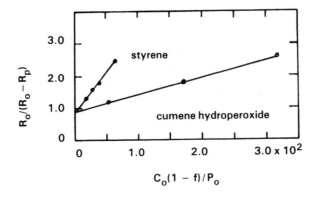

Figure 5.15P Inhibition of silica-alumina at 800°F. (Plank and Nace 1955. Reprinted with permission from *Industrial and Engineering Chemistry.* Copyright by American Chemical Society.)

5.16. Coke burning curves obtained by Weisz and Goodwin (1966) are shown in Figure 5.16P. The oxygen concentration and effective diffusivity are 3×10^{-6} mol/cm³ and 5×10^{-3} cm²/sec.

1. Check whether the criterion of negligible diffusion limitation is valid for their results.
2. Determine the kinetics of coke combustion to the extent you can.
3. Determine the 85% burnoff time for the two cases shown below.

Note that about 5 wt% carbon corresponds to monolayer coverage by coke.

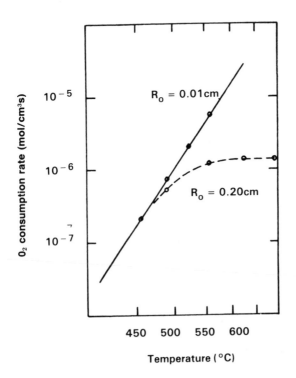

Figure 5.16P Coke burning rates of silica-alumina cracking catalyst. Initial carbon content is 3.4 wt%. (Weisz and Goodwin 1966. Reprinted with permission from *Journal of Catalysis.* Copyright by Academic Press.)

5.17. Explain for diffusion-limited reactions why the coke content is the highest at the pellet core for series coking whereas it is the highest at the surface for parallel coking. The work by Masamune and Smith (1966) may be useful.

REFERENCES

Baker, R.T.K., M.A. Barber, P.S. Harris, F.S. Feates and R.J. Waite, J. Catalysis, *26*, 51 (1972).

Baker, R.T.K. and J.J. Chludzinski, Jr., J. Catalysis, *64*, 464 (1980).

Bartholomew, C.H., G.D. Weatherbee and G.A. Jarvi, J. Catalysis, *60*, 257 (1979).

Becker, E.R. and J. Wei, J. Catalysis, *46*, 372 (1977).

Beeckman, J.W. and G.F. Froment, Chem. Eng. Sci., *35*, 805 (1980).

Beeckman, J.W. and G.F. Froment, Ind. Eng. Chem. Fund., *21*, 243 (1982).

Bischoff, K.B., Chem. Eng. Sci., *18*, 711 (1963).

Butt, J.B., Adv. Chem. Ser., *109*, 259 (1972).

Butt, J.B., D.M. Downing and J.W. Lee, Ind. Eng. Chem. Fund., *16*, 270 (1977).

Butt, J.B. and E.E. Petersen, *Catalyst Deactivation* (in press).

Carberry, J.J. and R.L. Gorring, J. Catalysis, *5*, 529 (1966).

Dadyburjor, D.B., AIChE J., *28*, 720 (1982).

Delmon, D. and G.F. Froment, eds., *Catalyst Deactivation,* Elsevier Scientific Publishing Co., New York (1980).

Dumez, F.J. and G.F. Froment, Ind. Eng. Chem. Proc. Des. Dev., *15*, 291 (1976).

Feng, C.F. and W.E. Stewart, Ind. Eng. Chem. Fund., *12*, 143 (1973).

Froment, G.F. and K.B. Bischoff, *Chemical Reactor Analysis and Design,* p 298, Wiley, New York (1979).

Haldemann, R.G. and M.C. Botty, J. Phys. Chem., *63*, 489 (1959).

Hegedus, L. and E.E. Petersen, Cat. Rev., *9*, 245 (1974).

Kehoe, P.G. and J.B. Butt, AIChE J., *18*, 347 (1972).

Khang, S.J. and O. Levenspiel, Ind. Eng. Chem. Fund., *12*, 185 (1973).

Klingman, K.J. and H.H. Lee, preprint annual AIChE meeting, San Francisco (1984).

Lee, J.W., J.B. Butt and D.M. Downing, AIChE J., *24*, 212 (1978).

Lee, H.H., Chem. Eng. Sci., *35*, 1149 (1980).

Lee, H.H., Chem. Eng. Sci., *35*, 905 (1980a).

Lee, H.H. and J.B. Butt, AIChE J., *28*, 405 (1982).

Levinter, M.E. G.M. Panchenkov and M.A. Tanatarov, Intern. Chem. Eng., *7*, 23 (1967).

Masamune, S. and J.M. Smith, AIChE J., *12*, 384 (1966).

Maxted, E.B., Adv. Cat., *3*, 129 (1951).

Mills, G.A., E.R. Boedekker and A.G. Oblad, J. Am. Chem. Soc., *72*, 1554 (1950).

Petersen, E.E., Chem. Eng. Sci., *37*, 669 (1982).

Plank, C.J. and D.M. Nace, Ind. Eng. Chem., *47*, 2374 (1955).

Szepe, S. and O. Levenspiel, Proc. European Fed. 4th Chem. React. Eng., Brussels, Pergamon Press (1970).

Varghese, P. and E.E. Wolf, AIChE J., *26*, 55 (1980).

Voorhies, A., Jr., Ind. Eng. Chem., *37*, 318 (1945).

Wakao, N. and J.M. Smith, Chem. Eng. Sci., *17*, 825 (1962).

Weekman, V.W., Jr., AIChE J., *16*, 397 (1970).

Weisz, P.B. and R.D. Goodwin, J. Catalysis, *6*, 226 (1966).

Wen, C.Y., Ind. Eng. Chem., *60*, 34 (1968).

Wheeler, A., in P.H. Emmett (ed.) *Catalysis,* Vol. 2, Reinhold, New York (1955).

CHAPTER 6

Physical Deactivation and Sintering

6–1 INTRODUCTION

Catalyst deactivation caused by physical processes rather than chemical reactions is termed *physical deactivation*. Reduction of the exposed surface area of the catalyst is often the result of physical deactivation. This reduction in surface area may be due to the growth of catalyst crystallites dispersed on a support, or to the dissolution of metal atoms into the bulk of the support. The growth of crystallites leads to a population of larger crystallites resulting in less surface area available for reaction per unit volume of catalyst pellet. When the metal atoms are dissolved into the support, the catalyst is in effect lost into the support. The exposed surface area of the catalyst is also reduced when particulates present in the feed deposit at the mouth of pores. This mode of physical deactivation is similar in nature to shell-progressive chemical deactivation: the pore mouth is first covered by the deposited particles and then the covered area grows inward with time, and finally the pore mouth is completely blocked, essentially disabling the catalyst since the reactants cannot reach the catalyst inside the pores. There is a fine line to be drawn between this mode of physical deactivation and chemical deactivation caused by the deposition of certain species: the particulates responsible for physical deactivation come with the feed, whereas the species responsible for chemical deactivation are formed by chemical reactions taking place in the pellet.

The physical deactivation processes can be reversed to a certain extent, but only after the damage has been done. Cyclic treatment of crystallites in an oxidizing atmosphere and then a reducing atmosphere has been found to redisperse large crystallites. Dissolved metal atoms can be made to migrate back to the support surface by proper heat treatment. The deposited particulates can be removed either by combustion or by dissolving them with a suitable solvent.

Our understanding of the physical deactivation of a catalyst is quite limited. In fact, the physical deactivation due to metal dissolution and particulate deposition has drawn interest only recently. On the other hand, some understanding of crystallite growth has been gained, even though there are still many questions to be resolved. The process of catalyst crystallites growing via the movement of small crystallites or atoms is called *sintering* (sometimes called *coarsening*). Large crystallites grow at the expense of smaller ones so as to minimize the chemical potential of the system. Since large crystallites have less exposed surface area per unit volume of catalyst, sintering leads to less surface area available for the reaction. For instance, it has been found (Dalla Betta et al. 1976) that the catalysts used in the catalytic converter for automobiles lose 95% of their original activity due to sintering in 50,000 miles of driving. There are also many important industrial reactions in which sintering is a major factor, a typical example being the silver catalyst used for ethylene oxidation. Major emphasis will be placed on sintering in this chapter because it is better understood than the other modes of physical deactivation.

Sintering takes place via the movement of either crystallites or atoms (or molecules) detached from the crystallites on the surface of a support. For the crystallites or atoms to move on the surface there has to be a source of energy whose level is high enough to overcome the binding energy between the crystallite and the support in the case of crystallite migration, or the metal to metal bond energy in that of atom migration. This energy is considered to be derived from the vibration of atoms, the energy of which is proportional to temperature. The presence of certain gaseous species affects the sintering since it can alter the state of the crystallite surface. The presence of oxygen, for instance, can alter the surface of metal crystallites into a metal oxide surface. Therefore, the nature of the interaction between a catalyst and its support, for instance, is altered from between a metal and a support, to between a metal oxide and a support. It is not surprising then that the sintering rate depends on temperature, types of metal and support, and the atmosphere in which the sintering takes place in addition to the exposed surface area. Depending on whether the movement of the crystallite or the ad-atom is responsible for the crystallite growth, sintering is said to take place through the mechanism of crystallite migration (Ruckenstein and Pulvermacher 1973) or atom migration (Flynn and Wanke 1974; Wynblatt and Gjostein 1975). The magnitude of the metal to metal bond energy relative to that of the binding energy between the metal and the support determines whether the sintering takes place through the mechanism of crystallite migration or atom migration. It is useful to note that for most of the metals the level of interaction between the metal and the support is in the order of alumina > silica > carbon.

While the understanding of the other modes of physical deactivation is quite limited, certain useful conclusions can be made on the effect of physical deactivation on the catalyst activity as related to reactor design, provided that the physical deactivation takes place uniformly throughout the pellet. Therefore, this general case will be treated first before proceeding to sintering.

6–2 EFFECT OF PHYSICAL DEACTIVATION OF CATALYST

It has been pointed out that the exposed surface area is reduced in a shell-progressive manner when the deposition of particles causes the physical deactivation. On the other hand, sintering and dissolution of atoms can be considered to take place uniformly within the pellet. The dissolution of the metal atoms into the support is largely determined by the density distribution of the atoms on the support, which is assumed to be uniform. Therefore, on average, the dissolution would be uniform. As discussed earlier, sintering depends on temperature and the atmosphere in which the catalyst is operating. Since the pellet can be considered isothermal, except when a pellet undergoes shell-progressive chemical deactivation, sintering should take place uniformly provided that the gas composition, in particular the oxygen content, does not change significantly during the course of a reaction. In many reactions of practical interest involving catalyst sintering, oxygen is present in excess, and therefore, sintering can be considered to take place uniformly. It is assumed in what follows that physical deactivation takes place uniformly.

Suppose that the exposed surface area of the catalyst can be measured. Given this measurement, an overall pellet effectiveness can be determined for the reactor design, which represents the combined effects of diffusion, chemical deactivation, and physical deactivation on the intrinsic rate of the reaction. The exposed surface area of a metal catalyst is typically measured using the chemisorption method. As pointed out in Chapter 1, however, this method cannot be used for catalysts of metal compounds. Furthermore, the chemisorption method fails to account for the whole exposed surface area of metal catalysts if a part of the exposed surface area is irreversibly oxidized or forms a compound with the support. Electron microscopy can also be used for the determination of the exposed surface area. While this method is applicable to metal catalysts as well as metal compounds, the size of crystallites counted, based on the micrographs of TEM (transmission electron microscopy), is limited to typically 1 nm for an alumina support and to 2 nm for a silica support, which gives a poorer contrast than alumina. SEM (scanning electron microscopy), which can also be used, gives a poorer resolution than TEM. Therefore, small crystallites cannot be detected. A typical SEM micrograph of silver supported on fused alumina is shown in Figure 6.1. The X-ray line broadening method for the average crystallite size does account for small crystallites even though its accuracy is rather limited in the nm size range. In view of these difficulties, it is desirable to use all three methods for the determination of the exposed surface area whenever possible. For metal compounds, selective poisoning reactions can also be used. It is important to recognize that any surface area measurement should not be taken to give an accurate absolute value, but rather to measure the change of surface area, i.e., the ratio of surface area normalized with respect to the initial surface area. Furthermore, smoothed, average data should be used to represent the change of surface area with time.

In order to examine the effect of sintering, poisoning, and diffusion, the overall pellet effectiveness E is defined as the ratio of the observed rate (global rate) to the intrinsic rate unaffected by diffusion and catalyst deactivation:

$$E = \frac{R_G}{r_c}$$

(6.1)

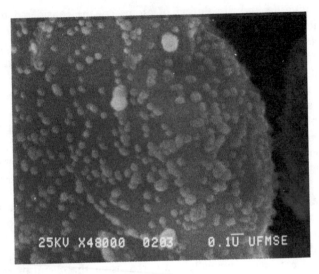

Figure 6.1   A SEM micrograph of silver supported on fused alumina.

It is assumed in what follows that sintering is uniform; poisoning can be uniform, shell-progressive, or in between these two extremes. For these various cases, the overall pellet effectiveness will be derived in this section. Since sintering can take place whether the surface is poisoned or not, it reduces the total surface area S_t. Furthermore, both the poisoned and the active surface areas sinter the same way under the assumption of uniform sintering. In contrast, poisoning reduces only the active surface area S_r. In the absence of poisoning, there is no exposed surface area poisoned and so $S_t = S_r$. This distinction of surface areas, between S_t and S_r, is important in determining the effects of sintering and poisoning on the rate of reaction. While the heat and mass transfer resistances, both within the catalyst pellet and across the pellet-fluid interface, affect the rate of reaction, it will be assumed that only diffusion within the pellet and heat transfer across the interface are important in accordance with the results obtained in Chapter 4.

It is straightforward to obtain the overall effectiveness factor when the reaction is free of diffusional effects. The global rate in this case is:

$$R_G = kh(S_r/S_0)f(C) \tag{6.2}$$

if the intrinsic rate is written as:

$$r_c = kf(C) \tag{6.3}$$

where S_0 is the total active surface area for fresh catalyst, and $h(S_r/S_0)$ represents the dependence of the catalytic activity on the surface area ratio. For structure-insensitive reactions, the activity is directly proportional to the surface area ratio and so $h(S_r/S_0)$ is simply S_r/S_0; for structure-sensitive reactions, the activity has to be correlated to the surface area ratio. Since $S_r = S_0$ for fresh catalyst, and the rate constant k in Eq. 6.3 is for fresh catalyst, it follows that $h(1) = 1$. Further, it follows from the definition of the overall effectiveness factor (Eq. 6.1 and Eqs. 6.2 and 6.3) that:

$$E = h(S_r/S_0)$$

In the absence of chemical deactivation, the overall effectiveness factor is simply $h(S_t/S_0)$ since then $S_r = S_t$.

For diffusion-affected reactions, the generalized internal effectiveness factor for fresh catalyst, η_G, can be used to arrive at the following overall pellet effectiveness for uniform chemical deactivation:

$$E = \eta_G[h(S_r/S_0)]^{1/2} \tag{6.5}$$

where
$$\eta_G = \frac{1}{Lf(C_b)}\left[\frac{2D_e I(C_b)}{k_s}\right]^{1/2}; \quad I(C_b) = \int_{C_c}^{C_b} f(C)dC \tag{6.6}$$

where L is the characteristic pellet length, D_e is the effective diffusivity, C_b is the bulk-fluid concentration of the main reactant, k_s is the rate constant evaluated at the pellet surface temperature, and C_c is the pellet center concentration. As was the case in previous chapters, the center concentration can be taken as zero when the following modified generalized modulus is greater than 3:

$$(\phi_G)_m = \phi_G[h(S_r/S_0)]^{-1/2} = \frac{Lf(C_b)}{\left[\frac{2D_e h(S_r/S_0)}{k_s}\int_0^{C_b} f(C)dC\right]^{1/2}} \tag{6.7}$$

Otherwise, the center concentration is estimated from:

$$\frac{C_c}{C_b} = \frac{1}{\cosh(\phi_G)_m} \quad \text{for } (\phi_G)_m < 3 \tag{6.8}$$

When the poisoning is nonuniform, there exists an activity distribution within the pellet, and the generalized effectiveness factor obtained in Chapter 4 for a nonuniform activity profile can be used to arrive at the following expression:

$$E = \eta_G[h(\bar{S}_r/\bar{S}_0)]^{1/2} \tag{6.9}$$

where the overbar denotes the quantities evaluated at the pellet surface. This approximate result gives accurate values of the pellet effectiveness when the surface activity is far from zero, say $h(\bar{S}_r/\bar{S}_0) > 0.5$. Another limitation is that it is applicable only to diffusion-limited reactions.

If the rate of poisoning is very high such that shell-progressive poisoning takes place, the surface activity is zero and Eq. 6.9 is not applicable. In order to treat this extreme case of poisoning, one needs to recognize that the poisoned catalyst in the completely deactivated outer shell sinters the same way as the active catalyst in the still active inner core of the pellet, but that only sintering affects this active inner core. At steady state, the rate of reactant transport through the deactivated outer shell should be equal to the rate at which the reactant is consumed by the reaction in the active inner core:

$$D_e \left(\frac{C_b - C_d}{l} \right) = \eta_d (L - l) k_d f(C_d) \tag{6.10}$$

where l is the length of the deactivated outer shell, and the subscript d denotes the boundary between the deactivated outer shell and the active inner core. Here, the generalized effectiveness factor for the active inner core is given by:

$$\eta_d = \frac{[k_d M(C_d)]^{1/2}}{L(1 - \gamma) k_d f(C_d)}; \qquad M(C_d) = 2 D_e h(S_t/S_0) \int_{C_c}^{C_d} f(C) dC \tag{6.11}$$

The fraction of catalyst deactivated, γ, is simply l/L for the slab-like pellet. Combining these two equations yields the following expression for the calculation of C_d:

$$C_b - C_d = \frac{L\gamma}{D_e} [k_d M(C_d)]^{1/2} \tag{6.12}$$

The definition of pellet effectiveness yields:

$$E = \frac{[k_d M(C_d)]^{1/2}}{L k_s f(C_b)} \tag{6.13}$$

Note that C_d appearing in Eq. 6.13 can be calculated from Eq. 6.12 and that the activity function h is given in terms of the total exposed surface area S_t, rather than the active area S_r since no poisoning takes place in the active inner core such that $S_t = S_r$.

The overall pellet effectiveness represents the combined effects of diffusion, poisoning, and sintering on the intrinsic rate. A simple multiplication of the intrinsic rate by the pellet effectiveness gives the actual rate at which the reactants are converted, i.e., the global rate. This pellet effectiveness can also be used to determine how much of the loss in catalytic activity is solely due to sintering. For this purpose, it is sufficient to know the pellet effectiveness in the absence of sintering, E_n, since the ratio of E to E_n is a measure of the loss of catalytic activity solely due to sintering. This ratio can readily be obtained from Eq. 6.4 for diffusion-free reactions:

$$\frac{E}{E_n} = \frac{h(S_r/S_0)}{h(S_{r_n}/S_0)} = \frac{h(S_r/S_0)}{h(S_r/S_t)} \tag{6.14}$$

where S_{r_n} is the active surface area that would exist in the absence of sintering. Poisoning reduces only active surface area. Therefore, the total surface area remains at its original area S_0 in the absence of sintering and the ratio of active surface area to the total surface area is given by S_{r_n}/S_0. Suppose sintering also takes place. Under the assumption of uniform sintering, both active and poisoned surface areas are reduced the same way so that the ratio of active to total surface area remains the same. Since the active surface area in this case is S_r, and the total surface area is S_t, then:

$$\frac{S_{r_n}}{S_0} = \frac{S_r}{S_t} \tag{6.15}$$

This relationship was used in Eq. 6.14. The ratio E/E_n has been obtained for various cases (Lee 1981) and these are summarized in Table 6.1. The overall pellet effectiveness in the absence of poisoning can be obtained by simply replacing S_r by S_t.

In order to use the information on surface area for reactor design, the change of the surface area with time has to be related to the reaction conditions. This is possible only for sintering because of limited understanding of the other modes of physical deactivation. Since surface diffusion plays a pivotal role in sintering, it is considered first for the development of the sintering kinetics to follow.

Table 6.1 Overall Pellet Effectiveness Ratio, E/E_n

	E/E_n
Diffusion-free reaction	$\dfrac{h(S_r/S_0)}{h(S_r/S_t)}$
Diffusion-affected reaction and uniform poisoning	$\left[\dfrac{h(S_r/S_0)}{h(S_r/S_t)}\right]^{1/2}$
Diffusion-affected reaction and general case of poisoning	$\left[\dfrac{h(\overline{S}_r/\overline{S}_0)}{h(\overline{S}_t/\overline{S}_0)}\right]^{1/2}$ for $\overline{S}_r/\overline{S}_0 > 0.5$
Diffusion-affected reaction and shell-progressive poisoning	$\left[h\left(\dfrac{S_t}{S_0}\right)\right]^{1/2}\left[\int_{C_c}^{C_d} f(C)dC \middle/ \int_{(C_c)_n}^{(C_d)_n} f(C)dC\right]^{1/2}$

C_d is given by Eq. 6.12 and $(C_d)_n$ by:

$$C_b - (C_d)_n = \frac{L\gamma}{D_e}\left[2D_e k_d \int_{(C_c)_n}^{(C_d)_n} f(C)dC\right]^{1/2}$$

6-3   SURFACE DIFFUSION

A catalyst is usually dispersed on a support in a discrete crystallite form as shown in Figure 6.1. These crystallites, which are composites of atoms or molecules, range in size anywhere from below 1 nm to about 100 nm. The crystallites can grow under typical reaction conditions due to the heat liberated by the chemical reaction, which causes the migration of atoms (molecules) or small crystallites. This migration is often aided by the presence of a gas, particularly oxygen for platinum group metals and chlorine for base metals. In order for sintering to occur via atom migration, an atom (molecule) has to be detached from the crystallite, forming an ad-atom, which then migrates over the support surface to other crystallites. A model of a crystallite is shown in Figure 6.2 (Somorjai 1972). For crystallite migration, on the other hand, a whole crystallite has to migrate over the support surface. The question of which mechanism is operative, then, is equivalent to whether the detachment of an atom followed by surface migration is energetically easier to take place than the movement of a crystallite. An understanding of surface diffusion provides a key to the question.

Surface diffusion via the movement of atoms is considered to be a multistep process in which atoms break away from their lattice position (e.g., a kink site at a ledge in Figure 6.2) and migrate over the surface until they find their new equilibrium sites. Consider an ad-atom held on the surface by a small binding energy. Let the energy of this interaction be ΔE_d. This energy is the difference between the energy of an ad-atom in its excited state, (i.e., in between two equilibrium sites) and that at its equilibrium condition. The thermal vibration of the ad-atom about the equilibrium site is the source of the energy for the movement. If the vibration frequency of the ad-atom is ν_0, the probable number of times that this atom jumps out of the energy well, or the jump frequency is given by:

$$f_r = \zeta \nu_0 \exp(-\Delta E_d / k_B T) \qquad (6.16)$$

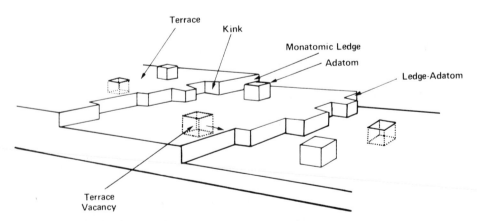

Figure 6.2   A model of a heterogeneous surface showing different surface sites. (From *Principles of Surface Chemistry*, Somorjai 1972.)

where k_B is the Boltzmann constant, and ζ is the number of equivalent neighboring sites. For instance, the ζ value of a (111) face of a face-cubic-centered metal is 6. If it is assumed that surface diffusion can occur only via the movement of ad-atoms, these species have to be created before any appreciable diffusion can occur. The fraction of ad-atoms, C/C_0, on the surface is given by:

$$C/C_0 = \exp(-\Delta E_f/k_B T) \tag{6.17}$$

where ΔE_f is the energy of formation of an ad-atom, C the concentration of ad-atoms, and C_0 the total concentration of surface sites from which the ad-atom can break away. The probabilities multiply and therefore the jump frequency is the product of the two events:

$$f_r = \zeta \nu_0 \exp[-(\Delta E_d + \Delta E_f)/k_B T] \tag{6.18}$$

This frequency is for a single jump of an ad-atom to a neighboring equilibrium site. Now, consider the random walk of an atom for long-distance motion. The question is how far a diffusing atom will move in time t and a very large number of jumps. For this net displacement of an atom, a one-dimensional random walk is considered. If the root mean square distance is used as the net displacement, then for the expected value of X^2, $\langle X^2 \rangle$:

$$\langle X^2 \rangle = n_j a^2 \tag{6.19}$$

where X is the net distance covered by the atom, n_j the number of jumps, and a the distance of one jump, which is taken as the distance between two nearest neighboring sites. If t is the time required to make n_j jumps, n_j is given by:

$$n_j = f_r t \tag{6.20}$$

which, when combined with Eq. 6.19, yields:

$$\langle X^2 \rangle = f_r t a^2 \tag{6.21}$$

It is customary to define the diffusion coefficient as:

$$D_s \equiv \frac{\langle X^2 \rangle}{2bt} = \frac{f_r a^2}{2b} \tag{6.22}$$

where b is the number of coordinate directions in which the jumps may occur with equal probability. For diffusion equally probable in two directions, use of Eq. 6.18 in 6.22 yields:

$$D_s = \frac{\zeta a^2 \nu_0}{4} \exp\left[-\frac{(\Delta E_d + \Delta E_f)}{k_B T}\right] \tag{6.23}$$

The root mean square distance can be obtained from Eqs. 6.21 and 6.22:

$$\langle X^2 \rangle^{1/2} = (2bD_s t)^{1/2} \tag{6.24}$$

This equation is usually used for the determination of the surface diffusivity D_s from the measurements of t and $\langle X^2 \rangle^{1/2}$. Determination of the activation energy is made in the usual manner using Eq. 6.23. For many cubic metals (Gjostein 1967), ΔE_f is of the order of 40 kcal/mole, and the sum of $(\Delta E_d + \Delta E_f)$ can often be approximated at high temperatures by $0.54 \Delta H_s$, where ΔH_s is the heat of sublimation. The correlations for surface self-diffusion coefficients are shown in Figures 6.3 and 6.4 for face-centered cubic metals and body-centered cubic metals, respectively. These correlations are for the diffusion of metal on the same metal surface. The neighboring site distance "a" is of the order of a few angstrom, and ν_0 is of the order of 10^{12} sec^{-1}.

The surface diffusion of crystallites is more complicated than the diffusion of ad-atoms due to the dependence of surface diffusivity on crystallite size. A review on this subject can be found in the article by Kashchiev (1979). Theoretical models available for crystallite migration predict power or exponential dependence of the diffusivity on the size r (or the number of atoms n_c making up the crystallite). The activation energy for the diffusion can be either size-dependent or size-indepen-

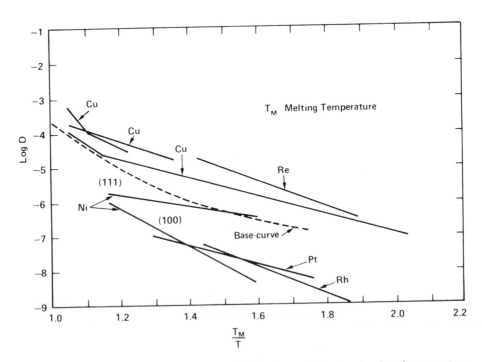

Figure 6.3 Correlations between surface self-diffusion coefficient and reduced temperature T_M/T for face-centered-cubic metals. (From *Principles of Surface Chemistry*, Somorjai 1972.)

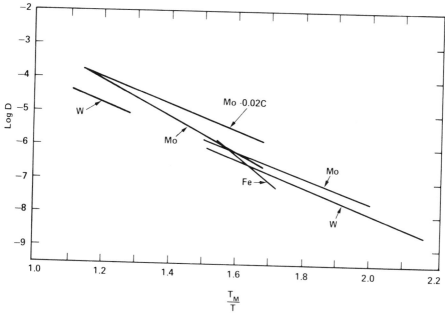

Figure 6.4   Correlations between surface self-diffusion coefficient and reduced temperature for body-centered-cubic metals. (From *Principles of Surface Chemistry,* Somorjai 1972.)

dent. The surface diffusivity also depends on the mode of crystallite migration. These factors complicate the treatment.

6–4   MECHANISMS AND KINETICS OF SINTERING

In this section, the mechanisms of sintering are treated in detail to arrive at sintering kinetics. The first mechanism to be considered is that of atom migration, sometimes termed *Oswald ripening* (Lifshitz and Slyozov 1961) by analogy to the growth of particles suspended in a gas or liquid by movement of individual atoms or molecules. The second is the mechanism of crystallite migration, which is the movement of individual crystallites along the surface to cause growth by coalescence, analogous to coagulation by Brownian motion.

Mechanism of Atom Migration and Kinetics

Consider individual crystallites dispersed on a flat substrate and assume that only single atoms migrate. Each crystallite will have a concentration of single atoms at the interface (line interface) between the crystallite and the substrate, which depends on the crystallite size. Smaller crystallites will have a higher concentration than larger crystallites because of their smaller curvature. At some distance away from the crystallites, there exists a "far field" concentration of single atoms (ad-atoms) on the substrate, representing the "bulk" of ad-atoms for the whole system at any given time. This concentration is higher than the interfacial concentration for the larger crystallites but lower than that for the smaller crystallites. Because of this concentration difference, the ad-atoms will move from the smaller crystallites to the bulk, and then to the larger ones, resulting in the growth of the larger crystallites at the expense of the smaller ones. This mechanism of sintering is termed the *mechanism of atom migration* or Ostwald ripening. In addition to this ripening, which takes place on a global scale involving the bulk ad-atoms, "direct ripening" (Ruckenstein and Dadyburjor 1978) can also occur locally between crystallites that are sufficiently close to each other, even though the bulk is not supersaturated with respect to the larger crystallites.

The atom migration is caused by the difference in the chemical potentials of the crystallites, since the level of concentration of ad-atoms at the interface is determined by the chemical potential. Consider partially wetting crystallites in the form of spherical segments, making a contact angle θ with a flat substrate as shown in Figure 6.5 (Wynblatt and Gjostein 1975). The chemical potential μ of any crystallite depends on its radius of curvature r:

$$\mu = \mu_0 + 2\sigma V_m/r \tag{6.25}$$

where μ_0 is μ for an infinite-sized crystallite, σ is the interfacial tension between the crystallite and gas, and V_m is the volume of an ad-atom. This relationship can be rewritten in terms of surface concentrations to give the Gibbs-Thompson relationship:

$$n(r)/n^{eq} = \exp(2\sigma V_m/k_B Tr) \tag{6.26}$$

where n^{eq} is the surface concentration (number of atoms/area) corresponding to the concentration of an infinite-sized crystallite. Consider the first of the series processes of surface diffusion of ad-atoms and eventual transfer at the interface of the crystallite (Figure 6.5). Taking the center of the crystallite of radius $r \sin \theta$ as the coordinate origin, a steady state mass balance in polar coordinate yields:

$$\frac{1}{R}\frac{d}{dR}\left[RD_1\frac{dn}{dR}\right] = 0 \tag{6.27}$$

where n is the local surface concentration of ad-atoms, and D_1 is the surface diffusivity of ad-atoms on the substance $[D_1 = D_{10} \exp(-\Delta E_d/k_B T)]$. The solution is:

$$n(R) = K_1 \ln R + K_2 \tag{6.28}$$

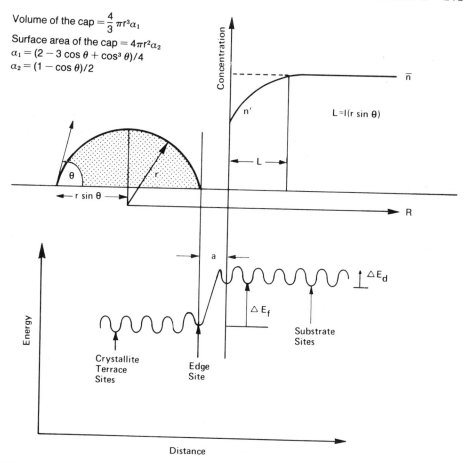

Volume of the cap $= \frac{4}{3}\pi r^3 \alpha_1$

Surface area of the cap $= 4\pi r^2 \alpha_2$

$\alpha_1 = (2 - 3\cos\theta + \cos^3\theta)/4$

$\alpha_2 = (1 - \cos\theta)/2$

Figure 6.5 Concentration-distance and energy-distance diagram for the surface diffusion of ad-atoms. (Wynblatt and Gjostein 1975. Reprinted with permission from *Progress in Solid State Chemistry.* Copyright by Pergamon Press, Inc.)

At the edge of the crystallite ($R = r\sin\theta$), there exists a certain surface concentration n'. At a "screening" distance L, which is taken as a multiple of the radius ($r\sin\theta$), i.e., $l(r\sin\theta)$ where l is the multiple, the surface concentration reaches its bulk value $\bar{n}$. These boundary conditions, when used in the above equation, give:

$$n(R) = \frac{\bar{n} - n'}{\ln l}\ln R + K_2 \tag{6.29}$$

The number of atoms attaching per second at the perimeter of the crystallite by surface diffusion is given by:

$$J_s = (2\pi r \sin \theta) \left[D_1 \frac{dn}{dR} \right]_{R = \sin \theta}$$

$$= \frac{2\pi D_1}{\ln l} (\bar{n} - n')$$

$$(6.30)$$

Consider now transfer at the interface. The rate at which the ad-atoms one interatomic distance a away are attached to the crystallite is equal to the circumferential area $(a2\pi r \sin \theta)$ times the rate at which these atoms (n' number of atoms/area) jump over the energy barrier ΔE_d, i.e., $[2\pi r \, a \sin \theta] n' \beta'$ where the jump frequency β' is given by:

$$\beta' = \nu_0 \exp(-\Delta E_d / k_B T)$$

$$(6.31)$$

In order for an edge atom to jump out to occupy a site on the substrate, it must surmount an energy barrier $\Delta E_f + \Delta E_d$, where ΔE_f and ΔE_d are the energy required to detach an atom from the crystallite and the energy required for an ad-atom to move on the substrate, respectively. Therefore, the jump frequency of these edge atoms is given by:

$$\beta = \nu_p \exp[-(\Delta E_f + \Delta E_d) / k_B T]$$

$$(6.32)$$

where ν_p and ν_0 are the vibrational frequencies on the crystallite edge and substrate sites. Then, the rate of atom detachment is given by $[2\pi r \, a \sin \theta] n_r \beta$, where n_r is the surface concentration of atoms on the crystallite edge sites. The net flux across the interface is equal to the difference between these two rates:

$$J_i = (2\pi r \, a \sin \theta)(\beta' n' - \beta n_r)$$

$$= (2\pi r \, a \sin \theta)\beta'(n' - \beta/\beta' n_r)$$

$$(6.33)$$

Since $J = J_s = J_i$ at steady state, Eqs. 6.30 and 6.33 give:

$$J = \frac{(2\pi r \, a \, \beta' \sin \theta)(2\pi D_1/\ln l)}{2\pi D_1/\ln l + 2\pi r \, a \, \beta' \sin \theta} (\bar{n} - \beta/\beta' n_r)$$

$$(6.34)$$

The growth rate (dr/dt) of the crystallite of interest is obtained from:

$$\frac{d}{dt}\left(\frac{4}{3}\pi r^3 \alpha_1\right) = J V_m$$

$$(6.35)$$

which leads to:

$$\frac{dr}{dt} = \frac{V_m}{4\pi r^2 \alpha_1} \frac{(2\pi r \, a \, \beta' \sin \theta)(2\pi D_1/\ln l)}{2\pi D_1/\ln l + 2\pi r \, a \, \beta' \sin \theta} (\bar{n} - \beta/\beta' n_r)$$

$$(6.36)$$

The Gibbs-Thompson relationship of Eq. 6.26 can be used to write n_r and $\bar{n}$ as

$$n_r = n_r^{eq} \exp(2\sigma V_m/k_B T r) \tag{6.37}$$

$$\bar{n} = \bar{n}^{eq} \exp(2\sigma V_m/k_B T \bar{r}) = (\beta/\beta') n_r^{eq} \exp(2\sigma V_m/k_B T \bar{r}) \tag{6.38}$$

where n_r^{eq} and $\bar{n}^{eq}$ are the atom surface concentrations on the crystallite edge sites and at the screening distance L, respectively, in equilibrium with an infinite-sized crystallite. The fact that $\bar{n}^{eq} = (\beta/\beta') n_r^{eq}$ has been used in Eq. 6.38. When these equations are used in Eq. 6.36, the growth rate expression becomes:

$$\frac{dr}{dt} = \frac{V_m n_r^{eq}(\beta/\beta')}{4\pi r^2 \alpha_1} \frac{(2\pi r a \beta' \sin\theta)(2\pi D_1/\ln l)}{2\pi D_1/\ln l + 2\pi r a \beta' \sin\theta}$$

$$\left\{ \exp\left(\frac{2\sigma V_m}{k_B T \bar{r}}\right) - \exp\left(\frac{2\sigma V_m}{k_B T r}\right) \right\} \tag{6.39}$$

This equation essentially defines the critical size $\bar{r}$. It is seen that the crystallites of size $\bar{r}$ neither grow nor shrink ($d\bar{r}/dt = 0$). This means that the surface concentration at the edges of these crystallites is the same as the average surface concentration prevailing at some distance away from the crystallites. This critical size changes with time as the size distribution of crystallites changes with time.

The relative importance of interface transfer with respect to surface diffusion can be determined by comparing the relative magnitude of resistances (Wynblatt and Gjostein 1975). If surface diffusion is the controlling step, one should have:

$$\frac{2\pi D_1}{\ln l} << 2\pi r a \beta' \sin\theta$$

The screening distance L is typically 2 to 3 times $r \sin\theta$, making ($\ln l$) close to unity. The surface diffusivity of ad-atoms on the substrate is given by:

$$D_1 = a^2 v_0 \exp(-\Delta E_d/k_B T)$$

Therefore, the condition becomes:

$$a << r \sin\theta$$

This condition is always met in physically realistic situations since $\sin\theta$ is close to unity and $a << r$. Thus, surface diffusion is most likely to be the controlling step in realistic situations. This conclusion is a result of the relationship, $\bar{n}^{eq} = (\beta/\beta') n_r^{eq}$.

When the surface diffusion is the controlling step, Eq. 6.39 reduces to:

$$\frac{dr}{dt} = \frac{V_m n_r^{eq}(\beta/\beta')2\pi D_1}{4\pi r^2 \alpha_1 \ln l} \left\{ \exp\frac{2\sigma V_m}{k_B T \bar{r}} - \exp\frac{2\sigma V_m}{k_B T r} \right\} \tag{6.40}$$

The exponential terms in this equation can be approximated as follows:

$$\exp\left(\frac{2\sigma V_m}{k_B T\bar{r}}\right) - \exp\left(\frac{2\sigma V_m}{k_B T r}\right) = -\left[\exp\left(\frac{2\sigma V_m}{k_B T r}\right)\right]\left[\frac{2\sigma V_m}{k_B T r}\left(1-\frac{r}{\bar{r}}\right)\right] \quad (6.41)$$

This approximation is accurate (Lee 1980) when:

$$\epsilon \equiv \frac{2\sigma V_m}{k_B T\bar{r}} \leq 1 \quad (6.42)$$

Under typical sintering conditions, the value of ϵ ranges from 0.3 to 0.8 for average crystallite sizes ranging from 5 nm to 20 nm. The dimensionless surface energy ϵ is readily calculable for a given system. Substitution of Eq. 6.41 into Eq. 6.40 results in the following expression:

$$\frac{dr}{dt} = -\frac{A_D}{r^3}\left(1-\frac{r}{\bar{r}}\right)\exp\frac{2\sigma V_m}{r k_B T} \quad (6.43)$$

where

$$A_D \equiv \frac{V_m^2 n_r^{eq}(\beta/\beta')D_1\sigma}{k_B T\alpha_1 \ln l} \quad (6.44)$$

The sintering kinetics for the mechanism of atom migration can be obtained from the expression for the growth rate (Eq. 6.43) and a continuity equation for the size distribution of crystallites:

$$\frac{\partial f}{\partial t} + \frac{\partial}{\partial r}\left(f\frac{dr}{dt}\right) = 0 \quad (6.45)$$

where $f(r,t)dr$ is the number of crystallites in the size range between r and $r + dr$ at any given time t. After a lengthy manipulation, one arrives at the following sintering kinetics (Lee 1980):

$$\frac{dS}{dt} = -KS^5\exp(mS) \quad (6.46)$$

where $K(T) = $ (constant) A_D
$= $ (constant) $\exp[-(\Delta E_f + \Delta E_d)/k_B T]$ (6.47)

$$m = \text{(constant)}\left(\frac{2\sigma V_m}{k_B T}\right) \equiv \text{(constant)}(\gamma) \quad (6.48)$$

Use of Eq. 6.23 in 6.47 results in a rate constant K that is essentially the surface diffusivity D_s. A weak temperature dependence of the form of $T^{-3/2}$ was neglected in Eq. 6.47. The quantity m can be taken constant if an average temperature is used for the range of temperatures of interest. The growth rate of the critical crystallites can be obtained in a similar manner (Lee 1980):

$$\frac{d\bar{r}}{dt} = \frac{K_r}{\bar{r}^3} \exp(\gamma/\bar{r}) \qquad (6.49)$$

where the temperature dependence of K_r is the same as that of K. This expression can be used to obtain the time dependence of the average size since in most cases the average crystallite size is very close to the critical size (Wynblatt 1976). The kinetics of sintering for the unlikely case of interface control can also be obtained in a similar manner. It should be noted that the exponential term appearing in the sintering kinetics can be set to unity when the dimensionless surface energy ϵ is less than 0.1. This would be true for a typical platinum system, for instance, if the average size is greater than approximately 60 nm.

While sintering can also take place via vapor phase diffusion, the likelihood of this is remote for metal catalysts because of their high melting points (Table 6.2). However, sintering can be considerably aided by the formation of oxides or chlorides. The melting point of metal compounds such as oxides is much lower than that of metals. Since the sublimation energy is a measure of ΔE_f, the energy required to detach an atom from a crystallite is much lower for metal compounds

Table 6.2 Properties of Metals (Jones 1971)

Metal	Solid/Vapour Surface Energy at T_m, erg/cm²	Atomic Volume of Solid at T_m, cm³/g atom	Melting Point (T_m), °C	Molar Surface Energy, kcal/g atom	Heat of Sublimation at T_m, kcal/g atom
Ag	1120 ± 65	10.94	961	11.1 ± 0.6	66.2
Al	1080 ± 200	10.80	660	10.7 ± 1.9	77.6
Au	1390 ± 70	10.74	1063	13.7 ± 0.7	85.5
Be	$900 \pm NA$	(5.09)	1287	$5.4 \pm NA$	75.3
Co	1950 ± 175	7.17	1495	14.6 ± 1.3	98.7
Cr	2300 ± 350	7.89	1857	18.4 ± 2.8	88.5
Cu	1720 ± 100	7.59	1084	13.4 ± 0.8	78.7
Fe(δ)	2080 ± 110	7.60	1536	16.2 ± 0.9	93.6
Mo	1960 ± 200	9.98	2617	18.3 ± 1.9	150.4
Nb	2260 ± 300	11.58	2467	23.3 ± 3.1	170.0
Ni	1810 ± 180	7.13	1453	13.5 ± 1.4	99.8
Pt	2280 ± 800	9.64	1769	20.9 ± 7.1	132.2
Sn	$680 \pm NA$	16.59	232	$8.9 \pm NA$	71.9
Ta	2500 ± 500	11.65	2977	26.0 ± 4.8	188.1
Ti	$1690 \pm NA$	10.99	1670	$16.9 \pm NA$	107.7
W	2650 ± 340	10.16	3407	25.1 ± 3.2	205.0
Zn	$830 \pm NA$	9.55	420	$7.5 \pm NA$	30.6

Table 6.3   Sintering Kinetics for the Mechanism of Atom Migration

$$\frac{dS}{dt} = -KS^n \exp(mS)$$

$$n = 3 \text{ or } 5$$

Controlling Step	n	K	ΔE
Interface transfer	3	$\sim \exp(-\Delta E/k_B T)$	$(\Delta E_f + \Delta E_d)$
Surface diffusion	5	$\sim \exp(-\Delta E/k_B T)$	$(\Delta E_f + \Delta E_d)$
Surface diffusion (Wynblatt 1976) aided by the formation of oxides or chlorides	5	$\sim (P_{gas})^2 K_{eq} \exp(Q/k_B T)^*$	$Q = \Delta E_s - \Delta E_f$

$$* \quad M(s) + q\, O_2\,(g) \overset{K_{eq}}{=\!=} MO_{2q}\,(g)$$
$K_{eq} = $ equilibrium constant
$\Delta E_s = $ energy required to transfer an oxide from the substrate to the vapor phrase

than for metals. The sintering kinetics for various cases are summarized in Table 6.3.

Mechanism of Crystallite Migration

According to the mechanism of crystallite migration, crystallites move as such on the substrate surface and then merge into larger ones when they are sufficiently close. Growth is assumed to depend on the rate of surface migration of the crystallites and the nature of their interaction when they are very close to each other. The probable driving force is assumed to be the thermal motion of the atoms at the metal-substrate interface. If the interaction between crystallites is so strong that they merge into a single unit upon contact within a time period which is short enough compared to the time necessary for the migration, sintering is considered diffusion-controlled. On the other hand, it is considered coalescence-controlled if the merging process is slow compared to the diffusion process.

Consider crystallites highly dispersed on a planar substrate (Ruckenstein and Pulvermacher 1973). Let C_k be the surface concentration of crystallites of k atoms, i.e., the number of crystallites of k atoms per unit surface area of substrate. The concentration C_k increases by the collisions of crystallites composed of i and $(k - i)$ atoms and decreases by the collisions of crystallites composed of k atoms with any other crystallites. In analogy with chemical kinetics, the rate of collision between crystallites composed of i and j units is written as:

$$b_{ij} = K_{ij} C_i C_j \tag{6.50}$$

where K_{ij} is a rate constant dependent on the mobility of crystallites over the substrate and the nature of the interaction between crystallites. The rate of change of C_k is thus given by:

$$\frac{dC_k}{dt} = \frac{1}{2} \sum_{i+j=k} K_{ij} C_i C_j - C_k \sum_{i=1}^{\infty} K_{ik} C_i \qquad (6.51)$$

In order to obtain information on K_{ij}, the process of crystallite diffusion must be examined in some detail. A mass balance during a small time interval θ_0 at or around a given real process time t is given by:

$$\frac{\partial C_k}{\partial \theta} = D_{kj} \frac{\partial^2 C_k}{\partial r^2} + \frac{1}{r} \frac{\partial C_k}{\partial r} \qquad (6.52)$$

where D_{kj} is the relative diffusion coefficient of a crystallite k with respect to that of a crystallite j. The cylindrical coordinate system is such that the origin is at the center of the surface of contact between a crystallite of j units and the substrate. This equation is used during a time interval θ_0, which must be small compared to the time in which an appreciable modification of the crystallite distribution takes place. Further, the rate of collision should be practically independent of θ during θ_0. Assuming that the thermal motion of the crystallite of j units is independent of the motion of the crystallite of k units, one can write:

$$D_{kj} = D_k + D_j \qquad (6.53)$$

The initial and boundary conditions are:

$$C_k = C_{k0} \qquad \theta = 0, \, r > R_{kj} \qquad (6.54)$$

$$\alpha_{kj} C_k = D_{kj} \frac{\partial C_k}{\partial r} \qquad \theta > 0, \, r = R_{kj} \qquad (6.55)$$

where α_{kj} is the rate constant for the merging process, and R_{kj} is the radius of interaction of the two colliding crystallites given by:

$$R_{kj} = r_k + r_j \qquad (6.56)$$

When the rate of merging is very high ($\alpha_{kj} \to \infty$), the process is diffusion-controlled and $C_k = 0$ for $R_{jk} = r_j + r_k$. When the rate is very small, on the other hand, the process is coalescence-controlled. The rate of collision ϕ_{kj} is equal to the diffusion flux times the perimeter $2\pi R_{kj}$:

$$\phi_{kj} = \left| 2\pi R_{kj} D_{kj} \left(\frac{\partial C_k}{\partial r} \right) \right|_{R_{kj}} \right| \qquad (6.57)$$

Since $\phi_{kj} = K_{kj} C_{kj}$, K_{ij} can be calculated from Eq. 6.57. The expression for K_{ij} thus obtained for large values of $(D_{ij}\theta/R_{ij}^2)$ or to be more precise, for $\ln (4D_{ij}\theta/R_{ij}^2) \gg 1$, is (Ruckenstein and Pulvermacher 1973):

$$K_{ij} = 4\pi D_{ij}/\ln (4D_{ij}\theta/R_{ij}^2) \qquad (6.58)$$

when the growth is diffusion-controlled. When it is coalescence-controlled:

$$K_{ij} = 2\pi R_{ij}\alpha_{ij} \tag{6.59}$$

It should be recognized that the diffusion equation (Eq. 6.52) ignores the collision between various crystallites. This is appropriate if the time interval needed to achieve the quasi-steady state is an infinitesimal fraction of the process time t.

In order to use these results in the rate expression of Eq. 6.54 for the derivation of sintering kinetics, D_{ij} and α_{ij} have to be expressed in terms of the crystallite size. These expressions have been suggested by Ruckenstein and Pulvermacher as:

$$D_{ij} = (\text{constant}) \left(\frac{1}{r_i^{n'}} + \frac{1}{r_j^{n'}} \right); \qquad n' = 0,1,2 \tag{6.60}$$

$$\alpha_{ij} = (\text{constant}) \left(\frac{r_i^{m'} + r_j^{m'}}{r_i + r_j} \right); \qquad m' = 1,2 \tag{6.61}$$

Using these results in the K_{ij} expressions, one obtains:

$$K_{ij} = \begin{cases} (\text{constant})(r_i^{-n'} + r_j^{-n'}) & \text{diffusion-controlled} \\ (\text{constant})(r_i^{m'} + r_j^{m'}) & \text{coalescence-controlled} \end{cases} \tag{6.62}$$

where $\ln(4D_{ij}\theta/R_{ij})$ is assumed to be relatively constant.

With these expressions, the continuous version of Eq. 6.51 can be written as:

$$\frac{\partial n(v,t)}{\partial t} = \frac{K}{2} \int_0^v n(\bar{v},t)n(v - \bar{v},t)[\bar{v}^{m'} + (v - \bar{v})^{m'}]d\bar{v}$$

$$- Kn(v,t) \int_0^\infty n(\bar{v},t)(v^{m'} + \bar{v}^{m'})d\bar{v} \tag{6.63}$$

where $n\,dv$ is the number of crystallites per unit substrate surface area having a volume between v and $v + dv$, and K is a constant, independent of size. Equation 6.62 has been used in 6.51 for K_{ij}. Integrating this equation over all sizes, one obtains:

$$\frac{dN_u}{dt} = -\frac{\bar{K}}{2} \int_0^\infty \int_0^\infty n(v,t)n(\bar{v},t)[\bar{v}^{m'} + v^{m'}]dv d\bar{v} \tag{6.64}$$

where N_u is the total number of crystallites per unit area. A similarity transformation of Eq. 6.64 yields the following expression for the rate of change of the total surface area of crystallites per unit substrate area:

$$\frac{dS}{dt} = -KS^{4-3m}$$

$$= -KS^n \tag{6.65}$$

where $m = -n'/3$ when sintering is diffusion-controlled, and $m = m'/3$ when sintering is coalescence-controlled. These are the sintering kinetics for the mechanism of crystallite migration. According to Eq. 6.61, m' can assume a value of 0, -1, or -2 when the growth is diffusion-controlled, giving an n value of 4, 5, or 6, respectively. When the growth is coalescence-controlled, n can assume a value of 2 or 3. It is seen then that depending on the expression used for the size dependence of K_{ij}, the value of n can be anywhere from unity to more than ten. In general, however, the value of n appears to be smaller for the case of coalescence-control than for diffusion-control. An Arrhenius temperature dependence can be assumed for K.

It is of interest at this point to compare the size-dependence of surface diffusivity for crystallite migration obtainable from the theoretical models with the expressions used by Ruckenstein and Pulvermacher for the size dependence of the diffusivity [m' of 0, -1, or -2]. According to the results of the theoretical models (Kaschiev 1979), m' can assume a value ranging from 0 to -5. On the other hand, the exponential size dependence of the surface diffusivity suggests that m' can assume a range of values. It is conceivable then that the value of m' and therefore that of n in Eq. 6.65 can be wide-ranging when the mechanism is crystallite migration. The mode of crystallite migration dictates the size-dependence of the diffusivity and this size-dependence complicates the *a priori* determination of the value of n in Eq. 6.65 (Lee and Ruckenstein 1983).

6–5   RELATIVE IMPORTANCE AND SINTERING BEHAVIOR

The basic question regarding sintering mechanisms is whether the crystallites migrate as such or in the form of ad-atoms. Energetically, the question is whether it takes more energy to detach an atom from the edges of a crystallite than to move a crystallite. A quantitative conclusion regarding the relative importance of the mechanism for a given system can be reached by examining the ratio of diffusivities, $D_s/D(r)$. Because of the dependence of $D(r)$ on the mode of crystallite migration, it is difficult to tell exactly under what conditions one mechanism is dominant over the other unless detailed knowledge of the mode of crystallite migration is available for the system under consideration. As the section on surface diffusivity reveals, the activation energy for the surface diffusion of ad-atoms is proportional to the sublimation energy. The activation energy for the surface diffusion of crystallite is dependent on crystallite size. While the mode of crystallite migration dictates the form of this size dependence, it is readily seen that the activation energy increases with increasing size. Therefore, sintering is more likely to be dominated by crystallite migration when the crystallite size is small. The activation energy for crystallite migration is proportional to the activation energy between an atom and the substrate multiplied by a function of crystallite size, which in turn depends on the mode of crystallite migration. It follows then that crystallite migration, as opposed to atom migration is more likely to take place when the level of interaction between catalyst and substrate is weaker. In fact, it has been observed experimentally that Pt crystallites as large as 20 nm migrate (Chu and Ruckenstein 1977) on a carbon substrate, although undoubtedly the migration was aided by gases evolving from the substrate.

Oxides and chlorides have a lower melting point and lower sublimation energy than the metals. Therefore, these compounds are more likely to sinter via the mechanism of atom migration than the metals. It is known that an oxidizing atmosphere is usually more conducive to sintering than inert or reducing atmospheres. This can be explained readily if the species being transported in the oxidizing atmosphere is the oxide, whereas it is the metal in the other atmospheres.

It should be recognized that the most substantive assumption made in deriving the sintering kinetics is that the surface is planar. In typical supported catalysts, however, the surface is not planar but rather curved, and there are many crevices and cracks. These nonideal surfaces and boundaries serve as physical barriers against the movement of atoms or crystallites. It has been shown experimentally (Wynblatt 1980) that sintering is localized when the radius of curvature for a curved surface is less than approximately four times the crystallite size. Experiments were carried out on a sinusoidal surface and when the radius of curvature was less than four times the crystallite size, sintering was localized in the valleys of the surface. It can be concluded therefore that the curved surface is essentially planar for crystallites whose size is less than one fourth of the radius of the curvature. Since the typical pore size distribution for supported catalysts is in the range of 1 to 100 nm, undoubtedly some crystallites "see" planar surfaces while others see curved surfaces. Those that see curved surfaces should grow faster in the valleys than those that see planar surfaces, while those in the intermediate range should grow at a rate intermediate between these two extremes. Therefore, the sintering kinetics obtained for planar surface, when applied to supported catalysts, describe sintering averaged over various nonideal surfaces.

6–6   REACTOR POINT EFFECTIVENESS

Now that the kinetics are known for the change of exposed surface area due to sintering, the results of Section 6–2 can be used to arrive at the reactor point effectiveness. The overall pellet effectiveness derived earlier can be used for this purpose. However, the term S_r/S_0 has to be expressed in terms of known quantities before it can be used for the reactor point effectiveness.

The active surface area can be reduced, not only by chemical deactivation, but also by sintering. While only the active surface area is affected by chemical deactivation, active as well as deactivated surface area is affected by sintering. In order to obtain an expression for the rate at which the active surface area is reduced by both chemical deactivation and sintering, the surface area ratio S_r/S_0 must be written as follows:

$$\frac{S_r}{S_0} = \left(\frac{S_t}{S_0}\right)\left(\frac{S_r}{S_t}\right)$$

(6.66)

The sintering kinetics provide the time dependence of (S_t/S_0):

$$\frac{d}{dt}\left(\frac{S_t}{S_0}\right) = -K_a \left(\frac{S_t}{S_0}\right)^n \exp[m'(S_t/S_0)] \qquad \text{atom migration}$$

$$= -K_c \left(\frac{S_t}{S_0}\right)^{n_c} \qquad \text{crystallite migration}$$

(6.67)

It has been established in Section 6–2, that under the assumption of uniform deactivation, the ratio of active surface area to total surface area should be the same whether or not chemical deactivation also reduces the surface area:

$$\frac{(S_r)_{\text{no sintering}}}{S_0}\left[=\left(\frac{S_r}{S_0}\right)_p\right] = \frac{S_r}{S_t}$$

(6.68)

where the term $(S_r/S_0)_p$ denotes the surface area ratio when the surface area is reduced only by chemical deactivation. The rate at which the active surface area is reduced only by chemical deactivation can be obtained from:

$$\frac{d}{dt}\left(\frac{S_r}{S_0}\right)_p = -\frac{1}{Q}r_p$$

(6.69)

where the rate of chemical deactivation r_p can be expressed in terms of $(S_r/S_0)_p$. Here, Q is the poisoning capacity of the pellet in moles of poisoning species per unit pellet volume. Using Eq. 6.68 in Eq. 6.66, one gets:

$$\frac{S_r}{S_0} = \left(\frac{S_t}{S_0}\right)\left(\frac{S_r}{S_0}\right)_p$$

(6.70)

where (S_t/S_0) and $(S_r/S_0)_p$ are given by Eqs. 6.67 and 6.69, respectively.

Take as an example, independent chemical deactivation due to the chemisorption of poisoning species. The kinetics (Chapter 5) can be represented by:

$$r_p = k_p'NC_v$$

$$= k_pN\left\{\frac{(C_r/C_{t0})_p}{[1 + G(c)]^n}\right\} = k_pN\left\{\frac{(S_r/S_0)_p}{[1 + G(c)]^n}\right\}$$

(6.71)

where $k_p = k_p' C_{t_0}$

C_{t_0} = total number of sites per pellet volume at time zero
C_r = number of active sites/pellet volume
k_p' = chemisorption rate constant
N = concentration of poisoning species
C_v = number of vacant active sites/pellet volume $(= C_r/[1 + G(c)]^n)$

Here, the number of vacant sites C_v has been re-expressed in terms of surface sites per unit volume of pellet, and the subscript p denotes that the active area reduction is solely due to poisoning.

The relationship of Eq. 6.70 is also applicable to the case of shell-progressive chemical deactivation since $(S_r/S_0)_p = 1 - \gamma$, where γ is the fraction of catalyst completely deactivated. The value of γ is the same under the assumption of uniform sintering whether or not the pellet is also sintered. Here, the time dependence of γ can be determined from:

$$\frac{d}{dt}(Q\gamma) = r_p \tag{6.72}$$

The reactor point effectiveness can now be readily determined from the overall pellet effectiveness developed earlier together with the known time dependence of the change of active surface area. Under the assumptions of negligible external mass transfer resistance and an isothermal pellet, the reactor point effectiveness is simply the pellet effectiveness multiplied by (k_s/k_b), where k_s and k_b are the rate constants evaluated at the pellet surface and bulk-fluid temperatures:

$$\Lambda = \left(\frac{k_s}{k_b}\right) E \tag{6.73}$$

The reactor point effectivenesses thus obtained for various cases are summarized in Table 6.4. Take as an example the first entry in Table 6.4, which is for diffusion-free reactions. Since the only transport resistance to consider is in heat transfer across the pellet-bulk fluid interface, the reactor point effectiveness is simply given by:

$$\Lambda = \left(\frac{k_s}{k_b}\right) h(S_r/S_0) \tag{6.74}$$

Since the ratio of rate constants can be approximated by:

$$\left(\frac{k_s}{k_b}\right)^{1/2} = 1 + \frac{1.2 E_a}{2 R_g T_b}(T_s/T_b - 1) \tag{6.75}$$

Table 6.4   Reactor Point Effectiveness: Uniform Sintering $r_c = kf(C)$

	Λ
I Diffusion-free reaction	$\dfrac{h(S_r/S_0)}{1 - \dfrac{1.2(-\Delta H)Lk_bf(C_b)h(S_r/S_0)}{2h_T R_g T_b^2}}$
II Diffusion-affected reaction and uniform chemical deactivation	$\dfrac{\dfrac{1}{Lf}\left[\dfrac{2D_eh(S_r/S_0)}{k_b}\right]^{1/2}\left[\displaystyle\int_{C_c}^{C_b} f(C)dC\right]^{1/2}}{1 - \dfrac{1.2E_a(-\Delta H)[2D_ek_bh(S_r/S_0)]^{1/2}}{2h_T R_g T_b^2}}$
III Diffusion-limited reaction and general case of chemical deactivation (for $\bar{S}_r/\bar{S}_0 > 0.5$)	Substitute $h(\bar{S}_r/\bar{S}_0)$ for $h(S_r/S_0)$ in the above expression.
IV Diffusion-affected reaction and shell-progressive chemical deactivation	$\Lambda = \dfrac{D_e(C_b - C_d)}{k_bf(C_b)L^2\gamma}$ $C_b - C_d = \left(\dfrac{L\gamma}{D_e}\right)\left[2D_eh(S_t/S_0)k_b\displaystyle\int_{C_c}^{C_d} f(C)dC\right]^{1/2}B$

$$B = \exp\left\{\left(\frac{E_a}{2R_gT_b}\right)\left[\frac{(-\Delta H)D_e(1+(\mathrm{Bi})_h\gamma)(C_b - C_d)}{h_T T_b L\gamma + (1+(\mathrm{Bi})_h\gamma)(-\Delta H)D_e(C_b - C_d)}\right]\right\}$$

The following relationships are used for γ and the surface area ratios:

$$\frac{S_r}{S_0} = \left(\frac{S_t}{S_0}\right)\left(\frac{S_r}{S_0}\right)_p$$

$$\frac{d}{dt}\left(\frac{S_r}{S_0}\right)_p = -\frac{1}{Q}r_p$$

$$\frac{d}{dt}\left(\frac{S_t}{S_0}\right) = \begin{cases} -K_a(S_t/S_0)^n\exp[m'(S_t/S_0)] & \text{atom migration} \\ -K_c(S_t/S_0)^{n_c} & \text{crystallite migration} \end{cases}$$

For the entry III, $\bar{S}_t/\bar{S}_0 = S_t/S_0$

$$\text{and } \frac{d}{dt}(\bar{S}_r/\bar{S}_0)_p = -\frac{1}{Q}\bar{r}_p$$

For the entry IV, $\dfrac{d\gamma}{dt} = \dfrac{r_p}{Q}$

$$\text{and } \gamma = 1 - (S_r/S_0)_p$$

and the temperature ratio in Eq. 6.75 is given by:

$$T_s/T_b - 1 = \frac{(-\Delta H)L\Lambda kf(C_b)}{h_T T_b} \tag{6.76}$$

which results from the heat balance across the pellet-bulk fluid interface, use of Eqs. 6.75 and 6.76 in Eq. 6.74 yields the first entry in Table 6.4. While the approximation of Eq. 6.75 has been used for the first three entries in the table, this approximation has not been made for the reactions affected by shell-progressive chemical deactivation. The only difference when compared with the reactor point effectiveness obtained in Chapter 5 for this case is the inclusion of $(S_t/S_0)^{1/2}$ in the expression for $(C_b - C_d)$. The expressions for the time dependence of γ and surface area ratios, which are necessary for the calculation of the reactor point effectiveness, are also given in Table 6.4.

Summary

With the possible exception of sintering, the physical deactivation of a catalyst is not well understood. Nevertheless, the effects of physical deactivation can still be quantified provided that the time progression of the change of active surface area is known. Because there is a better understanding of sintering compared with the other types of physical deactivation, emphasis has been placed on sintering. The theory behind the sintering kinetics is perhaps as sound as that behind the chemical kinetics of heterogeneous reactions. The weakness is in the application of the sintering kinetics derived for flat substrate to a porous supported catalyst. A similar weakness in chemical kinetics is the treatment of a heterogeneous surface as homogeneous.

In many reactions, sintering is as important as chemical deactivation. For those reactions affected by both sintering and chemical deactivation, reactor point effectivenesses have been developed for use in reactor design, which is treated in Part III.

NOTATION

a	distance for a single jump
b	number of coordinate directions in which jumps may occur
b_{ij}	rate of collision between crystallites composed of i and j units
$(Bi)_h$	Biot number for heat
C_b	bulk fluid concentration of key species
C_c	pellet center concentration
C_d	concentration at the boundary between completely deactivated outer region and fresh inner core in pellet

C_i, C_j	crystallites of i, j atoms per surface area
C/C_0	fraction of ad-atoms on surface
D	diffusivity of crystallite
D_e	effective diffusivity of key species in catalyst pellet
D_s	surface diffusivity of atom or molecule
D_1	surface diffusivity of ad-atom on substrate
D_{10}	pre-exponential factor for the surface diffusivity D_1
D_{ij}	relative diffusivity of a crystallite i with respect to that of a crystallite j
E	overall pellet effectiveness defined by R_G/r_c
E_n	E when sintering is absent
ΔE_d	binding energy between ad-atom and substrate
ΔE_f	energy required to detach an atom from crystallite
f	crystallite size distribution
f_r	jump frequency
$f(C)$	concentration dependence of intrinsic rate of reaction
$G(C)$	product inhibition term
$h(S_r/S_0)$	dependence of catalyst activity on surface area ratio
h_T	film heat transfer coefficient
$(-\Delta H)$	heat of reaction
I	integral defined in Eq. 6.6
J_i	number of atoms crossing the crystallite-substrate interface per time
J_s	number of atoms attaching per time at crystallite perimeter by surface diffusion
k	reaction rate constant
k_B	Boltzmann constant
k_d	reaction rate constant at the boundary between completely deactivated outer shell and fresh inner core in a pore-mouth poisoned pellet
k_p	rate constant for poisoning reaction
k_s	reaction rate evaluated at pellet surface temperature
K_{ij}	rate constant defined in Eq. 6.50
l	screening distance at which $n = \bar{n}$; length of the part of the pellet completely deactivated
L	characteristic dimension of catalyst pellet; screening distance
n	surface concentration of ad-atom; constant
n'	n at crystallite edge
$\bar{n}$	n for crystallites of critical size
n_c	number of atoms making up a crystallite
n_j	number of jumps
n_r	concentration of edge atoms associated with a crystallite
$\bar{n}^{eq}$	n in equilibrium with infinite-sized crystallite
N_u	total number of crystallites
Q	poison capacity of catalyst pellet
r	crystallite size
$\bar{r}$	critical crystallite size

r_c	intrinsic rate of chemical reaction
r_p	intrinsic rate of poisoning reaction
R	distance between two atoms; radial coordinate
R_g	gas constant
R_G	global rate (actual rate)
R_p	global rate of poisoning reaction
R_{ij}	$r_i + r_j$
S	exposed surface area of catalyst crystallite
S_0	S for fresh catalyst
S_r	active surface area
S_t	total exposed surface area at time t
t	time
T	temperature
T_b	bulk-fluid temperature
T_s	pellet surface temperature
V_m	volume of an ad-atom
X	net distance covered by atom
z	pellet coordinate

Greek Letters

α_1	shape factor for volume given by $(2 - 3 \cos \theta + \cos^3 \theta)/4$ (see Figure 6.5)
β	jump frequency for detachment of atoms from crystallites
β'	jump frequency for attachment of atoms over the crystallite-substrate interface
γ	fraction of catalyst deactivated
ϵ	quantity defined in Eq. 6.42
η_G	generalized effectiveness factor
Λ	reactor point effectiveness defined in Eq. 6.73
μ	chemical potential
μ_0	μ for infinite-sized crystallite
ν_p	vibration frequency at crystallite edge sites
ν_0	vibration frequency on substrate sites
ρ	fluid density
σ	surface energy
ϕ	Thiele modulus
ϕ_p	Thiele modulus for poisoning reaction
ϕ_{kj}	quantity defined by Eq. 6.57
ϕ_G	generalized modulus in Eq. 6.7
$(\phi_G)_m$	modified ϕ_G defined by Eq. 6.7

Subscripts

b	bulk fluid
c	pellet center

d   at the boundary between completely deactivated outer shell and fresh
           inner core in catalyst pellet

s   pellet surface

Superscript

‾   average; at pellet surface

PROBLEMS

6.1. The intrinsic rate of a catalytic reaction measured using fine powders is
10^{-7} mole/(min · volume of powder). These fine powders are compressed
into pellets and the observed rate changes to one tenth of the intrinsic rate
under identical reaction conditions. After 100 hrs of use, the observed rate
is 5×10^{-9} mole/(min · volume). What is the ratio of active catalyst surface
area to the initial active surface area? Is the assumption of negligible effects
of physical deactivation on the concentration profile justified? If so, what
is E/E_n?

6.2. Show for a structure-insensitive, first-order reaction that the last entry in
Table 6.1 reduces to:

$$\frac{E}{E_n} = \left(\frac{S_t}{S_0}\right)^{1/2} \left\{ \frac{1 + L\gamma(k_d/D_e)^{1/2}}{1 + L\gamma\left[\frac{k_d(S_t/S_0)}{D_e}\right]^{1/2}} \right\}$$

and Eq. 6.12 to:

$$C_d = \frac{C_b}{1 + L\gamma[k_d(S_t/S_0)/D_e]^{1/2}}$$

when the reaction is diffusion-limited. Show also that E/E_n approaches unity
when the reaction is severely diffusion-limited and the extent of shell-progressive poisoning is significant. Give physical reasons for this behavior. The
dominance of shell-progressive chemical deactivation over physical deactivation should be clear from this example.

6.3. If the root mean square distance is used for the distance that a crystallite
travels, Eq. 6.24 reduces to:

$$\bar{X} = (D_c t)^{1/2}; \qquad \bar{X} = \langle \bar{X}^2 \rangle^{1/2}$$

where D_c is the crystallite diffusivity. Suppose that the diffusivity of the
crystallite is related to the surface diffusivity of an ad-atom by:

$$D_c = 0.3\ D_s'\ (r/a)^4; \qquad a = \text{interatomic spacing},\ D_s' = 2bD_s$$

If D_s' is given by $D_s' = 300\ e^{-40\ \text{kcal}/RT}$ (cm^2/sec), how far does a 2-nm-radius crystallite travel in 1 hour at 770°K? The data are those used by Richardson and Crump (1979) for their study of nickel on silica.

6.4. Chen and Schmidt (1978) studied the sintering of Pt on amorphous silica. The average size was in the range of 5 to 15 nm at 1000°K. Calculate the range of ϵ (Eq. 6.42). Use σ of 2100 erg/cm^2.

6.5. Show that when $\epsilon < 0.1$, the approximation of Eq. 6.41 is almost the same as $\epsilon(1 - 1/\rho)$ in the range of $0.4 < \rho < 2.0$ where $\rho = r/\bar{r}$. The typical size distribution of crystallites belongs to this range of ρ.

6.6. Sintering data in terms of average crystallite size obtained at 920°K for Pt on silica are given below:

Average Size (nm)	Time (hr)
11.4	¼
13	½
13.5	1
16.5	3
20	8

If sintering takes place via the mechanism of atom migration and $\bar{r} = r_{\text{average}}/1.03$, determine the sintering kinetics for the average crystallite size. Can you tell the difference between the case of surface-diffusion controlling and that of interface-transfer controlling?

6.7. Derive the result given in Table 6.3 for the case of surface diffusion aided by the formation of oxides or chlorides. The growth rate, when the surface diffusion controls, is given by:

$$\frac{dr}{dt} = -\frac{A_D}{r^3}\left(1 - \frac{r}{\bar{r}}\right)\exp\left(\frac{2\sigma V_m}{\bar{r}k_B T}\right) \tag{1}$$

$$A_D = \frac{D_1 V_m^2 \sigma n_r^{eq}(\beta/\beta')}{\alpha_1 k_B T \ln l} \tag{2}$$

a. Suppose the metal atoms are transported over the substrate in the form of oxide or chloride molecules. The terms that might be significantly affected by this assumption are $(D_1\beta/\beta')$ and n_r^{eq}. The quantity n_r^{eq} for the molecules can be obtained by equating the vaporization and return fluxes of oxide or chloride molecules:

$$n'_{eq}v'_s\ \exp\{-\Delta E_{sv}/k_B T\} = X'P'_{eq}/(2\pi m'k_B T)^{1/2} \tag{3}$$

where primed quantities refer to the volatile oxides or chlorides and

X' = evaporation-condensation coefficient

ΔE_{sv} = energy required to transfer a molecule from the substrate to the vapor phase

m' = mass of an oxide or a chloride

$P'_{eq} = K_{eq}(P_{gas})^q$ for $M(s) + qO_2 \overset{k_{eq}}{=} MO_{2q}(g)$

The quantity $(D_1\beta/\beta')$ can be written as:

$$(D_1\beta/\beta') = a'^2 v'_p \exp(-\Delta E_f/k_B T) \tag{4}$$

Substituting Eqs. (3) and (4) into (2), one has:

$$A_D = \frac{a'^2 \sigma V_m^2 X' K_{eq}(P_{gas})^q}{a_1 k_B T \ln l(2\pi m' k_B T)^{1/2}} \exp\{(\Delta E_s - \Delta E_f)/k_B T\}$$

b. Show that Eq. (1), when used in the number balance equation, yields:

$$\frac{dS}{dt} = K'' S^5 \exp(m/S)$$

6.8. Show that Eq. 6.64 leads to:

$$\frac{dS}{dt} = -KS^{4-3m'}$$

a. Show that:

$$\frac{dN_u}{dt} = -b_1 N_u^{2-m'}$$

$$b_1 = \frac{K}{2} \phi^{m'} \frac{A^{m'-2}}{B^{m'}} \int_0^\infty \int_0^\infty \Psi(\eta)\Psi(\bar{\eta})(\eta^{m'} + \eta^{-m'}) d\eta d\bar{\eta}$$

$$\phi = \int_0^\infty vn(v,t) dv$$

Use the transformation $\eta = v/\bar{v}(t)$ and $n(v,t) = h(t)\Psi(\eta)$.

b. Show that:

$$S = b_3 N_u^{1/3}$$

since $S = b_2 \int_0^\infty v^{2/3} n(v,t) dv$

where b_2 is a geometric shape factor.

c. Show then that: $\qquad \dfrac{dS}{dt} = -KS^{4-3\,m'}$

where K is a composite of constants.

6.9. Determine the sintering kinetics for the following data:

$T(°C)$	S_t/S_0	Time (hr)
700	0.9	2
	0.81	4
	0.72	7
	0.59	11
750	0.6	2
	0.42	4
	0.32	6
	0.23	8
800	0.38	1
	0.23	3
	0.10	7
	0.07	9

Can you determine whether the sintering takes place via crystallite migration or atom migration?

6.10. Given the sintering kinetics and the kinetics of chemical deactivation, how can one calculate the time dependence of (S_r/S_0)?

6.11. Derive the reactor point effectiveness given in the second row of Table 6.4 for a reaction affected by diffusion, uniform chemical deactivation, and sintering.

a. Under the assumptions of an isothermal pellet and negligible external mass transfer resistance, show that the pellet effectiveness is:

$$\frac{1}{Lf(C_b)}\left[\frac{2D_e h(S_r/S_0)}{k_s}\right]^{1/2}\left[\int_{C_c}^{C_b} f(c)dc\right]^{1/2}$$

b. Use Eqs. 6.73 and 6.75 to arrive at the results.

c. Calculate the reactor point effectiveness as a function of S_r/S_0 for the following conditions:

$\phi = L(k_b/D_e)^{1/2} = 5$ $\qquad\qquad (-\Delta H) = 30$ kcal/mole
$T_b = 600°K$ $\qquad\qquad\qquad\quad C_b = 10^{-6}$ mole/cm^3
$E_a = 20$ kcal/mole $\qquad\qquad\quad r_c = kC$
$h_T = 3.2 \times 10^{-4}$ cal/sec cm$^{20°}$K $\qquad D_e = 10^{-2}$ cm^2/sec
$L = 1$ cm

Assume that the reaction is structure-insensitive.

d. Repeat the calculation for a Thiele modulus of 1 this time and plot both results on a Λ versus S_r/S_0 plane. What can you conclude (note that $C_c = C_b/\cosh(\phi_G)_m$)?

6.12. How does one calculate the reactor point effectiveness for a reaction affected by shell-progressive deactivation (fourth row in Table 6.4) at $t = 0$? Note that $\gamma = 0$ at $t = 0$ and therefore, Λ becomes infinite at time zero.

REFERENCES

Chen, M. and L.D. Schmidt, J. Catalysis, *55*, 348 (1978).

Chu, Y.F. and E. Ruckenstein, Surface Sci., *67*, 17 (1977).

Dalla Betta, R.A., R.C. McCune and J.W. Sprys, Ind. Eng. Chem. Prod. Res. Dev., *15*, 169 (1976).

Flynn, P.C. and S.E. Wanke, J. Catalysis, *34*, 390 (1974).

Gjostein, N.A., in Burke et al. (eds.), *Surfaces and Interfaces,* Syracuse University Press, Syracuse (1967).

Jones, H., Metal Sci. J., *5*, 15 (1971).

Kashchiev, D., Surface Sci., *86*, 14 (1979).

Lee, H.H., J. Catalysis, *63*, 129 (1980).

Lee, H.H., Chem. Eng. Sci., *36*, 950 (1981).

Lee, H.H. and E. Ruckenstein, Cat. Rev. (1983).

Lifshitz, I.M. and V.V. Slyozov, J. Phys. Chem. Solids, *19*, 35 (1961).

Richardson, J. T. and J.G. Crump, J. Catalysis, *57*, 417 (1979).

Ruckenstein, E. and B. Pulvermacher, AIChE J., *19*, 356 (1973).

Ruckenstein, E. and D.B. Dadyburjor, Thin Solid Films, *55*, 89 (1978).

Somorjai, G.A., *Principles of Surface Chemistry,* Prentice-Hall, Englewood Cliffs, N.J. (1972).

Wynblatt, P. and N.A. Gjostein, Prog. Solid State Chem., *9*, 21 (1975).

Wynblatt, P., Acta Metallurgica, *24*, 1175 (1976).

Wynblatt, P., in Bourdon (ed.), *Growth and Properties of Metal Clusters,* p 15, Elsevier Co., Netherlands (1980).

CHAPTER 7

Multiphase Reactions

7–1 INTRODUCTION

The reaction between gaseous and liquid reactants catalyzed by a solid is called a multiphase reaction in the context of heterogeneous catalytic reactions. A multiphase reactor is usually used when the liquid reactant is too nonvolatile to vaporize. For instance, a reactor of this type is used to hydrogenate petroleum feeds of wide boiling-point range for desulfurization, cracking, or straight hydrogenation. A fraction of these feeds has too high a boiling point to vaporize. The two most common reactors for multiphase reactions are trickle-bed and slurry reactors. The trickle-bed is a fixed bed packed with catalyst pellets to which both gaseous and liquid reactants are fed downward cocurrently, usually in the downward direction. The gas phase is continuous throughout the bed, whereas the liquid phase is dispersed as the liquid trickles down the bed. While many different configurations and flow regimes can result depending on the mode of feeding (countercurrent or cocurrent, upward or downward) and the feed rates, interest here is with the downward-flowing trickle-bed operated in the trickle-flow regime where the pellets are completely wetted and covered with a thin liquid film. A slurry reactor is one in which catalyst particles are suspended in a liquid reactant by agitation and gas flow, and the gaseous reactant is fed to the reactor through a suitable distributor located at the bottom. The catalyst particles are very small (~100 micron) and the heat capacity of the liquid is high so that isothermal conditions usually prevail.

The presence of a liquid phase and a liquid-solid interface in multiphase reactors results in added transport resistances. For instance, the effective diffusivity in liquid-filled pores (of the order of 10^{-4} to 10^{-5} cm²/sec) is much smaller than that in gas-filled pores (of the order of 10^{-2} cm²/sec). The solubility of the gaseous reactant is an important factor since the gaseous reactant has to be dissolved into the liquid reactant for the reaction to take place on the catalyst surface. As emphasized in Chapter 4, the Biot number for heat is much larger than the Biot number for mass for liquid-solid systems; the opposite is true for gas-solid systems. Therefore, the major external resistance lies in the mass transport, and the pellet is not necessarily isothermal. In many cases, however, the equilibrium gas concentration in the liquid is quite low and, thus, the heat evolved is small in spite of high heats of reaction. The pellet can be considered isothermal in such a case, as in the hydrodesulfurization of heavy petroleum feeds. On the other hand, high concentrations and heats of reaction as in the hydrocracking of petroleum fractions can result in nonisothermal pellets. The catalysts used for multiphase reactions are not immune from catalyst deactivation, typically caused by coking and metals deposition. High molecular weight organometallic compounds present in petroleum feeds, in particular petroleum residues, split off metals (typically Ni and V) upon cracking, eventually leading to pore-blocking.

In this chapter, the effects of transport resistances on the rate of reaction will be examined, so that the global rate suitable for direct inclusion in reactor conservation equations can be derived. Because of the important role that the mass transport plays in the overall kinetics, this subject will be taken up first.

7–2 MASS TRANSFER COEFFICIENTS

The mass transfer resistances in a multiphase reactor at the gas-liquid and liquid-solid interfaces are shown in Figure 7.1 (Smith 1981). In the case of a slurry reactor, the continuous gas-phase shown in the figure should be replaced by a gas bubble. If equilibrium exists at the gas-liquid boundary,

$$C_{ig} = F(C_{iL}) \tag{7.1}$$

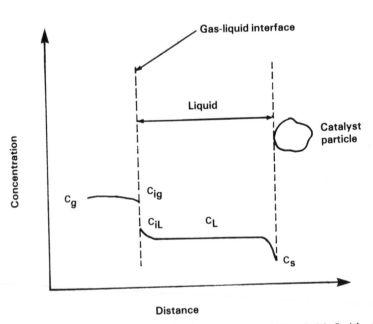

Figure 7.1 Concentration profiles in a multiphase reactor. (After J. M. Smith, *Chemical Engineering Kinetics*, 3rd ed., © 1981; with permission of McGraw-Hill Book Company, New York.)

where the function F relates C_{ig} to C_{iL}. If Henry's law applies, this reduces to:

$$C_{ig} = HC_{iL} \tag{7.2}$$

The resistance at the gas-liquid interface is dominated by the liquid side of the interface. When the gaseous reactant is pure, as in hydrogenation, there is no resistance to diffusion from the bulk gas to the gas-liquid interface. Even when the gas phase is a mixture, the gas-side transfer coefficient k_g is much larger than k_L/H. Also, for slightly soluble gases, the resistance to mass transfer is predominantly on the liquid side. Therefore, C_{ig} can be set equal to the bulk concentration C_g and the above relationship can be written as:

$$C_g = HC_{L_{eq}} \equiv HC_{eq} \tag{7.3}$$

Under these conditions the global rate based on catalyst volume can be written as:

$$R_G = \frac{k_L a_L}{\epsilon_s}(C_{eq} - C_L) = k_c a_c(C_L - C_s) \tag{7.4}$$

where a_L is the gas-liquid interfacial area per unit volume of liquid (bubble-free in the case of slurry reactor), and a_c is the reciprocal of the characteristic length of a pellet, ϵ_s is the ratio of catalyst volume to liquid volume, and k_L and k_c are the appropriate mass transfer coefficients.

Several correlations (Calderbank and Moo-Young 1961; Misic and Smith 1971; Juvekar and Sharma 1973) for the mass transfer coefficient k_L in a slurry reactor are available. One correlation (Calderbank and Moo-Young 1961) that accounts for the energy dissipation due to agitation by the impeller gives:

$$k_L = 0.592\, D_A^{1/2}(\sigma/\nu)^{1/4} \tag{7.5}$$

where D_A = molecular diffusivity of reactant in liquid, cm²/sec
  ν = kinematic viscosity, cm²/s
  σ = energy dissipation rate per unit mass of liquid, erg/s·g

$$\sigma = \frac{N_p \rho_L N^3 D_I^5}{W}\,\phi \tag{7.6}$$

where D_I = impeller diameter, cm
  N = impeller speed, rps
  W = mass of liquid in slurry, g
  $N_p = \dfrac{P}{\rho_L N^3 D_I^5}$ (~10) P = power input, erg/s

  ϕ = a correction factor ($0 < \phi < 1$) to account for the decrease in the energy dissipation rate due to gas bubbles. For $Q/ND_i^3 < 0.035$, ϕ $= 1 - 12.6\ Q/ND_i^3$ where Q is the gas flow rate (cm³/s)

ρ_L = liquid density

In the absence of mechanical agitation and for bubbles whose diameter is less than 2.5 mm (the usual size range for slurry reactors), the following correlation (Calderbank 1958) is available:

$$k_L \left(\frac{\mu_L}{\rho_L D_A} \right)^{2/3} = 0.31 \left(\frac{\Delta\rho\mu_L g}{\rho_L^2} \right)^{1/3} \tag{7.7}$$

where $\Delta\rho$ = difference in density between liquid phase and gas bubbles, g/cm³

μ_L = viscosity of liquid phase, g/cm·s

g = gravity, cm/s²

ρ_L = density of liquid, g/cm³

This correlation is for bubbles rising through the liquid phase because of a gravitational force. The basis for correlating k_c as a function of agitation speed and particle size is Kolmogoroff's theory (Levins and Glastonbury 1972) of isotropic turbulence. According to this theory, the Reynolds number (Re) is defined in terms of the energy dissipation rate, given by:

$$\text{Re} = \begin{cases} \left(\dfrac{\sigma d_p^4}{\nu^3} \right)^{1/2} & \zeta > d_p \\[4mm] \left(\dfrac{\sigma d_p^4}{\nu^3} \right)^{1/3} & \zeta < d_p \end{cases} \tag{7.8}$$

where the eddy size ζ is defined by:

$$\zeta = \left(\frac{\nu^3}{\sigma} \right)^{1/4} \tag{7.9}$$

Here, the energy dissipation rate σ is given by Eq. 7.6, and d_p is the particle size. The experimental data (Harriott 1962; Brian et al. 1969) for k_c are well correlated by plotting the Sherwood number ($k_c d_p/D_A$) against the Reynolds number given by Eq. 7.8, as shown in Figure 7.2.

For trickle-beds, correlations for $k_L a_L$ are in terms of the pressure drop in the reactor or the flow velocities. Illustrative of the latter type is the correlation by Goto and Smith (1975):

$$\frac{k_L a_L}{D_A} = \alpha_L \left(\frac{G_L}{\mu_L} \right)^{\eta_L} \left(\frac{\mu_L}{\rho_L D_A} \right)^{1/2} \tag{7.10}$$

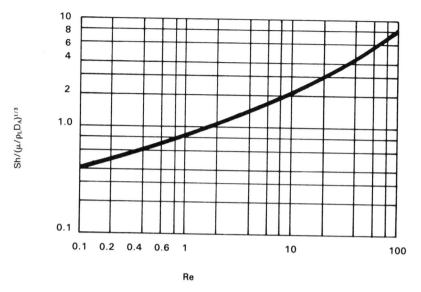

Figure 7.2   Correlation for the mass transfer coefficient, k_c. (After J. M. Smith, *Chemical Engineering Kinetics*, 3rd ed., © 1981; with permission of McGraw-Hill Book Company, New York.)

where D_A = molecular diffusivity of the diffusing component, cm^2/s
  G_L = superficial mass velocity of the liquid, $g/cm^2 \cdot s$
  μ_L = liquid viscosity, $g/cm \cdot s$
  $k_L a_L$ = liquid side mass transfer coefficient, $1/s$
  α_L = a correction factor ($\sim 7\, n_L^{-2}(cm)$), and n_L = 0.4 for granular particles.

This correlation applies to the trickle-flow regime; much higher values can be obtained at higher liquid and gas rates. The trickle-flow regime is usually for $G_L < 3 \times 10^3$ $g/cm^3 \cdot h$. A j factor correlation (Dharwadkar and Sylvester 1977) in the trickle-flow regime for the liquid-particle mass transfer coefficient k_c is:

$$j_D = 1.64\,(Re_L)^{-0.331}, \qquad 0.2 < Re_L < 2400 \qquad (7.11)$$

where $j_D = \dfrac{k_c a_c}{U_L a_t}\left(\dfrac{\mu_L}{\rho_L D_A}\right)^{2/3}$ (7.12)

  U_L = superficial velocity of liquid

  a_t = total external area of particles per unit volume of reactor
  $Re_L = d_p U_L \rho_L / \mu_L$

It is notable that both k_L and k_c in the trickle-flow regime are independent of the gas flow rate. Satterfield and coworkers (1978) also gave correlations independent of the gas flow rate in the trickle-flow regime. Further details on mass transfer coefficients can be found in the book by Smith (1981), from which the bulk of this section is derived.

7-3 TRANSPORT EFFECTS AND GLOBAL RATES

As pointed out earlier, the major external resistance is that of mass transfer, and therefore, the effect of external heat transfer can be neglected. Furthermore, internal (intraparticle) transport effects can be neglected in slurry reactors except under some unusual reaction conditions since the size of the catalyst particles is of the order of 100 microns. In trickle-beds, however, both the internal heat and mass transport effects can be important.

Consider first the slurry reactor. Since the internal resistances are negligible, the global rate can be written directly in terms of surface concentrations, and Eq. 7.4 becomes:

$$\frac{k_L a_L}{\epsilon_s} (C_{eq} - C_L) = k_c a_c (C_L - C_s) = kf(C_s) = R_G \tag{7.13}$$

where the intrinsic rate is represented by $kf(C)$. It follows from Eq. 7.13 that:

$$R_G = \left(\frac{1}{1/k_c a_c + \epsilon_s/k_L a_L}\right)(C_{eq} - C_s) \tag{7.14}$$

where C_s is the solution of:

$$\left(\frac{1}{1/k_c a_c + \epsilon_s/k_L a_L}\right)(C_{eq} - C_s) = kf(C_s) \tag{7.15}$$

Given the intrinsic rate expression $kf(C)$, Eq. 7.15 can be solved for the surface concentration C_s, which upon inserting into Eq. 7.14 yields the global rate in terms of C_{eq}, the rate constant, and the mass transfer coefficients. A pseudo-first-order rate expression is often adequate when the gaseous reactant dissolves slightly in the liquid reactant. In such a case, the global rate is simply:

$$R_G = \left(\frac{1}{1/k_c a_c + \epsilon_s/k_L a_L + 1/k}\right) C_{eq} \tag{7.16}$$

which follows from Eqs. 7.14 and 7.15. The relative magnitude of the reciprocals of the rate constant and mass transfer coefficients determines the relative importance of the three resistances. If one of the mass transfer coefficients is much smaller than the rate constant, for instance, the observed rate would appear to be almost independent of temperature, the controlling step for this case being the mass transfer resistance.

The existence of internal resistances complicates the analysis of transport effects for trickle-beds since the pellet cannot necessarily be assumed isothermal. Reactions in which the heat effect is negligible are considered first, and the case of a nonisothermal pellet will be treated in the following section. For arbitrary kinetics $kf(C)$, the internal, isothermal effectiveness factor (Chapter 4) is:

$$\eta_i(C_s) = \frac{1}{Lf(C_s)} \left[2D_e k \int_{C_c}^{C_s} f(C)dC \right]^{1/2} \tag{7.17}$$

where D_e is now the effective diffusivity of gas in liquid-filled pores. The global rate can be expressed as:

$$\frac{k_L a_L}{\epsilon_s} (C_{eq} - C_L) = k_c a_c (C_L - C_s) = \eta_i k f(C_s) = R_G \tag{7.18}$$

As in the slurry reactor, these expressions give for the global rate:

$$R_G = \left(\frac{1}{1/k_c a_c + \epsilon_s/k_L a_L} \right) (C_{eq} - C_s) \tag{7.19}$$

$$\left(\frac{1}{1/k_c a_c + \epsilon_s/k_L a_L} \right) (C_{eq} - C_s) = \eta_i(C_s) k f(C_s) \tag{7.20}$$

Equation 7.20 can be solved for C_s. This value of C_s in turn is used in Eq. 7.19 for the global rate. For a first-order reaction, Eqs. 7.19 and 7.20 can be combined to give:

$$R_G(C_{eq}) = \left[\frac{\tanh \phi/\phi}{1 + \phi \tanh \phi/B_{mL}} \right] k C_{eq} \tag{7.21}$$

where the Biot number for mass in terms of liquid-phase properties is given by:

$$B_{mL} = \frac{1}{D_e(1/k_c a_c + \epsilon_s/k_L a_L)} \left(= \frac{K_L L}{D_e} \right) \tag{7.22}$$

and the Thiele modulus ϕ is $L(k/D_e)^{1/2}$. Here, the characteristic pellet length L is equal to $1/a_c$. As detailed in Chapter 4, the center concentration C_c in Eq. 7.17 can be set to zero in the asymptotic region of strong diffusion effects, i.e., $\phi_G \geq 3$ where the generalized modulus is given by:

$$\phi_G = \frac{Lf(C_s)}{\left[\left(\frac{2D_e}{k} \right) \int_0^{C_s} f(C)dC \right]^{1/2}} \tag{7.23}$$

Otherwise, C_c is approximated by $C_s/\cosh \phi_G$.

While the treatment so far has been based on the assumption of completely wetted pellets, there are cases where the pellets are partially wetted. It has been found (Satterfield 1975; Herskowitz et al. 1979) that at low liquid velocities, a fraction of the pellet surface is not covered by liquid, i.e., partial wetting. This does not mean that the pores are free of liquid. The pores are still filled with liquid due to capillary forces, but the pore-mouth is in direct contact with gas rather than liquid. This partial wetting can take place when the Reynolds number defined by $Re = L\rho_L U_L / \mu_L$ is less than about 30. The fraction of the surface covered by liquid, q, is known to range from 0.6 to 1. It has been found (Tan and Smith 1980) that a linear combination of effectiveness factors gives a satisfactory result for the partially wetted pellet:

$$\eta_o = q\eta_L + (1 - q)\eta_G \qquad (7.24)$$

where the liquid-side overall effectiveness factor η_L and the gas-side counterpart η_G are given, for instance, for a first-order reaction by:

$$\eta_L = \frac{\tanh \phi / \phi}{1 + \phi \tanh \phi / B_{mL}} \qquad (7.25)$$

$$\eta_G = \frac{\tanh \phi / \phi}{1 + \phi \tanh \phi / HB_{mG}} \approx \frac{\tanh \phi}{\phi} \qquad (7.26)$$

The gas-side Biot number for mass (HB_{mG}) is usually very large so that its effect can be neglected as shown in Eq. 7.26. The net effect of partial wetting is a higher observed rate than would be expected from completely wetted pellets under identical conditions. Based on this fact, a criterion (Lee and Smith 1982) for partial wetting has been derived for a first-order reaction:

$$\frac{R'_G L^2 \rho_p}{D_e C_L} > \frac{\phi \tanh \phi}{1 + \phi \tanh \phi / B_{mL}} \qquad (7.27)$$

where R'_G = global rate per unit weight of pellet
C_L = bulk concentration of gaseous reactant in liquid
ρ_p = pellet density

If the indicated condition is satisfied, the pellets are partially wetted. The criterion, however, cannot say anything about the value of q. There appear to be no accurate correlations available for the prediction of q. An additional complication arises when a fraction of the liquid is vaporized due to the heat of reaction, as in hydro-cracking of petroleum feeds. In such a case, the pores are filled with both liquid and vapor. This problem has been considered by Dudukovic (1977).

The criteria of negligible transport effects are useful for the determination of intrinsic kinetics. Satterfield and coworkers (1969) give the following criterion for negligible mass transfer resistance:

$$\frac{R_G L}{K_L C_{eq}} < 0.15 \qquad (7.28)$$

where K_L is the overall liquid-side mass transfer coefficient. A tighter criterion by Lee and Smith (1982) for exothermic reactions is:

$$\frac{R'_G L \rho_p}{K_L C_L \epsilon_p} < \frac{\beta}{1-\beta} \tag{7.29}$$

where $\beta = (-\Delta H)D_e C_L/\lambda T_b$, and ϵ_p is the pellet porosity.

7–4 GLOBAL RATE FOR TRICKLE-BEDS: NONISOTHERMAL PELLETS

A pellet may not be isothermal when heat effects are significant as in hydrocracking. While complications may arise due to the vaporization of liquid, it will be assumed in what follows that the pellet is completely wetted. Steady-state, one-dimensional balance equations for a slab-like pellet can be written in dimensionless form as:

$$\frac{d^2u}{dx^2} = \phi^2 F(u,v) \tag{7.30}$$

$$\frac{d^2v}{dx^2} = -\beta\phi^2 F(u,v) \tag{7.31}$$

with $\dfrac{du}{dx} = \dfrac{dv}{dx} = 0$ at $x = 0$ $\tag{7.32}$

$$\frac{du}{dx} = B_{mL}(1-u), \quad \frac{dv}{dx} = B_{hL}(1-v) \qquad \text{at } x = 1 \tag{7.33}$$

where $u = C/C_L$, $v = T/T_b$, $x = z/L$, $B_{mL} = K_L L/D_e$
$B_{hL} = hL/\lambda$, $F(u,v) = r_c(C,T)/r_c(C_L,T_b) = f(u) \exp(\gamma(1-1/v))$
$\phi^2 = r_c(C_L,T_b)L^2/DC_L$, $\beta = (-\Delta H)D_e C_L/\lambda T_b$, $\gamma = E_a/R_g T_b$ $\tag{7.34}$

The notation can be found at the end of this chapter. Here, K_L is the overall mass transfer coefficient defined by:

$$\frac{1}{K_L a_c} = \frac{1}{k_c a_c} + \frac{\epsilon_s}{k_L a_L} \tag{7.35}$$

The major interest here is with an approximate expression for an overall effectiveness factor. Since a general result cannot be obtained for arbitrary kinetics, attention is restricted to a first-order reaction for which $F = u \exp(\gamma(1-1/v))$. The perturbation method of Pereira and co-workers (1979) is used for this purpose, but with a modification: the boundary condition of Eq. 7.33 for v is retained as such for the liquid-solid system under consideration, for B_{hL} is much larger than unity and β is much smaller than unity (0.1 ~ 0.001) such that B_{hL}/β will assume a large value. In the work of Pereira and co-workers, the boundary condition is transformed to $(B_{hL}/\beta)\epsilon(1-v)$ for the gas-solid system for which B_{hL}/β is finite.

Here, ϵ is a small perturbation parameter equal to β. Asymptotic expansions of the functions involved yield:

$$u \sim \sum_{i=0}^{\infty} \epsilon^i u_i(x), \qquad v \sim \sum_{i=0}^{\infty} \epsilon^i v_i(x) \tag{7.36}$$

$$F(u,v) \sim \sum_{i=0}^{\infty} \epsilon^i F_i(u_0, v_0, u_1, v_1, \ldots, u_i, v_i) \tag{7.37}$$

where

$$\begin{aligned}
F_0 &= F(u_0, v_0) \\
F_1 &= F_u(u_0, v_0)u_1 + F_v(u_0, v_0)v_2 + F_{uv}(u_0, v_0)u_1 v_1 \\
&\quad + [F_{uu}(u_0, v_0)u_1^2 + F_{vv}(u_0, v_0)v_1^2]/2
\end{aligned}$$

and $F_u(u_0, v_0)$ is $\partial F/\partial u$ evaluated at (u_0, v_0), etc. From the definition of effectiveness factor, it follows that:

$$\eta = \int_0^1 F(u,v)dx \tag{7.38}$$

and

$$\eta \sim \sum_{i=0}^{\infty} \epsilon^i \eta_i$$

$$\eta_i = \int_0^1 F_i(u_0, v_0, u_1 v_1, \ldots, u_i v_i)dx \tag{7.39}$$

When these expansions are substituted into Eqs. 7.30 and 7.31, the terms of equal order in ϵ gives for the ϵ^0 terms:

$$\frac{d^2 u_0}{dx^2} = \phi^2 F(u_0, v_0) \tag{7.40}$$

$$\frac{d^2 v_0}{dx^2} = 0 \tag{7.41}$$

$$u_0' = v_0' \qquad \text{at } x = 0 \tag{7.42}$$

$$u_0' = B_{mL}(1 - u_0), \qquad v_0' = B_{hL}(1 - v_0) \qquad \text{at } x = 1 \tag{7.43}$$

and for the ϵ^1 terms:

$$\frac{d^2u_1}{dx^2} = \phi^2 F_1 = \phi^2(u_1 + \gamma u_0 v_1) \tag{7.44}$$

$$\frac{d^2v_1}{dx^2} = -\phi^2 F_0(u_0, v_0) = -\phi^2 u_0 \tag{7.45}$$

$$u_1' = v_1' = 0 \qquad \text{at } x = 0 \tag{7.46}$$

$$u_1' = -B_{mL}u_1, \qquad v_1' = -B_{hL}v_1 \text{ at } x = 1 \tag{7.47}$$

where the prime denotes differentiation with respect to x. The zeroth-order solutions of Eqs. 7.40 and 7.41 are:

$$v_0 = 1 \tag{7.48}$$

$$u_0 = A \cosh \phi x; \qquad A = \frac{B_{mL}}{\phi \sinh \phi + B_{mL} \cosh \phi} \tag{7.49}$$

It follows from Eq. 7.39 that:

$$\eta_0 = \frac{\tanh \phi}{\phi(1 + \phi \tanh \phi/B_{mL})} \tag{7.50}$$

which is the isothermal effectiveness factor. The solutions of Eqs. 7.44 and 7.45 give the first-order corrections u_1 and v_1 which are:

$$u_1 = A_1 + A_2 \cosh \phi x + A_3 x \sinh \phi x + A_4 \cosh 2\phi x \tag{7.51}$$

$$v_1 = \frac{1 - \cosh \phi x/\cosh \phi + \phi \tanh \phi/B_{hL}}{1 + \phi \tanh \phi/B_{mL}} \tag{7.52}$$

where the constants A_i are defined in Notation. The normalized temperature can be written as:

$$v = 1 + \epsilon v_1 + O(\epsilon^2) \tag{7.53}$$

Eq. 7.52 gives an estimate of the maximum temperature difference in the pellet:

$$\Delta v = v - 1 = \epsilon \Delta v_1 \quad \text{or} \quad (\Delta v)_{max} = \beta(\Delta v_1)_{max}$$

$$(\Delta v_1)_{max} = v_1(x = 0) - v_1(x = 1) = (v_1)_{max} - (v_1)_{min}$$

$$= \frac{1 - 1/\cosh \phi}{1 + \phi \tanh \phi/B_{mL}} \tag{7.54}$$

It is seen that the maximum temperature deviation from the bulk-liquid temperature can be approximated by:

$$(\Delta v)_{max} = \frac{(\Delta T)_{max}}{T_b} = \frac{\beta(1 - 1/\cosh \phi)}{1 + \phi \tanh \phi / B_{mL}} \tag{7.55}$$

where $(\Delta T)_{max} = T_c - T_b$, and T_c is the pellet center temperature. The fractional maximum temperature rise in the pellet is at most β according to Eq. 7.55. Therefore, the value of β should be sufficient to tell whether the pellet can be treated as isothermal. For instance, if β is 0.01, the maximum possible temperature rise in the pellet is of the order of 1% of T_b. Typical range of β is 0.1 to 0.001 for liquid-solid systems.

A first-order correction for η can be obtained from:

$$\eta_1 \phi^2 = \frac{du_1}{dx}\bigg|_{x=1} = -B_{mL} u_1 \bigg|_{x=1} \tag{7.56}$$

which follows from Eqs. 7.39, 7.44, 7.46, and 7.47. The expression for u_1 at $x = 1$ is complex, however, this reduces considerably for $\phi \geq 3$ and η_1 becomes:

$$\eta_1 = \frac{\gamma(1/3 + \phi/B_{hL})}{2\phi(1 + \phi/B_{mL})^3} \qquad \text{for } \phi \geq 3 \tag{7.57}$$

Therefore, the overall effectiveness factor for $\phi \geq 3$ can be expressed as

$$\eta = \eta_0 + \epsilon\eta_1 + \Theta(\epsilon^2)$$

$$= \frac{1}{\phi(1 + \phi/B_{mL})}\left[1 + \frac{\beta\gamma(1/3 + \phi/B_{hL})}{2(1 + \phi/B_{mL})^2}\right] + \Theta(\epsilon^2) \qquad \text{for } \phi \geq 3 \tag{7.58}$$

In view of the fact that $B_{hL} \gg B_{mL}$ for liquid-solid systems, the maximum fractional error in using the isothermal effectiveness factor is of the order of $\beta\gamma/[2(1 + \phi/B_{mL})^2]$ for $\phi \geq 3$.

An approximate expression for the global rate of a first-order reaction can be written directly from Eq. 7.58:

$$R_G = \eta k(T_b)C_L$$

$$= \frac{k(T_b)C_L}{\phi(1 + \phi/B_{mL})}\left\{1 + \frac{\beta\gamma(1/3 + \phi/B_{hL})}{2(1 + \phi/B_{mL})^2}\right\}; \qquad \phi \geq 3 \tag{7.59}$$

For $\phi < 3$, Eqs. 7.51 and 7.56 can be used for η_1, which together with η_0 gives a near-accurate value of η, and therefore, of R_G.

While specific results have been restricted to a first-order reaction, the result regarding the maximum temperature rise in the pellet, which is of the order of

β, and that regarding the maximum fractional error in using the isothermal effectiveness factor, which is of the order of $\beta\gamma/[2(1 + \phi/B_{mL})^2]$, should be useful in assessing the suitability of the assumption of an isothermal pellet for complex reactions.

Summary

The global rates for multiphase reactions, in which both gaseous and liquid reactants are involved, can be obtained in a straightforward manner for slurry reactors owing to the lack of internal transport resistances. In contrast to gas-phase reactions, the external mass transfer resistance can have a significant effect on the rate of multiphase reactions. In addition, the internal transport resistances can be significant in trickle-beds. For the trickle-beds, similar results as in gas-phase reactions can be used with appropriate redefinition of dimensionless quantities, if the pellet is isothermal. On the other hand, the global rate cannot readily be obtained for a nonisothermal pellet. Nevertheless, an approximate expression for the global rate can be obtained for a first-order reaction. These results provide an order of magnitude estimate of the maximum temperature rise in a pellet and of the error involved in using the isothermal effectiveness factor for rate expressions more complex than first-order.

NOTATION

a_c	liquid-solid interfacial area per unit volume of catalyst, reciprocal of characteristic length L
a_L	gas-liquid interfacial area per unit volume of liquid
A	$1/(\cosh \phi + \phi \sinh \phi/B_{mL})$
A_1	$\gamma A^2/2$
A_2	$(B_{mL}(A_1 + A_3 \sinh \phi + A_4 \cosh 2\phi) + A_3(\sinh \phi + \phi \cosh \phi) + 2A_4(\phi \sinh 2\phi)/A_5$
A_3	$\phi^2 A\gamma(\cosh \phi + \phi \sinh \phi/B_{hL})/2$
A_4	$-A_1/3$
A_5	$-(\phi \sinh \phi + B_{mL} \cosh \phi)$
B_{hL}	hL/λ
B_{mG}	$k_g L/D_e$
B_{mL}	$K_L L/D_e$
C	concentration of gaseous reactant
C_g	gas-phase reactant concentration
C_c	pellet center concentration
C_L	bulk gaseous reactant concentration in liquid
C_s	gaseous reactant concentration at liquid-solid interface
C_{ig}	gas side interfacial concentration
C_{iL}	liquid side interfacial concentration
C_{eq}	gaseous reactant concentration in liquid in equilibrium with C_g

d_p	particle size or diameter
D_A	molecular diffusivity of gaseous reactant in liquid
D_e	effective diffusivity of gaseous reactant in liquid filled pore
E_a	activation energy
$f(u,v)$	$r_c(C,T)/r_c(C_L,T_b)$
G_L	superficial mass velocity of liquid
H	Henry's constant in concentration unit
k	rate constant
k_c	individual mass transfer coefficient at liquid-solid interface
k_g	gas-side individual mass transfer coefficient at gas-liquid interface
k_L	liquid-side individual mass transfer coefficient at gas-liquid interface
K_L	liquid-side overall mass transfer coefficient given in Eq. 7.22
L	characteristic length of pellet, volume/external surface area
q	fraction of external pellet surface covered by liquid
R_g	gas constant
R_G	global rate in volume unit
R'_G	global rate in weight unit
Re	Reynolds number defined by Eq. 7.8
Re_L	Reynolds number defined in Eq. 7.12
T	temperature
T_b	liquid bulk temperature
T_c	pellet center temperature
u	C/C_L
v	T/T_b
u_0, u_1	zeroth and first order value of u in asymptotic expansion of u
v_0, v_1	zeroth and first order value of v in asymptotic expansion of v
x	pellet coordinate normalized with respect to L

Greek Letters

α_L	a correction factor in Eq. 7.10
β	$(-\Delta H)D_e C_L/\lambda T_b$
γ	$E_a/R_g T_b$
ϵ	a small perturbation parameter equal to β
ϵ_s	ratio of catalyst volume to liquid volume
ϵ_p	pellet porosity
η_i	internal effectiveness factor given by Eq. 7.17
η_G	gas-side effectiveness factor
η_L	liquid-side effectiveness factor
η_o	overall effectiveness factor defined by Eq. 7.24
η_0, η_1	zeroth and first order value of η in asymptotic expansion of η
λ	effective heat conductivity
μ_L	liquid viscosity
ρ_L	liquid density

ρ_p	pellet density
σ	energy dissipation rate given by Eq. 7.6
ϕ	Thiele modulus for first-order reaction $(L(k/De)^{1/2})$ or $\phi^2 = r_c(C_L, T_b)L^2/DC_L$
ϕ_1	Thiele modulus for pseudo-first-order reaction
ϕ_G	generalized modulus given by Eq. 7.23
ν_L	kinematic viscosity of liquid

PROBLEMS

7.1. Obtain an expression for the global rate in terms of C_{eq} and transfer coefficients for a second-order reaction taking place in a slurry reactor.

7.2. Obtain an expression for the global rate in terms of C_{eq} and transfer coefficients for a second-order reaction taking place in a trickle-bed. The typical value of B_{mL} ranges from 1 to 10. Suppose that $k = 10^{-5}$ (s $\cdot$ mol/cm^3)$^{-1}$, $C_{eq} = 10^{-6}$ mol/cm^3, $L = 1$ cm, $D_e = 10^{-5}$ cm^2/s, and the pellet is isothermal. What are the values of the ratio, $R_G(C_{eq})/kC_{eq}^2$, for B_{mL} of 1 and 10? How would the results change if the characteristic length L is changed by a factor of 5 and 0.5? Use Eqs. 7.22 and 7.23 if necessary.

7.3. An estimate of the maximum fractional error e_f in using the isothermal effectiveness factor in a trickle-bed is given for a first-order reaction by:

$$e_f = \frac{\beta\gamma}{2(1 + \phi/B_{mL})^2}, \qquad \phi \geq 3$$

This result can be used for arbitrary kinetics for an approximate estimate if ϕ is replaced by ϕ_G given by Eq. 7.23. A typical value of the Arrhenius number γ is 30, and the upper bound of β is approximately 0.1. From the data given in Problem 7.2, first calculate C_s for the second-order reaction, and then calculate ϕ_G. For what value of B_{mL} does the fractional error become less than 0.1?

7.4. Derive a criterion of negligible transport effects for a first-order reaction taking place in trickle-bed. Note that one only needs to replace C_{eq} in Eq. 7.21 by C_L if kC_L is taken as the intrinsic rate. First obtain the criterion for an isothermal pellet and then for a nonisothermal pellet.

7.5. Derive the first-order correction of the effectiveness factor η_1 for $\phi \geq 3$ (Eq. 7.57). Also obtain η_1 when ϕ is less than 3, i.e., when the pellet center concentration is not close to zero.

REFERENCES

Brian, P.L.T., H.B. Hales and T.K. Sherwood, AIChE J., *15*, 419, 727 (1969).
Calderbank, P.H. and M.B. Moo-Young, Chem. Eng. Sci., *16*, 39 (1961).

Calderbank, P.H., Trans. Inst. Chem. Eng. (London), *36*, 443 (1958).

Dharwadkar, A. and N.D. Sylvester, AIChE J., *23*, 376 (1977).

Dudukovic, M.P., AIChE J., *23*, 940 (1977).

Goto, S. and J.M. Smith, AIChE J., *21*, 706 (1975).

Harriott, P., AIChE J., *8*, 93 (1962).

Herskowitz, M., R.G. Carbonell and J.M. Smith, AIChE J., *25*, 272 (1979).

Juvekar, V.A. and M.M. Sharma, Chem. Eng. Sci., *28*, 825 (1973).

Lee, H.H. and J.M. Smith, Chem. Eng. Sci., *37*, 223 (1982).

Levins, D.M. and J.R. Glastonbury, Chem. Eng. Sci., *27*, 537 (1972).

Misic, D. and J.M. Smith, Ind. Eng. Chem. Fund., *10*, 380 (1971).

Pereira, C.J., J.B. Wang and A. Varma, AIChE J., *25*, 1036 (1979).

Satterfield, C.N., AIChE J., *21*, 209 (1975).

Satterfield, C.N., M.W. van Eek and G.S. Bliss, AIChE J., *24*, 709 (1978).

Satterfield, C.N., A.A. Pelossof and T.K. Sherwood, AIChE J., *15*, 227 (1969).

Smith, J.M., *Chemical Engineering Kinetics*, 3rd ed., McGraw-Hill, New York (1981).

Tan, C.S. and J.M. Smith, Chem. Eng. Sci., *35*, 1601 (1980).

CHAPTER 8

Selectivity and Stability

The selectivity and yield of a desired product is of major interest in multiple reactions. In order to assess the effects of transport processes and catalyst deactivation on the selectivity and yield, discussion shall be confined to simple intrinsic kinetics, for the qualitative behavior of complex reactions is often similar to that of, for instance, first-order reactions. This qualitative behavior of the selectivity and yield as affected by transport resistances and catalyst deactivation will be treated in the first part of this chapter with the understanding that the same approach, when coupled with numerical methods, can lead to quantitative results for any system. An excellent treatment of multiple reactions can be found in the book by Aris (1975).

The second part of this chapter deals with the stability of a single pellet in a fluid stream. This stability represents the stability of a reaction at a point in the reactor. This part deals with two types of stability: one in which the stability around known steady states is of concern, and the other in which the knowledge of steady states is not required. For the first type, uniqueness and multiplicity of the solution to the steady state system equations is of interest; for the second, only the stability of the system is of concern. Perhaps, the most difficult part of the stability problem lies in determining whether the equations describing a system adequately represent the physical process. Two different sets of equations describing the same system could give conflicting conclusions on stability.

8-1 SELECTIVITY AFFECTED BY TRANSPORT PROCESSES

Consecutive first-order reactions provide a basis for assessing the effects of diffusion on selectivity. For the following isothermal, consecutive reaction network:

$$A \xrightarrow{k_A} B \xrightarrow{k_B} C$$

steady state mass balances give:

$$\nabla^2 C_A = \phi_A^2 C_A \tag{8.1}$$

$$\nabla^2 C_B = \phi_B^2 C_B - \phi_A^2 C_A / \gamma \tag{8.2}$$

where $\phi_i^2 = L^2(k_i/D_i)$ for $i = A,B$, $\gamma = D_B/D_A$, and the length coordinate is normalized. Here, L is the ratio of volume to external surface area. Boundary conditions are:

$$C_A = A_s, \quad C_B = B_s \quad \text{on the pellet surface} \tag{8.3}$$

The linear equations 8.1 and 8.2 can be solved using the principle of superposition:

$$C_A = a_1 u_A + a_2 u_B \tag{8.4}$$

$$C_B = a_3 u_A + a_4 u_B \tag{8.5}$$

in which u_i $(i = A,B)$ is the solution of:

$$\nabla^2 u_i = \phi_i^2 u_i \tag{8.6}$$

$$u_i = 1 \quad \text{on the pellet surface} \tag{8.7}$$

Substitution of Eqs. 8.4 through 8.7 into Eqs. 8.1 and 8.2 yields:

$$a_1 = A_s, \qquad a_2 = 0$$

$$a_3 = \frac{\phi_A^2 A_s}{\gamma(\phi_B^2 - \phi_A^2)} = \frac{sA_s}{1 - \gamma s}, \qquad a_4 = B_s - a_3 \tag{8.8}$$

where $s = k_A/k_B$. The point yield in the reactor is given by:

$$-\frac{(R_G)_B}{(R_G)_A} = -\frac{D_B \iint_{\text{surface}} (\partial C_B/\partial \mathbf{n})ds}{D_A \iint_{\text{surface}} (\partial C_A/\partial \mathbf{n})ds} \tag{8.9}$$

where $\mathbf{n}$ is the outward normal length coordinate. Substitution of Eqs. 8.4 and 8.5 into Eq. 8.9 with the known parameters yields:

$$-\frac{(R_G)_B}{(R_G)_A} = \frac{\gamma s}{\gamma s - 1} - \frac{\gamma \eta_B \phi_B^2}{\eta_A \phi_A^2}\left(-\frac{B_s}{A_s} + \frac{s}{\gamma s - 1}\right) \tag{8.10}$$

When the diffusional resistance is negligible, the internal effectiveness factors η_i approach unity, and Eq. 8.10 becomes:

$$-\frac{(R_G)_B}{(R_G)_A} = 1 - \frac{B_s}{sA_s}\left(= -\frac{r_B}{r_A}\right) \tag{8.11}$$

which is the point yield for homogeneous reactions as represented by the intrinsic rates r_i ($i = A,B$) indicated above. In the other extreme region of strong diffusion limitations, $\eta_i \phi_i$ approaches unity, and Eq. 8.10 reduces to:

$$-\frac{(R_G)_B}{(R_G)_A} = \frac{(s\gamma)^{1/2}}{(s\gamma)^{1/2} + 1} - \frac{B_s}{A_s}(\gamma/s)^{1/2} \tag{8.12}$$

A comparison between Eqs. 8.11 and 8.12 gives the effect of diffusion on yield. A catalyst would be chosen to maximize the yield of the desired product B. As apparent from Eq. 8.11, this implies that $s = (k_A/k_B)$ would be much greater than unity. For a large value of s, the first term in the right hand side of Eq. 8.12 approaches unity. Therefore, the relative magnitude of yields for this case can be expressed from Eqs. 8.11 and 8.12:

$$\frac{(-R_G)_B/(R_G)_A}{(-r_B/r_A)} = \frac{1 - B_s/[(s/\gamma)^{1/2}A_s]}{1 - B_s/(sA_s)} \tag{8.13}$$

which is less than unity for nominal values of the effective diffusivity ratio. It is seen then, that the point yield decreases with increasing diffusion limitations. On the other hand, in rare instances in which s is less than unity, the yield can increase with increasing diffusion limitations. It is noted in this regard that physically $-(R_G)_B/(R_G)_A \geq 0$ and therefore, the second term in Eqs. 8.11 and 8.12 is less than or equal to the first term.

The undersirable effect of diffusion when $s(= k_A/k_B) > 1$ is clearly demonstrated in Figure 8.1 for the consecutive dehydrogenation reactions of butene (Voge and Morgan 1972).

For a plug-flow reactor in which the effects of external transport resistances are negligible, the point yield is equal to $(-dB/dA)$, where A and B are bulk fluid concentrations. Therefore, an integration of Eq. 8.10 with respect to the dependent variables A and B gives the reactor yield, since $-(R_G)_B/(R_G)_A = (-dB/dA)$, and $A_s = A$ and $B_s = B$ for this plug flow reactor, yielding (Carberry 1966):

$$\frac{B_L}{A_f} = \frac{s(p-\gamma)}{(\gamma s - 1)(p-1)}\left[\left(\frac{A_L}{A_f}\right)^p - \frac{A_L}{A_f}\right] + \frac{B_f}{A_f}\left(\frac{A_L}{A_f}\right)^p$$

where the subscripts f and L denote the concentrations at the inlet and outlet of the reactor, respectively, and $p = \gamma \eta_B \phi_B^2/(\eta_A \phi_A^2)$.

For parallel reactions with a common reactant:

$$A \xrightarrow{k_B} B \quad (n^{\text{th}} \text{ order})$$

$$A \xrightarrow{k_C} C \quad (m^{\text{th}} \text{ order})$$

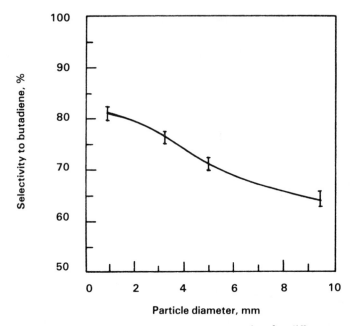

Figure 8.1 Selectivity to butadiene at 35% butene conversion for different particle sizes. (Voge and Morgan 1972; reprinted with permission from *Industrial and Engineering Chemistry, Process Design and Development.* Copyright by American Chemical Society.)

the selectivity is of interest only when the orders of the reactions are not the same, for the selectivity would not be affected by diffusion resistance if the orders were the same. For the above reactions, in which species B is the desired product, the point selectivity in a reactor when diffusional effects are absent is:

$$\frac{r_B}{r_A} = \frac{k_B}{k_C} A_s^{n-m} \tag{8.14}$$

When the concentration of species A depends on the pellet coordinate due to diffusion, the point selectivity is:

$$\frac{(R_G)_B}{(R_G)_A} = \frac{k_B}{k_C} A_s^{n-m} \frac{\displaystyle\int_V (C_A/A_s)^n dV}{\displaystyle\int_V (C_A/A_s)^m dV} \tag{8.15}$$

where V denotes volume integration. Then, the relative point selectivity defined by $[(R_G)_B/(R_G)_A]/(r_B/r_A)$ can be obtained directly from Eqs. 8.14 and 8.15:

$$\frac{(R_G)_B/(R_G)_A}{(r_B/r_A)} = \frac{\displaystyle\int_V (C_A/A_s)^n dV}{\displaystyle\int_V (C_A/A_s)^m dV} \tag{8.16}$$

It is clear from Eq. 8.16 that the relative point selectivity is smaller than unity when $n > m$ and larger than unity when $n < m$ since $(C_A/A_s) \leq 1$. Therefore, the point selectivity increases with decreasing diffusional resistance when $n > m$; it increases with increasing diffusional resistance when $n < m$.

The relative point selectivity for the following independent parallel reaction network in which B is the desired product:

$$A \xrightarrow{k_B} B \qquad (n^{\text{th}} \text{ order})$$

$$X \xrightarrow{k_C} C \qquad (m^{\text{th}} \text{ order})$$

can be written as:

$$\frac{(R_G)_B}{(R_G)_C} = \frac{\eta_A k_B A_s^n}{\eta_x k_C X_s^m} = \left(\frac{\eta_A}{\eta_x}\right)\left(\frac{r_B}{r_C}\right)$$

The diffusional effect on the selectivity is self-explanatory.

It has been shown by Wei (1962) that any complex network of first-order reactions consisting of combinations of consecutive and parallel paths including reversible paths can be decomposed into a system involving only independent parallel reaction paths of irreversible first-order reactions. Let r_c be a vector of rate expressions corresponding to a vector of concentrations C for all species. Then:

$$\nabla^2 C = \overline{K} C; \; r_c = K C, \overline{K} \equiv D^{-1} K / L^2$$

where K is a matrix of rate constants and the spatial coordinate is normalized with respect to the characteristic length L. If Λ is a diagonal matrix obtained from the matrix $\overline{K}$ by a similarity transformation, one has:

$$\Lambda = V^{-1} \overline{K} V$$

where V is the matrix whose columns are the coordinates of the eigenvectors given by:

$$\overline{K} v_m = -\lambda_m v_m$$

The eigenvalues are obtained from:

$$\det (\overline{K} + \lambda_m I) = 0$$

Then the original balance equation can be written as:

$$\nabla^2 C_t = \Lambda C_t$$

which is now decomposed into independent parallel reaction paths of irreversible first-order reactions. The transformed concentration C_t is related to the original concentration C through the eigenvectors:

$$C = V\, C_t \tag{8.17}$$

For such a decomposed system, the gradient at the pellet surface can be written in terms of the transformed concentrations C_t as:

$$\left.\frac{\partial C_t}{\partial \eta}\right| = \Lambda \eta C_t(1) \tag{8.18}$$

where η is a diagonal matrix whose elements are the effectiveness factors for the decomposed irreversible first-order reactions. If the global rates are written as $R_G = K_e C_t(1)$ in which the effective rate constant matrix K_e is defined by:

$$K_e = DV\Lambda\, \eta \tag{8.19}$$

it follows from Eqs. 8.17 and 8.18 that:

$$R_G(C) = K_e V^{-1} C(1) \tag{8.20}$$

The point selectivity can readily be obtained from Eq. 8.20. In fact, this technique has been used earlier in examining the effects of diffusion on the selectivity for a first-order consecutive reaction network. The result obtained by Wei (1962) for a three component system is shown in Figure 8.2. It is seen that the lengths between paths decrease when the reactions are affected by diffusion, indicating that selectivity differences are decreased. This result shows that diffusional effects are usually detrimental to selectivity in the sense that it is more difficult to have products differing in composition.

It has been shown in Chapter 4 that major transport resistances under typical operating conditions lie in external heat transport and internal mass transport.

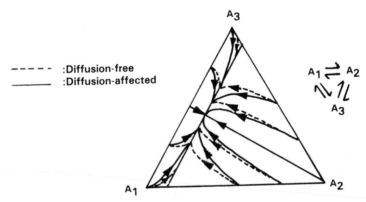

Figure 8.2   Effect of diffusion on the reaction paths in an integral reactor, equal diffusivities. (Wei 1962; reprinted with permission from Journal of Catalysis. Copyright by Academic Press.)

Let us examine the combined effects of these two resistances on selectivity. A simple way of accomplishing this is to compare the selectivity affected only by diffusion with that affected by both diffusion and external heat transport resistance. Let $(R_G)_B^s/(R_G)_A^s$ be the point yield of a consecutive reaction network $A \xrightarrow{k_A} B \xrightarrow{k_B} C$ based on pellet surface temperature and bulk fluid concentrations. In the absence of external resistance to heat transfer, the surface temperature is equal to the bulk fluid temperature. Therefore, $(R_G)_B^s/(R_G)_A^s$ is the point yield when only the diffusional resistance is present. For the rate expression $r_c = kf$ (composition), the point yield when the reactions are affected by both diffusion and external heat transport, $(R_G)_B/(R_G)_A$, is simply:

$$\frac{(R_G)_B}{(R_G)_A} = \left(\frac{(R_G)_B^s}{(R_G)_A^s}\right)\left(\frac{s_b}{s_s}\right) \tag{8.21}$$

where s_b and s_s are the rate constant ratios evaluated at bulk fluid and pellet surface temperatures, respectively. The rate constant ratio s_s can be expressed as:

$$
\begin{aligned}
s_s &= s_b \exp\{-(\epsilon_A - \epsilon_B)[1/(T_s/T_b) - 1]\} \\
&= s_b\{1 + 1.2(\epsilon_A - \epsilon_B)(T_s/T_b - 1)\}
\end{aligned}
\tag{8.22}
$$

where T_s and T_b are the surface and the bulk temperature and ϵ_A and ϵ_B are the Arrhenius numbers for k_A and k_B. Here, the factor 1.2 was introduced for better approximation of the exponential term as discussed in Chapter 4. The temperature ratio is obtained from a heat balance around the external film.

$$h(T_s/T_b - 1) = \frac{L}{T_b} \sum_i (-\Delta H)_i (R_G)_i \qquad (i = A, B) \tag{8.23}$$

Use of Eqs. 8.22 and 8.23 in Eq. 8.21 yields:

$$\frac{(R_G)_B/(R_G)_A}{(R_G)_B^s/(R_G)_A^s} = \frac{s_b}{s_s} = \frac{1}{1 + 1.2(\epsilon_A - \epsilon_B)(L/hT_b) \sum_i (-\Delta H)_i (R_G)_i} \tag{8.24}$$

This ratio for exothermic reactions is larger than unity when $\epsilon_B > \epsilon_A$ and smaller when $\epsilon_B < \epsilon_A$, meaning that the point yield is increased due to the external resistance when $\epsilon_B > \epsilon_A$ and decreased when $\epsilon_B < \epsilon_A$. It is therefore entirely possible when $\epsilon_B > \epsilon_A$ that the combined effects of transport resistances are in fact to increase the point yield. The same conclusions can be made for the point selectivity of the parallel reactions ($A \xrightarrow{k_B} B$, $A \xrightarrow{k_A} C$) in which species B is the desired product.

8–2   SELECTIVITY AFFECTED BY DEACTIVATION

The effect of catalyst deactivation on selectivity can be easily assessed for very small catalyst particles such as those used in fluidized-beds, for which the diffusional effect is absent and therefore, deactivation is uniform. A good example is provided by Weekman and Nace (1970) for the catalytic cracking of gas oil (A) into gasoline (G), and dry gas and coke (S) which deactivate the catalyst:

$$A \underset{k_3}{\overset{k_1}{\rightleftharpoons}} G \overset{k_2}{\searrow} S$$

This lumped version of the network of many reactions involved in the cracking was represented by the following phenomenological rate expressions:

$$(R_G)_1 = k_1 A^2 h_1$$

$$(R_G)_2 = k_2 G h_2$$

$$(R_G)_3 = k_3 A^2 h_3$$

where h_i ($i = 1, 2, 3$) is the fraction of catalyst still remaining active after deactivation, and k_i is the rate constant for fresh catalyst. Then the point yield can be expressed as:

$$-\frac{(R_G)_G}{(R_G)_A} = \frac{(R_G)_1 - (R_G)_2}{(R_G)_1 + (R_G)_3} = \frac{1}{1 + (k_3 h_3 / k_1 h_1)} - \frac{h_2 h_2}{k_1 h_1 + k_3 h_3} \left(\frac{G}{A^2}\right) \quad (8.25)$$

It is clear from Eq. 8.25 that the point yield is not affected by deactivation if the h_i's are the same. This implies that deactivation does not affect the point yield if all the reactions occur on the same sites. On the other hand, deactivation would affect the yield if different reactions take place on different sites. Since the above reaction network is a combination of consecutive and parallel reactions, the same conclusions can be made for simple consecutive and parallel networks. Weekman and Nace (1970) assumed that the h_i's are the same, which is represented by:

$$h_i = h = \exp(-\alpha t)$$

where α is a constant that would depend on bulk concentration and temperature. The data that they obtained from a moving bed, for which plug flow can be assumed, are compared with the model in Figure 8.3. For plug-flow, $-R_G/R_A = -dG/dA$ and Eq. 8.25 can be integrated to give the predicted curve in the figure. The good agreement shown is an indication that the assumption of $h_i = h$ is reasonable. On the other hand, De Pauw and Froment (1975) found for the n-

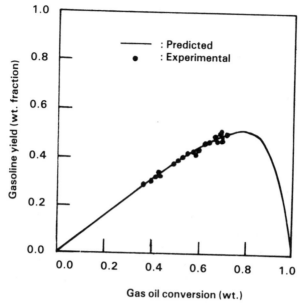

Figure 8.3 Instantaneous gasoline yield for the catalytic cracking of gas oil: experimental values are for various catalyst residence times. (Weekman and Nace 1970)

pentane isomerization over Pt/Al_2O_3 that two side reactions occur on sites different from those involved in the main reaction, resulting in a selectivity that varies with time.

While the effect of deactivation alone is rather straightforward to analyze, the combined effect of deactivation and diffusion is not, because of diffusional intrusion into deactivation even when all reactions occur on the same site. Therefore, the selectivity can vary with time regardless of the nature of the active sites. Attention here will be confined to the case in which all reactions occur on the same sites. It has been shown in Chapter 5 that the internal effectiveness factor for a pellet undergoing deactivation is given by:

$$\eta_i = \frac{\left[2\bar{g}D \int_0^{C_b} r_c(c)dc \right]^{1/2}}{L\bar{r}_c}$$

This equation applies in the asymptotic region of strong diffusion effects, where $\bar{g}$ is the fraction of external surface activity still remaining active after deactivation, and $\bar{r}_c$ is the intrinsic rate in terms of surface conditions. This internal effectiveness factor for a deactivated pellet is simply the internal effectiveness factor for a fresh pellet multiplied by $\bar{g}^{1/2}$. The restriction placed on the relationship was that the following condition be met:

$$\left|\frac{\bar{g}'D^{1/2}}{\bar{g}(2\bar{g})^{1/2}}\right| \left\{\frac{\int_0^{C_b}\int_0^C r_c(\alpha)d\alpha dC}{\left[\int_0^{C_b} r_c(C)dc\right]^{3/2}}\right\} \ll 1$$

For an nth-order reaction, this condition is equivalent to:

$$\left|\frac{\bar{g}'L}{\bar{g}^{3/2}}\right|\left(\frac{1}{\phi}\right)\left[\frac{(n+1)^{1/2}}{(2)^{1/2}(n+2)}\right] \ll 1$$

where ϕ is the usual Thiele modulus. This condition is usually met in the asymptotic region of strong diffusion effects, unless the fractional surface activity is small. Therefore, in the initial stage of deactivation when $\bar{g}$ is relatively large, the combined effect of diffusion and deactivation is to reduce the intrinsic rate constant k to an effective rate constant $(k\bar{g})^{1/2}$. It follows then that the point selectivity is independent of time in this initial stage of deactivation. For instance, s in Eq. 8.12 and k_B/k_C in Eq. 8.15 would remain the same in spite of catalyst deactivation. This time independence cannot hold when the surface activity becomes small or pore-mouth deactivation takes place. This is clear from Figure 8.4 obtained by Sada and Wen (1967) for simple networks of irreversible first-order reactions affected by independent pore-mouth poisoning. It is seen that the selectivities for consecutive and independent parallel reaction networks are time dependent, whereas the selectivity for the parallel reaction network with a common reactant is time invariant

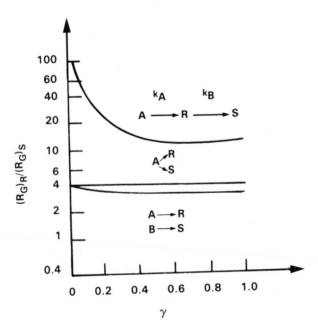

Figure 8.4 Selectivity behavior of simple multiple reactions affected by pore-mouth poisoning: equal diffusivities, $\phi_A = 1$, $k_A/k_B = 4$, $C_A = C_B$. (Sada and Wen 1967. Reprinted with permission from *Chemical Engineering Science*. Copyright by Pergamon Press, Inc.)

when all reactions are affected by deactivation. It is also seen that the selectivity for the consecutive network reaches a plateau after poisoning, although it improves slightly with increasing levels of the fraction of catalyst completely deactivated (γ) or after the plateau is reached. The selectivity for the independent parallel network deteriorates when it is affected by poisoning and decreases with time.

The selectivities for any complex reaction networks of first-order reactions can be compactly analyzed using the approach of Wei (1962) discussed earlier. All that is required for the analysis is the expressions for the effectiveness factor vector $\boldsymbol{\eta}$, which in this case is the overall effectiveness factor representing both deactivation and diffusion effects. In the domain of transformed concentration C_t, the elements of $\boldsymbol{\eta}$, η_k are given, for instance, for a slab-like pellet by:

$$\eta_k = \frac{1/\phi_k}{1 + \gamma\phi_k} \; ; \qquad \phi_k = L\left(\frac{k_k}{D_k}\right)^{1/2}$$

which was obtained in Chapter 5 for independent pore-mouth poisoning. Results such as those shown in Figure 8.4 can readily be obtained for any complex network of first-order reactions (see Problem 8.6).

Finally, it should be recognized that the time independence of selectivity is valid only for the same bulk-fluid conditions, i.e., when the conditions are the same whether or not deactivation takes place. This condition is not satisfied in reactor operation due to changing catalyst activity with time and therefore, the selectivity invariably changes with time except under unusual circumstances. Nevertheless, the time independence or dependence of selectivity is a useful piece of information in laboratory studies in which the bulk-fluid conditions can be held constant.

8-3 THE STABILITY PROBLEM FOR A SINGLE PELLET

The stability problem for a symmetric pellet in which a single reaction takes place can be stated as follows: Determine the stability of the system described by:

$$\frac{\partial C}{\partial t} = \nabla(D\nabla C) - r_c(C,T)$$

$$\rho_p C_p \frac{\partial T}{\partial t} = \nabla(\lambda \nabla T) + (-\Delta H)r_c(C,T)$$

when it is subjected to perturbations. The boundary and initial conditions are:

$$D\frac{\partial C}{\partial \mathbf{n}} = k_g(C_b - C)$$

$$\frac{12\lambda}{\partial \mathbf{n}}\frac{\partial T}{\partial \mathbf{n}} = h(T - T_b)$$

on the boundary and

$$C(\mathbf{r},0) = C_0(\mathbf{r})$$

$$T(\mathbf{r},0) = T_0(\mathbf{r})$$

For one-dimensional transport of heat and mass, the equations can be reduced in dimensionless form:

$$\frac{\partial z}{\partial \tau} = \frac{1}{\xi^q}\frac{\partial}{\partial \xi}\left(\xi^q \frac{\partial z}{\partial \xi}\right) - \phi^2 f(z)\exp[\epsilon(1 - 1/y)] \qquad (8.26)$$

$$\text{Le}\frac{\partial y}{\partial \tau} = \frac{1}{\xi^q}\frac{\partial}{\partial \xi}\left(\xi^q \frac{\partial z}{\partial \xi}\right) + \beta\phi^2 f(z)\exp[\epsilon(1 - 1/y)] \qquad (8.27)$$

with

$$\frac{\partial z}{\partial \xi} = \text{Bi}_m(1 - z) \qquad (8.28a)$$

$$\frac{\partial y}{\partial \xi} = \text{Bi}_h(y - 1) \qquad (8.28b)$$

on the boundary and

$$z(\xi,0) = z_0(\xi) \qquad (8.29a)$$

$$y(\xi,0) = y_0(\xi) \qquad (8.29b)$$

where $z = C/C_b$, $y = T/T_b$, $\tau = Dt/R^2$, $\xi = r/R$,
$\beta = (-\Delta H)DC_b/\lambda T_b$, $\epsilon = E_a/R_g T_b$, $\phi^2 = L^2 k_b f(C_b)/DC_b$
$\text{Le} = \rho_p C_p D/\lambda$, $\text{Bi}_m = k_g R/D$, $\text{Bi}_h = hR/\lambda$, $f(z) = f(C)/f(C_b)$ $\qquad (8.29c)$

Here, the intrinsic rate is represented by $r_c = k(T)f(C)$, and the subscript b denotes the bulk fluid conditions. The exponent q is a constant depending on pellet geometry: $q = 0$, 1, and 2 for a slab-like, cylindrical, and spherical pellets, respectively. It is rather difficult to establish stability for the general problem as posed since this requires an analysis of a very wide class of perturbations. It is much easier to establish instability conditions. As a result, most of the studies available are concerned with instability criteria.

Steady state operation is invariably concerned with stability since one must question whether or not a system will eventually converge to a steady state when it is subjected to perturbations, i.e., changes in operating conditions. Furthermore, steady state solutions do permit certain limited statements to be made about stability. For example, Gavalas (1968) has shown that, given multiple solutions of a steady state equation, there are $(2m + 1)$ steady states except under exceptional circumstances, and that at least m of them are unstable. It will also be seen that a certain claim on stability requires the resolution of uniqueness or multiplicity of the steady state.

A system represented by a set of equations describing a physical system is said to be stable around a steady state if the transients of the system are bounded when the system is subjected to small perturbations from the steady state. The system is said to be asymptotically stable if it is stable and it eventually returns to the steady state. If in addition, the steady state is approached exponentially, it is called exponentially asymptotically stable. If any magnitude of perturbation is allowed, it is said to be globally stable. The stability problem as related to reactor startup and control is concerned with the following questions (Luss 1977):

a) What are the realizable steady states?
b) What is the set of initial conditions from which a given steady state is attained?
c) Are the steady states stable or unstable? What is the nature of stability if stable?

As indicated earlier, it is rarely possible to fully resolve these questions. In particular, it is extremely difficult to prove global stability. Therefore, in what follows, the main concern will be with asymptotic stability with respect to small perturbations.

First, a single wire problem will be considered as an introduction. It brings out essential features of the stability problem and is similar in many respects to the more familiar stability problem with a CSTR. A simple and yet realistic case of negligible internal heat transfer and external mass transfer resistances will also be treated. The general problem will then be analyzed in a limited form for the case of unit Lewis number (Le) appearing in Eq. 8.27. This considerably simplifies the problem and yet the result can be extended to a certain extent to the general case of arbitrary Lewis numbers. Readers interested in more details on stability can refer to the book by Aris (1975) and the review by Luss (1977), from which the subsequent sections are derived.

8-4 A SINGLE WIRE PROBLEM AND SLOPE CONDITIONS

A single catalytic wire on the surface of which a first-order reaction takes place can be described by;

$$\frac{dA_s}{dt} = \bar{k}_g(A_b - A_s) - kA_s \tag{8.30}$$

$$\frac{(\rho C_p)_w R}{2} \frac{dT_s}{dt} = h(T_b - T_s) + (-\Delta H)kA_s \tag{8.31}$$

where the subscripts b and s denote the bulk and surface conditions, k is a rate constant, $\bar{k}_g$ is a mass transfer coefficient including the volume to surface area ratio, and R is the radius of the wire. At steady state, these equations reduce to:

$$\bar{k}_g(A_b - A_s) = kA_s \tag{8.32}$$

$$h(T_s - T_b) = (-\Delta H)kA_s \tag{8.33}$$

Using Eq. 8.32 in 8.33 yields:

$$Q_c = h(T_s - T_b) = (-\Delta H)\bar{k}_g A_b \left(\frac{k}{k_g + k}\right) \equiv Q_h \tag{8.34}$$

This relationship is similar to that for a CSTR with cooling: Q_c is the rate of heat removal, and Q_h is the rate of heat generation, which is a sigmoidal function of T_s due to the exponential temperature dependence of the rate constant k. A value of T_s that satisfies Eq. 8.34 is a steady state solution. This is illustrated in Figure 8.5, which shows Q_c and Q_h lines as a function of T_s. Any intersection of these two Q graphs is a steady state solution. It is seen that three steady states can exist.

It has been proved by Lyapounov that the local stability (as opposed to

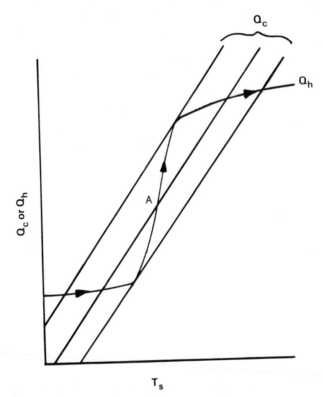

Figure 8.5 Possible steady states for the single wire problem. (Cordoso and Luss 1969; reprinted with permission from *Chemical Engineering Science.* Copyright by Pergamon Press.)

global stability) of a set of nonlinear ordinary differential equations is determined by the stability of the linearized equations about the steady state. Linearization of Eqs. 8.30 and 8.31 using a Taylor series expansion about the steady state yields:

$$\frac{d\mathbf{x}}{dt} = \mathbf{A}\mathbf{x} \tag{8.35}$$

where

$$\mathbf{x}^T = [A_s - A_{ss}, \ T_s - T_{ss}] \tag{8.36}$$

$$\mathbf{A} = \begin{bmatrix} -(\bar{k}_g + {}_{ss}) & -A_{ss}\left(\frac{\partial k}{\partial T}\right)_s \\ \dfrac{2(-\Delta H)k_{ss}}{(\rho C_p)_w R} & \dfrac{2}{(\rho C_p)_w R}\left\{(-\Delta H)A_{ss}\left(\dfrac{\partial k}{\partial T}\right)_s - h\right\} \end{bmatrix} \tag{8.37}$$

Here, the second subscript denotes the steady state value. The roots of the quadratic characteristic equation (eigenvalues) of the matrix $\mathbf{A}$ determine the stability of the equations: the system will converge exponentially to the steady state if all roots have a negative real part and, therefore, is asymptotically stable. It will show a limit cycle if the roots are imaginary with zero real parts. It is unstable if any of the roots has a positive real part. Since the perturbations will decay asymptotically if and only if all the eigenvalues of the matrix $\mathbf{A}$ have a negative real part, it follows that the necessary and sufficient conditions for local stability are:

$$\frac{dQ_c}{dT_s} > \frac{dQ_h}{dT_s} \quad \text{or} \quad h > \frac{(-\Delta H)\bar{k}_g A_{ss}}{k_{ss} + \bar{k}_g}\left(\frac{\partial k}{\partial T}\right)_s \tag{8.38}$$

$$\frac{h\bar{k}_g}{\bar{k}_g + k_{ss}} + \frac{R\bar{k}_g(\rho C_p)_w}{2} > \frac{dQ_h}{dT_s} \tag{8.39}$$

The first condition is known as the slope condition: a steady state solution is unstable whenever the slope of the heat generation curve exceeds that of the heat removal curve, e.g., the steady state A in Figure 8.5. It is seen from the figure that when three steady states exist, at most two are stable. The second condition has no simple physical meaning and is implied by the slope condition whenever:

$$\frac{\bar{k}_g(\rho C_p)_w R}{2h} \geq \frac{k_{ss}}{\bar{k}_g + k_{ss}} \tag{8.40}$$

For the more realistic case of internal mass transport and external heat transport resistances, the system equations written for a slab-like pellet are:

$$\frac{\partial C}{\partial t} = D\frac{\partial^2 C}{\partial x^2} - kC \tag{8.41}$$

$$V_p \rho_p C_p \frac{\partial T}{\partial t} = S_r h(T_b - T_s) + (-\Delta H)k \int_{V_P} C dV \qquad (8.42)$$

with the usual boundary conditions. Amundson and Raymond (1965) analyzed this problem and obtained the following necessary condition for stability:

$$Bi_h > \beta \frac{d(\eta \phi^2)}{dy} = \beta \frac{d}{dy} (\phi \tanh \phi) \qquad (8.43)$$

This is another form of the slope condition. Since this is a necessary condition, it is not sufficient to guarantee stability. In fact, Lee and coworkers (1972) found that a change in the ratio of Le/β that appears in the dimensionless version of Eq. 8.42 may destabilize a steady state even when Eq. 8.43 is satisfied. However, they found that instabilities exist only for unrealistically low values of Le/β.

8-5   THE CASE OF UNIT LEWIS NUMBER AND EXTENSIONS

The original problem posed is now analyzed in limited forms: first a sufficient condition for instability (Jackson 1973) when external resistances are negligible, and then the global stability for a unique steady state (Luss and Lee 1968) with an additional assumption on the relationship between temperature and concentration. In both cases, the Lewis number is set at unity for simplification. It shall soon be seen that the case of unit Lewis number is an important base case from which the questions raised in Section 8-3 can be resolved.

For the nth-order reaction considered by Jackson (1973), the Eqs. 8.26 through 8.29 reduce to:

$$\frac{\partial z}{\partial \tau} = \frac{1}{\xi^q \partial \xi} \left(\xi^q \frac{\partial z}{\partial \xi} \right) - \phi^2 z^n \exp[\epsilon(1 - 1/y)] \qquad (8.44)$$

$$Le \frac{\partial y}{\partial \tau} = \frac{1}{\xi^q \partial \xi} \left(\xi^q \frac{\partial y}{\partial \xi} \right) + \beta \phi^2 z^n \exp[\epsilon(1 - 1/y)] \qquad (8.45)$$

with

$$\frac{\partial z(0,\tau)}{\partial \xi} = \frac{\partial y(0,\tau)}{\partial \xi} = 0 \qquad (8.46)$$

$$y(1,\tau) = z(1,\tau) = 1 \qquad (8.47)$$

and with Eq. 8.29 remaining the same. For any steady state, Eqs. 8.44 and 8.45 give $y + \beta z = 1 + \beta$. *Residual enthalpy* shall be defined as:

$$E^* = y + \beta z - (1 + \beta) \qquad (8.48)$$

Utilizing this definition, Eqs. 8.44 through 8.47 can be rewritten for a unit Lewis number as:

$$\frac{\partial y}{\partial \tau} = \frac{1}{\xi^q \partial \xi} \frac{\partial}{\partial \xi}\left(\xi^q \frac{\partial y}{\partial \xi}\right) + \beta^{1-n}\phi^2(E^* + 1 + \beta - y)^n \exp[\epsilon(1 - 1/y)] \qquad (8.49)$$

$$\frac{\partial E^*}{\partial \tau} = \frac{1}{\xi^q \partial \xi}\frac{\partial}{\partial \xi}\left(\xi^q \frac{\partial E^*}{\partial \xi}\right) \qquad (8.50)$$

with
$$\frac{\partial y(0,\tau)}{\partial \xi} = \frac{\partial E^*(0,\tau)}{\partial \xi} = 0 \qquad (8.51)$$

$$y(1,\tau) = 1, \ E^*(1,\tau) = 0 \qquad (8.52)$$

$$y(\xi,0) = y_0(\xi), \ E^*(\xi,0) = E_0(\xi) \qquad (8.53)$$

The distributions y_0 and z_0 will give the initial distributions of y and z. However, because one can choose to investigate a very special class of perturbations when wishing to prove instability, it is permissible to insist that $E^*(\xi,\tau)$ never departs from its steady state value, the constant function $E^* = 0$ for all $\tau \geq 0$. The reasoning here is that it is sufficient to consider any chosen perturbations to demonstrate instability. Therefore, Eq. 8.49 is reduced to:

$$\frac{\partial y}{\partial \tau} = \frac{1}{\xi^q \partial \xi}\frac{\partial}{\partial \xi}\left(\xi^q \frac{\partial y}{\partial \xi}\right) + \phi^2\beta^{1-n}(1 + \beta - y)^n \exp[\epsilon(1 - 1/y)] \qquad (8.54)$$

and Eq. 8.50 is no longer needed for further consideration.

If y_s is the steady state under investigation, it satisfies:

$$\frac{1}{\xi^q d\xi}\frac{d}{d\xi}\left(\xi^q \frac{dy_s}{d\xi}\right) + \phi^2\beta G(\xi) = 0 \qquad (8.55)$$

with the usual boundary conditions, where:

$$G(\xi) = \beta^{-n}(1 + \beta - y_s)^n \exp[\epsilon(1 - 1/y_s)] \qquad (8.56)$$

The departure from steady state:

$$y^* = y(\xi,t) - y_s(\xi) \qquad (8.57)$$

thus satisfies:

$$\frac{\partial y^*}{\partial \tau} = \frac{1}{\xi^q \partial \xi}\frac{\partial}{\partial \xi}\left(\xi^q \frac{\partial y^*}{\partial \xi}\right) - F(\xi)y^* \qquad (8.58)$$

$$y*(1,\tau) = 0, \quad \frac{dy*}{d\xi}(0,\tau) = 0 \tag{8.59}$$

Here, the quantity F given by:

$$F(\xi) = -\phi^2\beta \left(\frac{dG}{dy_s}\right)_s \tag{8.60}$$

is the one term approximation of $G(\xi)$ about the steady state for small $y*$. Assuming a separable solution:

$$y* = \theta(\tau)w(\xi) \tag{8.61}$$

for Eq. 8.58, yields the following eigenvalue problem:

$$\frac{1}{\xi^q d\xi}\frac{d}{d\xi}\left(\xi^q\frac{dw}{d\xi}\right) - [F(\xi) + \lambda]w = 0 \tag{8.62}$$

$$w(1) = 0, \quad dw/d\xi|_0 = 0 \tag{8.63}$$

and
$$\theta(\tau) = e^{\lambda\tau} \tag{8.64}$$

If at least one positive eigenvalue of Eq. 8.62 is found, the steady state will be unstable.

The method Jackson (1973) uses to establish the condition for a positive eigenvalue is based on a comparison theorem of Sturm. Consider:

$$\frac{1}{\xi^q d\xi}\frac{d}{d\xi}\left(\xi^q\frac{dv}{d\xi}\right) - F(\xi)v = 0 \tag{8.65}$$

with
$$dv(0)/d\xi = 0, \quad v(1) = 1 \tag{8.66}$$

This is to be compared with Eq. 8.62. The Sturmian theorem states that if Eq. 8.62 has a positive eigenvalue, then the function v will have at least as many zeros in the interval $(0,1)$ as does v, and in fact the ith zero of w will occur at a smaller value of ξ than the ith zero of v. Indeed, as Jackson shows, the position of the zero will move continuously as λ varies. Hence, if v has at least one zero in the interval $(0,1)$, one will always find a positive eigenvalue for which the solution of Eq. 8.62 satisfies the boundary condition $w(1) = 0$. It is therefore a sufficient condition for instability that v should change its sign at least once in the interval $(0,1)$.

It is now desirable to relate this criterion to the geometry of the (η,ϕ)-curve. This curve can be separated into branches, as shown in Figure 8.6 which are bounded by branch points (or bifurcation points) at which $d\eta/d\phi = \infty$. Except

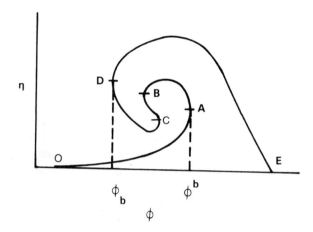

at a bifurcation value of ϕ the steady state solution $y_s(\xi,\phi)$ and the effectiveness factor are continuously differentiable functions of ϕ, and:

$$P(\xi) = \frac{\partial}{\partial \phi}\, y_s(\xi) \tag{8.67}$$

satisfies, when used in Eq. 8.55:

$$\frac{1}{\xi^q}\frac{d}{d\xi}\left(\xi^q \frac{dP}{d\xi}\right) - F(\xi)P = -2\beta\phi G(\xi) \tag{8.68}$$

with

$$P(1) = 0, \qquad dP(0)/d\xi = 0 \tag{8.69}$$

But if Eq. 8.68 is multiplied by $\xi^q v$, Eq. 8.65 by $\xi^q P$, and subtracted, one has on integrating over $(0,1)$:

$$\left(\frac{dP}{d\xi}\right)_{\xi=1} = -\frac{2\phi\beta}{v(1)}\int_0^1 \xi^q G(\xi)v(\xi)\,d\xi \tag{8.70}$$

On the other hand:

$$\phi^2\eta = -\frac{q+1}{\beta}\left(\frac{dy_s}{d\xi}\right)_{\xi=1} \tag{8.71}$$

and so:

$$\frac{d(\phi^2\eta)}{d\phi} = 2\phi(q+1)\int_0^1 \xi^q G(\xi)\frac{v(\xi)}{v(1)}\,d\xi \tag{8.72}$$

where Eq. 8.70 has been used. Since $G(\xi) > 0$, a sufficient condition for instability will be:

$$\frac{d}{d\phi} (\phi^2 \eta) < 0 \tag{8.73}$$

for this can only be the case if $v(\xi)$ changes its sign at least once in $0 \le \xi \le 1$.

The eigenvalues λ are displaced continuously as one moves along a branch. Therefore, when a positive eigenvalue crosses the real axis, $v(1) = 0$ and $d(\phi^2\eta)/d\phi \rightarrow -\infty$. It follows that a shift from instability to stability can occur only at a bifurcation point. Moreover, any branch for which $d\eta/d\phi \rightarrow -\infty$ at either one of the branching points is unstable. Shown in Figure 8.6 is an (η,ϕ)-curve having at most five steady states obtained by Aris and Copelowitz (1970) for an exothermic reaction in a spherical pellet. The branches of the solution are OA, AB, BC, CD and DE, the first and last corresponding to the minimal and maximal solutions in the interval $\phi_b < \phi < \phi^b$. Since all solutions on the same branch have the same stability character, any branch for which $d(\phi^2\eta)/d\phi$ is negatively infinite (Eq. 8.73) at an endpoint is unstable. If one traces the sign of this derivative, as Figure 8.7, it can be seen that the branches AB, BC, and CD must give unstable steady states. Therefore, only the maximal and minimal solutions may be stable.

Luss and Lee (1968) demonstrated that a unique steady state is stable when Le $= 1$ under the assumption of zero residual enthalpy. This assumption is not too much a restriction since a separable solution for E^* (Eq. 8.50) gives:

$$E^*(\xi,\tau) = \sum_{i=1}^{\infty} a_i v_i e^{-\lambda_i t} \tag{8.74}$$

where the coefficients a_i have to be chosen to satisfy the initial distribution $E_0(\xi)$, and $v_i(\xi)$ are eigenfunctions of:

$$\frac{1}{\xi^q d\xi} \left(\xi^q \frac{dv}{\xi} \right) + \lambda_i v = 0 \tag{8.75}$$

with
$$v(1) = 0, \qquad dv(0)/d\xi = 0 \tag{8.76}$$

Now that E^* decays exponentially with time, any time t at which E^* is sufficiently close to zero can be chosen as the starting point. Hence, the only equation of interest now is Eq. 8.54, written in the following form:

$$\frac{\partial y}{\partial \tau} = \nabla^2 y + f(y) \tag{8.77}$$

with
$$y = 1 \text{ on the boundary} \tag{8.78}$$

where $f(y)$ represents the heat generated by the reaction.

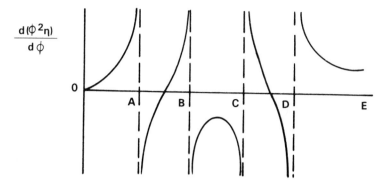

Figure 8.7 Slopes for the branches of the solution for the reaction in Figure 8.6. (Aris 1975)

The maximum principle for parabolic equations (Il'in et al. 1962), which is needed for the demonstration, can be stated as follows: if $u(\mathbf{r},\tau)$ is a continuous function satisfying:

$$\frac{\partial u}{\partial \tau} = \nabla^2 u + c(\mathbf{r})u$$

in the domain of interest, and $c(\mathbf{r})$ is bounded from above, then $u(\mathbf{r},0) \geq 0$ in the domain and $u(\mathbf{r},\tau) \geq 0$ on the boundary imply that $u(\mathbf{r},\tau) \geq 0$ in the domain for all $\tau > 0$.

Let y_s be the unique steady state solution of:

$$\nabla^2 y_s + f(y_s) = 0 \tag{8.79}$$

with

$$y_s = 1 \text{ on the boundary,}$$

then:

$$y^* = y(\mathbf{r},\tau) - y_s(\mathbf{r}) \tag{8.80}$$

satisfies:

$$\frac{\partial y^*}{\partial \tau} = \nabla^2 y^* + f'(y^*)y^* \tag{8.81}$$

with

$$(y^*)_s = 0 \text{ on the boundary} \tag{8.82}$$

where

$$f'(y^*) = \frac{f[y(\mathbf{r},\tau)] - f[y_s(\mathbf{r})]}{y(\mathbf{r},\tau) - y_s(\mathbf{r})} \tag{8.83}$$

Now $f'(y^*)$, though unknown, is certainly bounded from above so that the theorem applies to Eq. 8.81: if the small perturbation y^* is initially one-sided (i.e., either non-negative or non-positive) then it remains one-sided. Moreover, let:

$$v = \frac{\partial y(\mathbf{r},\tau)}{\partial \tau} \tag{8.84}$$

so that partial differentiation of Eq. 8.81 yields:

$$\frac{\partial v}{\partial \tau} = \nabla^2 v + f'(y^*)v \tag{8.85}$$

with $\qquad\qquad\qquad v = 0 \quad$ on the boundary. $\tag{8.86}$

Then the theorem also applies to this equation and v exhibits the same one-sided behavior.

Consider the solution of the transient equations for which $y(\mathbf{r},0) \equiv 1$, and denote the corresponding deviation by y_1^*. Then y_1^* satisfies Eq. 8.81, and:

$$y_1^*(\mathbf{r},0) = 1 - y_s(\mathbf{r}) \le 0 \tag{8.87}$$

and $\qquad\qquad\qquad v_1(\mathbf{r},0) = f(1) > 0 \tag{8.88}$

Thus, according to the maximum principle:

$$y_1^*(\mathbf{r},\tau) = y_1(\mathbf{r},\tau) - y_s(\mathbf{r}) \le 0 \tag{8.89}$$

$$v_1(\mathbf{r},\tau) = \frac{\partial y_1(\mathbf{r},\tau)}{\partial \tau} \ge 0 \tag{8.90}$$

Hence, $y_1(\mathbf{r},\tau)$ rises monotonically and with no oscillations to $y_s(\mathbf{r})$. An entirely similar argument can be used for $y_{1+\beta}^*(\mathbf{r},\tau)$, the solution of the transient equation whose initial value is constant at $(1 + \beta)$. It will therefore tend downwards toward the steady state. Calculations carried out by Luss and Lee (1968) for these two extreme cases, given in Figure 8.8, show the way in which these two solutions approach $y_s(\mathbf{r})$ in a spherical pellet; it is noticeable that $y_{1+\beta}^*$ approaches y_s much more rapidly than y_1^*. Finally one can show in the same way that no solution with initial values $y_s(\mathbf{r})$ between 1 and $1 + \beta$ ever crosses y_1^* or $y_{1+\beta}^*$ (Luss and Lee 1968), which means that:

$$1 \le y_s(\mathbf{r}) \le 1 + \beta \tag{8.91}$$

implies: $\qquad\qquad y_1^*(\mathbf{r},\tau) \le y(\mathbf{r},\tau) < y_{1+\beta}^*(\mathbf{r},\tau) \tag{8.92}$

Hence, any transient $y(\mathbf{r},\tau)$ must converge to the unique steady state, and uniqueness implies global stability, when $Le = 1$.

The problems considered so far in this section have been restricted to the

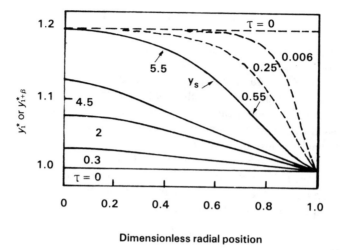

Dimensionless radial position

Figure 8.8 Approach to the steady state (y_s) of y_1^* and $y_{1+\beta}^*$ for $\beta = 0.2$, irreversible first-order reaction in a spherical pellet: - - - - $y_{1+\beta}^*$, —— y_1^*. (Luss and Lee 1968; reprinted with permission from *Chemical Engineering Science*. Copyright by Pergamon Press, Inc.)

case of unit Lewis number. This allowed a great simplification of the problems. The usual range of Lewis number is 0.5 to 1000 and in most practical applications it is greater than unity. When Le $\neq$ 1 the transients are described by the two coupled equations 8.44 and 8.45 instead of one, and the stability analysis becomes more difficult. Linearization of these two equations yields:

$$\frac{\partial \eta_1^*}{\partial \tau} = \nabla^2 \eta_1^* - \left(\frac{\partial r_c}{\partial z}\right) \eta_1^* - \left(\frac{\partial r_c}{\partial y}\right) \eta_2^*$$

$$\text{Le} \frac{\partial \eta_2^*}{\partial \tau} = \nabla^2 \eta_2^* + \beta \left(\frac{\partial r_c}{\partial z}\right) \eta_1^* + \beta \left(\frac{\partial r_c}{\partial y}\right) \eta_2^*$$

Assuming separable solutions of the form:

$$\eta_i^* = w_i e^{\lambda \tau}, \qquad i = 1, 2$$

one obtains:

$$\nabla^2 w_1 - \left(\frac{\partial r_c}{\partial z} + \lambda\right) w_1 - \left(\frac{\partial r_c}{\partial y}\right) w_2 = 0$$

$$\nabla^2 w_2 + \beta \left(\frac{\partial r_c}{\partial z}\right) w_1 + \left[\beta \left(\frac{\partial r_c}{\partial y}\right) - \lambda \text{Le}\right] w_2 = 0$$

$$w_1 = w_2 = 0 \quad \text{on the boundary.}$$

These non-selfadjoint equations have an infinite set of corresponding eigenvalues, some of which may be complex. Luss (1977) argues that changes in Lewis number can shift only pairs of eigenvalues across the imaginary axis and therefore, a steady state that has an odd number of positive eigenvalues for Le = 1 will have at least one positive eigenvalue for any Lewis number, implying instability. He also argues that the Lewis number may affect only the stability of those steady states that are candidates for stable steady states as determined by the criterion of Gavalas (1968) discussed earlier. It appears that the Lewis number has a detrimental effect on stability when it is less than unity.

The stability analysis becomes quite difficult when both the intra- and interparticle gradients have to be accounted for. Only when $(Bi)_h = (Bi)_m$ can the previous analysis be applied to obtain general criteria. The strongest available criterion is that of Gavalas (1968) regarding the stability of $(2m + 1)$ steady states.

Summary

It has been seen that transport resistances and catalyst deactivation can have a significant effect on selectivity. It has also been seen that the effect of transport resistances is not necessarily detrimental, even though the selectivity is, in general, adversely affected by diffusion for consecutive reactions. The selectivity is independent of deactivation for structure-insensitive reactions when the same sites are involved for all reactions, provided that the concentration and temperature do not change with deactivation. This condition is rarely met in reactor operation and therefore, the selectivity changes with the extent of deactivation.

The stability associated with a single pellet has been examined in some detail. This stability represents the stability at a point in the reactor. Perhaps the most significant results of the stability analyses are the slope condition regarding the rates of heat removal and heat generation and the criterion due to Gavalas (1968) regarding the number of unstable steady states. The slope condition, which essentially states that the reactor is unstable when the rate of heat generation by the reaction is larger than that of heat removal by the bulk-fluid, is an important criterion for the safe operation of reactors involving highly exothermic reactions.

NOTATION

a_i	$i = 1$ to 4, constants in Eqs. 8.4 and 8.5
A	bulk fluid concentration of species A
A_f	reactor inlet concentration of species A
A_L	reactor outlet concentration of species A
A_s	pellet surface concentration of species A
A_{ss}	pellet surface concentration of species A at steady state
$\mathbf{A}$	matrix in Eq. 8.37
B	bulk fluid concentration of species B
B_f	reactor inlet concentration of species B
B_L	reactor outlet concentration of species B

B_s	pellet surface concentration of species B
Bi_h	Biot number for heat defined in Eq. 8.29c
Bi_m	Biot number for mass defined in Eq. 8.29c
$\mathbf{C}$	concentration vector
$\mathbf{C}_t$	concentration vector transformed through Eq. 8.18
C	concentration
C_A	concentration of species A
C_B	concentration of species B
C_0	initial concentration
C_p	specific heat content
D	effective diffusivity
$\mathbf{D}$	effective diffusivity matrix
E^*	residual enthalpy defined by Eq. 8.48
E_a	activation energy
$f(C)$	concentration dependence of rate of reaction
$F(\xi)$	quantity defined in Eq. 8.60
$\bar{g}$	fractional pellet surface activity remaining after deactivation
$G(\xi)$	quantity defined by Eq. 8.56
h	film heat transfer coefficient; fraction of catalyst still remaining active after deactivation (h_i, $i = 1$ to 3)
$(-\Delta H)$	heat evolved per mole reacted
k, k_A, k_B, k_C, k_i	rate constant with subscript denoting the species involved
$k_g, \bar{k}_g$	film mass transfer coefficient, k_g including volume to surface area ratio
$\mathbf{K}$	rate constant matrix
$\mathbf{K}_e$	effective rate constant matrix in Eq. 8.19
L	pellet volume to external surface area ratio
Le	Lewis number defined in Eq. 8.29c
m, n	order of reaction
$\mathbf{n}$	outward normal vector
p	$\gamma \eta_B \phi_B^2 / (\eta_A \phi_A^2)$
P	quantity defined in Eq. 8.67
q	0 for slab, 1 for cylinder, 2 for sphere
Q_c	rate of heat removal given in Eq. 8.34
Q_h	rate of heat generation given in Eq. 8.34
$\mathbf{r}$	spatial coordinate vector
r_c	intrinsic rate of reaction given by $r_c = kf(C)$
$\bar{r}_c$	r_c evaluated at pellet surface
r_A, r_B	intrinsic rate of reaction for species A and B, respectively
R	length of spatial coordinate
R_G	global rate of reaction
$(R_G)_A, (R_G)_B$	global rate of reaction for species A and B, respectively
$(R_G)_A^s, (R_G)_B^s$	$(R_G)_A$ and $(R_G)_B$ evaluated at pellet surface temperature
$(R_G)_i$	global rate for i^{th} reaction path
s	k_A/k_B
s_b	s evaluated at bulk fluid temperature

S_x	external surface area
t	time
T	temperature
T_b	bulk-fluid temperature
T_s	pellet surface temperature
T_0	initial temperature
T_{ss}	pellet surface temperature at steady state
u_A, u_B	transformed concentration given in Eq. 8.4
v_m	columns of $\mathbf{V}$
V, V_p	pellet volume
$\mathbf{V}$	matrix in Eq. 8.18
W	variable given in Eq. 8.61
$\mathbf{x}$	vector in Eq. 8.35
X_s	pellet surface concentration of species X
y	T/T_b
y^*	variable defined in Eq. 8.57
z	C/C_b

Greek Letters

α	constant used in exponential deactivation function
β	constant defined in Eq. 8.29c
γ	fraction of catalyst deactivated; D_B/D_A
Λ	diagonal matrix in Eq. 8.19
ϵ	$E_a/R_g T_b$
ϵ_A, ϵ_B	ϵ for r_A and r_B, respectively
λ	effective thermal conductivity
λ_m	eigenvalue
ξ	r/R
η	effectiveness factor
η_i	internal effectiveness factor
η_1^*	$z - z_s$
η_2^*	$y - y_s$
η_A, η_B, η_X	η for r_A, r_B, and r_x, respectively
ρ_p	pellet density
τ	Dt/R^2
$(\rho C_p)_w$	ρC_p for single wire
ϕ	Thiele modulus defined in Eq. 8.29c
ϕ_A, ϕ_B	ϕ for r_A and r_B, respectively

Subscripts

b	bulk-fluid
i	species i
s	pellet surface; steady state
p	pellet
s_s	s evaluated at pellet surface temperature

PROBLEMS

8.1. Consider first-order consecutive reactions $A \xrightarrow{k_a} B \xrightarrow{k_b} C$. Show that the point selectivity $-(R_G)_B / (R_G)_A$ can be expressed as follows:

$$\frac{-(R_G)_B}{(R_G)_A} = \frac{s}{s-1} \left(1 - \frac{\eta_B}{\eta_A} \right) + \frac{\eta_B}{\eta_A} \left(\frac{-r_B}{r_A} \right)$$

when $\gamma = 1$. Plot $-(R_G)_B / (R_G)_A$ as a function of η_B / η_A for $s = 0.5, 2,$ and 10, with $(-r_B/r_A)$ set to 0.5. What can you conclude about the effect of diffusion? Note that when $s > 1$, $\eta_B / \eta_A > 1$.

8.2. Consider the following reaction network:

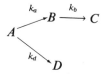

Find an expression for the point selectivity $-(R_G)_B / (R_G)_A$ when all reactions are first-order. When the reaction path $A \longrightarrow D$ is second-order, the network is known as a van de Vusse network. For the van de Vusse network, would you expect the yield to increase when k_d is decreased? What if the reactions are severely diffusion-limited? For detailed results, refer to Carberry (1976).

8.3. For the following reactions:

$$A \xrightarrow{k_B} B \ (n^{th} \text{ order})$$

$$A \xrightarrow{k_C} C \ (m^{th} \text{ order})$$

in which B is the desired product, Roberts (1972) has shown that:

$$\frac{(R_G)_B / (R_G)_A}{r_B/r_A} \simeq \frac{m+1}{2n-m-1}$$

in the asymptotic region of strong diffusion effect if:

$$r_B/r_C = k_B A_s^{n-m}/k_C \ll 1.$$

Derive this result.

8.4. Use the diagonalization technique to write the global rates as in Eq. 8.20 for the following first-order reactions:

$$A \underset{k_{-1}}{\overset{k_1}{\rightleftharpoons}} G$$

$$k_3 \searrow \quad \swarrow k_2$$

$$S$$

8.5. Show for the case of uniform deactivation that the point selectivity is not dependent on time and therefore, is independent of deactivation, provided that all reactions are first-order and take place on the same sites. Is the statement correct if the kinetics for some reactions are not first-order?

8.6. Use the approach of Wei (1962) to obtain global rate expressions R_G for a complex first-order reaction network when the reactions are also affected by catalyst deactivation. Assume pore-mouth poisoning and slab geometry. Note that η_k for each reaction path is given by:

$$\eta_k = \frac{1/\phi_k}{1 + \gamma\phi_k} ; \qquad \phi_k^2 = L^2 k_k/D_k$$

Based on the expressions obtained, construct the selectivity curve given in Figure 8.4 for the case of pore-mouth poisoning.

8.7. Show that Eq. 8.43 is a necessary condition for the stability of the system described by Eqs. 8.41 and 8.42. What is the physical meaning of the necessary condition. Why can the system become unstable even when Eq. 8.43 is satisfied? Give physical reasons.

8.8. For this problem, follow the steps given below to obtain an instability criterion.

Dimensionless mass and heat balances for a slab-like pellet in which a single reaction takes place can be written as follows:

$$\frac{\partial w}{\partial \theta} - \frac{\partial^2 w}{\partial z^2} = -g(w)U(y) \tag{1}$$

$$Le\frac{\partial y}{\partial \theta} - \frac{\partial^2 y}{\partial z^2} = \beta g(w)U(y) \tag{2}$$

where w and y are the dimensionless concentration and temperature, respectively. Here, the rate expression is given as a product of a temperature dependent function $U(y)$ and a concentration-dependent function $g(w)$. Le and β are Lewis number and Prater number, respectively.

a. Expand the equations about a steady state using:

$$w(z, \theta) = w_0(z) + \epsilon w_1(z, \theta) + (\epsilon^2) \tag{3}$$

$$y(z, \theta) = y_0(z) + \epsilon y_1(z, \theta) + (\epsilon^2) \tag{4}$$

so that upon substituting Eqs. (3) and (4) into (1) and (2), and collecting terms of ϵ^n of same order for $n = 0$ and $n = 1$, there result:

$$\left[\begin{array}{l} \dfrac{\partial^2 w_0}{\partial z^2} = g_0 U_0 \end{array}\right.$$ (5)

$$\qquad\qquad\qquad \text{for } \epsilon = 0$$

$$\dfrac{\partial^2 y_0}{\partial z^2} = -\beta g_0 U_0 \qquad\qquad$$ (6)

$$\left[\begin{array}{l} \dfrac{\partial w_1}{\partial \theta} - \dfrac{\partial^2 w_1}{\partial z^2} = -(U_0 g_0' w_1 + U_0' g_0 y_1) \end{array}\right.$$ (7)

$$\qquad\qquad\qquad\qquad \text{for } \epsilon = 1$$

$$\text{Le } \dfrac{\partial y_1}{\partial \theta} - \dfrac{\partial^2 y_1}{\partial z^2} = \beta(U_0 g_0' w_1 + U_0' g_0 y_1)$$ (8)

where $g_0' = \dfrac{dg}{dw}\bigg|_{w=w_0}$ and $U_0' = \dfrac{dU}{dy}\bigg|_{y=y_0}$ (9)

b. Assume separable solutions for w_1 and y_1:

$$w_1 = W e^{-\lambda \theta}$$ (10)

$$y_1 = Y e^{-\lambda \theta}$$ (11)

so that Eqs. (7) and (8) can be written as:

$$\lambda W + \dfrac{d^2 W}{dz^2} = U_0 g_0' W + U_0' g_0 Y$$ (12)

$$\lambda \text{Le } Y + \dfrac{d^2 Y}{dz^2} = -\beta(U_0 g_0' W + U_0' g_0 Y)$$ (13)

Define Y_1 by:

$$Y = \dfrac{-\beta}{\text{Le}} W + (\text{Le} - 1) Y_1$$ (14)

and show that Eqs. (12) and (13) can be transformed into:

$$\dfrac{d^2 W}{dz^2} = W(U_0 g_0' - \dfrac{\beta}{\text{Le}} U_0' g_0 - \lambda) + U_0' g_0(\text{Le} - 1) Y_1$$ (15)

$$Y_1 = \dfrac{1}{(\lambda \text{Le})^{1/2}} \int_0^z \sin[(\lambda \text{Le})^{1/2} (z - \xi)] \left(\dfrac{-\beta}{\text{Le}} \dfrac{d^2 W}{d\xi^2}\right) d\xi$$ (16)

Note that one is interested in the sign of λ for stability. Also note that now Eqs. (15) and (16) are sufficient for considerations of stability.

c. Seek a series solution for W in terms of an orthogonal set f_n over $(0,1)$, i.e.:

$$W(z) = \sum_{n=0}^{\infty} A_n f_n(z) \tag{17}$$

Substitute Eq. (17) into (15) and (16) and integrate over the interval to obtain the following:

$$\Sigma A_n \alpha_{mn} = -\lambda B A_m + \Sigma A_n \beta_{mn} - \frac{\beta(Le-1)}{Le(\lambda Le)^{1/2}} \Sigma A_n \gamma_{mn} \tag{18}$$

where $B = \int_0^1 f_n^2 dz$

$\alpha_{mn} = \int_0^1 f_m f_n'' dz$

$\beta_{mn} = \int_0^1 f_m f_n (U_0 g_0' - \frac{\beta}{Le} U_0' g_0) dz$

$\gamma_{mn} = \int_0^1 \int_0^z \sin[(\lambda Le)^{1/2} (z - \xi)] f_m f_n'' (\xi) d\xi dz$

d. For a nontrivial solution of Eq. (18) for A_m's, the determinant of the matrix given by the equation must be zero, which will yield solution(s) for λ. For stability, λ should be positive. Equation (18) is an infinite-dimensional matrix for the solution of A_m's. However, 1×1 ($n = m = 0$) case can be considered for an instability criterion. Obtain the instability criterion for the 1×1 case.

REFERENCES

Amundson, N.R. and L.R. Raymond, AIChE J., *11*, 339 (1965).

Aris, R. and J. Copelowitz, Chem. Eng. Sci., *25*, 909 (1970).

Aris, R., *Mathematical Theory of Diffusion and Reaction in Permeable Catalyst*, Vol. I and II, Oxford University Press, London (1975).

Carberry, J.J., Chem. Eng. Sci., *21*, 275 (1966).

Carberry, J.J., *Chemical and Catalytic Reaction Engineering*, pp 486–490, McGraw-Hill, New York (1976).

De Pauw, R. and G.F. Froment, Chem. Eng. Sci., *30*, 789 (1975).

Gavalas, G.R., *Nonlinear Differential Equations of Chemically Reacting System*, Springer-Verlag, Heidelberg (1968).

Il'in, A.M., A.S. Kalashnikov and O.A. Oleinik, Russ. Math. Surv., *17*, (3), 1 (1962).

Jackson, R., Chem. Eng. Sci., *28*, 1335 (1973).

Lee, J.C.M., L. Padmenabhan and L. Lapidus, Ind. Eng. Chem. Fund., *11*, 117 (1972).

Luss, D., in Lapidus and Amundson, eds., Chap. 4, Prentice-Hall, Englewood Cliffs, N.J. (1977).

Luss, D. and J.C.M. Lee, Chem. Eng. Sci., *23*, 1237 (1968).

Roberts, G., Chem. Eng. Sci., *27*, 1409 (1972).

Sada, E. and C.Y. Wen, Chem. Eng. Sci., *22*, 559 (1967).

Voge, H.H. and C.Z. Morgan, Ind. Eng. Chem. Proc. Des. Dev., *11*, 454 (1972).

Weekman, V.W., Jr. and D.M. Nace, AIChE J., *16*, 397 (1970).

Wei, J., J. Catalysis, *1*, 526, 538 (1962).

PART III

Reactor Design and Analysis

CHAPTER 9

Fixed-Bed Reactors

9–1 INTRODUCTION

As evident from the list of typical catalytic processes given in Table 1.1, most catalytic processes are carried out in fixed-bed reactors. These reactors are usually large capacity units, reaching in some cases, as in ammonia synthesis, capacities of half a million tons per year. The reactors are not single packed tubes. With all the auxiliary equipment for gas processing such as feeding, compressing, and heating or cooling, and the support units, the reactors are indeed complicated. Large capacity reactors can have tens of thousands of tubes operating in parallel. At the heart of these reactors is still a single tube packed with a supported catalyst in one form or another.

Many considerations go into the design of a fixed-bed reactor. These include the mode of heat exchange, bed pressure drop, safe operating temperature range, the mode of catalyst packing, and so on. Sizing a reactor, which is usually of primary concern, may turn out to be of secondary importance in the overall scheme of the design when economics and operability are considered. Rapid temperature rises in the bed caused by highly exothermic reactions may have a significant unfavorable effect on selectivity and equilibrium conversion. Or it may be altogether unsafe to operate the reactor under certain feed conditions. When effluent gases have to be recycled, the pressure drop in the bed has to be kept as low as possible to minimize the costs of recycling. It may be desirable to pack different parts of a reactor with catalyst particles of different size. These design considerations place limits within which a designer is free to choose the size and operating conditions of a reactor. Conservation of mass, energy, and momentum forms the basis on which these limits can be specified, as well as the size and operating conditions of the reactor.

Certain assumptions have to be made to write realistic and yet useful conservation equations for a fixed-bed reactor. Different models result depending on the assumptions made. The simplest model is the pseudo-homogeneous model in which the solid phase (catalyst particles) and the fluid phase as a whole are treated as homogeneous. In such cases, the usual conservation equations for homogeneous reactors, in which only one phase is involved, apply. In reality, however, there are certain transport resistances within the catalyst particles and also across the interface between solid and fluid phases. The reaction takes place within the catalyst

285

particles (solid phase) and the fluid phase is simply a medium that supplies the reactants and carries off the products. Heterogeneous models precisely represent this fact. The fluid and solid phases are treated separately, and the linkage between the two phases is the interfacial transport processes. Heterogeneous models are treated in this chapter. The treatment of homogeneous reactors (or equivalently pseudo-homogeneous models) can be found in a number of textbooks (e.g., Levenspiel 1972). It should be recognized that heterogeneous models reduce to pseudo-homogeneous models when all transport resistances are absent.

A rather complete treatment of heterogeneous models is presented first. It should become immediately evident that the conservation equations need not be solved in their entirety because of the simplifications that can be made for the design and analysis of fixed-bed reactors. The model used for the design and analysis should be as simple as the particulars of the reaction system under consideration will allow. In this regard, it is noteworthy that it is possible these days to tailor the design of a reactor according to the conservation equations. This capability, when fully utilized, should eventually lead to new types of reactors and to direct scale-up from experimental data to full-size plant reactors. Before this can be a reality, however, uncertainties regarding the accuracy of correlations for the transport properties (Chapter 14) have to be resolved, especially for highly exothermic reactions.

In this chapter, the classical approach to the design and analysis of fixed-beds is considered. Because of the complexity involved in using the classical approach, attention here is limited to reactions affected by diffusion, and the treatment of reactions affected by simultaneous diffusion and deactivation are postponed to Chapter 10, where the concept of reactor point effectiveness is used.

9-2   CONSERVATION EQUATIONS

The conservation of mass is considered first using fixed coordinates. The equation of continuity (Bird et al. 1960) for a multicomponent system can be written for each species i as:

$$\frac{\partial}{\partial t} \rho_i = -\nabla \cdot (\rho_i \mathbf{u} + \mathbf{j}_i) - r_i \qquad (9.1)$$

This equation can be rewritten in terms of concentration by dividing by the molecular weight M_i:

$$\frac{\partial}{\partial t} C_i = -\nabla \cdot (C_i \mathbf{u} + \mathbf{J}_i) - R_i \qquad (9.2)$$

where the observed rate of formation of species i due to chemical reaction, R_i, which is based on fluid volume, is now expressed in terms of concentration. The molar flux J_i can be due to gradients of concentration, pressure, and temperature. Here, however, the effective dispersion coefficient D_i is defined by the following relationship:

$$\mathbf{J}_i = -\mathbf{D}_i \nabla C_i \qquad (9.3)$$

For fixed-beds, the cylindrical coordinate system can be used. If only the velocity in the direction of flow (z) is taken, Eq. 9.2 can be rewritten with the aid of Eq. 9.3 as:

$$\epsilon_B \frac{\partial C_i}{\partial t} = -\frac{\partial}{\partial z}(vC_i) + \frac{\partial}{\partial z}\left\{(D_a)_i \frac{\partial C_i}{\partial z}\right\} + \frac{1}{r}\frac{\partial}{\partial r}\left\{(D_r)_i \, r \frac{\partial C_i}{\partial r}\right\} - (1 - \epsilon_B)R_{G_i} \qquad (9.4)$$

where D_a and D_r are the axial and radial effective dispersion coefficients, respectively, based on the total area including void space, v is the superficial velocity, and ϵ_B is the bed voidage. Here, the global rate R_{G_i} is multiplied by the factor $(1 - \epsilon_b)$ since R_{G_i} is based on pellet volume. For constant velocity and dispersion coefficients, this equation is further reduced to:

$$\epsilon_B \frac{\partial C_i}{\partial t} + v \frac{\partial C_i}{\partial z} - (D_a)_i \frac{\partial^2 C_i}{\partial z^2} - (D_r)_i \left(\frac{\partial^2 C_i}{\partial r^2} + \frac{1}{r}\frac{\partial C_i}{\partial r}\right) = -(1 - \epsilon_B)R_{G_i} \qquad (9.5)$$

The energy conservation equation can be written directly from Eq. 9.4 by simply replacing C_i by $\rho C_p T$, D by the effective thermal conductivity $\rho C_p K$, and R_{G_i} by $\Sigma(-\Delta H_i)R_{G_i}$, and by neglecting the kinetic and potential energy terms.

For a single reaction taking place in the fixed-bed, the conservation equations for constant v and ρC_p are:

$$\epsilon_B \frac{\partial C_b}{\partial t} + v \frac{\partial C_b}{\partial z} - D_a \frac{\partial^2 C_b}{\partial z^2} - D_r \left(\frac{\partial^2 C_b}{\partial r^2} + \frac{1}{r}\frac{\partial C_b}{\partial r}\right) = -(1 - \epsilon_B)R_G \qquad (9.6)$$

$$\epsilon_B \frac{\partial T_b}{\partial t} + v \frac{\partial T_b}{\partial z} - K_a \frac{\partial^2 T_b}{\partial z^2} - K_r \left(\frac{\partial^2 T_b}{\partial r^2} + \frac{1}{r}\frac{\partial T_b}{\partial r}\right) = \left(\frac{1 - \epsilon_B}{\rho C_p}\right)(-\Delta H)R_G \qquad (9.7)$$

where the subscript i has been dropped for the single reaction, and K_a and K_r are the axial and radial effective thermal dispersion coefficients, respectively. Here, the subscript b has been used for C and T to denote that these are bulk-fluid quantities. The radial boundary conditions are:

$$\frac{\partial T_b}{\partial r} = \frac{\partial C_b}{\partial r} = 0 \qquad \text{at the center} \qquad (9.8)$$

$$\frac{\partial C_b}{\partial r} = 0, \quad h_w(T_b - T_c) = -\rho C_p K_r \frac{\partial T_b}{\partial r} \qquad \text{at the wall} \qquad (9.9)$$

The first set of boundary conditions results from symmetry about the centerline of the tube. The second set results from the fact that no reactant transport takes place across the wall, and that heat transfer to the coolant, whose temperature is T_c, is equal to the heat conducted at the wall. The axial boundary conditions (Wehner and Wilhelm 1956) are:

$$v_{in}(C_b)_{in} = \left(vC_b - D_a \frac{\partial C_b}{\partial z} \right)_{z=0} \tag{9.10a}$$

at the inlet

$$v_{in}(T_b)_{in} = \left(vT_b - K_a \frac{\partial T_b}{\partial z} \right)_{z=0} \tag{9.10b}$$

$$\frac{\partial C_b}{\partial z} = \frac{\partial T_b}{\partial z} = 0 \qquad \text{at the outlet} \tag{9.11}$$

These boundary conditions are based on continuity of flux across a boundary, i.e., at the inlet and outlet of the bed. At the inlet continuity of flux requires that:

$$\text{Flux}|_{z=0-} = \text{flux}|_{z=0+}$$

$$\text{or} \quad \left(vC_b - D_a \frac{dC_b}{dz} \right)_{0-} = \left(vC_b - D_a \frac{dC_b}{dz} \right)_{0+}$$

If the feed is well mixed such that the dispersion term is zero, the boundary condition of Eq. 9.10a results. At the outlet $(vC)_{z-} = (vC)_{z+}$, and $(dC/dz)_{z+} = 0$ since no reaction takes place. These conditions lead to the boundary condition of Eq. 9.11. The initial conditions are the steady state solutions of Eqs. 9.6 and 9.7. The left hand side of Eqs. 9.6 and 9.7 represents the changes taking place in the fluid phase as a result of the reaction in the solid phase (catalyst pellets), which is given by the right hand side.

The global rate R_G is equal to the rate at which mass is transferred across the interface between fluid and solid phases, which in turn is related to the flux at the catalyst surface:

$$R_G = \frac{k_g}{L}(C_b - C_s) = \frac{D_e}{L} \frac{\partial C_p}{\partial x}\bigg|_{x=L} = \eta r_c \tag{9.12}$$

where k_g = film mass transfer coefficient
  L = characteristic length of catalyst pellet, volume/external surface; surface
  D_e = effective diffusivity
  C_s = fluid concentration at pellet surface
  η = effectiveness factor
  r_c = intrinsic rate of reaction

where the subscript p is for the pellet. In a similar manner, one gets:

$$(-\Delta H)R_G = \frac{h}{L}(T_s - T_b) = \frac{-\lambda_e}{L}\frac{\partial T_p}{\partial x}\bigg|_{x=L} \tag{9.13}$$

where $(-\Delta H)$ = heat of reaction
h = film heat transfer coefficient
T_s = pellet surface temperature
λ_e = effective thermal conductivity

In order to relate the intrinsic rate of reaction to the global rate, the effectiveness factor, or equivalently, the gradient at the pellet surface has to be calculated. This is obtained by solving the conservation equations for the pellet:

$$\epsilon_p \frac{\partial C_p}{\partial t} = D_e \frac{\partial^2 C_p}{\partial x^2} - r_c \tag{9.14}$$

$$(\epsilon \rho C_p)_p \frac{\partial T_p}{\partial t} = \lambda_e \frac{\partial^2 T_p}{\partial x^2} + (-\Delta H)r_c \tag{9.15}$$

where D_e and λ_e have been assumed constant, and ϵ_p is the pellet voidage. The usual boundary conditions at the center and the surface of pellet apply. When Eqs. 9.6 and 9.7 are made dimensionless, key reactor parameters emerge. Letting:

$$n = \frac{Z}{d_p}, \quad m = \frac{R}{d_p}, \quad A = \frac{Z}{R}, \quad \theta = \frac{Z}{v},$$

$$\bar{C} = \frac{C_b}{(C_b)_{in}}, \quad \bar{T} = \frac{T_b}{(T_b)_{in}}, \quad \bar{r} = \frac{r}{R}, \quad \bar{z} = \frac{z}{Z}, \quad \bar{t} = \frac{t}{\theta}$$

where d_p = pellet diameter
R = radius of fixed-bed
Z = length of fixed bed
C_{in}, T_{in} = reactor inlet concentration and temperature

the reactor conservation equations become:

$$\epsilon_B \frac{\partial \bar{C}}{\partial \bar{t}} + \frac{\partial \bar{C}}{\partial \bar{z}} - \frac{1}{n P_{ma}}\frac{\partial^2 \bar{C}}{\partial \bar{z}^2} - \frac{A}{m P_{mr}}\left(\frac{\partial^2 \bar{C}}{\partial \bar{r}^2} + \frac{1}{\bar{r}}\frac{\partial \bar{C}}{\partial \bar{r}}\right) = \frac{-R_G \theta}{(C_b)_{in}}(1-\epsilon_B) \tag{9.16}$$

$$\epsilon_B \frac{\partial \bar{T}}{\partial \bar{t}} + \frac{\partial \bar{T}}{\partial \bar{z}} - \frac{1}{n P_{ha}}\frac{\partial^2 \bar{T}}{\partial \bar{z}^2} - \frac{A}{m P_{hr}}\left(\frac{\partial^2 \bar{T}}{\partial \bar{r}^2} + \frac{1}{\bar{r}}\frac{\partial \bar{T}}{\partial \bar{r}}\right) = \frac{(-\Delta H)R_G \theta}{(\rho C_p)(T_b)_{in}}(1-\epsilon_B) \tag{9.17}$$

The axial Peclet number for mass dispersion P_{ma} and the radial Peclet number P_{mr} are given by:

$$P_{ma} = \frac{d_p v}{D_a}, \quad P_{mr} = \frac{d_p v}{D_r} \tag{9.18}$$

The Peclet numbers for heat dispersion are given by:

$$P_{ha} = \frac{vd_p}{K_a}, \qquad P_{hr} = \frac{vd_p}{K_r} \qquad (9.19)$$

It is seen then that the Peclet numbers and aspect ratios n, m, and A determine the extent to which the dispersion affects the reactor behavior.

Solving Eqs. 9.6 through 9.15 for the design and analysis of a fixed-bed reactor is not a simple matter by any means and may be a formidable numerical problem. The steady-state versions of these equations are used for design purposes. It shall soon be seen that considerable simplifications can be made for typical operating conditions.

9-3   SIMPLIFICATIONS

The nature of a reaction and the mode of reactor operation immediately suggest the kind of simplifications that can be made. If the heat of reaction is negligible so that a reactor is run isothermally, all heat balance equations can be eliminated. When the heat of reaction is not very high, the reactor is usually run as an adiabatic reactor, as many reactors are in practice. In such cases, radial gradients are negligible since only the velocity profile, which can be assumed to be flat, can give rise to gradients. Therefore, a one dimensional treatment of the conservation equations is sufficient for adiabatic reactors. For very exothermic reactions, however, external cooling is required to preserve the integrity of the catalyst pellets and to ensure the safety of reactor operation. In such cases, the whole equations as posed may have to be solved.

As indicated earlier, the Peclet numbers and aspect ratios determine the relative importance of dispersion terms. It is apparent from Eq. 9.16 that axial dispersion can be neglected compared to the convection term when (nP_{ma}) is large. The axial Peclet number for mass, P_{ma}, is about 2 for gases flowing through packed beds for Reynolds numbers (based on d_p) greater then 10. Since the axial aspect ratio n ($= Z/d_p$) is large, (nP_{ma}) for typical reactors is quite large and the axial mass dispersion term can be neglected. Much less is known about the axial and radial Peclet numbers for heat. However, P_{ha} is in general smaller than P_{ma}, as evidenced (Price and Butt 1977) by the necessity of including the axial heat dispersion term even when the axial mass dispersion term is negligible. Details on effective dispersion coefficients can be found in Chapter 14. Based on the mixing-cell dispersion analogy, in which the bed is considered to consist of n CSTRs in series, Carberry (1958) suggests that all axial dispersion terms can be neglected when n is greater than 150, except for unusual reactor configurations.

While a large value of (nP) for axial dispersion results in negligible dispersion effects compared to convection, the inverse is true for radial dispersion simply because the only transfer mechanism is conduction and a larger value of radial

Peclet number (a smaller value of K) means a higher resistance to conduction. The radial Peclet number for mass is typically 11 for Reynolds number greater than 30. The corresponding value for heat is generally smaller. On the other hand, the combination of radial and reactor aspect ratios m/A ($= R^2/d_p Z$) is quite small. It is readily seen then that the radial dispersion terms cannot readily be neglected unless the bed is adiabatic. If the tube radius is of the order of one pellet diameter and heat exchange is only at the tube wall, on the other hand, the radial gradients can be neglected since then the whole cross-section of the tube can be considered to be well-mixed. In general, the radial gradients can be neglected for all practical purposes if the radial aspect ratio m ($= R/d_p$) is 3 or 4 as suggested by Carberry (1976). As discussed earlier, the radial gradients exist when the bed is cooled externally, and this cooling is done only for very exothermic reactions: otherwise, the reactor is run adiabatically. Since the radial thermal gradient is caused by the external cooling and this in turn causes the radial concentration gradient, the major dispersion effect is in the radial direction, and axial dispersion can be neglected in comparison to radial dispersion. The conservation equations in such a case can be written as given in Table 9.1, and involve only axial convection and radial dispersion. Furthermore, the very exothermic nature of the reaction in such cases dictates that the radial temperature gradient be minimized to ensure safety of operation and to preserve the integrity of the catalyst pellets. Therefore, typical industrial reactors for highly exothermic reactions are designed to have a small radial aspect ratio, as practiced in the oxidation of naphthalene. In such cases, it is sometimes possible to neglect both axial and radial dispersion terms, yielding plug-flow conservation equations.

To sum up, it is sufficient to treat an adiabatic reactor as a plug-flow reactor. If the axial dispersion effect is to be included, only the heat dispersion term needs to be added. In the case of nonadiabatic, nonisothermal reactors, axial dispersion terms can be neglected in comparison to the radial dispersion terms. In addition, the radial dispersion terms can often be neglected if the radial aspect ratio is small. The conservation equations for various cases are summarized in Table 9.1.

The aspect ratios appearing in the conservation equations deserve further comment. The quantities appearing in the aspect ratios are important design parameters. The pellet size d_p is an important parameter determining the effects of inter- and intraphase transport, the tube radius is a major factor in heat transfer at the wall (in the form of $2/R$ in the overall heat transfer coefficient U), and the reactor length determines the contact time or the conversion. These parameters may then be determined *a priori* based on certain reactor requirements. This does not necessarily mean that the aspect ratios cannot be manipulated to achieve a desired design objective. If, for instance, it is desired to minimize radial gradients for a highly exothermic reaction without affecting the effectiveness factor and the conversion, R may be reduced, which results in a reduction of m and an increase in U, both of which reduce the radial gradients.

As discussed fully in Chapter 4, the Biot number is large enough under typical reaction conditions to allow isothermal treatment of catalyst pellets. Furthermore, the external mass transfer resistance (interphase mass transfer) can also be

Table 9.1 Conservation Equations for Design

Reactor Conditions		Left Hand Side of Eqs. 9.6 and 9.7
Adiabatic	$\left(\dfrac{Z}{d_p}\right)\left(\dfrac{d_p}{D_a}\right) > 300$	$v\dfrac{\partial C_b}{\partial z}$
	$\mathrm{Re} > 10$	$\rho C_p v\dfrac{\partial T_b}{\partial z}$
	$\left(\dfrac{Z}{d_p}\right)\left(\dfrac{d_p v}{D_a}\right) < 300$	$v\dfrac{\partial C_b}{\partial z}$
	$\mathrm{Re} > 10$	$\rho C_p v\dfrac{\partial T_b}{\partial z} - K_a\dfrac{\partial^2 T_b}{\partial z^2}$ if necessary.
		Otherwise, $\rho C_p v\dfrac{\partial T_b}{\partial z}$
Isothermal		$v\dfrac{\partial C_b}{\partial z}$
Nonadiabatic, nonisothermal	$R/d_p > 4$	$v\dfrac{\partial C_b}{\partial z} - D_r\left(\dfrac{\partial^2 C_b}{\partial r^2} + \dfrac{1}{r}\dfrac{\partial C_b}{\partial r}\right)$
		$\rho C_p v\dfrac{\partial T_b}{\partial z} - K_r\left(\dfrac{\partial^2 T_b}{\partial r^2} + \dfrac{1}{r}\dfrac{\partial T_b}{\partial r}\right)$
	$R/d_p \leq 4$	$v\dfrac{\partial C_b}{\partial z}$
	$\mathrm{Re} > 30$	$\rho C_p v\dfrac{\partial T_b}{\partial z}$

Pellet can be considered isothermal if $k_g \lambda_e/D_e h \gg 1$.
Pellet surface concentration can be considered to be equal to bulk fluid concentration when $\phi/(Bi)_m < 0.05$.

neglected under typical reaction conditions. Therefore, the heat balance of Eq. 9.15 can be eliminated and the pellet surface concentration can be taken to be equal to the bulk fluid concentration. These results are also summarized in Table 9.1. It is to be noted that the results given in Table 9.1 for the conservation equations are for guideline purposes only and that exceptions can always be found. Therefore, the results should be used with caution. The numbers used for pellet conditions are somewhat arbitrary. The condition of $\phi/(Bi)_m < 0.05$, for instance, ensures for a first-order reaction that the error in not including the interphase mass transfer resistance in the overall effectiveness factor will be less than 5% for an isothermal pellet.

9–4 OTHER CONSIDERATIONS

The pressure drop in a fixed-bed due to packed particles is an important design consideration, especially when a large amount of gas needs to be recycled. For very large capacity units, in fact, as in catalytic reforming and ammonia synthesis, a radial flow arrangement (Finneran et al. 1972) is used to reduce the pressure drop and thus enhance the recycle compressor capacity. The Fanning equation may be used for the estimation of the pressure drop:

$$F = \frac{\Delta P}{Z} = -f \frac{G_f^2}{\rho_f d_p} \tag{9.20}$$

where ΔP = pressure drop
G_f = fluid mass rate
f = friction factor
ρ_f = fluid density

which can be obtained from the Bernoulli equation by neglecting the kinetic and potential terms. The friction factor f is given by the Ergun relationship:

$$f = \frac{1}{Re} \left(\frac{1 - \epsilon_B}{\epsilon_B} \right)^2 \left(150 + 1.75 \frac{Re}{1 - \epsilon_B} \right) \tag{9.21}$$

Here, the Reynolds number is based on the particle diameter d_p. For a radial aspect ratio R/d_p less than 25, Mehta and Hawley (1969) suggest the following modified Ergun correlation, which includes wall effects:

$$f = \left(\frac{1 - \epsilon_B}{\epsilon_B^3} \right) \left(1 + \frac{d_p}{3(1 - \epsilon_B)R} \right)^2 \left(\frac{150(1 - \epsilon_B)}{Re} + \frac{1.75}{1 + d_p/(3(1 - \epsilon_B)R)} \right) \tag{9.22}$$

Hicks (1970) suggests the use of the following correlation for smooth spheres in the range of $300 < Re/(1 - \epsilon_B) < 60,000$:

$$f = 6.8 \left(\frac{Re}{1 - \epsilon_B} \right)^{0.8} \tag{9.23}$$

The conservation equations of 9.6 and 9.7 have been written with the assumption of constant fluid velocity. If there is a net change in the number of moles due to the reaction, or if there is a significant change in the reactor temperature or pressure, the velocity cannot be considered constant. This change in velocity can be accounted for with the aid of the ideal gas law and the reaction stoichiometry:

$$v = v_{in}(1 + \sigma y_A) \frac{P_{in}}{P} \frac{T_b}{(T_b)_{in}} \tag{9.24}$$

where $\sigma = \delta W_{Ain}$
     δ = change in the number of moles per one mole of species of interest, A
W_{Ain} = mole fraction of species A in feed
y_A = conversion of species A

This change has to be included in the conservation equations with the velocity appearing inside the differential operator as in Eq. 9.4. In addition, a momentum balance has to be added unless the bed pressure drop is negligible. In such a case, the momentum balance may take the form of the Bernoulli equation:

$$G_f \frac{dv}{dz} + \frac{dp}{dz} - F = 0 \qquad (9.25)$$

where F is given by the right hand side of Eq. 9.20. Use of Eq. 9.24 in 9.25 then yields an expression for dP/dz in terms of dT/dz and dy_A/dz. This expressiion may be solved with the mass and energy balance equations simultaneously for the design and analysis of fixed-beds. It should be recognized, on the other hand, that the effect of changing velocity is usually negligible in practice, even for highly exothermic reactions unless the reaction results in a significant change in the number of moles due to the reaction. For such reactions, the feed is usually quite diluted with inert gas for the reasons discussed below, and no excessive temperature rise should occur.

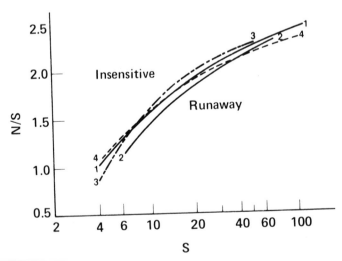

Figure 9.1 Runaway criteria for first-order reactions obtained by various authors. (1: Barkelew 1959; 2: Dente and Collina 1964; 3: Hlavecek et al. 1969; 4: Van Welsenaere and Froment 1970.) (After Froment and Bischoff, *Chemical Reactor Analysis and Design,* © 1979, Wiley. Reprinted by permission of John Wiley & Sons, Inc.)

Design of a fixed-bed for highly exothermic reactions requires additional consideration with regard to the integrity of the pellet and the safety of reactor operation. When a reaction is highly exothermic, a slight change in operating conditions such as the inlet temperature can lead to a situation where the reactor can literally explode. This is called runaway or parametric sensitivity. Several criteria have been developed that permit a selection of operating conditions and reactor dimensions for safe reactor operation. These criteria, shown in Figure 9.1, have been derived based on the pseudo-homogeneous plug-flow model for a first-order reaction. Runaway criteria for an nth-order reaction can be found in Morbidelli and Varma (1982). The dimensionless quantities involved are the dimensionless adiabatic temperature rise S given by:

$$S = \frac{T_{ad} - (T_b)_{in}}{(T_b)_{in}} = \frac{(-\Delta H)(C_b)_{in}}{\rho C_p (T_b)_{in}} \qquad (9.25a)$$

and N given by:

$$N = \frac{2}{R} \frac{U}{\rho C_p k} \qquad (9.26)$$

where k is the volumetric rate constant. A general treatment for the parametric sensitivity problem is given in Section 9–9.

The sensitivity problem considered above places constraints on the choice of operating parameters and reactor dimensions. The same is true when the bed pressure drop is significant, the constraint in this case being on the pellet size. The pellet size is also an important factor affecting inter- and intraphase transport. A smaller size reduces diffusional effects but causes a higher pressure drop and lower external transfer rates, resulting in poorer heat transfer. Therefore, a compromise has to be made regarding the choice of pellet size unless one of the factors is overriding. A similar situation occurs regarding the choice of reactor tube diameter. A smaller diameter ensures better heat transfer and better temperature control but results in a lower capacity per tube. The pellet size and reactor diameter are usually chosen before the reactor is sized. For highly exothermic reactions, not only is the feed diluted with inert gas but the bed packing is also diluted with inert pellets to reduce the heat liberated by the reaction on a unit volume basis. In addition, a cooling medium that gives a favorable reactor temperature profile can be chosen.

The design of fixed-bed reactors eventually involves solving the conservation equations (Eqs. 9.6 through 9.15). It is rare that analytical solutions are obtained because of the nonlinearity involving the Arrhenius expression and intrinsic kinetics. Therefore, the design invariably requires numerical solutions of the equations. This is dealt with in the next section.

9-5   NUMERICAL METHODS

As discussed in Section 9–3, the following conservation equations are sufficient for most practical reactor designs:

Reactor fluid side:

$$v\frac{\partial C_b}{\partial z} - D_r\left(\frac{\partial^2 C_b}{\partial r^2} + \frac{1}{r}\frac{\partial C_b}{\partial r}\right) = -(1 - \epsilon_B)R_G \tag{9.27}$$

$$v\frac{\partial T_b}{\partial z} - K_r\left(\frac{\partial^2 T_b}{\partial r^2} + \frac{1}{r}\frac{\partial T_b}{\partial r}\right) = \frac{(-\Delta H)}{\rho C_p}R_G(1 - \epsilon_B) \tag{9.28}$$

with
$$T_b(0,r) = T_0(r) \tag{9.29}$$

$$C_b(0,r) = C_0(r) \tag{9.30}$$

$$\frac{\partial T_b}{\partial r} = \frac{\partial C_b}{\partial r} = 0 \qquad \text{at } r = 0 \tag{9.31}$$

$$\frac{\partial C_b}{\partial r} = 0 \quad \text{and} \quad h_w(T_b - T_c) = -\rho C_p K_r\frac{\partial T_b}{\partial r} \qquad \text{at } r = R \tag{9.32}$$

Pellet side:

$$D_e\frac{\partial^2 C_p}{\partial x^2} = r_c; \qquad r_c = r_c[C_p, T_s = (T_p)_{x=L}] \tag{9.33}$$

with
$$\frac{\partial C_p}{\partial x} = 0 \qquad \text{at the center} \tag{9.34}$$

$$C_p = C_b \qquad \text{at the surface } (x = L)$$

Interface between pellet and reactor:

$$R_G = \frac{D_e}{L}\frac{\partial C_p}{\partial x}\bigg|_{x=L} = \eta r_c \tag{9.35}$$

$$(-\Delta H)R_G = \frac{h}{L}(T_s - T_b) \tag{9.36}$$

These equations are the steady-state version of Eqs. 9.6 through 9.15 with the assumptions of negligible axial dispersion, negligible external mass transfer resistance, isothermal pellets, and constant v and physical properties. Note that the external heat transfer resistance is still present as given by Eq. 9.36. Needless to say, the one-dimensional reactor (e.g., adiabatic reactor) is a special case of the general case given above.

Numerical solutions of these equations start at $z = 0$, at which T_b and C_b are known as a function of r. For the global rate R_G at each grid point in r, Eqs. 9.33 through 9.36 can be solved iteratively for $T_s(0,r)$ and $R_G(0,r)$. For the first iteration T_s is set equal to T_b, which allows the solution of Eq. 9.33 for the global rate R_G through Eq. 9.35. This R_G is used in Eq. 9.36 for the calculation of T_s. Based on this calculated T_s, the whole procedure is repeated until the new value of T_s is close to the value of T_s in the previous iteration. Based on all information at $z = 0$ now available, the radial profiles of T_b and C_b at the next grid point in z are obtained using a finite difference version of Eqs. 9.27 and 9.28. Using these newly calculated T and C profiles, the procedures described above at $z = 0$ are repeated to obtain R_G and T_s at this new grid point in z. These procedures can be repeated until $z = Z$ is reached. If an analytical expression for the internal effectiveness factor η is available, the whole procedure is greatly simplified since then all equations for the pellet and the interface can be replaced by:

$$R_G[T_s, C_b] = \eta r_c \tag{9.35}$$

$$\frac{h}{L}(T_s - T_b) = (-\Delta H)\eta r_c[T_s, C_b] \tag{9.37}$$

The nonlinear algebraic equation 9.37 can be solved for T_s, which in turn gives R_G in terms of the bulk fluid quantities C_b and T_b. No iteration procedures are necessary for the calculation of T_s in this case.

There are explicit and implicit numerical methods for the finite-difference version of Eqs. 9.27 and 9.28. There are also approximate numerical methods in which the radial derivatives are replaced by the functions obtained by differentiating assumed trial functions for the radial profiles, essentially transforming the equations into ordinary differential equations in z. The cell model (Hlavacek and Votruba, 1977) can also be used for the solution.

The explicit method gives the radial profiles at the new grid point in z explicitly, whereas the implicit method involves simultaneous solution of algebraic equations for the new radial profiles. The former replaces the axial derivative by a forward difference formula and the radial derivative by a central difference formula. The latter uses a backward difference formula for the axial derivative. Perhaps, the most important difference between the two is the stability of the numerical methods: the explicit method needs to satisfy much more stringent stability conditions than the implicit method. The grid system for explicit methods is shown in Figure 9.2. The dark circles represent the grid points from which the information for the open circle is obtained. The finite-difference formulas for the grid system of Figure 9.2 are given in Table 9.2. The finite-diffeence method corresponding

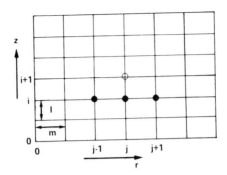

Figure 9.2a   Grid system for explicit method. (Hlavacek and Votruba 1977)

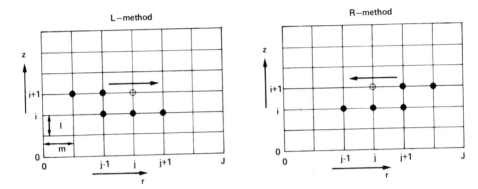

Figure 9.2b   Liu's grid system for explicit method. (Liu 1970)

Table 9.2   Finite-Difference Formulas for Various Methods

$$y = C_b/(C_b)_{in} \text{ or } T_b/(T_b)_{in}; \quad 0 \le z \le 1, 0 \le r \le 1$$
$$J: \text{ at the tube wall } (= (1/m))$$

a. Explicit method of Figure 9.2a
$$(j = 1, \dots , J - 1)$$

$$\frac{\partial y}{\partial z} = \frac{1}{l} [y(i + 1, j) - y(i, j)]$$

$$\frac{1}{r}\frac{\partial y}{\partial r} = \frac{1}{2jm^2} [y(i, j + 1) - y(i, j - 1)]$$

$$\frac{\partial^2 y}{\partial r^2} = \frac{1}{m^2} [y(i, j + 1) - 2y(i, j) + y(i, j - 1)]$$

b. *Explicit method of Figure 9.2b* (Liu 1970)

L-method ($j = 2, \ldots, J - 1$)   R-method ($j = 1, \ldots, J - 2$)

$$\frac{\partial y}{\partial z} = \frac{1}{l}[y(i + 1,j) - y(i,j)] \qquad \text{same}$$

$$\frac{1}{r}\frac{\partial y}{\partial r} = \frac{1}{4jm^2}[y(i + 1,j - 2) - 4y(i + 1,j - 1) \qquad = \frac{1}{4jm^2}[-y(i + 1,j + 2) + 4y(i + 1,j + 1)$$

$$+ 3y(i + 1,j) + y(i,j + 1) - y(i,j - 1)] \qquad - 3y(i + 1,j) + y(i,j + 1) - y(i,j - 1)]$$

$$\frac{\partial^2 y}{\partial r^2} = \frac{1}{2m^2}[y(i,j - 1) - 4y(i,j) \qquad = \frac{1}{2m^2}[-y(i + 1,j + 2) + 4y(i + 1,j + 1)$$

$$+ 3y(i,j + 1) - y(i + 1,j - 2) \qquad - 3y(i + 1,j) + y(i,j + 1) - 4y(i,j)$$

$$+ 4y(i + 1,j - 1)$$

$$- 3y(i + 1,j)] \qquad\qquad + 3y(i,j - 1)]$$

At $j = 1$, a hypothetical point, $(i + 1, -1)$, is necessary for the derivatives. For $y(i + 1, -1)$ use:

$$\frac{\partial^2 y}{\partial r^2} = \frac{1}{m^2}[y(i, 1) - y(i, 0) - y(i + 1, 0)$$

$$+ y(i + 1, -1)]$$

and other derivatives in the conservation equation, and then solve the resulting equation for $y(i + 1, -1)$.

At $j = J - 1$, a hypothetical point, $(i + 1, J + 1)$ is necessary for the derivatives. For this point, use:

$$\frac{\partial^2 y}{\partial r^2} = \frac{1}{m^2}[y(i + 1, J + 1) - y(i + 1, J)$$

$$- y(i, J) + y(i, J - 1)]$$

and other derivatives in the conservation equation and then solve the resulting equation for $y(i + 1, J + 1)$.

c. *Implicit method of Figure 9.3*

$$\frac{\partial y}{\partial z} = \frac{1}{l}[y(i + 1,j) - y(i,j) \qquad 0 \le \theta \le 1$$

$$\frac{\partial^2 y}{\partial r^2} = \theta \delta_r^2 y(i + 1,j) + (1 - \theta)\delta_r^2 y(i,j);$$

$$\delta_r^2 y(i + 1,j) = \frac{y(i + 1,j - 1) - 2y(i + 1,j) + y(i + 1,j + 1)}{m^2}$$

$$\frac{1}{r}\frac{\partial y}{\partial r} = \frac{1}{jm}\left\{\theta\frac{y(i + 1,j + 1) - y(i + 1,j - 1)}{2m} + (1 - \theta)\frac{y(i,j + 1) - y(i,j - 1)}{2m}\right\}$$

to Figure 9.2a is stable if the grid sizes are chosen in such a manner that:

$$\frac{l}{m^2} < \frac{1}{2\alpha} \qquad \alpha = \begin{cases} \dfrac{ZD_r}{R^2 v} & \text{for mass} \\[2ex] \dfrac{ZK_r}{R^2 v} & \text{for heat} \end{cases}$$

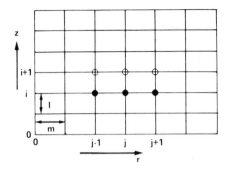

Figure 9.3 Grid system for implicit method.

where l and m are the step sizes of integration in the z and r directions, respectively ($l = 1/I$, $m = 1/J$). While this stability condition for numerical solution has been derived for linear equations, it can be used for nonlinear equations since the stability is mainly determined by the manner in which the derivatives are approximated by the finite-difference. While the scheme of Figure 9.2a has the stability limitation, it is by far the simplest method, involving straightforward marching in the z direction based on the information at previous grid points, i.e., at $z = li$ for the calculations at $z = l(i + 1)$. The finite-difference version for this scheme given in Table 9.2 can be applied for all j from $j = 1$ to $j = J - 1$ at any z, i.e., at any $(i + 1)$, where J ($= 1/m$) corresponds to the tube wall and $j = 0$ corresponds to the tube center. At $j = 0$ and $j = J$, the boundary conditions can be used to obtain T_b and C_b at the center and at the wall. The procedures for obtaining the values of T_b and C_b at these points can best be illustrated by re-writing the conservation equations as follows:

$$\frac{\partial y}{\partial z} + \gamma \left(\frac{\partial^2 y}{\partial r^2} + \frac{1}{r} \frac{\partial y}{\partial r} \right) = \beta R_G \qquad 0 \leq z \leq 1, 0 \leq r \leq 1 \qquad (9.38)$$

where y can be either $T_b/(T_b)_{in}$ or $C_b/(C_b)_{in}$, and γ and β are constants. For the value of $y(i + 1,0)$, y may be expanded around $r = 0$ in a Taylor series to give:

$$y(i + 1,1) = y(i + 1,0) + \left.\frac{\partial y}{\partial r}\right|_{i+1,0} m + \left.\frac{m^2}{2!} \frac{\partial^2 y}{\partial r^2}\right|_{i+1,0} + \theta(m^3)$$

which yields, when solved for $\partial^2 y/\partial r^2$:

$$\left.\frac{\partial^2 y}{\partial r^2}\right|_{i+1,0} = \frac{2}{m^2} \{y(i + 1,1) - y(i + 1,0)\} \qquad (9.39)$$

The first derivative disappears due to the boundary condition of Eq. 9.31. The term $(1/r)(\partial y/\partial r)$ is indefinite at $r = 0$. Applying the l'Hopital's rule, it becomes:

$$\frac{1}{r}\frac{\partial y}{\partial r}\bigg|_{r=0} = \frac{\partial^2 y}{\partial r^2}\bigg|_{i=0} \tag{9.40}$$

Upon substituting Eqs. 9.39 and 9.40 into 9.38, the equation becomes:

$$\frac{y(i+1,0) - y(i,0)}{l} + \gamma \frac{4}{m^2}\{y(i+1,1) - y(i+1,0)\} = \beta R_G(i,0)$$

This equation can be solved for $y(i+1,0)$ since it is the only unknown quantity. The basic idea here is that the differential equation has to be satisfied everywhere on the $z-r$ domain including the boundary. Similar procedures can be followed at $j = J$ for $y(i+1,J)$.

A five point numerical scheme was devised by Liu (1970) to alleviate the stability problem. Depending on whether the integration in r proceeds from left to right or from right to left, the scheme was termed L-method or the R-method. Here again, the integration procedures are straightforward with the finite-difference formulas given in Table 9.2 for $j = 2$ to $j = J - 1$ for the L-method and for $j = 1$ to $j = J - 2$ for the R-method. The calculation procedures at $j = 1$ for the L-method and at $j = J - 1$ for the R-method are detailed in the table. It is apparent that trial and error iterations are necessary to obtain the values at $j = 0$ and 1 (or $j = J - 1$ and J for the R-method), using the boundary condition at $j = 0$ and the suggested procedure for $j = 1$ given in the table. It is suggested by Liu (1970) that R_G evaluated at $z = i$ be used for the L-method and once the tube wall is reached the R-method be used with the value of R_G at $z = i + 1$, which is now available from the integration by the L-method, so that the average of the values obtained by the two methods can be used for the solution at each grid point.

The implicit method, while it requires additional computation, is better than the explicit method as evident from Table 9.3. The implicit method given in Table 9.2 is called the Crank-Nicholson method when the weighting factor θ is 0.5. As discussed earlier, this method requires simultaneous solution of algebraic equations at each i. When the expressions for the derivatives in Table 9.2 are used in Eq. 9.38, for instance, it is transformed into:

Table 9.3 Comparison of Numerical Methods (Carnahan et al. 1969)

Methods	Stability or Convergence Condition for Linear Equations	Order of Magnitude of Error
Explicit (Figure 9.2a)	$\dfrac{l}{m^2} < \dfrac{1}{2\alpha}$	$l + m^2$
Explicit (Figure 9.2b)	none	$\sim(l + m^2)$
Implicit	none	$l^2 + m^2$

$$\frac{\gamma l}{2m^2}\left(1-\frac{1}{2j}\right)y_{i+1,j-1}+\left(1-\frac{\gamma l}{m^2}\right)y_{i+1,j}+\frac{\gamma l}{2m^2}\left(1+\frac{1}{2j}\right)y_{i+1,j+1}$$

$$=-\frac{\gamma l}{2m^2}\left(1-\frac{1}{2j}\right)y_{i,j-1}+\left(1+\frac{\gamma l}{m^2}\right)y_{i,j}-\frac{\gamma l}{2m^2}\left(1+\frac{1}{2j}\right)y_{i,j+1}+l\beta R_{G\ i,j}$$

$$j=1,\ldots,J-1 \qquad\qquad (9.41)$$

The right hand side of Eq. 9.41 is a constant since all quantities are at $z = il$, which are known. This equation can be written $(J - 2)$ times and the number of unknowns is $(J - 2)$. Written in full for j, these equations form a tridiagonal matrix equation in the form of:

$$\mathbf{A}\mathbf{y}_{i+1} = \mathbf{b}_i$$

where $\mathbf{A}$ is a tridiagonal matrix of dimension of $(J + 1)$ including the points at the center and the wall, and $\mathbf{y}_{i+1}$ and $\mathbf{b}_i$ are vectors for y at $z = l(i + 1)$ and the right hand side of Eq. 9.41 evaluated at $z = li$, respectively. The Thomas method (Lapidus 1962) can be used to solve for $\mathbf{y}_{i+1}$. Similar procedures as in the explicit method can be used at the boundaries for y at $j = 0$ and $j = J$. The matrix equation is solved at each i until the reactor outlet is reached. Mihail and Iordache (1976) compared various numerical techniques and found Liu's scheme to be sensitive to step size and to lead to stability and convergence problems for severe operating conditions.

The method of lines lies midway between analytical and grid methods, and can also be used to solve the conservation equations. This method involves substituting finite differences for the derivatives with respect to one independent variable and retaining the derivatives with respect to the remaining variables. This approach replaces a given differential equation by a system of differential equations with a smaller number of independent variables, typically reducing a partial differential equation to a set of ordinary differential equations.

Consider Eq. 9.38. Suppose lines are drawn parallel to the r axis, assuming that the distance between two adjacent lines is constant and equal to h. Suppose that the $z-r$ region is intersected by the lines $z = z_0 + kh = z_k$ $(k = 0, 1, 2, \ldots, n)$. Since $y(0,r)$ is known, the z derivative can be replaced by a difference formula. For example, one can set:

$$\left.\frac{\partial y}{\partial z}\right|_{z=z_k} = \frac{y_k(r) - y_{k-1}(r)}{h}$$

This leads to the following sequence of equations:

$$\gamma\left(\frac{d^2y_k}{dr^2}+\frac{1}{r}\frac{dy_k}{dr}\right)+\frac{1}{h}y_k(r) = \beta R_G(y_k)+\frac{1}{h}y_{k-1}(r)$$

$$k = 0, 1, 2, \ldots, n; \quad n = 1/h$$

The set of differential equations resulting from the method of lines is solved analytically (Mkhlin and Smolitskiy 1967) or numerically (Carver 1976). Since the set of ordinary differential equations resulting from the application of the method of lines to Eq. 9.38 is not amenable to analytical solution, the set has to be solved numerically. Thus, the application of the method of lines leads to another grid method.

There are two useful approximate methods for the numerical integration of the conservation equations, approximate in the sense that radial profiles are approximated by trial functions. One of them is due to Villadsen and Stewart (1967), called the orthogonal collocation method, and the other is due to Ahmed and Fahien (1980) involving a three parameter function. It is instructive to examine typical radial profiles shown in Figure 9.4 since these methods involve an approximation of radial profiles. Because of the heat sink at the tube wall, the temperature gradient is steep at the wall. On the other hand, the concentration gradient at the wall is flat since no transport takes place there. The essence of these methods lies in approximating the profiles with suitable trial functions and then replacing the radial derivatives with the results obtained by differentiating the trial functions.

The collocation method is one of a general class of approximate methods known as the method of weighted residuals (Ames 1965). The method involves expanding the temperature and concentration in a series $\Sigma a_i(z)P_i(r)$ of known functions of radius, $P_i(r)$, multiplied by unknown functions a_i of z. The trial functions are substituted into the partial differential equations that are satisfied at discrete radial collocation points, r_j. This gives a set of ordinary differential equations governing $a_i(z)$. The trial functions are orthogonal polynomials, e.g.:

$$T(\bar{r},z) = T(1,z) + (1 - \bar{r}^2) \sum_{i=1}^{N} a_i(z)P_{i-1}(\bar{r}^2)$$

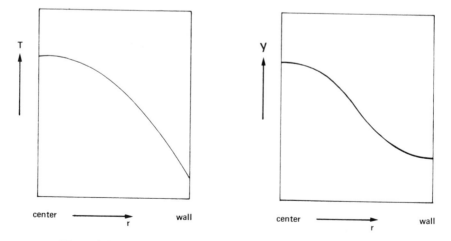

Figure 9.4   Typical radial profiles of temperature and conversion.

such that a minimization condition of the least-squared error integrated over $\bar{r}$ can be met. For the reactor simulation, a set of ordinary differential equations corresponding to the original conservation equations is obtained at each collocation point upon substitution, which are in turn solved for the quantities of interest. This application of the collocation method to reactor simulation is detailed by Finlayson (1971).

The averaging method of Ahmed and Fahien utilizes three-parameter radial profiles (1980) given by:

$$T = a_T + b_T \bar{r}^2 + c_T \bar{r}^{n_T}$$

$$y = a_c + b_c \bar{r}^2 + c_c \bar{r}^{n_c}$$

Use of boundary conditions leads to the expressions for the parameters a through c, which depend on centerline conversion and temperature, average conversion and temperature, and the conversion at the wall. Averaging over the cross section of the original conservation equations and evaluation of these equations at the centerline and wall gives five ordinary differential equations for the five quantities given above. This method is useful for the calculation of the hot-spot temperature.

The solution of the pellet-side conservation equation (Eq. 9.33) is a two-point boundary value problem but is not difficult to solve numerically so long as the intrinsic kinetics are simple and the reaction is moderately diffusion-limited. The problem becomes difficult to solve because of numerical sensitivity when the reaction is severely diffusion-limited and the accuracy of the result is not assured. However, an analytical expression can be obtained in this case for arbitrary intrinsic kinetics as shown in Chapter 4.

9–6   EXAMPLES AND PATTERNS

The simulation of an adiabatic reactor for the oxidation of SO_2 was carried out by Minhas and Carberry (1969). With the assumptions of negligible axial dispersion and constant fluid velocity, the conservation equations (Eqs. 9.27 and 9.28) reduce to:

$$v\frac{dC}{dz} = -(1 - \epsilon_B)R_G$$

$$\rho C_p v \frac{dT}{dz} = (-\Delta H)R_G(1 - \epsilon_B)$$

The boundary conditions are reactor inlet conditions. The intrinsic kinetics over a platinum catalyst are represented by:

$$r_c = \frac{k(SO_2)(O_2)^{1/2}}{[1 + [K_1(O_2)]^{1/2} + K_2(SO_3)]^2}\left[1 - \frac{(SO_3)}{K(SO_2)(O_2)^{1/2}}\right]$$

Instead of solving the pellet conservation equations, they used a definition of the Thiele modulus in terms of the limiting reactant SO_2 and the equilibrium nature of the reaction $SO_2 + 1/2\, O_2 = SO_3$ to obtain:

$$\phi = L \left[\frac{k(K_{eq} + 1)}{D_e K_{eq}} (SO_2)_b{}^{n-1} \right]^{1/2}$$

where the pellet was assumed to be isothermal. Simulation results are shown in Figure 9.5. These results are fairly close to actual plant conditions. For instance, a typical η value for the industrial SO_2 reactor is around 0.1.

The patterns shown in the figure are typical of an adiabatic reactor for exothermic reactions. The effectiveness factor generally decreases with reactor length, but more rapidly than shown in the figure. Because of the reaction equilibrium, the decrease is much more moderate in this case. The conversion and temperature usually increase rapidly near the inlet, but these increases are moderate toward the outlet, reaching plateaus as the reactant is depleted.

For a nonadiabatic reactor, consider the analysis by Carberry and White (1969) for the oxidation of naphthalene over V_2O_5. For the simplified network of first-order reactions:

$$\text{naphthalene} \xrightarrow[R_{G_1}]{k_1} \text{anhydride} \xrightarrow[R_{G_2}]{k_2} CO_2 + H_2O$$
$$\quad\;(1) \hspace{6.5em} (2)$$

the conservation equations are:

$$v\frac{\partial C_1}{\partial z} - D_r\left(\frac{\partial^2 C_1}{\partial r^2} + \frac{1}{r}\frac{\partial C_1}{\partial r}\right) = -(1 - \epsilon_B)R_{G_1} \qquad \text{for naphthalene}$$

$$v\frac{\partial C_2}{\partial z} - D_r\left(\frac{\partial^2 C_2}{\partial r^2} + \frac{1}{r}\frac{\partial C_2}{\partial r}\right) = -(1 - \epsilon_B)(R_{G_2} - R_{G_1}) \qquad \text{for anhydride}$$

$$v\frac{\partial T}{\partial z} - K_r\left(\frac{\partial^2 T}{\partial r^2} + \frac{1}{r}\frac{\partial T}{\partial r}\right) = (-\Delta H_1 R_{G_1} - \Delta H_2 R_{G_2})\left(\frac{1 - \epsilon_B}{\rho C_p}\right)$$

with the same boundary conditions as in Section 9–5. The pellet conservation equations were not solved. Instead, an analytical expression was used for the isothermal effectiveness factor for naphthalene. They found that the effect of radial mass dispersion was negligible, which makes it possible to write a reactor point yield ratio dC_1/dC_2 directly from the reactor conservation equations for naphthalene and anhydride. Results are shown in Figures 9.6 through 9.9. It is seen in Figure 9.6 that the centerline temperature goes through a maximum at a fractional length of about 0.1. This point is called a *hot spot*. For a given cross-section, the maximum temperature occurs at the centerline with shell-side cooling. The presence of a hot spot is a characteristic of nonadiabatic reactors with shell-side cooling. At the hot spot, the heat generated by the reaction is equal to the heat transferred

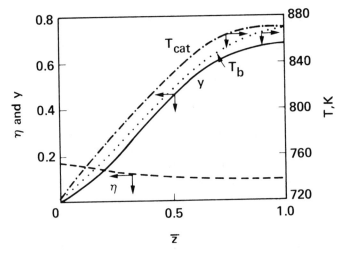

Figure 9.5 Simulation results for SO_2 oxidation. (Minhas and Carberry 1969)

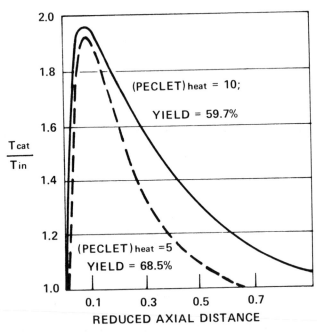

Figure 9.6 Normalized catalyst temperature profile at bed centerline: effect of radial Peclet number for heat on the profile, $P_{mr} = 10$. (Carberry and White 1969; reprinted with permission from *Industrial and Engineering Chemistry.* Copyright by American Chemical Society.)

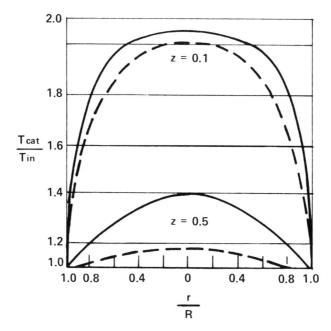

Figure 9.7 Radial catalyst temperature profiles: —— for $P_{hr} = 10$, ---- for $P_{hr} = 5$. (Carberry and White 1969; reprinted with permission from *Industrial and Engineering Chemistry.* Copyright by American Chemical Society.)

to the coolant through dispersion and heat conduction at the wall. The heat generation rate dominates the heat removal rate in the region close to the reactor entrance. Once the hot spot is passed, the heat removal rate dominates. This hot-spot temperature places a constraint on the reactor design since the viability of the catalyst is lost if the temperature exceeds a certain level: the catalyst as well as the support can be transformed into a different material at high temperatures. In industrial reactors, the hot spot migrates toward the outlet with the time on stream as the activity of the catalyst decreases due to deactivation. As expected, the hot-spot temperature as well as the catalyst temperature profile is lower for a smaller Peclet number for heat (better heat dispersion). This is also shown in Figure 9.7, in which the radial temperature profiles are given. The temperature patterns shown in these figures are typical of nonadiabatic reactors with shell-side cooling. The importance of good radial heat dispersion (approaching plug-flow behavior) is exemplified in Figure 9.9 where the yield of the desired intermediate product is given. The average yield is obtained from the mixing-cup average concentration:

$$\langle C_2(z)\rangle = \int_0^R rC_2(r,z)dr \bigg/ \int_0^R rdr$$

The overall effectiveness factor for the first reaction shows some interesting features (Figure 9.8). It is seen that the temperature effect dominates the overall effectiveness factor, causing a minimum near the hot spot. It should be recognized that the

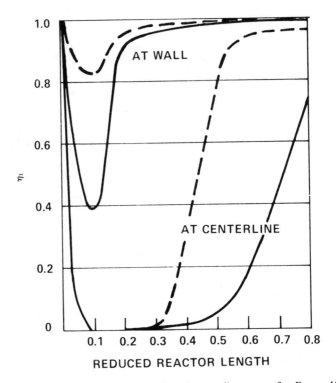

Figure 9.8 Effectiveness factor at reactor wall and centerline: —— for $P_{hr} = 10$, ---- for $P_{hr} = 5$. (Carberry and White 1969; reprinted with permission from *Industrial and Engineering Chemistry.* Copyright by American Chemical Society.)

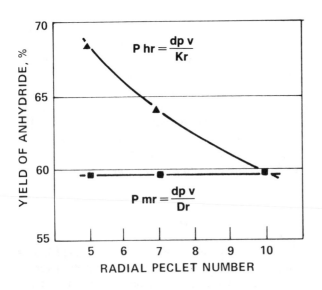

Figure 9.9 Effect of radial Peclet number on overall yield of anhydride, conversion = 99%. (Carberry and White 1969; reprinted with permission from *Industrial and Engineering Chemistry.* Copyright by American Chemical Society.)

effective rate constant at the hot spot is still quite high despite the very low overall effectiveness factor there ($10^{-3} \sim 10^{-4}$). This is due to the fact that the effective rate constant is proportional to $\eta_1 k_1$ ($\sim k_1^{1/2}$) and k_1 is quite high at high temperatures.

A successful application of the plug-flow heterogeneous model has been reported by Cappelli and coworkers (1972). Their simulation results show excellent agreement between the model predictions and the data that they obtained from a nonadiabatic industrial reactor for the synthesis of methanol from CO and H_2. The work by Dumez and Froment (1976) is another example of the application of the plug-flow heterogeneous model to an industrial reactor.

9-7 REACTOR/HEAT EXCHANGER COUPLING: AUTOTHERMIC REACTORS

When an exothermic reaction requires a high temperature, the heat of reaction can be used to preheat the feed. The coupling of the reactor with the heat exchanger through the use of reactor effluent fluid for the purpose of self-supporting the preheating requirement of the feed is termed autothermic operation (Van Heerden 1935). This can be accomplished externally, internally, or through recycling. These arrangements are shown in Figure 9.10. Ammonia synthesis reactors (Baddour et al. 1965) of the Nitrogen Engineering Co. and the TVA (Tennessee Valley Authority) are examples of the autothermic reactors having external heat exchange.

Consider the external arrangement. Since the reactor is adiabatic, it can be represented by the plug-flow model. Referring to Figure 9.10, the conservation equations are:

$$F_{in} \frac{dy}{dz} = (1 - \epsilon_B) R_G \qquad (9.42)$$

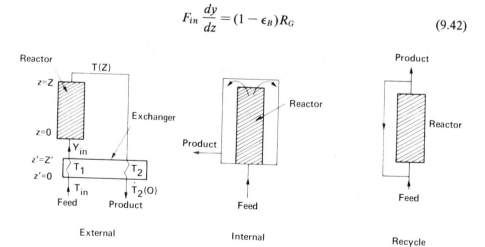

Figure 9.10 Autothermic reactor arrangements. (After J. M. Smith, *Chemical Engineering Kinetics,* 3rd ed., © 1981; with permission of McGraw-Hill Book Company, New York.)

$$v\rho C_p \frac{dT}{dz} = (1 - \epsilon_B)(-\Delta H)R_G \qquad (9.43)$$

$$(v\rho C_p)_1 \frac{dT_1}{dz'} = UA_c(T_2 - T_1) \qquad (9.44)$$

$$(v\rho C_p)_1 dT_1 = -(v\rho C_p)_2 dT_2 \qquad (9.45)$$

$$
\begin{aligned}
&\text{with } C = C_{in},\ T = T_1(Z') &&\text{at } z = 0 \\
&\qquad T_1(0) = T_{in} &&\text{at } z' = 0 \\
&\qquad T_2(Z') = T(Z) &&\text{at } z = Z \text{ or } z' = Z'
\end{aligned}
$$

where A_c = heat exchanger surface area per unit volume (2/radius)
F_{in} = inlet molar flow rate
y = conversion
U = overall heat transfer coefficient
Z = reactor outlet
Z' = heat exchanger outlet

Combining Eqs. 9.42 and 9.43 gives, when integrated from $z = 0$ to Z

$$\Delta y \equiv y(Z) - y_{in} = \frac{v\rho C_p}{(-\Delta H)F_{in}}[T(Z) - T_1(Z')] \qquad (9.46)$$

Assuming $(v\rho C_p)_1 = (v\rho C_p)_2$, Eq. 9.45 gives:

$$T_{in} - T_1(z') = T_2(0) - T_2(z') \qquad (9.47)$$

or

$$T_{in} - T_1(Z') = T_2(0) - T(Z) \qquad (9.48)$$

when evaluated at $z' = Z'$. Equation 9.47 shows that $(T_2 - T_1)$ in Eq. 9.44 is constant. Therefore, Eq. 9.44 gives, after integration:

$$T_1(Z') - T_{in} = \frac{UA_c Z'}{(v\rho C_p)_1}[T_2(0) - T_{in}] \qquad (9.49)$$

Adding $T(Z)$ to both sides of this equation and using Eq. 9.48 gives:

$$T(Z) - T_{in} = \left(1 + \frac{UA_c Z'}{(v\rho C_p)_1}\right)[T(Z) - T_1(Z')] \qquad (9.50)$$

This equation together with Eq. 9.46 yields:

$$y(Z) - y_{in} = \frac{T(Z) - T_{in}}{\left[1 + \dfrac{UA_c Z'}{(v\rho C_p)_1}\right]\left[\dfrac{F_{in}(-\Delta H)}{(\rho C_p v)_1}\right]} \equiv \xi\,[T(Z) - T_{in}] \qquad (9.51)$$

On a $y-T$ diagram this equation is a straight line with a slope of ξ. This is shown as dotted lines in Figure 9.11. A higher T_{in} shifts the line more to the right of the diagram. Another relationship between y and T can be obtained by inserting Eq. 9.46 into the right hand sides of Eqs. 9.42 and 9.43: Eq. 9.42 then gives $y(Z) = f_1[T_{in}]$ and Eq. 9.43 gives $T(Z) = f_2[T_{in}]$, combination of which yields a relationship between $y(Z)$ and $T(Z)$. For a given reaction system, $y(Z)$ is a function solely of $T(Z)$ if y_{in} is zero as T_{in} is eliminated in the process of linking the function f_1 to f_2 for the relationship. Because of the Arrhenius temperature dependence, this relationship gives a sigmoidal-shaped curve (solid line) as shown in Figure 9.11. Since the relationship is independent of T_{in} ($y_{in} = 0$), there exists only one curve for a given reaction system. The reaction system, on the other hand, has to satisfy not only this relationship, but also Eq. 9.51, meaning that the intersection points between the two curves are the steady state operating conditions.

If the choice of ξ and T_{in} is such that Eq. 9.51 crosses the sigmoidal curve at three points as in curve b, there exist three steady states, b_1, b_2 and b_3. Outside the region defined by lines a and c, there is only one steady state. These lines a and c themselves touch the sigmoidal curve at a tangent and two steady state operating points exist. For the line b, the points b_1 and b_3 represent stable operating points. The middle point, b_2, is unstable in the sense that a slight perturbation in operating conditions moves the operating point to either b_1 or b_3. For instance, a slight increase in $T(Z)$ at b_2 will cause a shift to b_1 since this increase has a

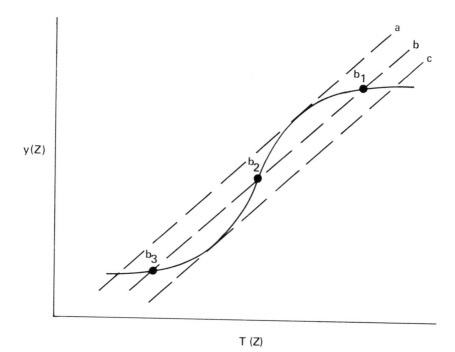

Figure 9.11   Possible steady state operating points for autothermic reactors.

much larger effect on the heat generated than the heat exchanged (the sigmoidal curve has a much larger slope at around b_2 than the line b). This is very similar to the phenomenon found for stirred-tank homogeneous reactors (e.g., Smith 1981). It is often undesirable to operate near the extinction point (where the heat of reaction cannot support the heating requirement) since a perturbation may lead to the quenching of the reaction (blowout). Nevertheless, the TVA ammonia reactors are operated near the extinction point to maximize throughput.

Similar results can be obtained for the other arrangements shown in Figure 9.10. Results for the recycle arrangement are given by Reilly and Schmitz (1966).

9–8   VARIATIONS OF FIXED BED REACTORS

Radial flow reactors are used in large capacity units where the pressure drop can become a limiting factor. Radial reactors are used in catalytic reforming and ammonia synthesis to eliminate problems of pressure drop associated with high throughput. A diagram for a radial flow reactor is given in Figure 9.12. Also given are idealized flow patterns [(a) and (b) in the figure] for a section (section A in the figure) of the catalyst bed. As shown in the figure, the fluid flow through the catalyst bed is either from a center pipe toward the wall or from the annulus to the center pipe. The void space in the vessel is for transportation of fluid and no reaction takes place.

According to the idealized flow patterns (Hlavacek and Kubicek, 1972), the slice of the bed cut in the radial direction can be considered as a fixed-bed with the inlet condition at $r = R_1$ and the outlet condition at $r = R_2$. Conservation equations written for this slice should be applicable to the whole length of the bed. Therefore, the conservation equations of Eqs. 9.27 through 9.36 are applicable to radial flow reactors with the convection terms $\partial c/\partial z$ and $\partial T/\partial z$ replaced by $\partial c/\partial r$ and $\partial T/\partial r$. This model is then identical to the one-dimensional model with axial dispersion as for the usual fixed-bed reactor. The same boundary conditions as those of Eqs. 9.10 and 9.11 apply, but this time Eq. 9.10 applies at $r = R_1$ and Eq. 9.11 at $r = R_2$, with z replaced by r for the flow pattern (a) in Figure 9.12. Unless the bed depth is quite shallow, the dispersion term can be neglected, resulting in a plug-flow model in the radial direction.

With the advent of automobile catalytic converters, monolithic catalyst beds have been developed to convert unburned hydrocarbons and carbon monoxide to combustion products. These oxidation reactions are highly exothermic. Thus, the catalyst is deposited only on the surface layers of the support. Other design considerations are a mechanically strong bed to withstand mechanical and thermal shock, and low pressure drop. These requirements led to monolithic catalyst beds of various configurations.

At the heart of the various configurations is an open channel, the inside wall of which is "coated" with the catalyst. As fluid passes through the open channel, it reacts on the catalyst coated on the inside wall. For the reactant i in the channel, one has:

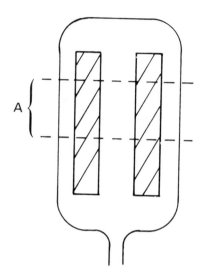

Figure 9.12 Radial flow reactor and idealized flow patterns. (Hlavacek and Votruba in *Chemical Reactor Theory: A Review,* ed. Lapidus and Amundson, © 1977. Reprinted by permission of Prentice-Hall, Inc., Englewood Cliffs, NJ.)

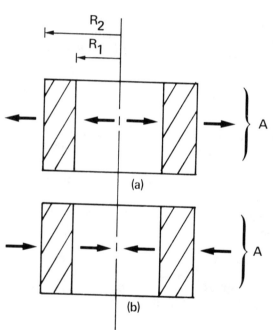

$$A_c v \frac{d(C_b)_i}{dz} + l\,(k_g)_i\,[(C_b)_i - (C_s)_i] = 0 \qquad i = 1, \ldots , N \qquad (9.52)$$

where A_c is the cross sectional area of the channel, l the circumference of the channel, $(C_s)_i$ the concentration of species i at the catalyst surface, and $(k_g)_i$

the film mass transfer coefficient for species i. Since the mass transfer rate should be equal to the reaction rate at the surface at steady state:

$$(k_g)_i\ [(C_b)_i - (C_s)_i] = (R'_G)_i\ \{(C_s)_i, T_s\} \tag{9.53}$$

where the global rate R'_G is based on the external surface area of the catalyst. Similarly, the heat balance for the reacting fluid is:

$$A_c v \rho C_p \frac{dT_b}{dz} + lh\ (T_b - T_s) = 0 \tag{9.54}$$

On the other hand, the heat balance for the solid wall, which is assumed to be at T_s, is:

$$-\delta\lambda \frac{d^2 T_s}{dz^2} + h\ (T_s - T_b) = \sum_i (-\Delta H_i R'_{Gi}) \tag{9.55}$$

where δ is the wall thickness, which is assumed to be small enough to give the area perpendicular to the heat flux as δl. The boundary conditions for the fluid are the channel inlet conditions. The boundary conditions for Eq. 9.55 are:

$$\frac{dT_s}{dz} = 0 \qquad \text{at the inlet and outlet} \tag{9.56}$$

Equations 9.52 through 9.56 represent the basic model for monolithic catalyst beds. More detailed modeling of these beds can be found in the articles by Hegedus (1975) and Votruba et al. (1975).

Design of a fixed-bed for highly exothermic reactions is perhaps the most demanding because of the sensitivity problem that these reactions can cause. In the next section the design of a countercurrent reactor/heat exchanger for highly exothermic reactions is considered.

9-9   DESIGN OF A COUNTERCURRENT REACTOR/HEAT EXCHANGER WITH PARAMETRIC SENSITIVITY

The parametric sensitivity problem associated with a countercurrent reactor/heat exchanger in which a highly exothermic reaction occurs has been discussed in Section 9-5. The overriding design concern is that no runaway situation should occur so that the reactor can be operated safely. Criteria developed for a first-order reaction have been presented in Figure 9.1 for the safe operation of a homogeneous reactor. This section develops a phase diagram of inlet reactant and coolant temperatures on which the regions of safe operation, negligible reaction, and possible ignition are clearly delineated in terms of the system parameters. Utilization of this phase diagram for the design is then examined.

As discussed earlier, reactors for highly exothermic reactions are usually designed to have a small ratio of pellet size to reactor radius so as to avoid undesirable effects of dispersion. It is therefore adequate to use a plug-flow model for such reactors. If f_a is the global rate expressed in terms of conversion, the conservation equations are:

$$\frac{1}{\pi R^2} \frac{dF}{dz} = -f_a(T_b, y) \tag{9.57}$$

$$\frac{1}{\pi R^2} \frac{d(\dot{M} C_{pr} T_b)}{dz} = (-\Delta H) f_a(T_b, y) - \frac{2U}{R} (T_b - T_c) \tag{9.58}$$

$$-\frac{1}{\pi R^2} \frac{d(\dot{m} C_{pc} T_c)}{dz} = \frac{2U}{R} (T_b - T_c) \tag{9.59}$$

where C_{pc} = mass specific heat of coolant
C_{pr} = mass specific heat of reacting fluid
$f_a = (1 - \epsilon_B) R_G$
F = molar flow rate (mol/time) of key species
$\dot{m}$ = mass flow rate of coolant
$\dot{M}$ = mass flow rate of reacting fluid
T_c = coolant temperature
y = conversion

Assuming that mass flow rates, specific heats, and the overall heat transfer coefficient U are constant, the above equations reduce to:

$$\frac{dy}{d\bar{z}} = a_1 f_a \tag{9.60}$$

$$\frac{dT_b}{d\bar{z}} = a_2 f_a - a_3(T_b - T_c) \tag{9.61}$$

$$\frac{dT_c}{d\bar{z}} = -a_4(T_b - T_c) \tag{9.62}$$

where $y = (F_{in} - F)/F_{in}$, $\bar{z} = z/Z$

$$a_1 = \frac{\pi R^2 Z}{F_{in}}, \quad a_2 = \frac{\pi R^2 Z(-\Delta H)}{\dot{M} C_{pr}}, \quad a_3 = \frac{2\pi R Z U}{\dot{M} C_{pr}}, \quad a_4 = \frac{2\pi R Z U}{\dot{m} C_{pc}} \tag{9.63}$$

F_{in} = inlet molal flow rate of key species

Given a reaction, the design problem is to choose the parameters involved in a_1 through a_4 and the reactant and coolant inlet temperatures that give the best possible performance of the reactor within the constraint of a no runaway situation.

For a given set of parameters, certain choices of inlet temperatures will result in ignition of the reaction and possible explosion. On the other hand, some other choices may be such that no appreciable amount of reaction can take place. The question at hand for the phase diagram is then to find "workable" or safe ranges of reactant (T_{in}) and coolant (T_{cin}) inlet temperatures, which lie between these two extremes, on a T_{in}-T_{cin} plane, as shown in Figure 9.13. The phase diagram is therefore divided into three regions: the region of safe operation, that of negligible reaction, and the ignition region, in which the runaway situation can occur. The line separating the safe region from the region of negligible reaction is to be called the quenching line. The line separating the safe region from the ignition region is to be called the ignition line. Referring to Figure 9.13, the ignition line in the region $T_{in} \geq T_{cin}$ is to be called the upper ignition line and that in the region $T_{in} < T_{cin}$ is called the lateral ignition line.

Approximate relationships for these lines have been derived from simple runaway criteria derivable from the system equations (Akella and Lee 1983). Procedures for constructing a phase diagram such as the one in Figure 9.13 have been developed based on the approximate relationships, summarized in Table 9.4. Consider a realistic problem involving a general hydrocarbon oxidation process. The reaction system given in Table 9.5 is similar to the one treated by Van Welsenaere and Froment (1970). A gaseous hydrocarbon mixed with oxygen and inerts is fed to a fixed-bed with a countercurrent heat exchanger. The reactor consists of 2500 tubes, 3 m long and 2.5 cm in diameter. The hydrocarbon concentration in the feed is 1.5 mol% in excess oxygen. Under these conditions a single pseudo-first-order irreversible reaction is assumed to take place:

$$R_G = r_c \rho_B = 8.5\,(10^7)\,\exp(-13{,}636/T)\,p\rho_B$$

where p is the partial pressure of the hydrocarbon. Since the concentration of the key reactant is very low, the total number of moles in the gas stream can be assumed constant to give:

$$R_G = 8.5\rho_B(10^7)\,\exp(-13{,}636/T)PW_{in}(1-y)$$

It follows then that the function f_a is:

$$f_a = k_0\exp(-13{,}636/T)PW_{in}(1-y) \tag{9.64}$$

where P is the total pressure and W_{in} is the feed mole fraction of the hydrocarbon. Since the reactor consists of 2500 tubes, a_1 through a_4 in Eqs. 9.60 through 9.62 should be rewritten appropriately for the system:

$$a_1 = \frac{Z(MW)}{\dot{M}W_{in}}, \quad a_2 = \frac{Z(-\Delta H)}{\dot{M}C_{pr}}, \quad a_3 = \frac{2ZU}{\dot{M}C_{pr}R}, \quad a_4 = \frac{2N\pi RZU}{\dot{m}C_{pc}} \tag{9.65}$$

where $\dot{M}$ and (MW) are the mass velocity and average molecular weight of the reacting fluid, respectively, and N is the number of tubes. In order to construct the phase diagram of Figure 9.13, one has to know the relationships for various lines. According to step (7) in Table 9.4, the quenching line is defined by:

$$T_1 = T_{c_{in}} + (T_{in} - T_{c_{in}}) \exp(-33.6)$$

and
$$1.956(10^9) \exp(-13{,}636/T_1) = 0.01 \qquad (9.68)$$

when the parameters in Table 9.5 are used. Since $e^{-33.6}$ is negligibly small, $T_1 = T_{c_{in}}$ practically, and hence, the quenching line is given by a straight line parallel to the ordinate at $T_{c_{in}} = 525°K$. Consequently, the quenching line for the upper operating zone is given by $T_{c_{in}} = 525°K$. The quenching line for the lower operating

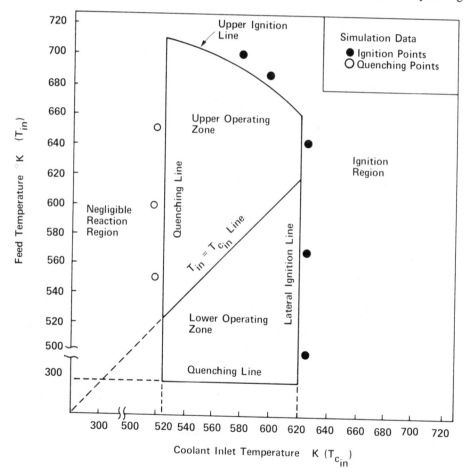

Figure 9.13 Phase diagram of inlet temperatures for a hydrocarbon oxidation reaction. (Akella and Lee 1983)

Table 9.4   Procedures for Constructing Phase Diagram

Global rate $f_a = f_a(T_b, y)$

(1)   The phase diagram is plotted on the $T_{in} - T_{c_{in}}$ plane with $T_{c_{in}}$ as the abscissa and T_{in} as the ordinate.

(2)   The phase diagram is divided diagonally along the $T_{in} = T_{c_{in}}$ line. The upper section where $T_{in} \geq T_{c_{in}}$ has an upper operating zone and the lower section where $T_{in} < T_{c_{in}}$ has a lower operating zone.

(3)   The upper operating zone is bounded on the left by the quenching line, on the top by the upper ignition line, and by the diagonal (see Figure 9.13).

(4)   The lower operating zone is bounded on the right by the lateral ignition line, at the bottom by the quenching line, and by the diagonal.

(5)   The upper ignition line is defined as the locus of the lowest values of $(T_{in}, T_{c_{in}})$ satisfying the conditions

$$\left.\frac{dT_b}{d\bar{z}}\right|_{\bar{z}=0} \text{ and } \left.\frac{d^2 T_b}{d\bar{z}^2}\right|_{\bar{z}=0} > 0$$

with $T_c|_{\bar{z}=0} = T_{c_{in}} + C_1 + C_2(T_{in} - T_{c_{in}});$ $T_{c_{in}} = T_c|_{\bar{z}=1}$

(6)   The lateral ignition line is the vertical line at $T_{c_{in}}$ defined by:

$$T_{c_{in}} = T_m - \frac{a_2}{a_3} f_a(T_m, 0) \tag{9.66}$$

such that: $\dfrac{dT_{c_{in}}}{dT_m} = 0.$

(7)   The quenching line is defined by:

$$a_1 f_a(T_1, 0) = 0.01 \tag{9.67}$$

with $T_1 = T_{c_{in}} + (T_{in} - T_{c_{in}})e^{-a_3};$ $T_1 = T_b|_{\bar{z}=1}$

where T_m = maximum temperature
$C_1 = F_{in}(-\Delta H)/\dot{m}C_{pc}$
$C_2 = \dot{M}C_{pr}/(\dot{m}C_{pc})$

a_1 through a_4 are given by Eq. 9.63

For $f_a(T_m, 0)$ in Eq. 9.66 above, simply set $T_b = T_m$, and $y = 0$ in the global rate expression, $f_a(T_b, y)$.

---

zone is given when T_{in} is set to the minimum available feed temperature. In this case it is assumed that $T_{in} = 300°\text{K}$ is the quenching line for the lower operating zone. These lines are shown in Figure 9.13. According to step (5) in Table 9.4, the upper ignition line is obtained from:

$$\frac{dT_b}{d\bar{z}} = 1.27(10^{12}) \exp(-13{,}636/T_{in}) - 33.6(T_{in} - T_c|_{\bar{z}=0}) > 0 \tag{9.69a}$$

Table 9.5 Reaction System of Hydrocarbon Oxidation (After Froment and Bischoff, *Chemical Reactor Analysis and Design,* © 1979, Wiley. Reprinted by permission of John Wiley & Sons, Inc.)

Reactor: 2500 packed tubes of 3 m long and 2.5 cm diameter
Parameters

Reactor length (Z)	3 m
Reactor radius (R)	2.5 cm
Mole fraction of hydrocarbons in feed (W_{in})	0.015
Molecular weight of feed (MW)	29.5 kg/kmol
Feed mass velocity or mass flow rate per area ($M/N\pi R^2$)	5000 kg/m²hr
Number of tubes (N)	2500
Coolant mass rate ($\dot{m}$)	1.0(10⁶) kg/hr
Heat of reaction ($-\Delta H$)	1.28(10⁶) kJ/kmol
Average specific heat of gas mixture (C_{pr})	1.0 kJ/kg°K
Average specific heat of coolant (C_{pc})	4.0 kJ/kg°K
Overall heat transfer coefficient (U)	3.5(10²) kJ/m²hr°K
Pre-exponential factor (k_0)	8.5(10⁷) kmol/kg hr
Activation energy divided by gas constant	1.3636(10⁴)°K
Catalyst bulk density (ρ_B)	1.3(10³) kg/m³
Total pressure (P)	1 atm
Pellet diameter (d_p)	3 mm
Feed temperature (T_{in})	variable ($\geqslant 300°$K)
Coolant inlet temperature ($T_{c_{in}}$)	variable

$$\frac{d^2 T_b}{d\bar{z}^2} = \exp(-27{,}272/T_{in})\{2.21(10^{28})/T_{in}^2 - 2.49(10^{21})\}$$

$$- \exp(-13.636/T_{in})\{5.83(10^{17})(T_{in} - T_c|_{\bar{z}=0})/T_{in}^2$$

$$+ 4.28(10^{13})\} + 1.11(10^3)(T_{in} - T_c|_{\bar{z}=0}) > 0 \tag{9.69b}$$

and $T_c|_{\bar{z}=0} = T_{c_{in}} + 13.31 + 0.015\,(T_{in} - T_{c_{in}})$ (9.69c)

For a given value of $T_{c_{in}}$, Eq. 9.69c gives $T_{c_{in}}$ in terms of T_{in}, which upon insertion into Eqs. 9.69a and 9.69b yields T_{in} values. The lower value of T_{in} (one from 9.69a and the other from 9.69b) is then chosen for the value of $T_{c_{in}}$. Since the quenching line is given by $T_{c_{in}} = 525°$K, pairs of ($T_{in} - T_{c_{in}}$) corresponding to $T_{c_{in}} = 525, 550, 575, 600$, etc. have been calculated. A continuous curve drawn through these points is shown in Figure 9.13. According to step (6) in Table 9.4, the lateral ignition line is given by:

$$T_{c_{in}} = T_m - 3.79\,(10^{10})\exp(-13{,}636/T_m) \tag{9.70}$$

such that $dT_{c_{in}}/dT_m = 0$. Therefore, the lateral ignition line is the vertical line at $T_{c_{in}} = 621°$K. This line is also shown in Figure 9.13. Using these relationships, the procedures of Table 9.4 were followed to construct the phase diagram of Figure 9.13 for the present problem. The dark and open circles in the figure are the temperatures corresponding to the onset of ignition and to the quenching region numerically calculated using the original system equations for comparison. It is seen that the phase diagram predicts the true nature of the reactor behavior very closely.

It is evident from Figure 9.13 that the phase diagram not only tells whether a set of inlet temperatures, which are most susceptible to unexpected fluctuations, would guarantee safe operation, but also shows the margin of safety. The boundaries surrounding the safe operation region are defined uniquely by the system parameters. It is thus possible to expand the boundaries of the safe operating region by proper selection of the system parameters, which in turn simplifies the design problem considerably.

The design problem for a reactor/heat exchanger is to choose reactor length (Z), reactor diameter (R), overall heat transfer coefficient (U), coolant and feed rates ($\dot{m}$ and $\dot{M}$), feed composition (W_{in}), and inlet temperatures (T_{in} and $T_{c_{in}}$) in such a way that design objectives are achieved, yet the reactor is safe from any possible runaway due to uncontrollable fluctuations in the operating variables. Due to the large number of design and operation parameters involved, the design problem is generally complicated. Since all these parameters are lumped into the coefficients a_1 through a_4 of the system equations and these coefficients in turn define the boundaries of the safe operating region, the design problem can be considered from the viewpoint of the phase diagram. One broad objective of the design is the proper choice of parameters such that the reactor can be operated safely over a wide range of temperature, since higher temperatures will result in higher conversions. Thus, the design problem can be condensed into a single problem of expanding the boundaries of the safe operation region by moving the quenching and ignition lines outward or upward, subject to the requirements of throughput.

Consider the example of hydrocarbon oxidation. The quenching line given by Eq. 9.67 is exclusively determined by a_1. This line is to be pushed to the left in Figure 9.13, allowing high conversion even at low coolant temperatures. In accordance with Eq. 9.65, this can be done by increasing Z, or decreasing $\dot{M}$ and W_{in}, which results in either a larger reactor or lower throughput. There should be an optimum search. In addition, the effect of changing these variables on other system parameters should also be taken into account. Nevertheless, since W_{in} is not involved in any other system parameters, in principle a_1 can be varied independently. The second quenching line does not pose a design problem since it is always represented by the lowest allowable feed temperature. The upper ignition line, on the other hand, is to be elevated in Figure 9.13. From Eq. 9.69, it is not possible to make any *a priori* comments as to the effect of a_1 through a_4. Nevertheless, a search can be made using Eq. 9.69 for moving the line upward. The movement of the lateral ignition line on the phase diagram is mainly controlled by the ratio of a_2/a_3. This line, given by Eq. 9.66, can be moved to the right of Figure 9.13 by decreasing a_2/a_3 or by decreasing the ratio R/U since $a_2/a_3 = (-\Delta H)R/U$. While decreasing R causes a reduction in throughput, increasing U (choice of material of construction and coolant) does not have any adverse effect on throughput. One interesting observation emerges from this analysis: fewer design variables are involved for reactor operation in the lower operating zone.

The phase diagram of Figure 9.13 reveals an important point. For a given coolant inlet temperature, any point in the upper operating zone is more sensitive to ignition than a point in the lower operating zone. This sensitivity is shown in Figure 9.14. It is seen that a slight increase in feed temperature in the upper operating zone can lead to ignition, whereas the sensitivity is not as severe in the lower operating zone. Therefore, from the sensitivity viewpoint, it is always better to operate the reactor in the lower operating zone than the upper. This fact leads to an interesting design alternative: a countercurrent reactor/heat exchanger without preheater, as opposed to the usual arrangement of a preheater followed by a reactor/heat exchanger. This alternative eliminates the sensitivity problem with respect to changes in the reactant inlet temperature and allows elimination of the preheater, reducing fixed costs. The additional reactor length required of this design for the same conversion is minimal when compared with the usual design with preheater (Akella and Lee 1983).

The sensitivity with respect to coolant inlet temperature remains almost the same for both types of reactor configuration, as shown in Figure 9.15. This sensitivity problem is much less severe than that with respect to feed temperature since there is much better control over the coolant inlet temperature. However, this precludes the attainment of exit conversions in excess of a certain level, for instance, 60% in Figure 9.15, since around this region even a slight increase in the coolant temperature leads to runaway, as shown in Figure 9.13, no matter what the feed temperature is. One way of circumventing this predicament is to realize that faster heat removal from the reactor flattens the reactor temperature profile, and thus, the reactor becomes operable even at higher coolant temperatures. This can be accomplished by increasing the value of a_3 in Eq. 9.61, which can be varied independently of the other a's. The results when a_3 is increased by 50% to 150% are shown in

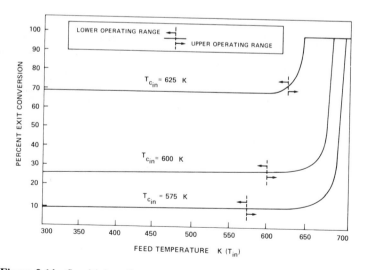

Figure 9.14 Sensitivity of conversion with respect to feed temperature.

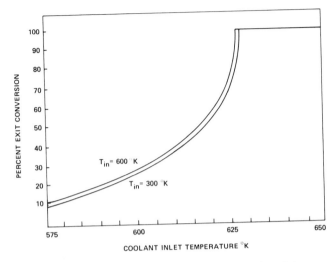

Figure 9.15 Sensitivity of conversion with respect to coolant inlet temperature.

Figure 9.16 for the reactor configuration without a preheater; slightly worse results are obtained for the configuration with a preheater. In view of the advantages discussed, the reactor configuration with no preheater should be considered as a viable alternative to the usual configuration for highly exothermic reactions.

The results obtained in this section are also applicable to the case of constant wall temperature, in which case the parameter a_4 can be set to zero for the construction of the phase diagram. Since a plug-flow model was used for the phase diagram, radial dispersion should have the effect of reducing the range of safe operating conditions; axial dispersion should tend to broaden the range.

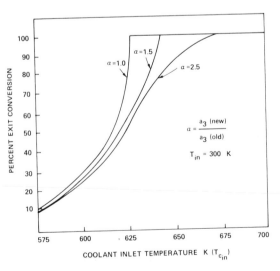

Figure 9.16 Effect of increasing the parameter a_3 on reactor sensitivity.

Summary

Design and analysis of heterogeneous fixed-bed reactors requires solving conservation equations of mass and energy for both the fluid phase (reactor) and the solid phase (pellet). These equations in their full form are extremely difficult to solve. The mode of reactor operation and the nature of the reaction, however, allow considerable simplification. Even with the simplifications that can be made, numerical solutions are not easy to obtain. In addition, there are design requirements that have to be satisfied. There are usually limitations placed on the pressure drop and operating conditions for safe reactor operation. Special types of fixed-beds are sometimes necessary as in radial flow reactors and monolithic catalyst beds. The desire to utilize the heat of reaction motivates the coupling of the reactor with a heat exchanger. If the reaction is highly exothermic, the reactor becomes quite sensitive to changes in operating conditions. For such cases, the phase plane diagram and the phase plane analysis can be utilized for the design.

When axial dispersion is significant, the question of multiple steady states (Luss and Amundson 1967) becomes important for the reactor start-up. Unless the bed is quite shallow, however, this should not pose a problem. Not considered in this chapter was the design and analysis of fixed-beds under conditions of catalyst deactivation. This is treated in the following chapter based on the approach of reactor point effectiveness.

NOTATION

a_1, a_2, a_3, a_4	parameters defined in Eq. 9.63 (or in Eq. 9.65)
a_i	parameters in orthogonal polynomial
a_c, b_c, c_c	z-dependent parameters for radial concentration profiles
a_T, b_T, c_T	z-dependent parameters for radial temperature profiles
$(Bi)_m$	Biot number for mass
A	Z/R
A_c	heat exchange surface area per unit volume
C	concentration
C_b	bulk fluid concentration
C_i	concentration of species i
$\bar{C}$	$C_b/(C_b)_{in}$
C_1, C_2	constants defined in Table 9.4
C_p	specific heat
C_{pc}	mass specific heat of coolant
C_{pr}	mass specific heat of reaction fluid
d_p	pellet size (pellet diameter)
D	dispersion coefficient for mass
D_a	axial dispersion coefficient
D_e	effective diffusivity
D_r	radial dispersion coefficient
E_a	activation energy

f	friction factor
f_a	global rate in terms of bulk fluid temperature and conversion
F	molar flow rate of key species
G_f	mass rate of fluid/area
h	film heat transfer coefficient
h_w	wall heat transfer coefficient for two-dimensional model
$(-\Delta H)$	heat of reaction
i	grid point in z-direction
j	grid point in r-direction
$\mathbf{j}_i$	mass flux vector for species i
$\mathbf{J}_i$	molar flux vector for species i
k	rate constant
k_g	film mass transfer coefficient
K	heat dispersion coefficient ($\rho C_p K = \lambda$)
K_a	axial dispersion coefficient for heat
K_r	radial dispersion coefficient for heat
l	grid interval in z-direction; circumference of an open channel
L	characteristic pellet length
m	grid interval in r-direction; R/d_p
$\dot{m}$	mass flow rate of coolant
$\dot{M}$	mass flow rate of reaction fluid
n	Z/d_p
n_c, n_T	z-dependent parameters for radial profiles of concentration and temperature, respectively
N	parameter defined by Eq. 9.26; number of tubes
p, P	partial pressure and total pressure, respectively
P_{ha}, P_{hr}	axial and radial Peclet numbers for heat, defined in Eq. 9.19
P_{ma}, P_{mr}	axial and radial Peclet numbers of mass, defined in Eq. 9.18
r	radial coordinate (0 at tube center)
r_i	rate of formation for species i in mass unit per unit volume of fluid
r_c	intrinsic rate in moles per unit volume of pellet
R	radius of tube
R_i	r_i in mole unit
R_G	global rate of formation by key species
R'_G	global rate based on external surface area
Re	Reynolds number based on d_p
S	parameter defined by 9.25
t	time
T	temperature
T_1	reactor outlet temperature used in Eq. 9.67
T_{ad}	adiabatic temperature at reactor outlet
T_b	bulk-fluid temperature
T_c	coolant temperature
T_{in}	T_b at reactor inlet (at $z = 0$)
$T_{c_{in}}$	coolant inlet temperature (at $z = Z$)

T_m	maximum temperature in reactor
$\overline{T}$	$T_b/(T_b)_{in}$
u	fluid velocity vector
U	overall heat transfer coefficient based on average temperature
v	superficial fluid velocity
W	mole fraction
x	pellet coordinate
y	conversion of key species; C_b or T_b in difference equations (Eq. 9.38)
z	axial coordinate ($z = 0$ at reactor inlet)
$\bar{z}$	z/Z
z'	heat exchanger coordinate
Z	reactor length
Z'	heat exchanger length

Greek Letters

α	ZD_r/R^2v for mass, and zK_r/R^2v for heat
β, γ	constants in Eq. 9.38
δ	quantity defined in Eq. 9.26; wall thickness
ϵ_B	bed porosity
η	effectiveness factor
θ	Z/v; weighting factor for implicit numerical method in Table 9.2
λ	thermal conductivity
λ_e	effective thermal conductivity
ρ_B	catalyst bulk density
ρ_i	mass density of species i
σ	quantity defined in Eq. 9.24
ϕ	Thiele modulus

Subscripts

b	bulk-fluid
c	coolant
f	fluid
i	species i
in	reactor inlet
s	pellet surface
p	pellet

PROBLEMS

9.1. List the assumptions made in writing Eq. 9.5 from Eq. 9.1.

9.2. Suppose that the pellet size is such that the diffusional effects are negligible. What equations can be deleted from Eqs. 9.6 through 9.15? Does this necessarily mean that the global rate (R_G) is equal to the intrinsic rate (r_c)? Under what conditions is R_G equal to r_c? If $R_G \neq r_c$, what additional conditions are neeeded in addition to Eqs. 9.6 and 9.7?

9.3. According to a dispersion model (Carberry 1958), the dispersion coefficient is given by:

$$D_a = \frac{l^2}{2\theta_D} \qquad \begin{array}{l} l: \text{mixing length} \\ \theta_D \approx l/v \end{array}$$

If the axial mixing length l in a packed bed is defined as the bed length divided by the number of perfect mixers n, $l = Z/n$. Therefore, a reactor with axial dispersion may be considered equivalent to n CSTR's in series where:

$$n = \frac{Zv}{2D_a}$$

Show then that a packed bed is equivalent to Z/d_p perfectly mixed, stirred tanks in series. Use the typical P_{ma} value of 2. By extending this analogy to radial dispersion, show that the number of perfectly mixed cells in the radial direction is $5(R/d_p)$ when $P_{mr} = 10$.

9.4. Consider a bed packed with spherical pellets. Show that the relative pressure drop is given by:

$$\frac{\Delta P}{P} = -(\text{constant}) \frac{ZG_f^{2.8}}{P^2 d_p^{0.2}}$$

for an isothermal reactor. This relationship should be applicable to the radial flow reactor with the idealization discussed in Section 9–8. If a fixed-bed with a length to depth ratio of 50 is operated as a radial flow reactor, how much of an increase in throughput would you expect if the relative pressure drop is the same for both modes of operation? Assume P to be the same for both modes.

9.5. Write the finite-difference version of Eqs. 9.27 through 9.32 using the explicit method of Figure 9.2a and the implicit method of Figure 9.3. Draw flow sheet diagrams for both methods. Make the equations dimensionless first.

9.6. Show for the averaging method of Ahmed and Fahien (1980) that the use of boundary conditions leads to:

$$y = y_0 - c_x \left(\frac{n_x}{2} \bar{r}^2 - \bar{r}^{n_x} \right); \qquad c_x = \frac{4(n_x + 2)}{(n_x + 4)(n_x - 2)}(y_0 - \bar{y})$$

$$T = T_0 + (T_w - T_0 - c_T)\bar{r}^2 + c_T \bar{r}^{n_T}; \qquad c_T = \left(\frac{n_T + 2}{n_T - 2} \right)(T_w - T_0 - 2\bar{T})$$

$$0 \leq \bar{r} \leq 1$$

when the wall temperature is constant at T_w. The subscript 0 is for the centerline quantities. The cup-mixing quantities are denoted by the overbar. From the expressions for y and c_x, show that:

$$n_x = \frac{4(\bar{y} - y_w)}{y_0 + y_w - 2\bar{y}}$$

The value of n_T can be found from the fact that T_w is constant: $dT_w/dz = 0$. Obtain 5 ordinary differential equations from Eqs. 9.27 and 9.28 for y_0, T_0, y_w, $\bar{y}$, and T. Note that the equations for the centerline quantities can be obtained by evaluating the radial derivatives at $\bar{r} = 0$ and using the assumed profiles.

9.7. Determine the length of a reactor for 35% yield (B formed/A_{in}) of phthalic anhydride by partial oxidation of naphthalene in air over V_2O_5. Assume each reaction to be of first-order and that the coolant temperature is constant.

$$\begin{array}{llll} P_{mr} = 8 & T_{in} = 320°C & Re = 184 & R = 4 \text{ cm} \\ P_{hr} = 6 & T_c = 310°C & d_p = 0.5 \text{ cm} & y_{in} = 0.5\% \\ P = 1 \text{ atm} & D_e = 0.1 \text{ cm}^2/\text{sec} & \epsilon_B = 0.5 \end{array}$$

$$C_p = 7 \text{ cal/mole°C}, \quad (kg/v)(Sc)^{2/3} = 1.15(Re)^{1/2} = \frac{h}{\rho C_p v}(Pr)^{2/3}$$

$$h_w = 10^{-3} \text{ cal/cm}^2 \, S°K$$

$$\text{napthalene} \xrightarrow[k_1]{-\Delta H_1} \text{anhydride} \xrightarrow[k_2]{-\Delta H_2} CO_2 \text{ and } H_2O$$

$$-\Delta H_1 = 43 \text{ kcal/mole}$$
$$-\Delta H_2 = 87 \text{ kcal/mole}$$

$T(°C)$	k_1	k_2
320	0.25	0.0095
370	3.5	0.035

Use the kinetic viscosity of air for the reaction mixture. Use the two-dimensional model. Assume a spherical, isothermal pellet.

9.8. For the following reaction carried out in an adiabatic reactor, calculate the reactor length required for 40% conversion.

$$A \xrightarrow{k} \text{products}$$

$$-r_A = \frac{1.8 \times 10^7 \exp(-29,000/R_g T)P_A}{1 + 18,000 \, P_A}$$

$L = 0.5$ cm, bed cross-sectional area $= 10^4$ cm^2
$\epsilon_B = 0.5$, $P = 1$ atm, y_{in} (mole fraction) $= 0.06$
Molar flow rate $= 1.5$ mole/hr cm^2, $D_e = 0.05$ cm^2/sec
$(-\Delta H) = 20$ kcal/mol, $T_{in} = 580°C$

Use the physical properties of air for the mixture. Also use the same j factor correlations as given in Problem 9.7.

9.9. Construct a diagram similar to the one given in Figure 9.11 for the following conditions:

ξ in Eq. 9.51 $= 125°K^{-1}$
$R_G = 10^{17} \exp(-18,000/RT)$ mol/hr $\cdot$ m^3
Reactor length $= 7$ m
$F_{in} = 1.5 \times 10^4$ mole/hr $\cdot$ m^2
$\epsilon_B = 0.5$ $v\rho C_p/(-\Delta H)F_{in} = 2.5 \times 10^{-3}$ (1/K)

Obtain the extinction temperature, i.e., the T_{in} that results in extinction of the reaction. Also calculate the T_{in} that gives a conversion of 0.5. Expand the Arrhenius exponential around $T(Z)$ to rewrite it as $A \exp(bT)$.

9.10. For the reaction system of Table 9.5, calculate the reactor lengths for 29.5% conversion, first for a reactor with a preheater ($T_{in} = 600°K$) and then for the reactor without a preheater ($T_{in} = 300°K$). Use $(T_c)_{in}$ of 600°K. Also calculate the preheater length for the reactor with a preheater by assuming that the exit stream from the reactor is used to exchange heat with the cold feed in a preheater geometrically and thermally similar to the main reactor. Assume that the main temperature difference across the heat exchanger is 10°K.

	Reactor Length (m)	Preheater Length (m)
With preheater	3	2.65
Without preheater	3.25	0

9.11. Construct a phase diagram of inlet temperatures for the following reaction system:

$a_1 = 5.92 \times 10^9$ $a_2 = 1.215 \times 10^9$
$a_3 = 10.2$ $a_4 = 7.7 \times 10^{-4}$
$R_G = 5.8 \times 10^8 \exp(-28,000/R_g T)$
$C_1 = 15$ $C_2 = 2 \times 10^{-4}$

All quantities are in cgs unit.

REFERENCES

Akella, L.M. and H.H. Lee, AIChE J., *28*, 87 (1983).

Ahmed, M. and R.W. Fahien, Chem. Eng. Sci., *35*, 897 (1980).

Ames, W.F., *Nonlinear Partial Differential Equations in Engineering*, Academic Press, New York (1965).

Baddour, R.F., P.L.T. Brian, B.A. Logeais and J.P. Eymery, Chem. Eng. Sci., *20*, 281 (1965).

Barkelew, C.R., Chem. Eng. Progr. Symp. Ser., *55*(25), 38 (1959).

Bird, B.R., W.E. Stewart and E.N. Lightfoot, *Transport Phenomena*, Wiley, New York (1960).

Carver, M.B., in *Numerical Methods for Differential Systems*, p 243, eds. Lapidus and Schiesser, Academic Press, New York (1976).

Cappelli, A., A. Collina and M. Dente, Ind. Eng. Chem. Proc. Des. Dev., *11*, 184 (1972).

Carberry, J.J. and D. White, Ind. Eng. Chem., *61*, 27 (1969).

Carberry, J.J., *Chemical and Catalytic Reaction Engineering*, McGraw-Hill, New York (1976).

Carberry, J.J., Can. J. Chem. Eng., *36*, 207 (1958).

Carnahan, B., H.A. Luther and J.O.Wilkes, *Applied Numerical Methods*, Chap. 7, Wiley, New York (1969).

Dente, M., and A. Collina, Chim. et Industrie, *46*, 752 (1964).

Dumez, F.J. and G.F. Froment, Ind. Eng. Chem. Proc. Des. Dev., *15*, 291 (1976).

Finlayson, B.A., Chem. Eng. Sci., *26*, 1081 (1971).

Finneran, J.A., L.J. Buividas and N. Walen, Hydrocarbon Processing, *51*, 127 (1972).

Froment, G.F. and K.B. Bischoff, *Chemical Reactor Analysis and Design*, Wiley, New York (1979).

Hegedus, L.L., AIChE J., *21*, 849 (1975).

Hicks, R.E., Ind. Eng. Chem. Fund., *9*, 500 (1970).

Hlavacek, V. and J. Votruba in *Chemical Reactor Theory*, Chap. 6, eds. Lapidus and Amundson, McGraw-Hill, New York (1977).

Hlavacek, V. and M. Kubicek, Chem. Eng. Sci., *27*, 177 (1972).

Hlavacek, V., M. Marek and T.M. John, Coll. Czechoslov. Chem. Comm., *34* 3868 (1969).

Lapidus, L., *Digital Computation for Chemical Engineers*, McGraw-Hill, New York (1962).

Levenspiel, O., *Chemical Reaction Engineering*, Wiley, New York (1972).

Liu, S.L., AIChE J., *16*, 501 (1970).

Luss, D. and N.R. Amundson, Chem. Eng. Sci., *22*, 253 (1967).

Mehta, D. and M.C. Hawley, Ind. Eng. Chem. Proc. Des. Dev., *8*, 280 (1969).

Mihail, R. and C. Iordache, Ind. Eng. Chem. Proc. Des. Dev., *31*, 83 (1976).

Minhas, S. and J.J. Carberry, Br. Chem. Eng., *14*, 799 (1969).

Mkhlin, S.G. and K.C. Smolitskiy, *Modern Analytic and Computational Methods in Science and Mathematics*, eds. Bellman and Kalaba, American Elsevier Co., New York (1967).

Morbidelli, M. and A. Varma, AIChE J., *28*, 705 (1982).

Price, T.H. and J.B. Butt, Chem. Eng. Sci., *32*, 393 (1977).

Reilly, M.H. and R.A. Schmitz, AIChE J., *12*, 153 (1966).

Smith, J.M., *Chemical Engineering Kinetics*, 3rd ed., Chap. 5, McGraw-Hill (1981).

Van Heerden, C., Ind. Eng. Chem., *45*, 1242 (1935).

Van Welsenaere, R.J. and G.F. Froment, Chem. Eng. Sci., *25*, 1503 (1970).

Villadsen, J.V. and W.E. Stewart, Chem. Eng. Sci., *22*, 1483 (1967).

Votruba, J., O. Mikus, K. Nguen, V. Hlavacek and J. Skrivanek, Chem. Eng. Sci., *30*, 117, 201 (1975).

Wehner, J.F. and R.H. Wilhelm, Chem. Eng. Sci. *6*, 89 (1956).

CHAPTER 10

Design and Analysis of Fixed-Beds: Approach of Reactor Point Effectiveness

10–1 INTRODUCTION

Rigorous procedures for the design and analysis of fixed-bed reactors based on heterogeneous models have been treated in Chapter 9. While such procedures can be used to solve the design and analysis problem in principle, the time and effort involved is often too costly. This is particularly so when the kinetics are too complex to yield an analytical expression for the effectiveness factor. The stability of the numerical scheme may also pose a problem. This chapter describes the use of reactor point effectiveness as an alternative to the rigorous design procedures of Chapter 9. For a fixed-bed reactor, the reactor conservation equations are:

$$L_c(C_b) = -R_G \text{(pellet and film conditions)}$$

$$L_T(T_b) = (-\Delta H)R_G \text{(pellet and film conditions)}$$

where the linear operators L_c and L_T operate on the bulk concentration and temperature, respectively. The linear operators take on different forms depending on whether a plug-flow model or dispersion model is used. The conservation equations for the pellet in which the reaction takes place are given by:

$$\bar{L}_c(C_p) = r_c$$

$$\bar{L}_T(T_p) = -(-\Delta H)r_c$$

where the linear operators $\bar{L}_c$ and $\bar{L}_T$ operate on the pellet concentration and temperature, respectively. The intrinsic rate r_c is related to the global rate R_G through the transport processes across the pellet-bulk fluid interface. For the design and analysis of fixed-beds, therefore, the pellet conservation equations have to be solved first, which involves solving nonlinear two-point boundary value problems. These solutions are then related to the bulk conditions through heat and mass transfer across the pellet-bulk fluid interface. The reactor conservation equations can then be solved for the reactor concentration (C_b) and temperature (T_b). This heterogeneous reactor problem can be transformed into a homogeneous one, in which only the reactor conservation equations are involved, through the use of the reactor point effectiveness. The reactor point effectiveness developed in Part II enables the global rate to be expressed in terms of the bulk-fluid conditions. Therefore, the design and analysis problem reduces to solving:

$$L_c(C_b) = -R_G(C_b, T_b)$$

$$L_T(T_b) = (-\Delta H)R_G(C_b, T_b)$$

which are equivalent to the conservation equations for homogeneous reactors. The usefulness of this approach will be fully appreciated when reactions affected by catalyst deactivation are considered.

The approach of reactor point effectiveness, however, does require some approximations. The adequacy of these approximations will be examined in detail. This will be followed by a section on the equivalence between the plug-flow model and the axial dispersion model, and that between the plug-flow and the radial dispersion model. With these equivalences established, it is possible to concentrate on the plug-flow model. Using this model, the design and analysis problem for reactions affected by diffusion, those affected by both diffusion and chemical deactivation, and those affected by catalyst sintering will be treated in detail. Detailed design and analysis procedures result from this treatment.

The simplification made possible by the use of the reactor point effectiveness allows a close examination of the design of a reactor affected by catalyst deactivation, the characteristic of which is time-dependence. This leads to the optimal design for a reactor whose performance is time-dependent. Finally, reactor design involving multiple reactions is considered, using essentially the approach of reactor point effectiveness.

10-2   ADEQUACY OF APPROXIMATIONS

The basic approximations made in arriving at the reactor point effectiveness are: (1) isothermal pellet, (2) negligible external mass transfer resistance, and (3) estimation of the pellet center concentration by a simple relationship when the reaction is not severely diffusion-limited. The first two approximations are quite adequate in view of the fact that the mass Biot number is of the order of hundreds under realistic reaction conditions. Both theoretical and experimental justifications for these approximations have been given in Chapter 4. The first approximation will be relaxed when reactions affected by pore-mouth poisoning are considered since a definite temperature gradient then exists within the pellet. An additional approximation is the representation of the difference between the Arrhenius exponentials evaluated at the pellet surface and the bulk-fluid temperatures by a linear relationship. This is quite adequate since the ratio of the bulk-fluid to the pellet surface temperature is close to unity. This approximation can be eliminated if desired, as will be seen in the treatment of reactors affected by catalyst deactivation. This leaves the third approximation regarding the pellet center concentration.

For the intrinsic kinetics, $r_c = kf(C)$, the internal effectiveness factor (Chapter 4) is:

$$\eta_i = \frac{2^{1/2}}{Lr_c(C_b)} \left[\int_{C_c}^{C_b} D_e(C)r_c(C)dC \right]^{1/2} \tag{4.73}$$

and the generalized modulus is:

$$\phi_G = \frac{Lr_c(C_b)}{2^{1/2}} \left[\int_0^{C_b} D_e(C)r_c(C)dC \right]^{-1/2} \tag{4.74}$$

where the pellet surface concentration has been replaced by the bulk concentration in view of the second approximation. As was shown in Chapter 4, the pellet center concentration C_c can be set to:

$$C_c/C_b = \begin{cases} 0 & \text{for } \phi_G \geqslant 3 \\ \dfrac{1}{\cosh \phi_G} & \text{for } 0.3 < \phi_G < 3 \end{cases} \tag{4.93}$$

The approximation for the pellet center concentration when $0.3 < \phi_G < 3$ is based on the fact that the value of the integral in Eq. 4.73 is quite insensitive to an error in the estimated value of C_c. Because of this tolerance, the approximation is adequate in the region of moderate diffusion effect ($0.3 < \phi_G < 3$). On the other hand, the error in the calculated value of η_i can be significant as ϕ_G approaches zero. This should not pose any problem since the reaction can be considered diffusion-free for all practical purposes when the value of ϕ_G is small, say less than 0.3.

As discussed in some detail in Chapter 4, the approach of reactor point effectiveness cannot be used when the intrinsic kinetics exhibit an apparent negative-order behavior.

10–3   EQUIVALENCE OF FIXED-BED MODELS

The reactor conservation equations given in Chapter 9 contain dispersion terms. Depending on the relative importance of dispersion effects, there results a two-dimensional model with radial dispersion, a one-dimensional model with axial dispersion, or a plug-flow model. If plug-flow is taken as the basic model, the dispersion models can be viewed as corrections to the plug-flow model, resulting from mixing effects. If the deviation from the plug-flow model due to axial dispersion is small, which is usually the case, this deviation can be accounted for by a first-order correction term added to the solution of the plug-flow model. A perturbation technique (Cole 1968) will be used to show the equivalence between the plug-flow model and the one-dimensional model with axial dispersion. When there exists a radial gradient, the process of averaging over the cross-section usually leads to simpler results. In fact, the quantities of interest are the average concentration and temperature rather than the local quantities. It will be shown that the averaging process leads to an equivalence between the plug-flow model and the two-dimensional model with radial dispersion under certain simplifying conditions.

For adiabatic fixed-bed reactors with axial dispersion, the steady state conservation equations are:

$$-\epsilon \frac{d^2 C_b}{dy^2} + \frac{dC_b}{dy} = -\tau R_G \qquad y \in (0,1) \tag{10.1}$$

$$-\beta \frac{d^2 T_b}{dy^2} + \frac{dT_b}{dy} = \left(\frac{-\Delta H}{\rho C_p}\right) \tau R_G \tag{10.2}$$

where $\epsilon = 1/P_m = 1/(Zv/D_a)$
$\beta = 1/P_h = 1/(Zv/K_a)$
$\tau = Z(1 - \epsilon_B)/v$
ϵ_B = bed porosity
ρC_p = molar heat content
R_G = global rate
y = normalized reactor coordinate
Z = reactor length
v = superficial fluid velocity
D_a = axial dispersion coefficient for mass
K_a = axial dispersion coefficient for heat

The dimensionless quantities ϵ and β are much smaller than unity since the Peclet number for heat (P_h) and the Peclet number for mass (P_m) are large for most reactors of interest. The boundary conditions are:

$$C_{in} = C_b - \epsilon \frac{dC_b}{dy}, \quad T_{in} = T_b - \beta \frac{dT_b}{dy} \qquad \text{at } y = 0 \tag{10.3}$$

$$\frac{dC_b}{dy} = \frac{dT_b}{dy} = 0 \qquad \text{at } y = 1 \tag{10.4}$$

Nonadiabatic reactors in which heat exchange occurs with the environment are not considered here, since then radial dispersion would dominate over axial dispersion. For an isothermal reactor for which τR_G is a function only of concentration, i.e., $\tau R_G = F(C_b)$, Burghardt and Zaleski (1968) obtained an approximate solution of Eq. 10.1 using a perturbation technique:

$$C_b = C_0 + \epsilon(\ln F_i/F - 1)F + (\epsilon^2) \tag{10.5}$$

where $C_0 =$ solution for the plug flow model, $\epsilon = 0$
   $F = \tau R_G(C_0)$
   $F_i = F$ evaluated at reactor inlet

This result can be used directly in Eq. 10.1 if a relationship between T_b and C_b can be found such that $\tau R_G(T_b, C_b)$ is expressible solely in terms of concentration. This relationship has been obtained by Lee (1981) using a perturbation technique:

$$T_b - T_{in} = \left(\frac{-\Delta H}{\rho C_p}\right)(C_{in} - C_b) + \left(\frac{-\Delta H}{\rho C_p}\right)(\epsilon - \beta)\frac{dC_0}{dy} \tag{10.6}$$

where dC_0/dy is the concentration gradient obtained from the solution of the plug-flow model. It can be seen that the second term in the right hand side of Eq. 10.6 is a first-order correction for the effect of dispersion on the relationship between T_b and C_b, since the first term represents the temperature change for a plug-flow reactor. Now that Eq. 10.6 gives T_b in terms of the solution of the plug-flow model and C_b, use of this relationship in Eq. 10.1 allows an expression for τR_G in terms of C_b and known quantities. Therefore, the solutions for the reactor with axial dispersion follow from Eqs. 10.5 and 10.6. It is seen that the plug-flow model leads directly to near-accurate solutions of the axial dispersion model. Details of using Eqs. 10.5 and 10.6 will be given when design and analysis are considered in the following section.

Consider now the radial dispersion model. The steady state conservation equations are:

$$\frac{\partial C_b}{\partial y} - \frac{\tau_a D_r}{R_t^2}\left(\frac{\partial^2 C_b}{\partial r^2} + \frac{1}{r}\frac{\partial C_b}{\partial r}\right) = -\tau R_G; \qquad r \in (0,1) \tag{10.7}$$

$$\frac{\partial T_b}{\partial y} - \frac{\tau_a K_r}{R_t^2}\left(\frac{\partial^2 T_b}{\partial r^2} + \frac{1}{r}\frac{\partial T_b}{\partial r}\right) = \left(\frac{-\Delta H}{\rho C_p}\right)\tau R_G \tag{10.8}$$

with the boundary conditions:

$$\frac{\partial C_b}{\partial r} = \frac{\partial T_b}{\partial r} = 0 \qquad \text{at } r = 0 \tag{10.9}$$

$$\frac{\partial C_b}{\partial r} = 0, \quad -\frac{\partial T_b}{\partial r} = \frac{h_w R_t}{\rho C_p K_r}(T_b - T_c) \qquad \text{at } r = 1 \tag{10.10}$$

where $\tau_a = Z/v$
$\quad D_r$ = radial dispersion coefficient for mass
$\quad K_r$ = radial dispersion coefficient for heat
$\quad r$ = normalized radial distance, $r = 0$ at the center
$\quad R_t$ = tube radius
$\quad h_w$ = overall heat transfer coefficient based on $(T_1 - T_c)$
$\quad T_c$ = coolant temperature
$\quad T_1 = T_b$ at $r = 1$

Multiplying Eqs. 10.7 and 10.8 by r and integrating the results with respect to r from zero to unity for the average values gives:

$$\frac{\partial \bar{C}}{\partial y} = -2\tau \int_0^1 rR_G dr \tag{10.11}$$

$$\frac{\partial \bar{T}}{\partial y} = -2\tau \left(\frac{-\Delta H}{\rho C_p}\right) \int_0^1 rR_G dr - \frac{2\tau_a h_w}{R_t \rho C_p} (T_1 - T_c) \tag{10.12}$$

where the boundary conditions have been used in the integration, and T_1 is T_b at $r = 1$. The mixing-cup averages are:

$$\bar{C} = \int_0^1 rC_b dr \Big/ \int_0^1 rdr = 2\int_0^1 rC_b dr \tag{10.13}$$

$$\bar{T} = \int_0^1 rT_b dr \Big/ \int_0^1 rdr \tag{10.14}$$

A comparison between Eqs. 10.11 and 10.12 and the conservation equations for the plug-flow model reveals that the radial dispersion model is the same as the plug-flow model in terms of average concentration and temperature if:

$$\int_0^1 rR_G dr \Big/ \int_0^1 rdr = R_G|_{T_b=\bar{T}, C_b=\bar{C}} \tag{10.15}$$

and if the overall heat transfer coefficient is based on the average temperature (U) instead of the fluid temperature at the wall (h_w) such that the heat transferred from the reactor to the coolant is written as $2U(\bar{T} - T_c)/R_t$ instead of $2h_w (T_1 - T_c)/R_t$. In general, the identity of Eq. 10.15 does not hold. The goodness of the plug-flow model, therefore, hinges on a proper choice of the value of U when radial gradients exist.

It has been shown that the solutions of the plug-flow model lead directly to near-accurate solutions of the axial dispersion model and that the plug-flow model is equivalent to the radial dispersion model with some simplifying assumptions. Radial gradients usually exist when there is a significant heat effect (Chapter 9). For highly exothermic reactions, the usual design practice is to select a small ratio of tube diameter to pellet size such that the undesirable radial gradients can be eliminated. Therefore, the design and analysis of reactors can be carried out in most cases based on the plug-flow model.

10–4 DESIGN AND ANALYSIS OF FIXED-BED REACTORS

The approach of reactor point effectiveness is used in this section to obtain detailed design and analysis procedures for reactions affected by diffusion, those affected by both diffusion and chemical deactivation, and those affected, in addition, by catalyst sintering. Results obtained in Chapters 4 through 6 are used to arrive at the procedures.

10–4–1 Reactions Affected by Diffusion

The steady state conservation equations for a plug-flow reactor with countercurrent cooling are:

$$\frac{dC_b}{dy} = -\tau R_G \tag{10.16}$$

$$\frac{dT_b}{dy} = \left(\frac{-\Delta H}{\rho C_p}\right) \tau R_G + \frac{v_c (\rho C_p)_c}{v \rho C_p} \frac{dT_c}{dy} \tag{10.17}$$

$$-\frac{dT_c}{dy} = \frac{2 U \tau_c}{(\rho C_p)_c R_t} (T_b - T_c) \tag{10.18}$$

where v_c = coolant velocity
$(\rho C_p)_c$ = heat content of coolant
$\tau_c = Z/v_c$

The boundary conditions are reactor inlet conditions (at $y = 0$) for C_b and T_b and the coolant inlet temperature (at $y = 1$) for T_c. It has been assumed that the velocity of the reacting fluid is constant. The case of a variable velocity will be treated later. The global rate is given by:

$$R_G = \Lambda(C_b, T_b) \, r_c \, (C_b, T_b); \qquad r_c = kf(C) \tag{10.19}$$

which can be rewritten as:

$$R_G = \frac{\dfrac{(2 D_e k_b)^{1/2}}{L} \left[\displaystyle\int_{C_c}^{C_b} f(\alpha) d\alpha \right]^{1/2}}{1 - \dfrac{1.2 E_a (-\Delta H)(2 D_e k_b)^{1/2}}{2 h R_g T_b^2} \left[\displaystyle\int_{C_c}^{C_b} f(\alpha) d\alpha \right]^{1/2}}$$

where the expression for the reactor point effectiveness developed in Chapter 4 (Eq. 4.83) has been used. For the reactor design, Eq. 10.16 can be integrated from the inlet to the outlet of the reactor with R_G given above to yield:

$$\tau = \int_{C_{out}}^{C_{in}} \left\{ \frac{1}{\frac{(2D_e k_b)^{1/2}}{L} \left[\int_{C_c}^{C_b} f(\alpha)d\alpha \right]^{1/2}} - \frac{1.2L\, E_a(-\Delta H)}{2hR_g T_b^2} \right\} dC_b \quad (10.21)$$

The right hand side of this equation can be integrated analytically or numerically to give the reactor size (τ), given the desired conversion (C_{out}). In order to integrate this equation with respect to C_b, however, T_b has to be expressed in terms of C_b. Combining Eqs. 10.16 and 10.17, and integrating the resulting equation from the reactor inlet to any point in the reactor gives:

$$T_b = T_{in} - \left(\frac{-\Delta H}{\rho C_p} \right)(C_b - C_{in}) + \frac{(\rho C_p)_c v_c}{\rho C_p v} \left[T_c - (T_c)_{out} \right] \quad (10.22)$$

where T_{in} is T_b at $y = 0$ and $(T_c)_{out}$ is T_c at $y = 0$. This equation provides the relationship necessary for the integration of Eq. 10.21. Because of the split boundary conditions on T_b and T_c, however, iterations on T_c are necessary as detailed below. Design equations for this general case are summarized in Table 10.1. The design of an adiabatic reactor is quite simple since then Eq. 10.22 reduces to:

$$T_b = T_{in} - \left(\frac{-\Delta H}{\rho C_p} \right)(C_b - C_{in}) \quad (10.23)$$

When Eq. 10.23 is used in Eq. 10.21, there results a simple design equation for adiabatic reactors:

$$\tau = \int_{C_{out}}^{C_{in}} \left\{ \frac{1}{\frac{(2D_e k_b)^{1/2}}{L} \left[\int_{C_c}^{C_b} f(\alpha)d\alpha \right]^{1/2}} \right.$$
$$\left. - \frac{1.2LE_a(-\Delta H)}{2hR_g \left[T_{in} - \left(\frac{-\Delta H}{\rho C_p} \right)(C_b - C_{in}) \right]^{1/2}} \right\} dC_b \quad (10.24)$$

with

$$k_b = k_0 \exp\left\{ (-E_a/R_g) \Big/ \left[T_{in} - \left(\frac{-\Delta H}{\rho C_p} \right)(C_b - C_{in}) \right] \right\} \quad (10.25)$$

where k_0 is the preexponential factor. It is seen that a straightforward integration as a function of C_b gives the reactor volume for adiabatic reactors. A simpler design equation results for isothermal reactors. These results are summarized in

Table 10.1 Design Equations for Reactions Affected by Diffusion

$$r_c(C) = kf(C); \qquad y \in (0,1)$$

a. *Nonadiabatic, nonisothermal reactor*

$$\int_{C_{in}}^{C_b} \left\{ \frac{1}{\frac{(2D_e k_b)^{1/2}}{L} \left[\int_{C_c}^{\xi} f(\alpha)d\alpha \right]^{1/2}} - \frac{1.2LE_a(-\Delta H)}{2hT_b^2 R_g} \right\} d\xi = -\tau y \tag{A}$$

$$k_b = k_0 \exp(-E_a/R_g T_b) \tag{B}$$

$$T_b = T_{in} - \left(\frac{-\Delta H}{\rho C_p}\right)(C_b - C_{in}) + \frac{(\rho C_p)_c}{\rho C_p} \frac{v_c}{v} [T_c - (T_c)_{out}] \tag{C}$$

$$(T_c)_j = (T_c)_{j-1} - \frac{2\tau_c U \Delta y}{(\rho C_p)_c R_t} [T_b - T_c]_{j-1} \tag{D}$$

$$(T_c)_{j=0} = (T_c)_{out} = (T_c)_{y=0}$$

b. *Adiabatic, nonisothermal reactor*

$$\int_{C_{out}}^{C_{in}} \left\{ \frac{1}{\frac{(2D_e k_b)^{1/2}}{L} \left[\int_{C_c}^{C_b} f(\alpha)d\alpha \right]^{1/2}} \right.$$
$$\left. - \frac{1.2LE_a(-\Delta H)}{2hR_g[T_{in} - (-\Delta H/\rho C_p)(C_b - C_{in})]^2} \right\} dC_b = \tau \tag{E}$$

$$k_b = k_0 \exp\left\{ (-E_a/R_g) \Big/ \left[T_{in} - \left(\frac{-\Delta H}{\rho C_p}\right)(C_b - C_{in}) \right] \right\} \tag{F}$$

c. *Isothermal reactor*

$$\int_{C_{out}}^{C_{in}} \frac{dC_b}{\frac{(2D_e k_b)^{1/2}}{L} \left[\int_{C_c}^{C_b} f(\alpha)d\alpha \right]^{1/2}} = \tau \tag{G}$$

$$T_s/T_b = 1 + \frac{(-\Delta H)L\Lambda k_b f(C_b)}{hT_b} \tag{H}$$

$$\Lambda = \frac{\dfrac{(2D_e k_b)^{1/2}}{Lk_b f} \left[\int_{C_c}^{C_b} f(C)dC \right]^{1/2}}{1 - \dfrac{1.2E_a(-\Delta H)(2D_e k_b)^{1/2}}{2hR_g T_b^2} \left[\int_{C_c}^{C_b} f(C)dC \right]^{1/2}} \tag{I}$$

$$\phi_G = \frac{L(k_b/2D_e)^{1/2} f(C_b)}{\left[\int_0^{C_b} f(C)dC \right]^{1/2}} \tag{J}$$

If $\phi_G \geq 3$, $C_c = 0$ for Eqs. (A) through (G).
If $0.3 < \phi_G < 3$, obtain C_c by:

$$C_c/C_b = \frac{1}{\cosh \phi_G} \tag{K}$$

If $\phi_G < 0.3$, the reaction is diffusion-free.

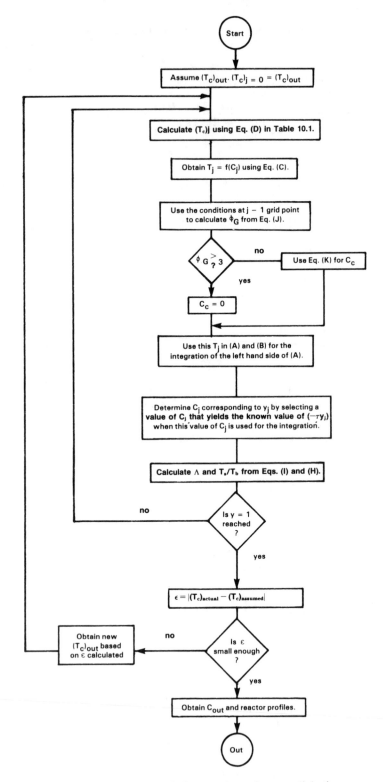

Figure 10.1 Design and analysis procedures for nonadiabatic, nonisothermal reactors for a given value of τ.

Table 10.1. While the design procedures for isothermal reactors and adiabatic reactors are straightforward to use, those for nonadiabatic reactors are complicated by the split boundary conditions on temperature. Procedures for nonadiabatic reactors are summarized in Figure 10.1 in the form of a flow chart. Note that Eq. (D) in Table 10.1 is the Euler version of Eq. 10.18 for numerical integration. Using the procedures given in Figure 10.1, a table of τ versus C_{out} can be generated, from which the value of τ corresponding to the desired conversion (C_{out}) can be selected.

In order to illustrate the use of the design and analysis procedures, the reaction system detailed in Table 10.2 has been completely analyzed. For the adiabatic reactor under consideration, the design equations (Eqs. (E) and (F) in Table 10.1) reduce to:

$$\int_C^{C_{in}} \left\{ \frac{1}{(2D_e k_b/\xi)^{1/2} \left[C_b - \frac{1}{\xi} \ln (1 + \xi C_b) \right]^{1/2}} - \frac{5.326 \times 10^{11}}{T_b^2} \right\} dC_b = \tau y \quad (10.26)$$

$$k_b = \exp\{14.6 - 12{,}000/[T_{in} + 4 \times 10^7 (C_{in} - C_b)]\} \quad (10.27)$$

with $T_b = T_{in} + 4 \times 10^7 (C_{in} - C_b)$

In view of the fact that ϕ_G at the reactor inlet is 5.6, the pellet center concentration was set equal to zero. Eq. (J) in Table 10.1 was used to evaluate the value of ϕ_G. The bulk temperature was used for the evaluation of the equilibrium constant ξ. The relationship between conversion and residence time (equivalently the reactor

Table 10.2 Parameters for the Example Reaction (Lee and Butt 1982)

$$r_c(C) = kf(C) = k \left[\frac{C}{1 + \xi C} \right]$$

$$k = \exp \left(-\frac{12{,}000}{T} + 14.6 \right) (1/\text{sec}), \quad E_a = 23.76 \text{ kcal/mole}$$

$$\xi = \exp \left(\frac{3600}{T} + 3.86 \right) (\text{cm}^3/\text{mole})$$

$T_{in} = 673°\text{K}$

$C_{in} = 1.81 \times 10^{-5} \text{ mole/cm}^3$

$(Bi)_m = 2500 \ (D_e = 1 \times 10^{-3} \text{ cm}^2/\text{sec}, \quad L = 1 \text{ cm}, \quad k_m = 2.5 \text{ cm/sec})$

$\left(\dfrac{-\Delta H}{\rho C_p} \right) = 4 \times 10^7 \, °\text{K cm}^3/\text{mole}$

$h = 3.245 \times 10^{-4} \text{ cal/sec cm}^2 \, °\text{K}$

$\phi_G = 5.63$ at the reactor inlet

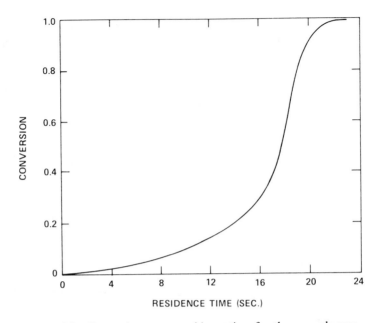

Figure 10.2 Conversion versus residence time for the example reaction of Table 10.2.

size) shown in Figure 10.2 was obtained by integrating Eq. 10.26. Simpson's method was used for the integration. It is seen that there exists a critical size of reactor where an incremental increase in reactor size significantly increases the conversion. This size corresponds to a residence time of approximately 17 seconds. Detailed calculations have been carried out for this case. The ratio of pellet surface temperature to bulk fluid temperature calculated from Eq. (H) in Table 10.1 is shown in Figure 10.3 along with the reactor point effectiveness calculated from Eq. (I). Reactor profiles of bulk fluid concentration and temperature are shown in Figure 10.4. It is seen from these figures that the ratio (T_s/T_b) increases rather rapidly as the conversion increases, whereas the reactor point effectiveness decreases monotonically. This example clearly demonstrates how simple it is to design and analyze a fixed-bed reactor using the approach of reactor point effectiveness, in sharp contrast with the usual approach given in Chapter 9.

Consider another example in which there are more than one species involved in the reaction:

$$A + 2B \rightarrow 3D + \text{other products}$$

Suppose that the intrinsic rate based on species A is given by:

$$r_A = \frac{kC_A C_B^2}{(1 + KC_D)}$$

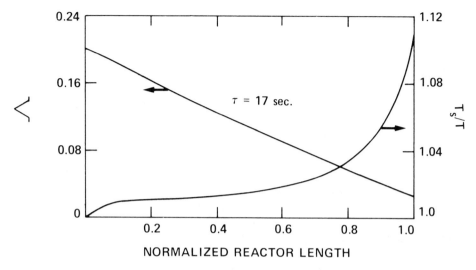

Figure 10.3 Reactor profiles of reactor point effectiveness and T_s/T_b.

Suppose further that the effective diffusivities are 10^{-2}, 4×10^{-1}, and 2×10^{-1} cm²/s for the species A, B, and D, respectively, and that the reactor inlet concentrations are 10^{-5} mol/cm³ for A and 4×10^{-5} mol/cm³ for B with no products in the feed. Here, the rate r_A is sought, expressed solely in terms of C_A such that it

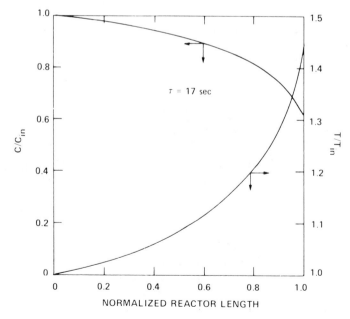

Figure 10.4 Reactor profiles of bulk-fluid concentration and temperature.

can be used for the concentration dependency $f(C)$ in Table 10.1. As detailed in Chapter 4, the relationships between the key species and the other species can be obtained by combining the pellet conservation equations for the species involved, which in this example yields:

$$C_B = (C_B)_b + \frac{2D_A}{D_B}[C_A - (C_A)_b] = (C_B)_b + 0.05[C_A - (C_A)_b]$$

$$C_D = (C_D)_b - \frac{3D_A}{D_D}[C_A - (C_A)_b] = (C_D)_b - 0.15[C_A - (C_A)_b]$$

where the D_i's ($i = A, B, D$) are the effective diffusivities, and the subscript b denotes the bulk fluid value. If the reactor dispersion coefficients are for the fluid mixture, which is usually the case, the stoichiometric relationship holds and in addition one has the following relationships:

$$(C_B)_b = (C_B)_{in} + 2[(C_A)_b - (C_A)_{in}]$$
$$(C_D)_b = (C_D)_{in} - 3[(C_A)_b - (C_A)_{in}]$$

where the subscript in denotes the reactor inlet value. The pellet local concentrations C_B and C_D then become:

$$C_B = (C_B)_{in} + 2[(C_A)_b - (C_A)_{in}] + 0.05[C_A - (C_A)_b]$$
$$C_D = -3[(C_A)_b - (C_A)_{in}] - 0.15[C_A - (C_A)_b]$$

since $(C_D)_{in} = 0$. Therefore, the rate expression for the key species A is:

$$r_A = k\left(\frac{C_A\{4 \times 10^{-5} + 2[(C_A)_b - 10^{-5}] + 0.05[C_A - (C_A)_b]\}^2}{1 + K\{3[10^{-5} - (C_A)_b] - 0.15[C_A - (C_A)_b]\}}\right)$$

Since the concentration dependency $f(C)$ for the rate expression in Table 10.1 is for the key species, which is A in this example, one has:

$$f(C) = f(C_A) = \frac{C_A\{4 \times 10^{-5} + 2[(C_A)_b - 10^{-5}] + 0.05[C_A - (C_A)_b]\}^2}{1 + K\{3[10^{-5} - (C_A)_b] - 0.15[C_A - (C_A)_b]\}}$$

This expression can be used directly in the design equations in Table 10.1 since the integration of $f(C)$ in Eq. (E) in Table 10.1, for instance, is with respect to C_A, and thus $(C_A)_b$ is constant for the integration of $f(C)$. This example clearly demonstrates the use of the design equations for a single reaction involving many species. As was the case in the earlier example, the bulk fluid temperature can be used for the value of the equilibrium constant K.

When the temperature change in the reactor and the volume change due to the reaction are significant, Eqs. 10.16 and 10.17 need to be rewritten to account for these changes. These are:

$$\frac{dx_b}{dy} = \frac{\tau_r}{(C_b)_r} R_G \left\{ \left[\left(\frac{T_r}{T_b} \right) \left(\frac{(C_b)_r(1 - x_b)}{1 + \delta x_b} \right) \right], T_b \right\} \tag{10.28}$$

$$\frac{dT_b}{dy} = \left(\frac{-\Delta H}{(\rho C_p)_r} \right) \tau_r R_G \left\{ \left[\left(\frac{T_r}{T_b} \right) \left(\frac{(C_b)_r(1 - x_b)}{1 + \delta x_b} \right) \right], T_b \right\} + \frac{(v\rho C_b)_c}{(v\rho C_p)_r} \frac{dT_c}{dy} \tag{10.29}$$

where $(C_b)_r$ = concentration evaluated at reference temperature and pressure
$\tau_r = \tau$ evaluated at reference velocity
T_r = reference temperature
δ = mole fraction times the change in the number of moles due to reaction per mole of species of interest
x_b = conversion of species of interest

The subscript r is for the reference conditions, which can be taken as the inlet conditions if so desired. The ideal gas law was used to rewrite Eqs. 10.16 and 10.17. Note that these equations are in the same form as Eqs. 10.28 and 10.29, C_b being replaced by the quantity within the bracket in the function R_G. Since these equations are in the same functional form as Eqs. 10.16 and 10.17, the same design and analysis procedures developed earlier can be applied. Combining Eqs. 10.18, 10.28, and 10.29, a relationship similar to Eq. 10.22 results.

It has been shown in Section 10.3 that the solution of the plug-flow model leads directly to that of the one-dimensional model with axial dispersion. These results can be applied to reactions affected by diffusion. Use of the results leads to the design equations for the fixed-beds with axial dispersion given in Table 10.3. Note that C_0 is the solution for the plug-flow model and that Eq. (E) in Table 10.3 is used for k_b in Eq. (B), whereas Eq. (G) is in Eq. (F) (see Problem 10.9 for the case of $\epsilon = 0$).

10–4–2 Reactions Affected by Diffusion and Chemical Deactivation

Design and analysis are treated separately in this section, for the analysis procedures do not lead directly to the design because of the time dependence of conversion caused by catalyst deactivation. The analysis procedures are treated first.

Analysis Procedures

It has been shown in Chapter 5 that the chemical deactivation of a catalyst can be uniform or shell-progressive (pore-mouth deactivation) depending on the relative speeds of the reaction and the transport of the species causing deactivation. Detailed analysis procedures are developed here. While deactivation reactions can cause shell-progressive or pseudo-shell-progressive deactivation in many cases, the activity profile of a pellet undergoing chemical deactivation can also exhibit a certain gra-

Table 10.3   Design Equations for Adiabatic Fixed-Beds with Axial Dispersion

$$C_b = C_0 + \epsilon\{\ln F_i/F - 1\}F \tag{A}$$

$$F = \tau \frac{\dfrac{(2D_e k_b)^{1/2}}{L}\left(\displaystyle\int_{C_c}^{C_0} f(\alpha)d\alpha\right)^{1/2}}{1 - \dfrac{1.2 E_a(-\Delta H)(2D_e k_b)^{1/2}}{2hT_b^2 R_g}\left(\displaystyle\int_{C_c}^{C_0} f(\alpha)d\alpha\right)^{1/2}} \tag{B}$$

$$F_i = \tau \frac{\dfrac{(2D_e k_{in})^{1/2}}{L}\left(\displaystyle\int_{(C_c)_{in}}^{C_{in}} f(\alpha)d\alpha\right)^{1/2}}{1 - \dfrac{E_a(-\Delta H)(2D_e k_{in})^{1/2}}{2hT_{in}^2 R_g}\left(\displaystyle\int_{(C_c)_{in}}^{C_{in}} f(\alpha)d\alpha\right)^{1/2}} \tag{C}$$

$$k_b = k_0 \exp(-E_a/R_g T_b), \qquad k_{in} = k_0 \exp(-E_a/R_g T_{in}) \tag{D}$$

$$T_b = T_{in} - (C_0 - C_{in})\left(\frac{-\Delta H}{\rho C_p}\right) - \left(\frac{-\Delta H}{\rho C_p}\right)(\beta - \epsilon)\frac{dC_0}{dy} \tag{E}$$

$$\frac{dC_0}{dy} = \tau \frac{\dfrac{(2D_e k_b)^{1/2}}{L}\left(\displaystyle\int_{C_c}^{C_0} f(\alpha)d\alpha\right)^{1/2}}{1 - \dfrac{1.2 E_a(-\Delta H)(2D_e k_b)^{1/2}}{2hT_b^2 R_g}\left(\displaystyle\int_{C_c}^{C_0} f(\alpha)d\alpha\right)^{1/2}} \tag{F}$$

$$k_b = k_0 \exp\left\{(-E_a/R_g)\Big/\left[T_{in} - \left(\frac{-\Delta H}{\rho C_p}\right)(C_0 - C_{in})\right]\right\} \tag{G}$$

dient that cannot be well approximated by the shell-progressive model. This general case of deactivation with an arbitrary profile is treated later.

For a plug-flow reactor in which the catalyst undergoes either uniform or shell-progressive deactivation, an additional mass balance for the species causing deactivation is required along with Eqs. 10.16 through 10.18:

$$\frac{dN_b}{dy} = -\tau R_p \tag{10.30}$$

where R_p is the global rate of the deactivation reaction and N_b is the bulk concentration of the species causing deactivation. The steady state pellet conservation equations are:

$$D_e \frac{d^2 C}{dz^2} = k(1 - \gamma)f(C) \tag{10.31}$$

$$-\lambda_e \frac{d^2 T}{dz^2} = (-\Delta H)k(1 - \gamma)f(C) \tag{10.32}$$

The boundary conditions are the usual ones at the pellet center and surface. Here, the fraction of catalyst deactivated, γ, is dependent on the pellet coordinate z. Without loss of generality, it may be assumed that the reaction is structure-insensitive, as indicated by the effective rate constant $k(1 - \gamma)$ in Eqs. 10.31 and 10.32. The rate of deactivation is much slower than the rate at which concentration and temperature reach their steady state values. This condition leads to the pseudo-steady state assumption with respect to the reactor conservation equations. With this assumption, the fraction of catalyst deactivated can be expressed (Chapter 5) as:

$$\frac{d}{dt}(Q\gamma) = r_p[N,C,\gamma] \tag{10.33}$$

where the intrinsic rate of deactivation can depend on concentrations of the main reactant and the poisoning species. For independent deactivation, this reduces (Chapter 5) to:

$$\frac{d\gamma}{dt} = \begin{cases} \dfrac{(k_p)_s N_s}{Q}(1-\gamma) & \text{uniform} \\[4mm] \dfrac{(k_p)_d}{Q}\left(\dfrac{N_s}{1+(\phi_p)_d^2\gamma}\right) & \text{shell-progressive} \end{cases} \tag{10.34} \tag{10.35}$$

where k_p = rate constant for deactivation reaction
$\quad Q$ = deactivation capacity of catalyst in moles of the species causing deactivation per apparent pellet volume
$\quad (\phi_p)_d = L[(k_p)_d/D_p]^{1/2}$
$\quad D_p$ = effective diffusivity of the species causing deactivation
$\quad s$ = subscript denoting pellet surface
$\quad d$ = subscript denoting the boundary between the fresh inner core and the completely deactivated pellet
$\quad N$ = concentration of the species causing deactivation

The pellet conservation equations are related to their reactor counterparts through transport across the pellet-bulk fluid interface:

$$h(T_s - T_b) = (-\Delta H)R_G L \tag{10.36}$$

$$k_g(C_b - C_s) = R_G L \tag{10.37}$$

$$(k_g)_p(N_b - N_s) = R_p L \tag{10.38}$$

where $(k_g)_p$ is the mass transfer coefficient for the species causing deactivation. The global rate of deactivation is given by:

$$R_p = \eta_p r_p \tag{10.39}$$

where η_p is the internal effectiveness factor for the deactivation reaction. For independent and uniform deactivation this reduces to:

$$R_p = (k_p)_s N_s (1 - \gamma) \tag{10.40}$$

In the case of shell-progressive deactivation, the global rate is given by the solution (Chapter 5) to:

$$D_p \frac{d^2 N}{dz^2} = 0 \qquad 0 \leq z \leq z_d$$

$$LR_p = D_p \frac{dN}{dz}\bigg|_{z=0} = \frac{L(k_p)_d N_s}{1 + (\phi_p)_d^2 \gamma} \tag{10.41}$$

Under the conditions of negligible external mass transfer resistance and an isothermal pellet, Eqs. 10.32, 10.37, and 10.38 can be removed from consideration for the case of uniform deactivation. However, Eq. 10.32 still needs to be retained in a limited form to account for the temperature drop in the deactivated outer shell in the case of shell-progressive deactivation.

The problem posed by Eqs. 10.16 through 10.18 and Eqs. 10.30 through 10.41 can be solved in principle for the analysis of a fixed-bed reactor. However, it is quite complicated and the numerical problem involved is not simple by any means, especially when it involves the design accounting for the time dependence of conversion. The reactor point effectiveness in Chapter 5 is now used to reduce this problem to a simpler one.

Uniform, Independent Deactivation. Since dependent deactivation can be treated in conjunction with the general case of deactivation considered later, attention here is restricted to independent deactivation. For a uniformly deactivated pellet, Eq. 10.30 can be combined with 10.40 to give:

$$\frac{dN_b}{dy} = -\tau(k_p)_s N_b (1 - \gamma) \tag{10.42}$$

The external mass transfer resistance has been neglected such that N_s is set to N_b. This equation can be integrated with the pseudo steady state assumption to give:

$$N_b[y;t] = (N_b)_{in} \exp\left\{-\int_0^y (k_p)_s \tau (1 - \gamma) dy\right\} \tag{10.43}$$

where $(N_b)_{in}$ is the reactor inlet concentration of the species causing deactivation. The global rate (Chapter 5, Table 5.3) is:

$$R_G = \Lambda k_b f(C_b) = \frac{[2D_e k_b(1-\gamma)I]^{1/2}/L}{1 - \dfrac{1.2E_a(-\Delta H)}{2hR_g T_b^2}[2D_e k_b(1-\gamma)I]^{1/2}} \qquad (10.44)$$

$$I \equiv \int_{C_c}^{C_b} f(C)dC$$

Equations 10.16, 10.34, and 10.43 then form a set of equations for the analysis procedures along with Eqs. 10.22 and 10.44:

$$\left\{ \frac{L}{\left[2D_e k_b \displaystyle\int_{C_c}^{C_b} f(C)dC\right]^{1/2}} - \frac{1.2E_a(-\Delta H)L(1-\gamma)^{1/2}}{2hR_g T_b^2} \right\}dC_b$$

$$= -\tau(1-\gamma)^{1/2}dy \qquad (10.45)$$

$$N_b = (N_b)_{in}\, \exp\left\{ -\int_0^y (k_p)_s \tau(1-\gamma)dy \right\} \qquad (10.43)$$

$$\gamma_j = \gamma_{j-1} + \Delta t[(k_p)_s N_b(1-\gamma)/Q]_{j-1} \qquad (10.46)$$

$$T_b = T_{in} - \left(\frac{-\Delta H}{\rho C_p}\right)(C_b - C_{in}) + \frac{(\rho C_p v)_c}{\rho C_p v}[T_c - (T_c)_{out}] \qquad (10.22)$$

Equation 10.45 has been obtained by combining Eq. 10.43 with Eq. 10.16 while Eq. 10.46 is the Euler version of Eq. 10.34 in time for numerical calculation. Equation 10.18 is also required for T_c. Let us examine in some detail the analysis procedures for an adiabatic reactor since the procedures are similar for nonadiabatic reactors. At time $t = 0^+$ ($\gamma = 0$), an integration of Eq. 10.45 yields all information on the reactor since the left hand side of this equation is a function only of C_b when Eq. 10.22 is used in 10.45. The last term in Eq. 10.22 disappears for an adiabatic reactor. The pellet surface temperature is then calculated for the evaluation of $(k_p)_s$. The initial profile ($t = 0^+$) of N_b is calculated from Eq. 10.43. The reactor profile of γ is obtained from Eq. 10.46 for the next time interval. This new γ profile (γ_j) is used for the integration of Eq. 10.45 for C_b corresponding to the new time, i.e., at $t_j = t_{j-1} + \Delta t$. These procedures are repeated until the final time of interest is reached. Equations necessary for the analysis are summarized in Table 10.4 for various modes of operation. The equation equivalent to Eq. 10.45 for an isothermal reactor is obtained by simply setting $(-\Delta H)$ equal to zero. The analysis procedures for an adiabatic reactor based on these equations are given in Figure 10.5 in the form of a flow chart. A similar chart can be obtained for nonadiabatic reactors.

Simulation results for an adiabatic reactor undergoing uniform deactivation (Lee and Butt 1982) are shown in Figures 10.6 and 10.7 for the reaction system given in Table 10.5. These results were obtained according to the procedures of

Table 10.4 Equations for Analysis of Plug-Flow Reactors: Uniform Deactivation and Diffusion Limitation

$$r_c(C,T) = kf(C)$$

a. *Adiabatic, nonisothermal reactors*

$$\left\{\frac{L}{\left[2D_ek_b\int_{C_c}^{C_b}f(\alpha)d\alpha\right]^{1/2}} - \frac{1.2E_a(-\Delta H)L(1-\gamma)^{1/2}}{2hR_gT_b^2}\right\}dC_b = -\tau(1-\gamma)^{1/2}dy \qquad (A)$$

$$T_b = T_{in} - \left(\frac{-\Delta H}{\rho C_p}\right)(C_b - C_{in}) \qquad (B)$$

$$N_b = (N_b)_{in}\exp\left\{-\int_0^y (k_p)_s\tau(1-\gamma)dy\right\} \qquad (C)$$

$$\gamma_j = \gamma_{j-1} + \Delta t\left[\frac{(k_p)_s}{Q}N_b(1-\gamma)\right]_{j-1} \qquad (D)$$

$$\Lambda k_bf = \frac{(1-\gamma)^{1/2}}{\dfrac{L}{\left[2D_ek_b\int_{C_c}^{C_b}f(\alpha)d\alpha\right]^{1/2}} - \dfrac{1.2E_a(-\Delta H)L(1-\gamma)^{1/2}}{2hR_gT_b^2}} \qquad (E)$$

$$T_s/T_b = 1 + \frac{(-\Delta H)\Lambda k_bfL}{hT_b} \qquad (F)$$

$$k_b = k_{00}\exp(-E_a/R_gT_b) \qquad (G)$$

$$(k_p)_s = k_{p0}\exp(-E_p/R_gT_s)$$

b. *Nonadiabatic, nonisothermal reactors*

$$T_b = T_{in} - \left(\frac{-\Delta H}{\rho C_p}\right)(C_b - C_{in}) + \frac{(v\rho C_p)_c}{v\rho C_p}[T_c - (T_c)_{out}] \qquad (B')$$

$$-\frac{dT_c}{dz} = \frac{2U}{R_t(\rho C_p)_c}\tau_c(T_b - T_c) \qquad (H)$$

[Equations (A) through (H) with (B) replaced by (B')]

c. *Isothermal reactors*

$$\int_C^{C_{in}}\frac{LdC_b}{\left[2D_ek_b\int_{C_c}^{C_b}f(\alpha)d\alpha\right]^{1/2}} = \tau\int_0^y (1-\gamma)^{1/2}dy \qquad (I)$$

$$N_b = (N_b)_{in}\exp\left\{-\int_0^y (k_p)_b\tau(1-\gamma)dy\right\} \qquad (J)$$

$$\gamma_j = \gamma_{j-1} + \Delta t\left[\frac{(k_p)_b}{Q}N_b(1-\gamma)\right]_{j-1} \qquad (K)$$

$$\Lambda k_bf = \frac{1}{L}\left\{(1-\gamma)2D_ek_b\int_{C_c}^{C_b}f(\alpha)d\alpha\right\}^{1/2} \qquad (L)$$

C_c is calculated as given in Table 10.1.

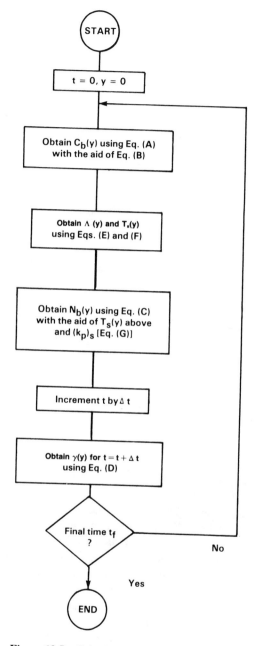

Figure 10.5 Calculation procedures for adiabatic, plug-flow reactors affected by diffusion and uniform poisoning.

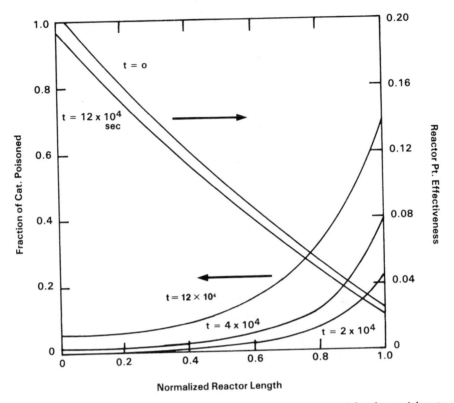

Figure 10.6 Reactor profiles of Λ and fraction of catalyst deactivated for the model system of Table 10.5: uniform poisoning and $\tau = 17s$.

Figure 10.5 for a residence time of 17 seconds. The reactor profiles for the fraction of catalyst deactivated, shown in Figure 10.6, reveal that the fraction deactivated increases rather sharply as the reactor outlet is approached, and that the time progression slows down with time. The concentration N_b turns out to remain relatively constant up to 12×10^4 seconds. The reactor profiles for γ are expected to show the trend in Figure 10.6 in view of the relatively constant concentration of the species causing deactivation (N_b) throughout the reactor, and in view of the fact that the maximum reactor temperature is usually reached at the outlet for an adiabatic reactor. The reactor point effectiveness, also shown in Figure. 10.6, does not depend on time as much as γ. This can be attributed to the compensating effect between the fraction of catalyst deactivated and the internal effectiveness factor: a larger fraction of catalyst deactivated results in a smaller catalyst activity while an increase in γ, on the other hand, causes an increase in the internal effectiveness factor. The outlet concentrations shown in Figure 10.7 reveal that the conversion decreases rather sharply initially, but then moderately with the time on stream. As expected, the reactor temperature decreases with increasing level of catalyst deactivation as shown in Figure 10.7.

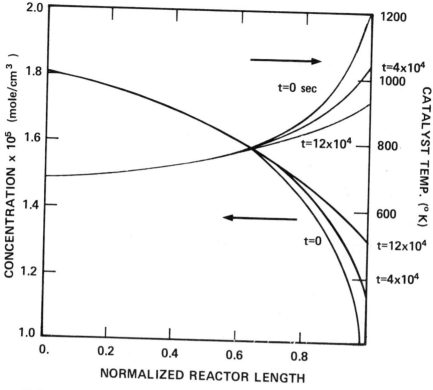

Figure 10.7 Profiles of concentration and temperature for the model system of Table 10.5: uniform poisoning and $\tau = 17$s.

In view of the fact that the reactor profile for N_b appears to have a strong effect on the other profiles, a τ value of 20 seconds has been chosen to examine the effect of a change in the N_b profile. The simulation results are shown in Figures 10.8 through 10.10. All the other parameters are the same as those given in Table 10.5, the only difference being the change of τ from 17 to 20 seconds. It is seen in Figure 10.8 that the species causing deactivation is almost completely depleted initially at the reactor outlet. This condition leads to reactor profiles for γ with the maximum near $y = 0.8$, as shown in Figure 10.9. As discussed earlier, the fraction of the catalyst deactivated is determined by the concentration of the species causing deactivation (N_b) and the local temperature. Initially, the species causing deactivation is nearly completely depleted at the outlet. Because of this depletion, the catalyst is not deactivated the most at the outlet even though the temperature is the highest there. As the time on stream increases, less of the species causing deactivation is consumed in the front part of the reactor because of the presence of partially deactivated pellets, making more of the poisoning species available toward the outlet. This results in an increase in N_b toward the outlet as shown in Figure 10.8, eventually resulting in a flat profile. The existence of a maximum

Table 10.5   Parameters for a Model Reaction System (Lee and Butt 1982)

$$r_c(C,T) = kf(C) = k\left(\frac{C}{1+\xi C}\right)$$

$$k = \exp\left(-\frac{12{,}000}{T} + 14.6\right)(1/\text{sec}), \quad E_a = 23.76 \text{ kcal/mol}$$

$$\xi = \exp\left(\frac{3600}{T} + 3.86\right)(\text{cm}^3/\text{mol})$$

$T_{in} = 673°\text{K}$

$C_{in} = 1.81 \times 10^{-5} \text{ mole/cm}^3$

$B_m = 2500, \quad D_e = 10^{-3} \text{ cm}^2/\text{sec}, \quad L = 1 \text{ cm}, \quad k_g = 2.5 \text{ cm/sec}$

$\left(\dfrac{-\Delta H}{\rho C_p}\right) = 4 \times 10^7 \text{ °K cm}^3/\text{mole}; \quad (-\Delta H) = 24.1 \text{ kcal/mol}$

$h = 3.245 \times 10^{-4} \text{ cal/sec cm}^2 °\text{K}$

$\phi_G = 5.63$ at the reactor inlet

$(N_b)_{in} = 5 \times 10^{-8} \text{ mole/cm}^3$

$Q = 10^{-4} \text{ mole/cm}^3 \text{ cat. pellet}$

$$k_p = \exp\left(-\frac{7000}{T} + 3.27\right), \quad E_p = 13.86 \text{ kcal/mol}$$

$D_p = 10^{-3} \text{ cm}^2/\text{sec}$

$\tau = 17 \text{ sec}$

$(Bi)_h = 20$

γ at some point other than the reactor outlet gives rise to some unusual behavior as shown in Figure 10.10, in which the time progression of observed rate, catalyst temperature, and bulk temperature at the outlet are shown. The most important fact revealed in this figure is that the maximum catalyst temperature can increase with increasing time. This fact has also been observed experimentally by Price and Butt (1977). This finding is important in view of the fact that adiabatic reactors are often designed with a constraint on the maximum allowable temperature. A design that disregards the effect of deactivation may not adequately meet the constraint. It is interesting to observe in Figure 10.10 that the bulk-fluid temperature decreases monotonically with time, in sharp contrast with the behavior of the catalyst temperature. This difference can be explained by the way in which the temperatures are determined: the bulk-fluid temperature is determined by the amount of reactant consumed up to the point of interest in the reactor (Eq. 10.23), whereas the catalyst temperature is determined by the local conditions (Eq. 10.36) of R_G and T_b. If the local global rate increases faster than the rate at which the

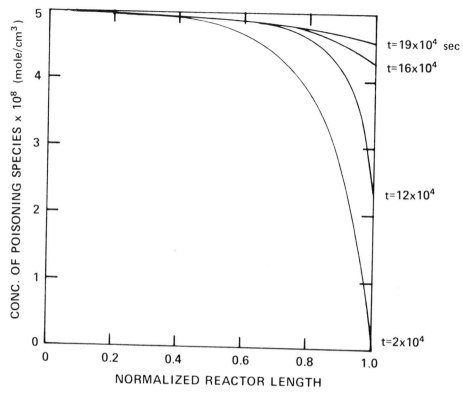

Figure 10.8 Reactor profiles of the concentration of the species causing deactivation: uniform deactivation and $\tau = 20s$.

bulk-fluid temperature decreases with time, the catalyst temperature can increase with increasing time despite the fact that the bulk-fluid temperature decreases with increasing time (Eq. 10.36). It is seen in Figure 10.10 that the local global rate (at $y = 1$) does increase with increasing time initially. The outlet concentration, on the other hand, decreases with increasing time, resulting in the monotonic decrease of the bulk-fluid temperature as shown in the figure.

Shell-Progressive, Independent Deactivation.   Consider the analysis procedures for reactors undergoing shell-progressive deactivation. The profile of N_b can be obtained by combining Eqs. 10.30 and 10.41 and then solving the resulting equation:

$$N_b[z;t] = (N_b)_{in} \exp\left\{-\int_0^y \frac{(k_p)_d \tau}{1 + (\phi_p)_d^2 \gamma} \, dy\right\} \tag{10.47}$$

The reactor point effectiveness can be expressed (Chapter 5) as:

$$\Lambda = \frac{D_e(C_b - C_d)}{k_b f(C_b) L^2 \gamma} \tag{10.48}$$

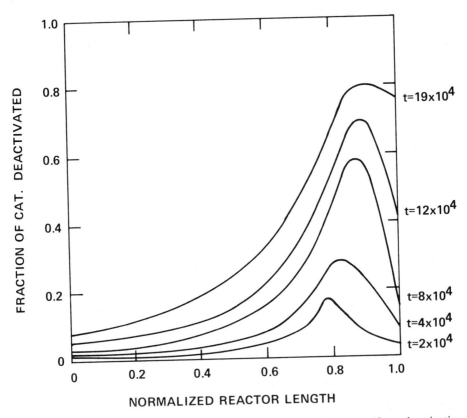

Figure 10.9 Reactor profiles of the fraction of catalyst deactivated: uniform deactivation and $\tau = 20$s.

When this is combined with Eq. 10.16, there results:

$$\frac{dC_b}{dy} = \frac{-\tau D_e(C_b - C_d)}{L^2\gamma} \tag{10.49}$$

At time zero, the value of γ is zero. Therefore, Eq. 10.45 should be used at time zero in place of Eq. 10.49 with γ set equal to zero. A set of equations necessary for the analysis of an adiabatic reactor can then be written as follows:

$$(C_b) = (C_b)_{i-1} - \frac{\tau D_e}{L^2\gamma} \Delta y \{(C_b)_{i-1} - (C_d)_{i-1}\} \tag{10.50}$$

$$C_b - C_d = \left(\frac{L\gamma}{D_e}\right)\left(2D_ek_b\int_{C_c}^{C_d} f(C)dC\right)^{1/2}$$
$$\exp\left\{\left(\frac{E_a}{2R_gT_b}\right)\left[\frac{(-\Delta H)D_e(1 + (Bi)_h\gamma)(C_b - C_d)}{hT_b\gamma + (1 + (Bi)_h\gamma)(C_b - C_d)(-\Delta H)D_e}\right]\right\} \tag{10.51}$$

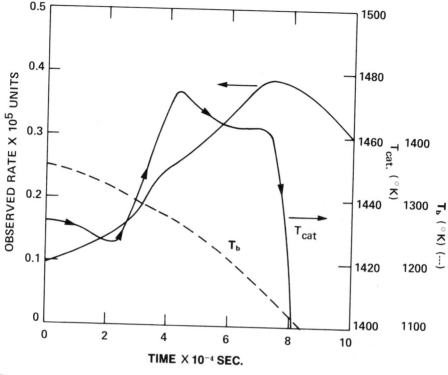

Figure 10.10   Time dependence of global rate, catalyst and bulk-fluid temperatures at the reactor outlet: uniform deactivation and $\tau = 20s$.

$$T_b = T_{in} - \left(\frac{-\Delta H}{\rho C_p}\right)(C_b - C_{in}) \qquad (10.23)$$

$$N_b = (N_b)_{in}\ \exp\left\{-\int_0^y \frac{(k_p)_d \tau}{1 + (\phi_p)_d^2 \gamma}\ dy\right\} \qquad (10.47)$$

$$\gamma_j = \gamma_{j-1} + \Delta t \left[\frac{(k_p)_d}{Q}\left(\frac{N_b}{1 + (\phi_p)_d^2 \gamma}\right)\right]_{j-1} \qquad (10.52)$$

Equation 10.51 has been obtained in Chapter 5 by equating the diffusion-limited transport rate through the completely deactivated outer shell to the diffusion-limited reaction within the fresh core of the pellet. The temperature drop in the deactivated outer shell has been accounted for in terms of the heat Biot number $(Bi)_h$ and γ. Equation 10.50 is the Euler version of Eq. 10.49 while Eq. 10.52 is that of Eq. 10.35 for numerical calculation. For nonadiabatic reactors, Eq. 10.18 needs to be added to the above set of equations with Eq. 10.22 in place of Eq. 10.23. These equations are summarized in Table 10.6 for various modes of operation while detailed analysis procedures for adiabatic reactors are given in Figure 10.11 in the form of a flow chart. The structure of the procedures shown in Figure

Table 10.6   Analysis Equations for Plug-Flow Reactors: Shell-Progressive, Independent Deactivation, and Diffusion Limitation

$$r_c(C,T) = kf(C)$$

a. *Adiabatic, nonisothermal reactors*

Use Eqs. (A), (B), (E), and (F) In Table 10.4 for the profiles of C_b, T_b, Λ, and T_s at $t = 0$ ($\gamma = 0$). Based on these initial profiles, proceed with the equations given below.

$$C_b - C_d = \left(\frac{L\gamma}{D_e}\right)\left[2D_ek_b\int_{C_C}^{C_d} f(\alpha)d\alpha\right]^{1/2}$$

$$\exp\left\{\left(\frac{E_a}{2R_gT_b}\right)\left[\frac{(-\Delta H)D_e(1+(Bi)_h\gamma)(C_b-C_d)}{hT_b\gamma+(1+(Bi)_h\gamma)(C_b-C_d)(-\Delta H)D_e}\right]\right\} \tag{A}$$

$$(C_b)_i = (C_b)_{i-1} - \frac{\tau D_e}{L^2\gamma}\,\Delta y\{(C_b)_{i-1}-(C_d)_{i-1}\} \tag{B}$$

$$T_b = T_{in} - \left(\frac{-\Delta H}{\rho C_p}\right)(C_b - C_{in}) \tag{C}$$

$$N_b = (N_b)_{in}\,\exp\left\{-\int_0^y \frac{(k_p)_d\tau}{1+(\phi_p)_d^2\gamma}\,dy\right\} \tag{D}$$

$$\gamma_j = \gamma_{j-1} + \Delta t\left[\frac{(k_p)_d}{Q}\left(\frac{N_b}{1+(\phi_p)_d^2\gamma}\right)\right]_{j-1} \tag{E}$$

$$\Lambda = \frac{D_e(C_b - C_d)}{k_bf(C_b)L^2\gamma} \tag{F}$$

$$T_s/T_b = 1 + \frac{(-\Delta H)\Lambda k_bf(C_b)L}{hT_b} \tag{G}$$

$$T_d/T_b = 1 + (1+(Bi)_h\gamma)(T_s/T_b - 1) \tag{H}$$

$$\begin{aligned} k_b &= k_{00}\exp(-E_a/RT_b) \\ (k_p)_d &= k_{p0}\exp(-E_p/RT_d) \end{aligned} \tag{I}$$

b. *Nonadiabatic, nonisothermal reactors*

$$T_b = T_{in} - \left(\frac{-\Delta H}{\rho C_p}\right)(C_b - C_{in}) + \frac{(\rho C_p)_c}{\rho C_p}\frac{v_c}{v}[T_c - (T_c)_{out}] \tag{C'}$$

$$-\frac{dT_c}{dz} = \frac{2U}{R_t(\rho C_p)_c}\,\tau_c\,(T_b - T_c) \tag{J}$$

[Equations (A) through (J) with (C) replaced by (C')]

c. *Isothermal reactors*

Use Eqs. (I) and (L) in Table 10.4 for the profiles of C_b and Λ at $t = 0$. Based on these initial profiles, proceed with the equations given below.

$$C_b - C_d = \left(\frac{L\gamma}{D_e}\right)\left[2D_ek_b\int_{C_C}^{C_d} f(\alpha)d\alpha\right]^{1/2} \tag{K}$$

Table 10.6 (cont.)

$$(C_b)_i = (C_b)_{i-1} - \frac{\tau D_e}{L^2 \gamma} \Delta y \{(C_b)_{i-1} - (C_d)_{i-1}\} \tag{L}$$

$$N_b = (N_b) \exp \left\{ \int_0^y \frac{-(k_p)\tau}{1 + (\phi_p)_d^2 \gamma} \, dy \right\} \tag{M}$$

$$\gamma = -\frac{1}{(\phi_p)_d^2} + \left[\frac{1}{(\phi_p)_d^4} + \frac{2D_p}{QL^2} \int_0^t N_b dt \right]^{1/2} \tag{N}$$

$$C_c/C_d = \begin{cases} 0 & \text{for } \phi_G \geq 3 \\ \dfrac{1}{\cosh \phi_G} & \text{for } \phi_G < 3 \end{cases}$$

$$\phi_G = Lk_b(1-\gamma)f(C_d)/\left[2\int_0^{C_d} D_e k_b(1-\gamma)f(C)dC \right]^{1/2}$$

10.11 is easy to understand if it is recognized that Eq. 10.23 is an auxiliary equation for Eq. 10.51 while Eq. 10.47 is the same for Eq. 10.52. For a given time then, simultaneous solutions of Eqs. 10.50 and 10.51 yield the reactor concentration profile. To proceed to the next time interval, Eq. 10.52 is solved for the γ profile for use in Eqs. 10.50 and 10.51, which yields the reactor concentration profile at time $(t + \Delta t)$. In order to initiate the numerical calculations, the initial concentration and temperature profiles are obtained using Eq. 10.45. Based on these profiles at $t = 0^+$, the N_b profile is obtained from Eq. 10.47 which in turn is used in Eq. 10.52 for the γ profile at time $(0^+ + \Delta t)$. Given the γ profile, Eqs. 10.50 and 10.51 are solved simultaneously. These procedures can be repeated at each successive time interval with the aid of Eq. 10.52 until the final time of interest is reached. The design equation (K) in Table 10.6 can be obtained directly from Eq. (A) by setting $(-\Delta H)$ equal to zero. Equation (N) has been obtained by integrating Eq. 10.35.

Simulation results for an adiabatic reactor undergoing shell-progressive deactivation (Lee and Butt 1982) are shown in Figures 10.12 through 10.17 for the model reaction system in Table 10.5. The procedures of Figure 10.11 have been followed to obtain these results for τ of 20 seconds. The relative speeds of transport and reaction of the species causing deactivation determine whether the catalyst is uniformly or shell-progressively deactivated. Therefore, the intrinsic rate of deactivation given in Table 10.5 should have been changed for the shell-progressive deactivation being considered. However, the same conditions have been used in the simulation to contrast the results with those obtained for uniform deactivation. The most striking thing is that the fate of the reactor is literally determined in the initial stage of deactivation, although this is not unexpected considering the nature of shell-progressive deactivation. This is most evident in Figure 10.12 in which the outlet concentration is shown to go through a drastic change initially with time, in sharp contrast with the concentration behavior shown in Figure

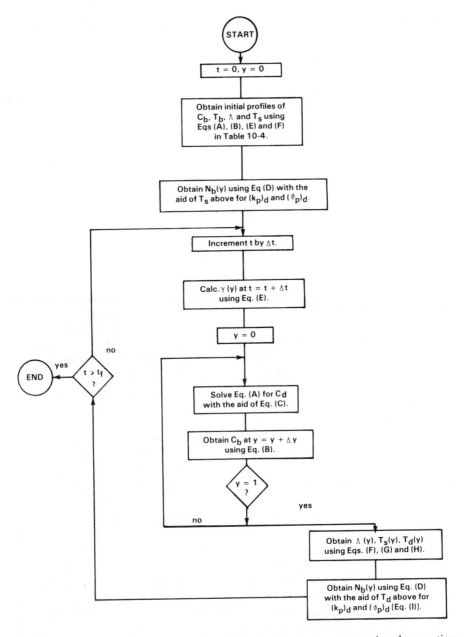

Figure 10.11   Calculation procedures for adiabatic, plug-flow reactors using the equations in Table 10.6.

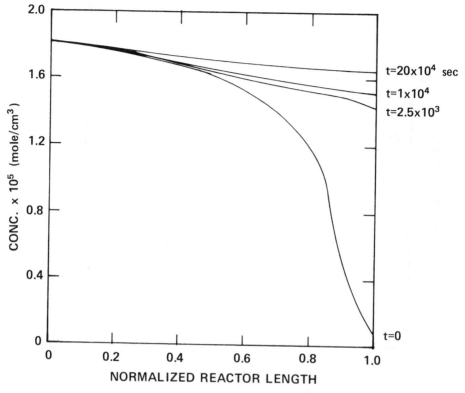

Figure 10.12   Profiles of the bulk-fluid concentration: shell-progressive deactivation.

10.7 for uniform deactivation under identical conditions. The γ profiles again go through a maximum as shown in Figure 10.14. However, the peaks are sharper than those for uniform deactivation. The species causing deactivation is also depleted at the outlet initially as was the case with uniform deactivation (Figure 10.13). Similar reasoning as for uniform deactivation applies for the existence of a maximum in the γ profile. The existence of a maximum causes aberrations in the profiles of the reactor point effectiveness and the concentration at the forefront of the fresh inner core, apparent in Figures 10.15 and 10.16. The profiles of reactor point effectiveness shown in Figure 10.15 are quite different from those obtained for uniform deactivation (Figure 10.6). It is seen that the reactor point effectiveness does decrease more rapidly with time than for uniform deactivation. This example clearly shows the devastating effect of shell-progressive deactivation on reactor performance as evident from a comparison of Figures 10.7 and 10.12 for the outlet concentration. If the rate of deactivation is higher than the rate given in Table 10.5, which should be the case for shell-progressive deactivation, the effect will be even more devastating than shown. This fact has an important bearing on the design of a reactor undergoing shell-progressive deactivation, as will be seen shortly.

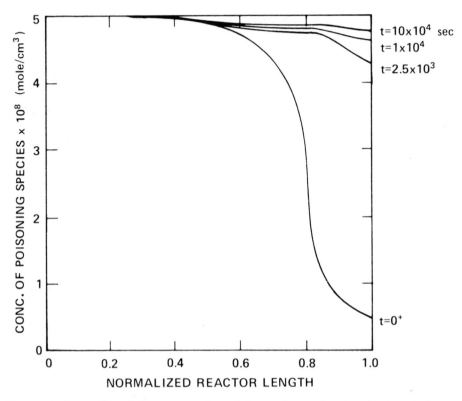

Figure 10.13 Profiles of the concentration of the species causing deactivation: shell-progressive deactivation.

General Case of Deactivation. As discussed earlier, the activity profile of a pellet undergoing chemical deactivation cannot always be adequately represented by either uniform or shell-progressive deactivation. Deactivation intermediate between uniform and shell-progressive deactivation results in a smooth activity profile within the pellet with its activity decreasing toward the pellet surface. Therefore, the result obtained in Chapter 4 for a pellet with a nonuniform activity distribution can be used for this general case. Utilizing the global rates derived in Chapter 5 for this case (Table 5.3), for independent deactivation:

$$R_G = \frac{J^{1/2}/L}{1 - \dfrac{1.2\epsilon(-\Delta H)}{2hT_b} J^{1/2}} \qquad (10.53)$$

$$J = 2k_b \, (1 - \bar{\gamma}) \, D_e \int_0^{C_b} f(C)dC \qquad (10.54)$$

$$R_p = \frac{[(k_p)_b(1 - \bar{\gamma})D_pH(C_b)]^{1/2}}{L[1 + G(C_b)]^n} \left\{ 1 + \frac{1.2\epsilon_pL(-\Delta H)}{2hT_b} R_G \right\} \qquad (10.55)$$

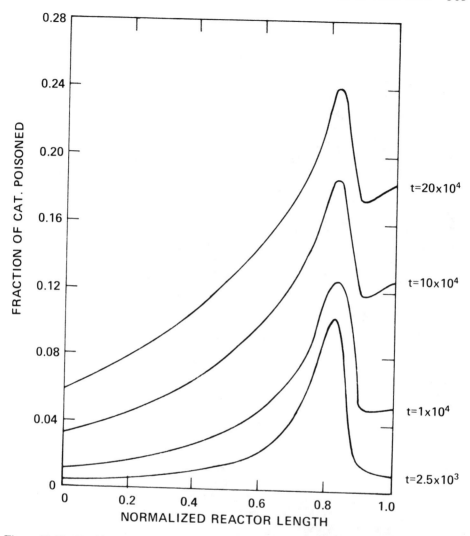

Figure 10.14 Profiles of the fraction of catalyst deactivated: shell-progressive deactivation.

$$H(C_b) = \frac{[1 + G(C_b)]^n + [1 + G(C = 0)]^n}{2} \tag{10.56}$$

$$\frac{d}{dt}(Q\bar{\gamma}) = \frac{(k_p)_s N_b (1 - \bar{\gamma})}{[1 + G(C_b)]^n} \tag{10.57}$$

where $\bar{\gamma}$ = fraction of catalyst deactivated at pellet surface

ϵ_p = Arrhenius number for deactivation reaction, $(E_p)/R_g T_b$

$[1 + G(C_b)]^n$ = product inhibition term.

For dependent deactivation:

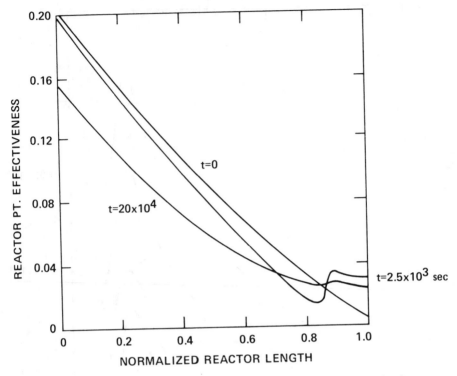

Figure 10.15   Profiles of reactor point effectiveness: shell-progressive deactivation.

$$R_G = \frac{J^{1/2}/L}{1 - \dfrac{1.2\epsilon(-\Delta H)}{2hT_b} J^{1/2}} \tag{10.53}$$

$$\frac{d}{dt}(Q\bar{\gamma}) = \begin{cases} \dfrac{(k_p)_s(C_A)_b(1-\bar{\gamma})}{\left[1 + \sum\limits_i (K_iC_i)^{m_i}\right]^n_{bulk}} & : \text{parallel} \qquad (10.58a) \\[2em] \dfrac{(k_p)_s(C_B)_b(1-\bar{\gamma})}{\left[1 + \sum\limits_i (K_iC_i)^{m_i}\right]^n_{bulk}} & : \text{series} \qquad (10.58b) \end{cases}$$

where C_A is the concentration of the main reactant, and C_B is that of the intermediate product in the dependent deactivation models:

$$\text{Parallel: } A + S \xrightarrow{\ k\ } B + S$$

$$A + S \xrightarrow{\ k_p\ } A \cdot S$$

$$\text{Series: } A + S \xrightarrow{\ k\ } B + S$$

$$B + S \xrightarrow{\ k_p\ } B \cdot S$$

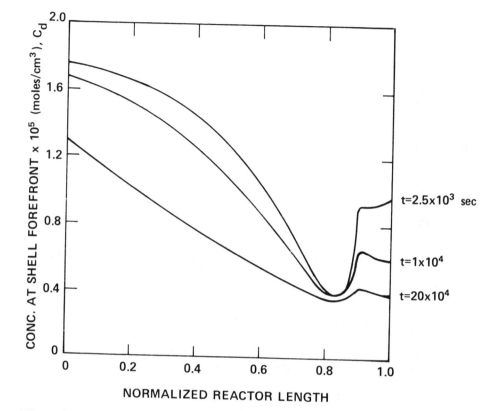

Figure 10.16   Profiles of concentration at the forefront of fresh inner core: shell-progressive deactivation.

In Eq. 10.54, the pellet center concentration has been set to zero for the diffusion-limited case being considered. As pointed out in Chapter 5, the global rates are usually valid for $\bar{\gamma}$ less than 0.5 (refer to Sections 4–8 and 5–6 for details).

The analysis equations and procedures given in Table 10.4 and Figure 10.6 for uniform deactivation can be utilized for independent deactivation provided that γ is replaced by $\bar{\gamma}$, and Eqs. (C) and (J) in Table 10.4 are replaced by:

$$(N_b)_i = (N_b)_{i-1} - (\Delta y)\tau \left[\frac{[(k_p)_b(1-\bar{\gamma})D_p H(C_b)]^{1/2}}{L[1+G(C_b)]^n} \left\{ 1 + \frac{1.2\epsilon_p L(-\Delta H)}{2hT_b} R_G \right\} \right]_{i-1}$$

(10.59)

$i =$ grid point in y

which is the Euler version of the reactor conservation equation:

$$\frac{dN_b}{dy} = -\tau R_p$$

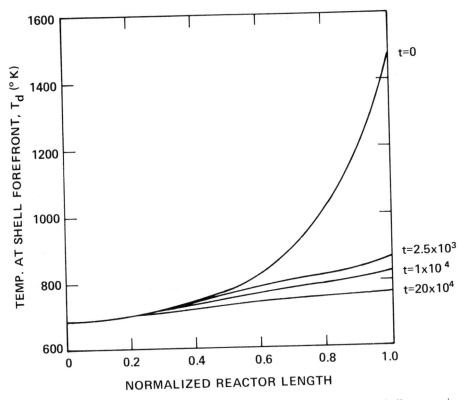

Figure 10.17   Profiles of temperature at the forefront of fresh inner core: shell-progressive deactivation.

The same equations and analysis procedures are also applicable to dependent deactivation provided that γ is again replaced by $\bar{\gamma}$, and Eqs. (C) and (J) in Table 10.4 are eliminated. It is seen that the analysis procedures for the general case are straightforward for both dependent and independent deactivation. On the other hand, their utility is limited to small values of $\bar{\gamma}$, say less than 0.5. More specifically, the value of R given by Eq. 4.102 (Section 4–8) should be much less than unity for accurate results. Further, ϕ_G has to be much larger than unity. While closed-form equations for global rates cannot be written when these restrictions are not met, similar analysis procedures as for $\bar{\gamma}$ less than 0.5 can be developed by adding more correction terms to the calculation of the flux at the pellet surface (Section 4–10) and relating this flux to the global rate.

While specific kinetics have been specified for dependent deactivation according to the model given above, arbitrary deactivation kinetics can be handled without problem. For instance, if the dependent deactivation kinetics are represented by:

$$r_p = k_p f_p(C)$$

Eq. 10.58 can be replaced by:

$$\frac{d}{dt}(Q\bar{\gamma}) = (k_p)_s f_p (C_b)(1 - \bar{\gamma})$$

and the same analysis procedures discussed above can be applied. One assumption made in deriving the global rates for the case of dependent deactivation is that the rate of the deactivation reaction is much smaller than that of the main reaction. This must be true for dependent deactivation in which the main reactant or product itself is the species causing deactivation. If not, the rate of deactivation reaction becomes of the same order of magnitude as the rate of the main reaction, and such a catalyst cannot be useful in the form of a pellet. This is precisely the reason why small particles, instead of pellets, of silica-alumina catalyst are used in a fluidized-bed for the catalytic cracking of hydrocarbons, for which the coking rate is high.

Design

The time dependence of conversion can complicate the design of a reactor affected by catalyst deactivation. In practice, however, this complication is avoided: the usual practice is to design the reactor as if there were no deactivation taking place and then operate the reactor at the desired conversion by constantly raising the reactor temperature to compensate for the declining catalytic activity. When the conversion drops to an unacceptable level despite adjustments in reactor temperature or the temperature reaches too high a value, the catalyst is regenerated. Therefore, a typical operating cycle consists of start-up, operation, and catalyst regeneration. An obvious problem with this design approach is that it does not take advantage of the knowledge of catalyst deactivation during the design phase.

Consider the choice of reactor size τ when the reactor is subject to catalyst deactivation. A measure of the performance of the reactor is the average conversion per unit volume of reactor in a given time period or some reference time period t_r. If t_f is defined as the duration of the reaction phase of the cycle, and b as the time required for catalyst regeneration, the number of cycles is:

$$n = \frac{t_r}{t_f + b} \tag{10.60}$$

There are additional costs incurred due to the reactor shut-downs caused by deactivation, which can be represented by a constant M. A performance index suitable for the choice of reactor size τ can then be written as:

$$J(\tau) = \frac{1}{\tau t_r}\left[n\int_0^{t_f}(C_{in} - C_{out})\,dt - Mn\right] \tag{10.61}$$

In essence, this index represents the net value of the product produced per unit volume of reactor, averaged over a reference time period t_r. This index, when Eq. 10.60 is utilized, reduces to:

$$\underset{\tau}{\text{Max }} J(\tau) = \underset{\tau}{\text{Max }} \left\{ \frac{1}{\tau(t_f + b)} \left[\int_0^{t_f} (C_{in} - C_{out}) dt - M \right] \right\}$$

(10.62)

subject to $C_{out}(t_f) > (C_{out})_s$

where $(C_{out})_s$ is the reactor outlet concentration corresponding to the conversion below which the catalyst is regenerated. When the duration of the reaction phase is relatively short due to rapid deactivation, the major concern has to be the longevity of the operation. In such cases, a suitable index for the choice of τ may be written as

$$\underset{\tau}{\text{Max }} J(\tau) = \underset{\tau}{\text{Max }} \frac{1}{\tau} \int_0^{t_f} (C_{in} - C_{out}) dt$$

(10.63)

subject to $C_{out}(t_f) > (C_{out})_s$

This index represents the amount of the product produced per one cycle on a unit reactor volume basis.

The specification of τ based on the criteria of Eq. 10.62 or 10.63 is relatively simple with the analysis procedures developed in the previous section. A value of τ can be chosen to calculate $J(\tau)$, subject to any additional constraints placed. From a plot of τ versus J, the value of τ that gives the maximum J can be chosen. We consider here the use of Eq. 10.63 for the specification of τ and then that of Eq. 10.62 in the next section.

For the reactor system in Table 10.5, values of J versus τ have been calculated for the case of uniform poisoning. Constraints are the maximum allowable catalyst temperature (or the maximum reactor inlet temperature) and the minimum allowable conversion. For the adiabatic reactor being considered, the reactor inlet tem-

Table 10.7 Performance of the Model Reactor in Table 10.5

	$(T_{cat})_{max} \leq 1183°K$	$x \geq 0.296$	
τ (sec)	T_{in} (°K)	$t_f \times 10^{-5}$ (sec)	
15	681.6	7.00	(8.1 days)
17	673.0	8.65	
20	661.8	11.00	
23	652.2	13.38	
30	644.2	18.80	(21.8 days)

perature (T_{in}) solely determines the maximum catalyst temperature. The inlet temperatures that satisfy the constraint on the maximum catalyst temperature are given in Table 10.7 along with the duration of operation t_f. It is clear from the table that the duration of the operation before catalyst regeneration is relatively short. On this basis alone, one may decide on the reactor size choosing, say τ of 30 seconds to have a relatively long reaction phase. It is nevertheless useful to determine how effective the reactor is on a unit volume basis, using the criterion of Eq. 10.63. A plot of J (Eq. 10.63) versus τ is given in Figure 10.18. The J values in the figure are normalized with respect to the J value corresponding to τ of 15 seconds. Although no maximum is reached under the conditions specified, it is clear that a larger reactor should be favored over a smaller reactor for higher conversion per unit reactor volume. Also given in the figure as the dotted line is

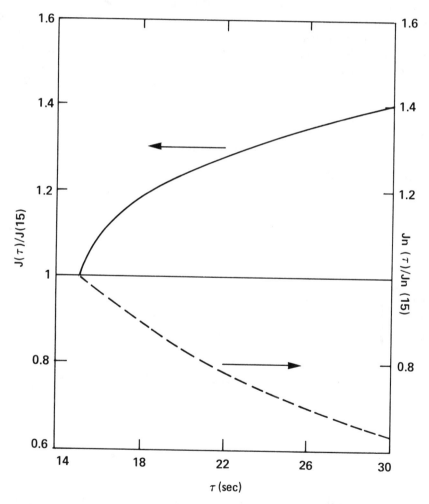

Figure 10.18   Performance index of Eq. 10.63 for a reactor affected by uniform deactivation.

the value of J (J_n in the figure) when catalyst deactivation is not taken into consideration, i.e., the usual design practice. This can be obtained by simply dividing the conversion at $t = 0$ by τ. It is seen that the usual design practice leads to a completely wrong conclusion that a smaller reactor should be favored over a larger one. Another pitfall that one can come across when catalyst deactivation is not taken into consideration has already been pointed out in a previous section: the fact that the maximum catalyst temperature can increase rather than decrease with an increasing level of deactivation, which will not be accounted for if the usual design practice is followed.

Similar conclusions can be made for shell-progressive deactivation regarding the choice of τ. Some important design considerations also emerge from the dominant feature of shell-progressive deactivation, i.e., the rapid initial change of conversion with time. This is shown in Figure 10.19, which has been obtained with a constraint on the maximum allowable catalyst temperature. It is seen that the conversion decreases exponentially with time initially, followed by an almost linear decrease with time. It is also seen that the level of conversion in this linear range is largely determined by the reactor size τ. This dominant role of the reactor size in determining the level of conversion in the linear range is also apparent in Figure 10.20, where the conversion for τ of 20 seconds is given as a function of time when the constraint on the maximum catalyst temperature is removed. Even with a change in the reactor inlet temperature from 662°K to 673°K, which can double the rate of reaction as indicated by the conversions at time zero in the

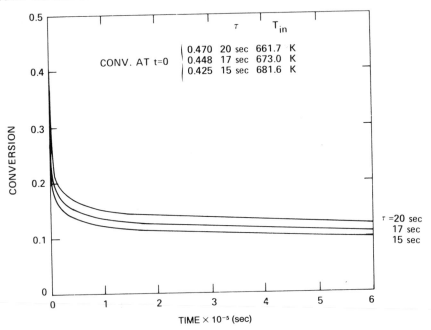

Figure 10.19 Time dependence of conversion: shell-progressive deactivation with a constraint on the maximum catalyst temperature.

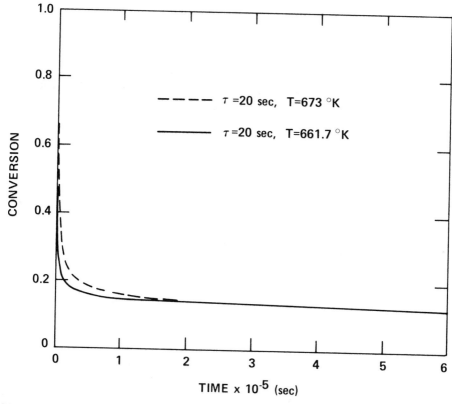

Figure 10.20 Time dependence of conversion: shell-progressive deactivation without the constraint on the maximum catalyst temperature.

figure, the conversions converge to the same level in the linear range. A few conclusions can be made from these observations on a reactor undergoing shell-progressive deactivation. The reactor may be sized according to a desired conversion in the linear range. Furthermore, the initial inlet temperature may be set as high as the constraint on the maximum catalyst temperature allows. Given the desired conversion x_d, then, the value of τ that yields x_d in the linear range can be found using the analysis procedures developed for the case of shell-progressive deactivation.

A more sophisticated design criterion may be used for the choice of τ. For instance, the choice of τ based on Eq. 10.62 or 10.63 may be subject to an optimal temperature policy during the operation so that J is maximized with regard to both τ and the temperature policy. This optimization problem will be treated in Chapter 13. The point here is that much better reactor performance can be attained by including the effects of catalyst deactivation in the design, as will be exemplified in the next section, and that such complicated design problems, whatever the performance index of the design may be, can be handled with relative ease using the analysis procedures developed in the previous sections.

10-4-3 Reactions Affected by Sintering

Design and analysis procedures based on reactor point effectiveness are developed in this section for reactions affected by diffusion, chemical deactivation, and catalyst sintering. Since these procedures closely parallel those developed in the previous section, only the general case of catalyst deactivation will be treated in detail.

For an adiabatic plug-flow reactor, the conservation equations are:

$$\frac{dC_b}{dy} = -\tau R_G = -\tau k_b(T_b)f(C_b)\Lambda \tag{10.64}$$

$$T_b = T_{in} - \left(\frac{-\Delta H}{\rho C_p}\right)(C_b - C_{in}) \tag{10.23}$$

where the reactor point effectiveness is given by:

$$\Lambda = \frac{\dfrac{1}{Lf}\left[\dfrac{2D_e(\bar{S}_r/\bar{S}_0)}{k_b}\right]\left[\displaystyle\int_{C_c}^{C_b} f(C)dC\right]}{1 - \dfrac{1.2E_a(-\Delta H)[2D_e k_b(\bar{S}_r/\bar{S}_0)]^{1/2}}{2hT_b^2 R_g}\left[\displaystyle\int_{C_c}^{C_b} f(C)dC\right]^{1/2}} \qquad \bar{S}_r/\bar{S}_0 > 0.5 \tag{10.65}$$

Here, S_r and S_0 are the active surface area and total surface area, respectively, for fresh catalyst, and the overbar denotes evaluation at the pellet surface. Without loss of generality, the functional dependence of the rate constant on surface area $h(S_r/S_0)$ can be set to S_r/S_0 for structure-insensitive reactions. This reactor point effectiveness for a certain smooth activity profile has been developed in Chapter 6 and is given in Table 6.5 along with the reactor point effectivenesses for the other cases. Combining Eqs. 10.64 and 10.65 and rearranging, there results:

$$\left\{\frac{L}{\left[2k_b D_e(\bar{S}_r/\bar{S}_0)\displaystyle\int_{C_c}^{C_b} f(C)dC\right]^{1/2}} - \frac{1.2LE_a(-\Delta H)}{2hT_b^2 R_g}\right\} dC_b = -\tau dy \tag{10.66}$$

For the species causing independent chemical deactivation:

$$\frac{dN_b}{dy} = -\tau R_p \tag{10.41}$$

The pellet mass balance for this species is:

$$D_p\frac{d^2 N}{dz^2} = r_p = \frac{k_p N(1-\gamma)}{[1+G(C)]^n} = \frac{k_p N}{[1+G(C)]^n}\left(\frac{\bar{S}_r}{\bar{S}_0}\right) \tag{10.67}$$

The factor $(1 - \gamma)/[1 + G(C)]^n$ is the fraction of active sites that are vacant, as shown in Chapter 5. This factor represents the profile of sites within the pellet that are available for both the main and deactivation reactions. If this profile is treated as the activity profile for the rate constant k_p, the result obtained in Section 4–8 for the internal effectiveness factor for a pellet with a nonuniform activity distribution can be applied (Section 5–6):

$$(\eta_p)_{in} = \frac{\tanh(\phi_p)_a}{(\phi_p)_a} \tag{10.68}$$

$$(\phi_p)_a = L \left\{ \frac{(k_p)_s(1 - \gamma)}{[1 + G(C_b)]^n} \frac{1}{D_p} \right\}^{1/2} \tag{10.69}$$

where the external mass transfer resistance has been neglected by setting $C_s = C_b$ in Eq. 10.69. Therefore, the mass balance of Eq. 10.41 can be rewritten as:

$$\frac{dN_b}{dy} = -\tau r_p (\eta_p)_{in}$$

$$= -\tau \frac{(k_p)_s N_b (1 - \bar{\gamma})}{[1 + G(C_b)]^n} \frac{\tanh(\phi_p)_a}{(\phi_p)_a} \tag{10.70}$$

where $\bar{\gamma}$ is related to $\overline{S}_r/\overline{S}_0$ by:

$$1 - \bar{\gamma} = \overline{S}_r/\overline{S}_0 \tag{10.71}$$

Using the pseudo-steady state assumption, the rate at which the active surface area changes with time is obtained from Table 6.4 as:

$$\frac{d}{dt}\left(\frac{\overline{S}_r}{\overline{S}_t}\right) = -\frac{(k_p)_s N_b}{Q[1 + G(C_b)]^n}\left(\frac{\overline{S}_r}{\overline{S}_t}\right) \tag{10.72}$$

$$\frac{d}{dt}\left(\frac{\overline{S}_t}{\overline{S}_0}\right) = \begin{array}{ll} -K_a(\overline{S}_t/\overline{S}_0)^n \exp[m'(\overline{S}_t/\overline{S}_0)] & \text{atom migration} \\ -K_c(\overline{S}_t/\overline{S}_0)^{n_c} & \text{crystallite migration} \end{array} \tag{10.73}$$

with $(\overline{S}_r/\overline{S}_0) = (\overline{S}_r/\overline{S}_t)(\overline{S}_t/\overline{S}_0)$ \hfill (10.73a)

where $\overline{S}_t$ is the total catalyst surface area at time t at the pellet surface, and K_a and K_c are the rate constants for sintering. Here, n, n_c, and m' are the constants for the sintering kinetics. For the design and analysis of the reactor affected by diffusion, chemical deactivation, and sintering, then, Eqs. 10.66 and 10.70 need to be solved with the aid of Eqs. 10.23, 10.68, and 10.71 through 10.73. The solution of Eq. 10.70 is:

$$N_b[y;t] = (N_b)_{in} \exp\left\{-\int_0^y \frac{\tau(k_p)_s(\overline{S}_r/\overline{S}_0)\tanh(\phi_p)_a}{[1 + G(C_b)]^n(\phi_p)_a}\,dy\right\} \tag{10.74}$$

Table 10.8 Design Equations for Plug-Flow Reactors: Diffusion-Limitation and General Case of Independent Catalyst Deactivation (Reprinted from Lee and Ruckenstein 1983, by courtesy of Marcel Dekker, Inc.)

$$r_c = kf(C); \qquad \bar{S}_r/\bar{S}_0 > 0.5$$

a. *Adiabatic, nonisothermal reactors*

$$\left\{ \frac{L}{\left[2k_b D_e(\bar{S}_r/\bar{S}_0) \int_0^{C_b} f(C)dC \right]^{1/2}} - \frac{1.2LE_a(-\Delta H)}{2hT_b^2 R_g} \right\} dC_b = -\tau dy \tag{A}$$

$$\begin{aligned}
k_b &= k_0 \exp(-E_a/R_g T_b) \\
(k_p)_b &= k_{p_0} \exp(-E_p/R_g T_b) \\
(k_p)_s &= k_{p_0} \exp(-E_p/R_g T_s) \\
(K_a)_s &= (K_a)_0 \exp(-E/R_g T_s)
\end{aligned} \tag{B}$$

$$T_b = T_{in} - \left(\frac{-\Delta H}{\rho C_p}\right)(C_b - C_{in}) \tag{C}$$

$$N_b[y;t] = (N_b)_{in} \exp\left\{ -\int_0^y \frac{\tau(k_p)_s \, \bar{S}_r/\bar{S}_0 \tanh(\phi_p)_a}{[1 + G(C_b)]^n (\phi_p)_a} \, dy \right\} \tag{D}$$

$$(\phi_p)_a = L \left\{ \frac{(k_p)_s (\bar{S}_r/\bar{S}_0)}{[[1 + GC_b)]^n D_p} \right\}^{1/2} \tag{E}$$

$$(\bar{S}_r/\bar{S}_t)_j = (\bar{S}_r/\bar{S}_t)_{j-1} - \left\{ \frac{(k_p)_s N_b}{Q[1 + G(C_b)]^n} \left(\frac{\bar{S}_r}{\bar{S}_t}\right) \right\}_{j-1} \Delta t \tag{F}$$

$$(\bar{S}_t/\bar{S}_0)_j = \begin{matrix} -\{(K_a)_s(\bar{S}_t/\bar{S}_0)^n \exp[m'(\bar{S}_t/\bar{S}_0)]\}_{j-1}\Delta t + (\bar{S}_t/\bar{S}_0)_{j-1} \\ \text{atom migration} \\ -\{(K_c)_s(\bar{S}_t/\bar{S}_0)^{n_c}\}_{j-1}\Delta t + (\bar{S}_t/\bar{S}_0)_{j-1} \\ \text{crystallite migration} \end{matrix} \tag{G}$$

$$\bar{S}_r/\bar{S}_0 = (\bar{S}_r/\bar{S}_t)(\bar{S}_t/\bar{S}_0) \tag{H}$$

$$\Lambda = \frac{(\bar{S}_r/\bar{S}_0)^{1/2}}{\dfrac{L}{\left[2D_e k_b \int_0^{C_b} f(C)dC \right]^{1/2}} - \dfrac{1.2E_a(-\Delta H)L(\bar{S}_r/\bar{S}_0)^{1/2}}{2hR_g T_b^2}} \tag{I}$$

$$T_s/T_b = 1 + \frac{(-\Delta H)L}{hT_b}(\Lambda k_b f) \tag{J}$$

b. *Nonadiabatic, nonisothermal reactors*

$$T_b = T_{in} - \left(\frac{-\Delta H}{\rho C_p}\right)(C_b - C_{in}) + \frac{(v\rho C_p)_c}{v\rho C_p}[T_c - (T_c)_{out}] \tag{C'}$$

$$-\frac{dT_c}{dz} = \frac{2U\tau_c}{R_t(\rho C_p)_c}(T_b - T_c) \tag{K}$$

[Equations (A) through (J) with (C) replaced by (C')]

Table 10.8 (*cont.*)

c.   *Isothermal reactors*

$$\frac{L}{\left[2k_bD_e(\bar{S}_r/\bar{S}_0)\displaystyle\int_0^{C_b} f(C)dC\right]^{1/2}} dC_b = -\tau dy$$

$$N_b = (N_b)_{in}\ \exp\left\{-\int_0^y \frac{\tau(k_p)_b(\bar{S}_r/\bar{S}_0)\tanh(\phi_p)_a}{[1+G(C_b)]^n(\phi_p)_a}\ dy\right\}$$

$$(\phi_p)_a = L\left\{\frac{(k_p)_b(\bar{S}_r/\bar{S}_0)}{[1+G(C_b)]^n D_p}\right\}^{1/2}$$

$$(\bar{S}_r/\bar{S}_t)_j = (\bar{S}_r/\bar{S}_t)_{j-1} - \left\{\frac{(k_p)_b N_b}{Q[1+G(C_b)]^n}\left(\frac{\bar{S}_r}{\bar{S}_t}\right)\right\}_{j-1}\Delta t$$

$$(\bar{S}_t/\bar{S}_0)_j = -\{(K_a)_b(\bar{S}_t/\bar{S}_0)^n\exp[m'(\bar{S}_t/\bar{S}_0)]\}_{j-1}\Delta t + (\bar{S}_t/\bar{S}_0)_{j-1}$$

$$\bar{S}_r/\bar{S}_0 = (\bar{S}_r/\bar{S}_t)(\bar{S}_t/\bar{S}_0)$$

$$\Delta K_b f = \frac{1}{L}\left\{(\bar{S}_r/\bar{S}_0)\,2D_e k_b\int_0^{C_b} f(C)dC\right\}^{1/2}$$

where Eq. 10.71 has been used. This solution, along with the solutions of Eqs. 10.72 and 10.73, is required for the integration of the reactor conservation equation, Eq. 10.66. The complete set of equations necessary for design and analysis is given in Table 10.8. Equations (F) and (G) in the table are the Euler versions of Eqs. 10.72 and 10.73 for numerical calculations. Design and analysis procedures for adiabatic reactors are summarized in Figure 10.21 in the form of a flow chart. At time zero, the catalyst is free from deactivation and therefore $\bar{S}_r/\bar{S}_0 = 1$. Integration of Eq. (A) in the table yields $C_b(y)$ at time zero when Eqs. (B) and (C) are used in Eq. (A). The temperature profile at time zero then follows directly from Eq. (C). The initial pellet surface temperature $T_s(y)$ follows from Eqs. (I) and (J). With this $T_s(y)$, the initial profile of $N_b(y)$ can be obtained from Eqs. (D) and (E). Having obtained all necessary information at time zero, the new surface area ratio for the next time interval ($t = 0^+ + \Delta t$, or next j) can be obtained from Eqs. (F) and (G) using the calculated $T_s(y)$ and Eqs. (B) and (H), which in turn are used for the integration of Eq. (A) for the next time interval. These procedures can be repeated until the final time of interest (t_f) is reached.

Design and analysis procedures for the limiting cases* of uniform and shell-progressive chemical deactivation coupled with sintering closely parallel those given for the general case and are very similar to those already developed in the previous section on reactions affected by diffusion and chemical deactivation. Likewise, the

---

* See problems at the end of this chapter.

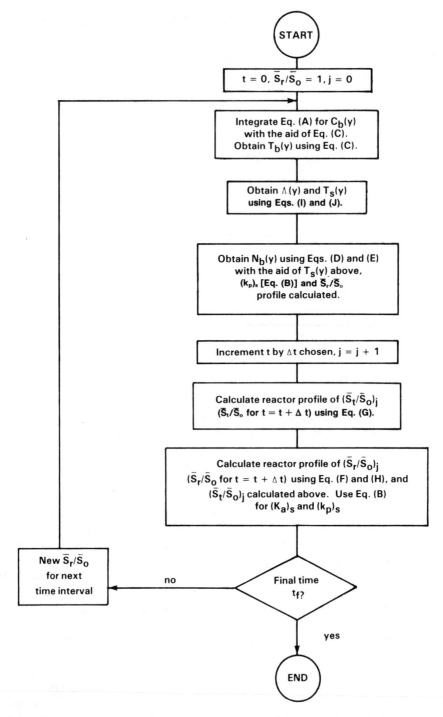

Figure 10.21 Design and analysis procedures for adiabatic, nonisothermal plug-flow reactors affected by diffusion, chemical deactivation, and sintering. (Reprinted from Lee and Ruckenstein 1983, by courtesy of Marcel Dekker, Inc.)

case of dependent deactivation can be handled in a similar way (see Section 10-4-2).

When the diffusional effect is totally absent, the reactor conservation equation reduces to:

$$\frac{dC_b}{dy} = -\tau R_G = -\tau \Lambda k_b f(C_b)$$

$$= \frac{-\tau(S_r/S_0)k_b f(C_b)}{1 - \dfrac{1.2(-\Delta H)E_a k_b(S_r/S_0)f(C_b)}{hR_g T_b^2}} \tag{10.77}$$

where the reactor point effectiveness given in Table 6.4 has been used. The conservation equation for the poisoning species is:

$$\frac{dN_b}{dy} = -\frac{\tau(k_p)_s N_b(S_r/S_0)}{[1 + G(C_b)]^n} \tag{10.78}$$

The design and analysis procedures based on Eqs. 10.77 and 10.78, and Eqs. 10.72 and 10.73 with the overbar removed can readily be obtained. In the absence of poisoning, all that is required is to set $(N_b)_{in}$ equal to zero in the flow chart given in Figure 10.21.

As an example, consider the design of an adiabatic reactor for the following reaction:

$$A \rightleftharpoons B$$

$$\begin{aligned}
A + S &\rightleftharpoons A \cdot S \ (K_A) \\
A \cdot S &\rightleftharpoons B \cdot S \ (k_s', \text{ controlling step}) \\
B \cdot S &\rightleftharpoons B + S \ (K_B) \\
N + S &\rightarrow N \cdot S \ (K_p, \text{ poisoning})
\end{aligned} \tag{10.79}$$

where S represents active sites. These elementary steps lead to:

$$r_c = \frac{k_s' C_{t_0} K_A(C_A - C_B/K)(S_r/S_0)}{1 + K_A C_A + K_B C_B} ; \qquad K = K_A K_s/K_B \tag{10.80}$$

$$r_p = \frac{k_p N(S_r/S_0)}{1 + K_A C_A + K_B C_B}$$

where $S_r/S_0 = C_r/C_{t_0}$. Here, C_r and C_{t_0} are surface concentrations of active sites and initial total sites, respectively. For this independent poisoning, the inhibition term $(1 + G)^n$ is:

$$(1 + G)^n = 1 + G = 1 + K_A C_A + K_B C_B = 1 + C_A(K_A - K_B) + K_B C_{A_0}$$

where the effective diffusivities are assumed to be the same for both A and B and C_{A_0} is the reactor inlet concentration of species A. In Table 10.9, $C = C_A$, and $C_{A_0} = C_{in}$ for the key species A. It is also assumed that the reaction is facile such that $h(S_r/S_0) = S_r/S_0$, and that the catalyst crystallite size is sufficiently large that sintering takes place via atom migration. For large metal crystallites, surface diffusion is the controlling step and thus the value of n in the sintering

Table 10.9 A Model System of the Reaction $A \rightleftharpoons B$ (reprinted from Lee and Ruckenstein 1983, by courtesy of Marcel Dekker, Inc.)

$$r_c = kf(C); \qquad k = k_s' C_{t_0} K_A$$

$$f(C) = f(C_A) = \frac{C_A - (C_{A_0} - C_A)/K}{1 + K_B C_{A_0} + C_A(K_A - K_B)}$$

$$[1 + G(C)]^n = 1 + C_A(K_A - K_B) + K_B C_{A_0}$$

$$h(S_r/S_0) = S_r/S_0$$

Sintering: $\dfrac{d}{dt}(S_t/S_0) = -K_a(S_t/S_0)^5 \exp[m'(S_t/S_0)]$

$$K_a = 4.1 \times 10^4 \exp(-40 \text{ kcal}/R_g T)(\text{min}^{-1})$$

$$m' = 3$$

$k = \exp\left(-\dfrac{12,000}{T} + 14.6\right)(\text{cm}^3/\text{sec}\cdot\text{cm}^3 \text{ pellet volume}); \qquad E_a = 23.76 \text{ kcal/mole}$

$k_p = \exp(-5000/T + 3.27)$
$K_A = \exp(3600/T + 3.86)(\text{cm}^3/\text{mol})$
$K_B = \exp(2800/T + 3.70)$
$K = \exp(6400/T + 7.56)$
$T_{in} = 673°K; \qquad D_e = D_p = 5 \times 10^{-3} \text{ cm}^2/\text{sec}$
$C_{in} = C_{A_0} = 1.8 \times 10^{-5} \text{ mol/cm}^3; \qquad h = 3.25 \times 10^{-4} \text{ cal/sec cm}^2°K$
$N_{in} = 5 \times 10^{-8} \text{ mol/cm}^3; \qquad (-\Delta H/\rho C_p) = 1.1 \times 10^7°K \text{ cm}^3/\text{mol}$
$L = 1 \text{ cm}; \qquad Q = 10^{-4} \text{ mol/cm}^3 \text{ cat. pellet}; \qquad (-\Delta H) = 30 \text{ kcal/mol}$

$$\phi_G = \left\{ \left[\frac{1}{Lf(C_b)}\right]\left[\frac{2D_e(S_r/S_0)}{k_s}\right]\int_0^{C_b}\left(\frac{C - (C_{in} - C)/K}{1 + K_B C_{in} + C_b(K_A - K_B)}\right)dC\right]^{1/2}\right\}^{-1}$$

$$\int_0^{C_b} f(C)dC = \int_0^{C_b}\frac{C - (C_{in} - C)/K}{1 + K_B C_{in} + C(K_A - K_B)}dC$$

$$= \left(\frac{1 + 1/K}{K_A - K_B}\right)C_b + \left[\frac{C_{in}}{K(K_A - K_B)} + \frac{(1 + K_B C_{in})(1 + 1/K)}{(K_A - K_B)^2}\right]$$

$$\times \ln\left(\frac{1 + K_B C_{in}}{1 + K_A C_{in} + (K_A K_B)C_b}\right)$$

Maximum catalyst temperature allowed $= 853°K$
$(C_{out})_s = 0.75 \times 10^{-5} \text{ mol/cm}^3$

$$\Delta t = 10^3 \text{ (min)}; \qquad \Delta y = 0.005; \qquad \tau: \text{varied}$$

kinetics given in Table 10.9 has been set to 5. Other pertinent parameters and conditions are also given in the table. The problem is to choose the reactor size τ that maximizes the performance index J of Eq. 10.62, subject to:

$$(T_{cat})_{max} \leq 853°K \text{ and } (C_{out})_s \leq 0.75 \times 10^{-5} \text{ mole/cm}^3 \text{ [conversion} > 58.3\%]$$

In order to calculate J as a function of τ, the T_{in} that yields the maximum catalyst temperature allowed at $t = 0$ for each of τ values selected was first calculated. This allows calculation of J with the procedures in Figure 10.21. The smallest value of ϕ_G was 2.5, and therefore, C_c can be set to zero. Values of the performance index J were then calculated using the procedures of Figure 10.21 for the selected τ values. The values of J normalized with respect to that of J for τ of 20 seconds are plotted in Figure 10.22 for $b = 2 \times 10^3$ min (1.4 days) and various values

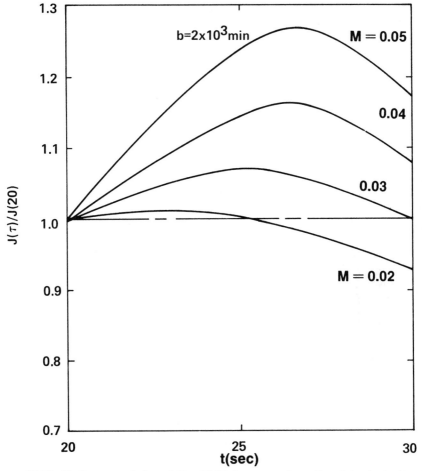

Figure 10.22 Performance index of Eq. 10.62 for a reaction affected by both sintering and poisoning. (Reprinted from Lee and Ruckenstein 1983, by courtesy of Marcel Dekker, Inc.)

of M. It is seen from the figure that a maximum in J exists between τ of 20 and 30 seconds, and that this maximum point moves to higher values of τ as M increases. An M value of 0.05 for a reactor with τ of 30 seconds means that the costs incurred due to catalyst regeneration are approximately 1% of the value of the product. Perhaps, more important than the specific values shown in the figure is the fact that the performance of a reactor undergoing catalyst deactivation can be significantly enhanced by properly sizing the reactor, and that this enhancement

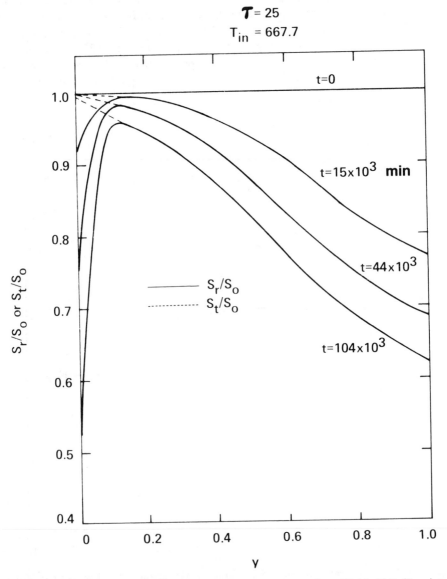

Figure 10.23 Reactor profiles of surface area ratios for the reaction of Table 10.9. (Reprinted from Lee and Ruckenstein 1983, by courtesy of Marcel Dekker, Inc.)

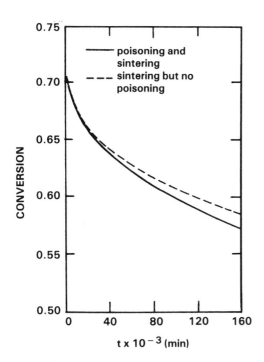

CONVERSION

— poisoning and
   sintering
--- sintering but no
   poisoning

$t \times 10^{-3}$ (min)

Figure 10.24 Effect of poisoning on conversion. (Reprinted from Lee and Ruckenstein 1983, by courtesy of Marcel Dekker, Inc.)

can only be realized by taking the effects of catalyst deactivation into consideration during the design phase.

Some interesting behavior emerges from the simulation results obtained in the process of calculating the performance index. Shown in Figures 10.23 through 10.25 are the simulation results for a reactor with τ of 25 seconds. The surface area ratios shown in Figure 10.23 for the active and total surface areas indicate that deactivation is dominated by catalyst sintering: the reduction of active surface area is almost entirely due to the sintering except for the entrance region. The poisoning species is almost entirely consumed in the entrance region so that its concentration is too small in the rest of the reactor to cause appreciable poisoning. On the other hand, a higher temperature toward the reactor outlet causes a higher rate of sintering, resulting in the decrease of the active surface area toward the outlet shown in the figure. Most of the conversion occurs near the outlet in an adiabatic reactor, and therefore, it is expected that the reduction of active surface area due to poisoning in the entrance region will not have a significant impact on the conversion, as verified by the results shown in Figure 10.24. It is seen from this figure that poisoning does not affect the conversion significantly at low times on stream. As the time on stream increases, however, the catalyst is deactivated too much near the outlet, and the reactor is eventually affected by the poisoning in the entrance region. This is reflected in the difference in the conversions shown in Figure 10.24 for $t > 40 \times 10^3$ min. The conversion corresponding to

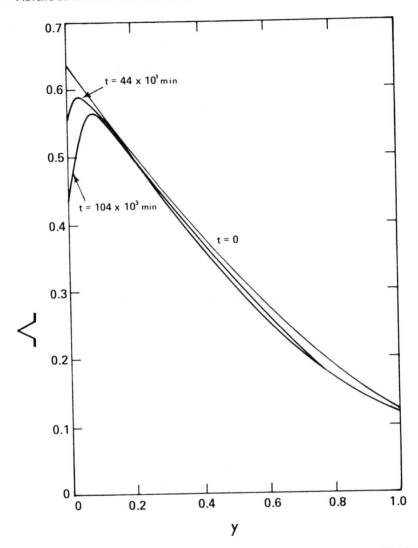

Figure 10.25 Profiles of reactor point effectiveness for the reaction system of Table 10.9. (Reprinted from Lee and Ruckenstein 1983, by courtesy of Marcel Dekker, Inc.)

the case of no poisoning was obtained by setting $(N_b)_{in}$ equal to zero in the procedures given in Figure 10–21. The profiles of the reactor point effectiveness shown in Figure 10.25 reveal that a maximum exists, which is due to the rapid change of the active surface area caused by poisoning in the entrance region. Based on the simulation results and the reasoning given earlier, it can be concluded for adiabatic reactors that the conversion is mainly affected by sintering initially, but then poisoning dominates as time increases. This will not be true, however, when poisoning takes place in a shell-progressive manner.

10-5   MULTIPLE REACTIONS

The reactor point effectiveness for multiple reactions cannot be expressed explicitly in terms of bulk concentrations and temperature. Nevertheless, the approach of using the reactor point effectiveness to obtain the global rates in terms of concentrations still applies as detailed in Section 4–9.

Consider multiple reactions taking place in an adiabatic reactor. If A_0 is designated as the main reactant from which all key species A_i ($i = 1,2, \ldots ,$ N) at each node shown in Figure 10.26 originate, the reactor conservation equations are:

$$\frac{d(A_0)_b}{dy} = \tau(R_G)_0 \tag{10.81}$$

$$\frac{d(A_i)_b}{dy} = \tau(R_G)_i \qquad i = 1,2, \ldots , N \tag{10.82}$$

$$\frac{dT_b}{dy} = \left(\frac{\tau}{\rho C_p}\right) \sum_{i=0}^{N} \{(-\Delta H_f)_i (R_G)_i\} \tag{10.83}$$

where $(R_G)_i$ is the net global rate for the key species A_i such that:

$$(R_G)_i = -\frac{D_i}{L} \frac{dA_i}{dz}\bigg|_{z=L} \qquad i = 0,1, \ldots , N \tag{10.84}$$

A few comments are in order at this time. It is understood in Figure 10.26 that there can be more than one species at each node. However, only the key species is required at each node as shown since the other species can be related to the key species through the diffusivity ratios and stoichiometric coefficients as detailed in Chapter 4. The use of the net global rate in place of the sum of individual global rates for all reaction paths involving the key species A_i necessitates the use of the heat of formation $(\Delta H_f)_i$ in Eq. 10.83 in place of the heat of reaction.

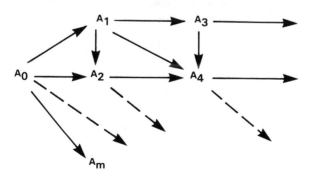

Figure 10.26   A reaction network for multiple reactions.

The pellet conservation equations are:

$$D_i \frac{d^2 A_i}{dz^2} = R_i \qquad i = 0, 1, \ldots, N \qquad (10.85)$$

where

$$R_i = \sum_{l=0}^{N} (r_{il} - \gamma_{li} r_{li}) \qquad (10.86)$$

Here, r_{il} is the intrinsic rate for the reaction path from species A_i to A_l, and γ_{il} is the stoichiometric coefficient per mole of A_i with the usual sign convention that R_i is the net intrinsic rate of consumption of species A_i. The intrinsic rate r_{il} may be expressed as:

$$r_{il} = k_{il}(T) f_{il}(A_0, A_1, \ldots, A_N)$$

where k_{il} is the rate constant for the reaction path between species A_i and A_l. With the assumptions of an isothermal pellet and negligible film mass transfer resistance that have been used throughout this chapter, one only needs to consider the film heat transfer to relate the reactions in the pellet to mass and heat transport in the bulk fluid:

$$h(T_s - T_b) = L \sum_{i=0}^{N} \{(-\Delta H_f)_i (R_G)_i\} \qquad (10.87)$$

According to the results obtained in Chapter 4, the net global rates are given by:

$$(R_G)_0 = -\frac{D_0}{L} \frac{dA_0}{dz}\bigg|_L = -\frac{1}{L} \left\{ 2D_0 \int_{(A_0)_c}^{(A_0)_b} R_0 dA_0 \right\}^{1/2} \qquad (10.88)$$

$$(R_G)_i = -\frac{D_i}{L} \frac{dA_i}{dz}\bigg|_L = \frac{1}{L} \left\{ 2D_i \int_{(A_i)_c}^{(A_i)_b} R_i dA_i \right\}^{1/2}, \qquad i = 1, 2, \ldots, N \quad (10.89)$$

where $(A_i)_c$ is the pellet center concentration of species A_i. The relationships between A_0 and A_i in the pellet, which are required for the numerical integration of R_i in Eqs. 10.88 and 10.89, are obtained in the process of solving the following equations at each grid point j in the pellet:

$$\frac{(\Delta A_i)_{j+1}}{(\Delta A_0)_{j+1}} = \left(\frac{D_0}{D_i}\right) \left\{ \frac{R_i(A_0, A_1, \ldots, A_N)|_j + R_i(A_0, A_1, \ldots, A_N)|_{j+1}}{R_0(A_0, A_1, \ldots, A_N)|_j + R_0(A_0, A_1, \ldots, A_N)|_{j+1}} \right\}$$

$$(10.90)$$

$$i = 1, 2, \ldots, N$$
$$j = 0 \text{ at the pellet surface}$$

For a selected increment ΔA_0, Eq. 10.90 represents N algebraic equations for N unknown increments ΔA_i at each grid point j in the pellet. The concentration increments are given by:

$$(A_i)_{j+1} = (A_i)_j + (\Delta A_i)_{j+1} \tag{10.91}$$

where (ΔA_i) is negative for $i = 0$ and positive for $i > 0$ in view of the fact that A_0 is "reactant" and the A_i's are "products." If $(\Delta A_0)_{j+1}$ is chosen to be the same for all intervals, $(\Delta A_0)_{j+1}$ may be replaced by the constant increment ΔA_0, but $(\Delta A_i)_{j+1}$ will still depend on the position j.

The design and analysis procedures follow from Eqs. 10.81 through 10.83 and Eqs. 10.87 through 10.91. At the reactor inlet at which $(A_0)_b = A_{in}$ and $(A_i)_b = 0$, which are the values of A_i at the pellet surface ($j = 0$) in Eq. 10.90, Eq. 10.90 can be solved for the pellet concentration profiles A_i ($i = 1, 2, \ldots, N$) with the aid of Eq. 10.91 using the chosen value of ΔA_0. As indicated earlier, $(\Delta A_0)_j$ may be chosen to vary with each increment. These concentration profiles are used in Eqs. 10.88 and 10.89 for the global rates. In order to solve Eq. 10.90, however, the pellet surface temperature, T_s, has to be used to calculate the rate constants in R_i. This is unknown, but by trial and error solution of Eqs. 10.87 and 10.90 the correct value of T_s can be calculated. Based on the correct global rates obtained from the trial and error solution, Eqs. 10.81 through 10.83 can then be used to numerically calculate $(A_0)_b$, $(A_i)_b$, and T_b for the next grid point in the reactor. These procedures can be repeated until the reactor outlet is reached. In order to calculate the global rates from Eqs. 10.88 and 10.89, the pellet center concentrations $(A_0)_c$ and $(A_i)_c$ have to be known. As detailed in Chapter 4, $(A_0)_c$ is estimated from:

$$\frac{(A_0)_c}{(A_0)_b} = \frac{1}{\cosh \phi_G} \tag{10.92}$$

where

$$\phi_G = \frac{LR_0[(A_0)_b, (A_1)_b, \ldots, (A_N)_b]}{\left[2D_0 \int_0^{(A_0)_b} R_0 dA_0 \right]^{1/2}} \tag{10.93}$$

This means that Eq. 10.90 is solved until $(A_0)_j$ reaches zero for use in Eq. 10.93. The grid index j corresponding to the value of $(A_0)_c$ calculated from Eq. 10.92 can be located for the values of $(A_i)_j$ corresponding to this j, which are the approximate pellet center concentrations.

The procedures outlined above can be quite time-consuming due to the iterations involving Eqs. 10.87 and 10.90. One way of overcoming the trial and error iteration problem is to combine Eq. 10.83 with 10.87:

$$\frac{dT_b}{dy} = \beta(T_s - T_b); \qquad \beta = \frac{h\tau}{\rho C_p L} \tag{10.94}$$

This can be integrated from any point y_m to a point Δy away (y_{m+1}) to give:

$$\int_{y_m}^{y_{m+1}} dT_b = \int_{y_m}^{y_{m+1}} \beta(T_s - T_b)dy \qquad (10.95)$$

If the trapezoid rule is used for the integration of the right hand side of Eq. 10.95, one has:

$$(T_s)_1 = (T_s)_0 + \left[\frac{1 + \beta \Delta y/2}{\beta \Delta y/2}\right](T_b)_1 - \left[\frac{1 - \beta \Delta y/2}{\beta \Delta y/2}\right](T_b)_0 \qquad (10.96)$$

$$(T_s)_{m+1} = (T_s)_m + \left[\frac{1 + \beta \Delta y/2}{\beta \Delta y/2}\right](T_b)_{m+1} - \left[\frac{1 - \beta \Delta y/2}{\beta \Delta y/2}\right](T_b)_m \qquad (10.97)$$

$$m = 1, 2, \ldots, M$$

Here, m is the index for the grid point in y, and Δy is the grid interval. Consider the use of Eq. 10.97 in place of 10.87. If the pellet surface temperature is known at the reactor inlet, Eqs. 10.81 through 10.83 allow the calculation of T_b and $(A_i)_b$ at the next grid point in y, i.e., $m = 1$. Then, the only unknown in Eq. 10.96 is $(T_s)_1$, which is the pellet surface temperature at the next grid point. This value of $(T_s)_1$ is then used in Eq. 10.90 for the global rates at $m = 1$, thereby eliminating trial and error iterations for all m except at the reactor inlet since the same procedures as those for $m = 1$ apply to all m due to the recursive relationship of Eq. 10.97. It is seen that the use of Eq. 10.97 in place of Eq. 10.87 results in a numerical scheme that requires iteration only at the reactor inlet. As discussed in Chapter 4, the rectilinear rule can be used in place of the trapezoid rule in deriving Eq. 10.90 with the following result:

$$\frac{(\Delta A_i)_{j+1}}{(\Delta A_0)_{j+1}} = \left(\frac{D_0}{D_i}\right) = \left\{\frac{R_i(A_0, \ldots, A_N)|_j}{R_0(A_0, A_1, \ldots, A_N)|_j}\right\} \qquad (10.98)$$

The use of Eq. 10.98 requires solving linear algebraic equations as opposed to the nonlinear algebraic equations of Eq. 10.90 and thus considerably simplifies the numerical problem. Furthermore, accurate results can be obtained using a predictor-corrector method, as detailed in Section 4.9.

Take as an example the following multiple reactions, which are the basic unit in any complex reaction network:

$$A_0 \overset{\longrightarrow A_1}{\underset{\longrightarrow A_2}{\big\downarrow}}$$

Suppose that the reaction stoichiometry and rates are given as follows:

$$A_0 + 2B_0 \longrightarrow 2A_1 + B_1 \qquad r_{01} = k_{01}A_0B_0^2$$
$$A_1 + B_1 \longrightarrow 2A_2 + 3B_2 \qquad r_{12} = k_{12}A_1B_1 \qquad (10.99)$$
$$A_0 + B_0 \longrightarrow 2A_2 + B_2 \qquad r_{02} = k_{02}A_0B_0$$

It follows then that the net intrinsic rates for species A_0, A_1, and A_2 are:

$$R_0 = -r_{01} - r_{02}$$
$$R_1 = -r_{12} + 2r_{01} \qquad (10.100)$$
$$R_2 = 2r_{02} + 2r_{12}$$

The pellet conservation equations (Eq. 10.85) for all the involved species can be combined two at a time and integrated twice to express the non-key-concentrations in terms of the key concentrations:

$$(B_0)_b - B_0 = -\frac{1}{2}\frac{D_2}{D_{B_0}}[(A_2)_b - A_2] - \frac{D_1}{D_{B_0}}[(A_1)_b - A_1]$$
$$(10.101)$$
$$(B_1)_b - B_1 = -\frac{D_0}{D_{B_1}}[(A_0)_b - A_0] - \frac{1}{2}\frac{D_2}{D_{B_1}}[(A_2)_b - A_2]$$

where the subscript b denotes the bulk-fluid concentration and the D's are effective diffusivities. The bulk-fluid concentrations can be related to reactor-inlet concentrations by combining the reactor conservation equations (Eq. 10.82) written for all the involved species two at a time and then integrating once. Equations 10.81 and 10.82 with the assumption of no products in the feed, yield:

$$(B_0)_b = (B_0)_{in} - \frac{1}{2}(A_2)_b - (A_1)_b$$
$$(10.102)$$
$$(B_1)_b = (A_0)_{in} - (A_0)_b - \frac{1}{2}(A_2)_b$$

Here the subscript in denotes the reactor inlet concentration. Equations 10.101 and 10.102 can be used in Eq. 10.100 to give:

$$R_0 = -k_{01}A_0\left[(B_0)_{in} - \frac{1}{2}A_2 - A_1\right]^2 - k_{02}A_0\left[(B_0)_{in} - \frac{1}{2}A_2 - A_1\right]$$

$$R_1 = -k_{12}A_1\left[(A_0)_{in} - A_0 - \frac{1}{2}A_2\right] + 2k_{01}A_0\left[(B_0)_{in} - \frac{1}{2}A_2 - A_1\right]^2 \quad (10.103)$$

$$R_2 = 2k_{02}A_0\left[(B_0)_{in} - \frac{1}{2}A_2 - A_1\right] + 2k_{12}A_1\left[(A_0)_{in} - A_0 - \frac{1}{2}A_2\right]$$

where the effectiveness diffusivities have been assumed the same for all species for illustration purposes.

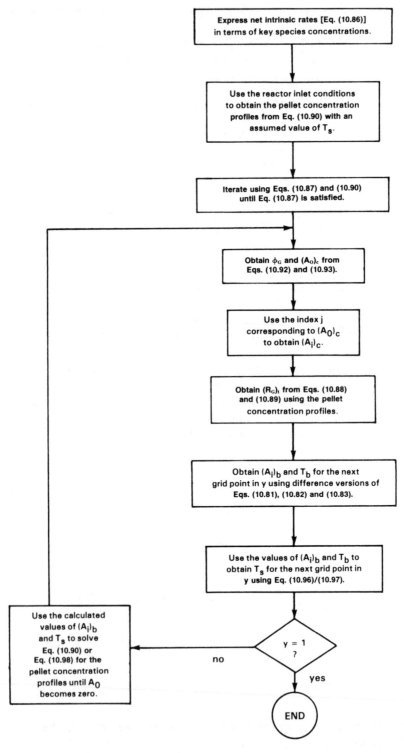

Figure 10.27 Design and analysis procedures for multiple reactions in an adiabatic reactor.

It is seen that the net intrinsic rates are now expressed solely in terms of the concentrations of the key species. Upon substituting Eqs. 10.91 and 10.103 into 10.90, the nonlinear algebraic equations of Eq. 10.90 become:

$$F_1[(\Delta A_1)_{j+1}, (\Delta A_2)_{j+1}] = f_1[(\Delta A_0)_{j+1}, (A_0)_j, (A_1)_j, (A_2)_j]$$
$$F_2[(\Delta A_1)_{j+1}, (\Delta A_2)_{j+1}] = f_2[(\Delta A_0)_{j+1}, (A_0)_j, (A_1)_j, (A_2)_j] \qquad (10.104)$$

where F_1 and F_2 are functions of the unknowns, $(\Delta A_1)_{j+1}$ and $(\Delta A_2)_{j+1}$, and f_1 and f_2 are constants for a selected value of $(\Delta A_0)_{j+1}$. These equations can be solved for the two unknowns at each grid point j to give the pellet concentration profiles of A_1 and A_2 versus A_0. In order to generate the pellet concentration profiles, a value for $(\Delta A_0)_{j+1}$, which can be a constant or a variable for each grid point, is first chosen. The concentrations at $j = 1$ can then be calculated from Eq. 10.104 with $(A_i)_0 = (A_i)_b$ at the pellet surface. The calculated values of $(\Delta A_1)_1$ and $(\Delta A_2)_1$ are used in Eq. 10.91 to obtain values for $(A_0)_1$, $(A_1)_1$, and $(A_2)_1$, which in turn are used in Eq. 10.104 to obtain $(\Delta A_1)_2$ and $(\Delta A_2)_2$ with a chosen value of $(\Delta A_0)_2$. These procedures can be repeated until A_0 reaches a value of zero, at which point the generalized modulus ϕ_G (Eq. 10.93) is calculated for the pellet center concentration of A_0 (Eq. 10.92). The index j corresponding to $(A_0)_c$ is then located to obtain the center concentrations $(A_1)_c$ and $(A_2)_c$. Now that all quantities in Eqs. 10.88 and 10.89 are known, the net global rates can be calculated.

The above example is just for the calculations of net global rates at a point in the reactor. These global rates have to be used in conjunction with the reactor conservation equations for design and analysis. The complete design and analysis procedures for the reactor are outlined in Figure 10.27.

Even with the simplifications made possible using the reactor point effectiveness, the design and analysis problem is still not simple. It should be clear, however, that the alternative of solving nonlinear differential equations with split boundary conditions on each is much less attractive.

Summary

The approach of reactor point effectiveness allows a general treatment of reactor design, whether the reaction is affected by transport processes, catalyst deactivation, or both. Relatively simple procedures have been developed for the design and analysis of heterogeneous reactors. These procedures, applicable to general forms of rate expressions for the main reaction, poisoning, and sintering, allow a rather straightforward treatment of complex design and analysis problems for reactors affected by catalyst deactivation. Including the effects of catalyst deactivation during the design phase has been shown to deliver better reactor performance. Design and analysis for multiple reactions is a very complicated problem. Use of the reactor point effectiveness, however, considerably simplifies it.

NOTATION

A_i	concentration of species A_i
$(Bi)_h$	Biot number for heat
$(Bi)_m$	Biot number for mass
C	concentration
C_b	bulk concentration for a species of interest
$(C_b)_r$	C_b evaluated at reference conditions
C_c	pellet center concentration
$\overline{C}$	cup-mixing concentration
C_i	concentration of species i
C_{in}	reactor inlet concentration for a species of interest
C_0	solution for plug-flow reactor in perturbation method
C_{out}	reactor outlet concentration
C_p	concentration of the species causing deactivation; specific heat content of reaction fluid
C_s	pellet surface concentration
D_a	axial dispersion coefficient for mass
D_e	effective diffusivity of main reactant
D_i	effective diffusivity for species A_i
D_r	radial dispersion coefficient for mass
D_p	effective diffusivity of the species causing deactivation
E	activation energy for sintering
E_a	activation energy for the main reaction
E_p	activation energy for the deactivation reaction
f	concentration dependence of intrinsic rate of the main reaction
f_p	concentration dependence of intrinsic rate of the deactivation reaction
F	$\tau R_G(C_0)$ in Eq. 10.5
F_i	F evaluated at reactor inlet
$[1 + G(C_b)]^n$	product inhibition term in the deactivation kinetics
h	film heat transfer coefficient
h_w	wall heat transfer coefficient for two-dimensional model
$(-\Delta H)$	heat of reaction
$H(C_b)$	$\{[1 + G(C_b)]^n + [1 + G(C = 0)]^n\}/2$
ΔH_f	heat of formation
i	grid point in space
I	integral defined in Eq. 10.44
j	grid point in time; grid point in z
J	integral defined by Eq. 10.54; integral defined in Eq. 10.62
k	rate constant for main reaction
k_1	k for pseudo-first-order reaction
k_b	k evaluated at T_b
k_g	film mass transfer coefficient
$(k_g)_p$	k_g for the species causing deactivation

k_p	rate constant for deactivation reaction
$(k_p)_d$	k_p evaluated at T_d
k_0	preexponential factor for k
k_{p0}	preexponential factor for k_p
K_a	axial dispersion coefficient for heat; rate constant for sintering kinetics in Eq. 10.73
K_c	rate constant for sintering kinetics in Eq. 10.73
K_r	radial dispersion coefficient for heat
K_s	equilibrium constant for the surface reaction in Eq. 10.79
L	characteristic pellet size
m'	constant for sintering kinetics in Eq. 10.73
n, n_c	exponents for sintering kinetics in Eq. 10.73
N	concentration of the species causing deactivation
N_b	bulk-fluid value of N
N_{in}	N at reactor inlet
P_h	Peclet number for heat
P_m	Peclet number for mass
Q	deactivation capacity of catalyst in moles of the poisoning species per pellet volume
r	normalized radial coordinate
r_c	intrinsic rate of the main reaction [$r_c = kf(C)$]
r_p	intrinsic rate of the deactivation reaction [$r_p = k_p f_p(C)$]
r_{il}	intrinsic rate for the reaction path between species A_i and A_l
R_g	gas constant
R_i	net intrinsic rate of consumption of species A_i
R_G	global rate of the main reaction
$(R_G)_i$	net global rate for species A_i
R_p	global rate of deactivation reaction
R_t	tube radius
S	surface site
S_0	total catalyst surface area at time zero or for fresh catalyst
S_r	active catalyst surface area at time t
S_t	total catalyst surface area at time t including the surface area made inactive by chemical deactivation
t	time
t_f	time at which catalyst is regenerated
T	temperature
T_b	bulk fluid temperature
T_c	coolant temperature
$(T_c)_{out}$	T_c at $y = 1$
T_d	temperature at the boundary separating fresh inner core from the completely deactivated outer shell of a pellet
T_r	reference temperature
T_s	pellet surface temperature
T_1	T_b at tube wall

$\overline{T}$	cup-mixing temperature
T_{in}	reactor inlet temperature
u	interstitial fluid velocity
U	overall heat transfer coefficient based on $\overline{T}$
v	superficial fluid velocity
v_c	coolant velocity
x	conversion
x_b	conversion of main reactant in bulk fluid
y	normalized reactor coordinate
z	pellet coordinate
Z	reactor length

Greek Letters

β	$1/P_h$; $h\tau/(\rho C_p L)$
γ	fraction of catalyst pellet deactivated
Λ	reactor point effectiveness
δ	mole fraction times the change in number of moles due to reaction per mole of species of interest
ϵ	Arrhenius number ($E_a/R_g T_b$); $1/P_m$
ϵ_B	bed porosity
ϵ_p	ϵ for deactivation reaction
η_i	internal effectiveness factor
η_f	η_i for fresh pellet
η_p	internal effectiveness factor for deactivation reaction
$(\eta_p)_{in}$	effectiveness factor given in Eq. 10.68
$(\eta_{in})_d$	η_i for the fresh inner core of a shell-progressively deactivated catalyst pellet
λ_e	effective thermal conductivity
ξ	equilibrium constant
ρ	molar density
τ	$Z(1 - \epsilon_B)/v$
τ_a	Z/v
τ_c	Z/v_c
τ_r	τ calculated at reference velocity
ϕ	Thiele modulus defined by Eq. 4.44
ϕ_G	generalized modulus defined by Eq. 4.74
ϕ_p	$L(k_p/D_p)^{1/2}$
$(\phi_p)_a$	Thiele modulus defined by Eq. 10.69
$(\phi_p)_d$	$L[(k_p)_d/D_p]^{1/2}$

Subscripts

b	bulk fluid
c	pellet center; coolant

d	at the boundary between fresh inner core and completely deacti-vated outer shell of a pellet
p	deactivation
r	reference condition
s	pellet surface
in	reactor inlet

Superscript

$^{-}$ evaluated at pellet surface; cup-mixing average

PROBLEMS

10.1. Consider an isothermal reactor for which:

$$\frac{dC}{dy} - \frac{1}{P_m}\frac{d^2C}{dy^2} = -\tau R_G(C) \qquad y \in (0,1)$$

For a large mass Peclet number, the above equation can be rewritten as:

$$\epsilon \frac{d^2C}{dy^2} - \frac{dC}{dy} - g(C) = 0 \qquad \text{(a)}$$

where $\epsilon = 1/P_m$ and $g(C) = \tau R_G(C)$.

The boundary conditions are:

$$C_{in} = C - \frac{dC}{dy} \qquad \text{at } y = 0 \qquad \text{(b)}$$

$$\frac{dC}{dy} = 0 \qquad \text{at } y = 1 \qquad \text{(c)}$$

Using perturbation, with ϵ as a small parameter, $C(y, \epsilon)$ for $\epsilon \neq 0$ can be expressed as a superposition of C_0 (solution for C when $\epsilon = 0$) and some perturbation terms:

$$C(y, \epsilon) = \sum_{i=0}^{\infty} \epsilon^i C_i \qquad \text{(d)}$$

Normally, a first-order correction or $C(y, \epsilon) = C_0 + \epsilon C_1$ is accurate enough.

a. Show that $g(C)$, when expanded in a Taylor series, can be expressed as:

$$g(C) = g(C_0) + \epsilon C_1 g'(C_0) + \epsilon^2 \left(\frac{1}{2} C_1^2 g''(C_0) + C_2 g'(C_0) \right) + \theta(\epsilon^3) \quad \text{(e)}$$

Use Eq. (d).

b. Show that the original conservation equation can be transformed into the following set of equations:

$$\frac{dC_0}{dy} + g(C_0) = 0 \quad \text{(f)}$$

$$\frac{dC_1}{dy} + C_1 g'(C_0) - \frac{d^2 C_0}{dy^2} = 0 \quad \text{(g)}$$

$$\frac{dC_2}{dy} + C_2 g'(C_0) + \frac{1}{2} C_1^2 g''(C_0) - \frac{d^2 C_1}{dy^2} = 0 \quad \text{(h)}$$

Use Eqs. (d) and (e) in (a) in matching the terms of the same order of ϵ^i ($i = 0,1,2$). Note that Eq. (f) is the conservation equation for the plug-flow model, and that Eqs. (g) and (h) can be solved in a successive manner, i.e., a solution for C_1 based on C_0 can be obtained, and so on.

c. Show that Eqs. (f) and (g) can be combined to give:

$$\frac{dC_1}{dC_0} - C_1 \frac{g'}{g} + g' = 0 \quad \text{(i)}$$

whose solution is:

$$C_1 = (A_1 - \ln g)g \qquad A_1 = \text{integration constant} \quad \text{(j)}$$

where $g = g(C_0)$, $g'(C_0) = g'$. Show that Eq. (h) can be written as:

$$\frac{dC_2}{dC_0} - C_2 \frac{g'}{g} - \frac{1}{2} \frac{C_1^2 g''}{g} + \frac{d(g dC_1/dC_0)}{dC_0} = 0 \quad \text{(k)}$$

Show also that the solution is:

$$C_2 = \left\{ A_2 - \int \left[\frac{1}{g} \frac{d(g dC_1/dC_0)}{dC_0} - \frac{1}{2} C_1^2 \frac{g''}{g^2} \right] dC_0 \right\} g \quad \text{(l)}$$

Using Eq. (j) in (l) and integrating by parts several times, the equation becomes:

$$C_2 = \left\{ A_2 + \frac{2 + (A_1 - \ln g)(A_1 - 2 \ln g)}{2} g' + \int_1^{C_0} \frac{g''^2}{g} \, dC_0 \right\} g \qquad \text{(m)}$$

d. Show that one set of the boundary conditions is:

$$C_1(0) = \left(\frac{dC_0}{dy} \right)_{y=0} \qquad \text{(n)}$$

$$C_2(0) = \left(\frac{dC_1}{dy} \right)_{y=0} \qquad \text{(o)}$$

Use the boundary condition of C for this purpose with the aid of Eq. (d). Using these conditions, one gets:

$$A_1 = \ln g_i - 1 \qquad g_i = g(C_0)|_{y=0}$$

$$A_2 = -\frac{1}{2} g_i' \qquad g_i' = g'(C_0)|_{y=1} \qquad \text{(p)}$$

e. The other boundary condition dictates that

$$\frac{dC}{dy}\bigg|_{y=1} \doteq \frac{dC_0}{dy}\bigg|_{y=1} + \epsilon \left(\frac{dC_1}{dy} \right)\bigg|_{y=1} + \epsilon^2 \left(\frac{dC_2}{dy} \right)\bigg|_{y=1}$$

$$\doteq -f_e + \epsilon \left(\frac{dC_1}{dy} \right)\bigg|_{y=1} + \epsilon^2 \left(\frac{dC_2}{dy} \right)\bigg|_{y=1} = 0$$

Since $f_e \neq 0$, each of ϵ^i's cannot be set to zero. To satisfy the condition, write $C(y, \epsilon) = C_0 + \epsilon C_1 + \epsilon^2 C_2 + f(y)$ \hfill (q)

Show that:

$$\epsilon \frac{d^2 f}{dy^2} - \frac{df}{dy} = 0$$

or

$$f = d_1 + d_2 e^{y/\epsilon}$$

when Eq. (q) is used in Eq. (a). According to the first boundary condition $d_1 = 0$, and therefore:

$$d_2 = \epsilon g_e \left[1 + \epsilon g_e' \left(\ln \frac{g_i}{g_e} - 2 \right) \right] e^{-1/\epsilon} \qquad \text{(r)}$$

where $g_e = g(C_0)|_{y=1}$.

f. Finally, show that:

$$C = C_0 + \epsilon \left(\ln \frac{g_i}{g_e} - 1 \right) g$$

$$+ \frac{\epsilon^2}{2} g \left\{ \left[2 + \left(\ln \frac{g_i}{g} - 1 \right) \left(\ln \frac{g_i}{g} - 3 \right) \right] g' - g'_i + 2 \int_1^{C_0} \frac{g''^2}{g} \, dC_0 \right\}$$

$$+ \epsilon g_e \left[1 + \epsilon g'_e \left(\ln \frac{g_i}{g_e} - 2 \right) \right] e^{y - 1/\epsilon} \tag{s}$$

g. Show that for a first-order reaction with $k\tau = 1$, use of the first-order term is sufficiently accurate for $P_m > 30$.

10.2. Derive Eq. 10.6 by letting:

$$C_b = \sum_i \epsilon^i C_i$$

$$T_b = \sum_i \beta^i T_i$$

and using the fact that:

$$dT_0/dC_0 = (-\Delta H)/\rho C_p \quad \text{or} \quad T_0 = T_{in} + \left(\frac{-\Delta H}{\rho C_p} \right) (C_0 - C_{in})$$

First arrive at the following result:

$$T_1 - \frac{dT_0}{dy} = (-\Delta H) \left(\frac{\epsilon}{\beta} \right) \left(C_1 - \frac{dC_0}{dy} \right); \qquad \beta = -\Delta H/\rho C_p$$

10.3. It is desired to examine the adequacy of the plug-flow model for an adiabatic reactor that is in operation. This is to be done based on temperature and concentration measurements at the inlet and outlet of the reactor. The following are known:

$T_{in} = 673°K$	$T_{out} = 1073°K$
$C_{in} = 5 \times 10^{-5}$ gmol/cm³	$C_{out} = 0.5 \times 10^{-5}$ gmol/cm³
$-\Delta H/\rho C_p = 10^7$ (°K cm³/mol)	$P_m = 100 \qquad P_h = 50$
$\tau = 10$ sec	$R_G|_{y=1} = 5 \times 10^{-5}$ gmol/(cm³sec)

What can you conclude on the adequacy of PFR model? What is the outlet concentration difference between the value calculated from the PFR model and that from the axial dispersion model if the observed rate at the inlet is 10 times the rate at the outlet?

10.4. Show that the inclusion of the radial velocity profile in Eqs. 10.7 and 10.8 also leads to Eqs. 10.11 and 10.12 if the cup-mixing concentration and temperature are used for $\bar{C}$ and $\bar{T}$ (Ahmed and Fahien 1980).

10.5. Determine whether the equality of Eq. 10.15 holds for a first-order reaction. Expand R_G in a Taylor series around $\bar{T}$ and $\bar{C}$.

10.6. For a first-order reaction taking place in an isothermal, plug-flow reactor, calculate τ:

   a.   assuming that no diffusional resistance exists

   b.   taking diffusional effects into account.

   Use 5 as the Thiele modulus. $C_{out}/C_{in} = 0.2$. Compare the two τ values to recognize the importance of diffusional effects in sizing a reactor properly.

10.7. Calculate the length of an adiabatic reactor required to convert 65% of SO_2 to SO_3. Obtain the reactor profiles of T_b, C_b, T_s, and Λ.

   $T_{in} = 643°K$. Feed at atmospheric pressure (7.8 mole % SO_2, 10.8% O_2 and 81.4% N_2) at 5700 kg/hr m². $(-\Delta H) = 21.4$ kcal/mole, mean specific heat $= 0.221$ kcal/kg°C.

$$r_c(\text{kmole } SO_2/\text{kg cat} \cdot \text{hr}) = \frac{k_1 P_{O_2} P_{SO_2}\left(1 - \dfrac{P_{SO_3}}{P_{SO_2} P_{O_2}^{1/2} K_p}\right)}{22.414\,(1 + K_2 P_{SO_2} + K_3 P_{SO_3})^2}$$

$\epsilon_B = 0.5$
$L = 0.8$ cm
$k_1 = \exp(12.16 - 5473/T)$
$K_2 = \exp(-9.953 + 8619/T)$
$K_3 = \exp(-72.745 + 52596/T)$
$K_p = \exp(11300/T - 10.68)$
$h = 3.3 \times 10^{-4}$ cal/sec cm² °K
$\rho_s = 1.5$ g/cm³
$D_e = 0.5 \times 10^{-2}$ cm²/sec

Use bulk temperature for K's. Use the equations given in Table 10.1 and the design procedures in Figure 10.1. Express r_c in terms of P_{SO_2} only using the stoichiometry and inlet conditions. The solution obtained using a hand-held calculator for $T_{in} = 643°K$ is 145 m. Note that small intervals have to be used for integration in the entrance region because of very slow conversion in that region. Redo the problem with $T_{in} = 720°K$. Assume that the effective diffusivities are the same for all species.

10.8. Derive Eqs. 10.28 and 10.29 from Eqs. 10.16 and 10.17. Use these equations in Problem 10.7 and resolve it. Show that no significant difference can be found for the exit conversion. Discuss the reasons.

10.9. Consider an adiabatic reactor for which $P_m = 200$ and $P_h = 20$. The axial Peclet number for mass dispersion is typically in the range of 200 ~ 500. The Peclet number for heat dispersion is in general smaller than that for mass dispersion. Even when mass dispersion can be neglected, heat dispersion may be significant. Obtain the reactor profiles of temperature and concentration for the following conditions:

$C_{in} = 10^{-5}$ mole/cm³ $T_{in} = 620°$K
$r_c = kC$ $k = \exp(-12,000/T + 14.6)$ (1/sec)
$D_e = 10^{-2}$ cm²/sec $L = 1$ cm
$-\Delta H = 20$ kcal/mole
$h = 4 \times 10^{-4}$ cal/sec cm² °K
$\tau = 10$ sec. $\rho C_p = 3.8 \times 10^{-4}$ cal/cm³ °K

Plot the results for the following cases:
a.   Assume plug-flow, neglecting dispersion of both heat and mass.
b.   Include heat dispersion.
c.   Include both mass and heat dispersion.
d.   Repeat for $P_h = 2$.

What can you conclude from the results? Neglecting the mass dispersion does not mean that $C_0 = C_b$. Rather, Eq. 10.1 should be solved as such instead of Eq. (A) in Table 10.3 with the aid of Eq. (E) in Table 10.3 but with ϵ set to zero for both Eqs. 10.1 and (A).

10.10. For the reaction system given in Table 10.5, calculate the generalized modulus ϕ_G at the reactor inlet.

10.11. For the reaction system given in Table 10.5, plot the reactor point effectiveness against γ at the reactor inlet.

10.12. Consider a first-order reaction taking place in an isothermal reactor. If the packed pellets are uniformly poisoned and the concentration of the poisoning species is constant throughout the reactor, what is the outlet concentration at $t = 10^7$ sec?

$L(k/D_e)^{1/2} = 3$ $L = 1$ cm
$C_{in} = 10^{-5}$g mole/cm³ $D_e = 0.1$ cm²/s
$\tau = 10$ seconds
$(k_p) = 10^{-4}$ sec. $N_b = 10^{-8}$ g moe/cm³
$Q = 10^{-4}$ g mole/cm³ cat. vol.

Suppose that the catalyst is regenerated when the outlet concentration (C_{out}) becomes 50% higher than C_{out} calculated above for $\tau = 10$ sec. Specify τ based on Eq. 10.63.

10.13. For the reaction system of Table 10.5, obtain reactor profiles of T_b, C_b, T_s, Λ, γ, and N_b as a function of time on stream for the case of uniform chemical deactivation. Use the following rate constant:

$k = \exp(-16,000/T + 20.0)$
$h = 8 \times 10^{-4}$ cal/cm² °Ks

10.14. Repeat Problem 10.13 for the case of shell-progressive deactivation. Use the following rate constant for the species causing deactivation.

$$k_p = \exp(-6000/T + 3.3)$$

In addition, obtain C_d and T_d profiles.

10.15. Repeat Problem 10.12 for the case of shell-progressive deactivation. Use the following effective diffusivity:

$$D_e = 0.01 \text{ cm}^2/\text{s}$$

10.16. Consider a pellet undergoing uniform chemical deactivation and sintering. If the main reaction is diffusion-free, the global rate is:

$$R_G = r_c \, (S_r/S_0) \tag{a}$$

and the observed rate of the deactivation reaction is given by:

$$R_p = \frac{(k_p)_s N_b (S_r/S_0)}{[1 + G(C_b)]^n} \tag{b}$$

For an adiabatic plug-flow reactor, then, the conservation equations are:

$$\frac{dC_b}{dy} = -\tau r_c(S_r/S_0) = -\tau k_s f(C_b)[S_r/S_0] \tag{c}$$

$$\frac{dN_b}{dy} = -\tau R_p = -\frac{\tau (k_p)_s N_b [S_r/S_0]}{[1 + G(C_b)]^n} \tag{d}$$

Based on Eqs. (c) and (d), and Eqs. (F) and (G) in Table 10.8, procedures very similar to those in Figure 10.21 can be obtained.

For the case of a diffusion-limited main reaction, show that the results given in Table 10.8 can be used as such by simply replacing $(\bar{S}_r/\bar{S}_0)$ by S_r/S_0 and $\bar{S}_t/\bar{S}_0$ by S_t/S_0. Also note (Chapter 4) that $[1 + G(C_b)]$ is close to unity (or $G(C_b) \ll 1$) for a uniformly deactivated, diffusion-limited reaction. Therefore, $G(C_b)$ in the equations of Table 10.8 can be set to zero. Further, show that Eq. (D) in Table 10.8 becomes for this case:

$$N_b[y;t] = (N_b)_{in} \, \exp\left\{-\int_0^y \tau(k_p)_s (S_r/S_0) dy\right\} \tag{e}$$

10.17. Consider a pellet undergoing shell-progressive chemical deactivation and uniform sintering. Show that the design equations given in Table 10.6 apply to this case. The only change is in Eq. (A) of Table 10.6:

$$C_b - C_d = \left(\frac{L\gamma}{D_e}\right)\left[2D_e(S_t/S_0)k_b \int_{C_c}^{C_d} f(C)dC\right]^{1/2}$$

$$\times \exp\left\{\left(\frac{E_a}{2R_g T_b}\right)\left[\frac{(-\Delta H)D_e(1+(Bi)_h\gamma)(C_b-C_d)}{hT_bL\gamma + (1+(Bi)_h\gamma)(-\Delta H)D_e(C_b-C_d)}\right]\right\}$$

for which the sintering kinetics given in Table 6.4 can be used for (S_t/S_0).

10.18. Derive Eqs. 10.101 and 10.102.

REFERENCES

Ahmed, M. and R.W. Fahien, Chem. Eng. Sci., 35, 897 (1980).

Burghardt, A. and T. Zaleski, Chem. Eng. Sci., 23, 575 (1968).

Cole, J.D., *Perturbation Methods in Applied Mathematics,* Blaisdell, New York (1968).

Lee, H.H., AIChE J., 27, 557 (1981).

Lee, H.H. and J.B. Butt, AIChE J., 28, 410 (1982).

Lee, H.H., and E. Ruckenstein, Cat. Rev., 25, 475 (1983).

Price, T.H. and J.B. Butt, Chem. Eng. Sci., 32, 393 (1977).

CHAPTER 11

Fluidized-Bed Reactors

11-1 INTRODUCTION

A fluidized-bed for catalytic reactions is one in which small catalyst particles (typically in the range of 20 to 300 μm) are suspended by fluid reactants. The earliest application of fluidization was for the catalytic cracking of gas oil into gasoline. Rapid deactivation of the catalyst due to coking necessitated the change from fixed-beds to fluidized-beds for the cracking reactions. In the continuous reactor-regenerator system (Froment and Bischoff 1979), shown in Figure 11.1, the oil is fed at the bottom of the reactor through a perforated plate distributor and the reaction products are removed at the top. Fines entrained in the product stream are separated in cyclones and are then fed back to the reactor. Catalyst particles are allowed to leave the reactor through a bottom standpipe in which they are steam-stripped to remove adsorbed hydrocarbons. They then plow under a static head through the transfer line to the regenerator. The lower static pressure in the riser leading to the regenerator is attained by aeration of particles with the air required for burning off the coke. The regenerated catalyst flows into a downcomer and back to the reactor. The pressure difference required for this transport is attained by injection of the oil into the riser of the reactor. The heat produced by the regeneration is carried to the reactor by the catalyst particles and there they evaporate, heat, and crack the oil. With the advent of zeolite catalysts, the traditional fluidized-bed has been modified to the riser cracker. These zeolite catalysts are so active that the cracking takes place entirely in the riser, leaving the traditional reactor vessel only for housing cyclones for rapid separation of the catalyst from the product gas and a stripper.

Significant catalyst losses by entrainment and attrition can occur because of the fluidization of small particles. On the other hand, the high turbulence created by fluidization leads to much better heat and mass transfer than is possible in a fixed-bed. In fact, fluidized-beds are now used for highly exothermic reactions for which close temperature control is important since the fluidized-bed can be operated isothermally. Fluidized-beds are an alternative to fixed-beds for highly exothermic reactions and when frequent regeneration of the catalyst is required.

A quantitative description of a bubbling fluidized-bed is complicated because of the presence of gas bubbles that form at the feed distributor, rise, and then rupture when they reach the free surface at the top of the bed, releasing downward the particles drawn into the wake of the bubbles as they rise. Despite the complexity

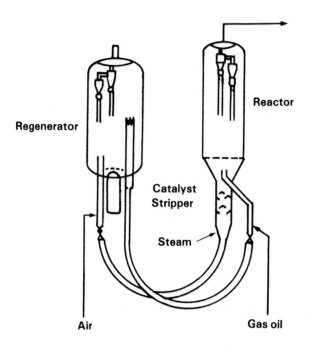

Figure 11.1 Reactor-regenerator system for catalytic cracking of gas oil. (Zenz and Othmer 1960)

Regenerator

Reactor

Catalyst
Stripper

Steam

Air   Gas oil

involved, simple-minded approaches have been used: one of them is to treat the
bed as a continuously-stirred tank reactor. In an analysis of a fluidized-bed for
the regeneration of coked catalyst, Pansing (1956) assumed the coked catalyst to
be well-mixed but for the gas phase to be in plug-flow. His work contains the
essential elements of two-phase models to evolve later: the presence of two distinct
phases, gas-rich and solid-rich, and the degree of mixing and contacting within
and between the phases. For coke burning, the rate of reaction is equated to the
mass transfer between the gas and the catalyst particles:

$$R_G' = kC_c\, p_i = K(p - p_i)$$

or

$$R_G' = \frac{p}{1/k + 1/KC_c} \tag{11.1}$$

where C_c = concentration of carbon on catalyst
$\quad p_i$ = oxygen partial pressure at catalyst surface
$\quad p$ = bulk gas-phase oxygen partial pressure
$\quad k,\ K$ = rate and mass transfer constants
$\quad R_G'$ = observed rate based on catalyst weight

A mass balance on oxygen in the plug-flow gas-phase gives:

$$F\frac{dN}{dw} = -R_G' \tag{11.2}$$

which can be integrated with the aid of Eq. 11.1 to yield:

$$-\frac{PC_c}{S \ln f} = \frac{C_c}{K} + \frac{1}{k}$$

(11.3)

where $f = N/N_0$, $S = FN_0/w$, and P is total pressure. Here, F is the volumetric feed rate, N is oxygen concentration, N_0 is inlet oxygen concentration, and w is the weight of the catalyst. Pansing correlated K to the mass flow rate and particle size ($K \propto G^m/d_p$) to fit pilot plant date and found the activation energy of k to be 35 kcal/mol, which is close to the 36 kcal/mol value found by Weisz and Goodwin (1966) for coke combustion. Pansing's concept of separating the solid-rich phase (emulsion phase) from the gas-rich phase (bubble phase) forms the basis for many two-phase models.

The understanding of fluidized-beds is far from satisfactory, particularly regarding the fluid mechanics. The value of various models for fluidized-beds, however, lies in providing the framework within which each specific application can be considered. This chapter examines fluidization characteristics, the role that gas bubbles play, and the rationale behind the two-phase theory, which naturally leads to the models based on the two-phase theory. A full section will be devoted to the catalytic cracker, in particular the riser cracker, since it represents the most important application of fluidized-beds to catalytic reactions. This chapter starts with the understanding that the intrinsic rate is essentially the same as the global rate in fluidized-beds, unless the catalyst is deactivated.

11–2 FLUIDIZATION

The flow regimes that can exist when a mass of fine particles is subjected to a fluid flowing upward are shown in Figure 11.2, along with a diagram of pressure drop versus superficial fluid velocity. When the superficial fluid velocity is just high enough to support the solid weight of the particles, incipient fluidization occurs and this velocity is called the incipient fluidization velocity or minimum fluidization velocity U_{mf}. This situation is represented by the flow regime (a). The incipient velocity is usually determined by drawing the lines shown in the lower diagram and finding their point of intersection. The voidage at minimum fluidization is calculated from:

$$\Delta P_{b,mf} = (\rho_s - \rho)(1 - \epsilon_{mf})gH_{mf}$$

(11.4)

where $\Delta P_{b,mf}$ = pressure drop at U_{mf}

ρ_s = solid density

ρ = fluid density

ϵ_{mf} = voidage at incipient fluidization

H_{mf} = bed height at U_{mf}

g = gravity

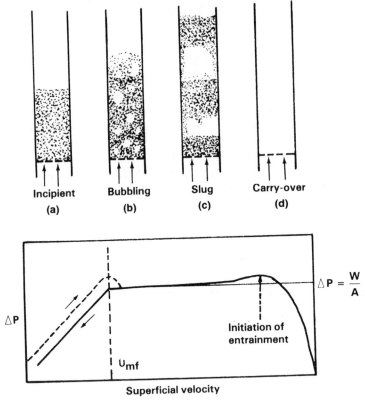

Figure 11.2 Flow regimes and pressure drop versus gas velocity for fluidized-beds (Kunii and Levenspiel 1969). (Reprinted from *Fluidization Engineering,* © Wiley, by permission of John Wiley & Sons, Inc.)

In the absence of experimental data, it is a good approximation to take $\epsilon_{mf} = 0.4$ (Davidson et al. 1977) for uniform spheres. The minimum fluidization velocity can be estimated by combining the Ergun equation with Eq. 11.4:

$$(1 - \epsilon_{mf})(\rho_s - \rho)g = \frac{150(1 - \epsilon_{mf})}{d_p^2 \epsilon_{mf}^3} \mu\, U_{mf} + 1.75 \frac{\rho\, U_{mf}^2}{d_p \epsilon_{mf}^3} \qquad (11.5)$$

which is an implicit expression for U_{mf}. Here, d_p is the particle size, and μ is the fluid viscosity. While the fixed-bed Ergun equation has been used, it should give a reasonable approximation of U_{mf}. In fact, the pressure drop through a fixed-bed has been shown (Richardson 1971) to be slightly more than through a fluidized-bed with the same particles and voidage fraction.

As the velocity is increased beyond U_{mf}, gas bubbles start to appear, as shown in flow regime (b) of Figure 11.2. This bubbling regime, in which the largest bubbles are smaller than the bed diameter D, is the one most commonly found in industrial bubbling beds. When the velocity is further increased, slugs appear as shown in the slugging regime (c) of Figure 11.2. In this regime, the bubble size is comparable to the diameter of the bed and the bubble motion is restricted by the walls. At a still higher pressure, slugs occupy the whole cross section and the particles rain through the low-voidage regions. This extreme case is often found in small diameter beds and is really a breakdown of proper fluidization. As the velocity exceeds the terminal velocity, solid particles are carried over, approaching the behavior of a transport reactor. The pressure drop remains relatively independent of the velocity until entrainment takes place. The terminal velocity can be estimated from:

$$U_t = d_p^2 \frac{(\rho_s - \rho)g}{18\mu} \tag{11.6}$$

As seen from the lower diagram of Figure 11.2, a certain hysteresis exists on decreasing the velocity. In the range of practical interest ($U > 2U_{mf}$), the bubbling regime prevails since industrial bubbling beds often have a height to diameter ratio close to unity, and thus, the bubble diameter is likely to be less than D.

For a bubbling fluidized-bed, the "excess flow" above the value needed for incipient fluidization is ($U - U_{mf}$) (area). The two-phase theory is based on the supposition that much of the excess flow passes through the bed as voids or bubbles. In this way the bed is divided into two phases, a bubble phase and an emulsion phase, which is like the bed at incipient fluidization. The emulsion phase is assumed to have the same voidage as at incipient fluidization. The presence of bubbles, shown in Figure 11.3, in the bubbling regime then leads to the two-phase theory. The essence of this theory is the hypothesis that the bubbles behave as if they were in a liquid. This analogy is often used to obtain an insight into the fluid mechanics of bubbles. In the absence of bubbles, a fluidized-bed can be treated as a simple CSTR, or a PFR. Fluidization behavior unique to fine particles will be discussed in Section 11–4, which involves nonbubbling beds at high gas velocities.

The operating range of superficial gas velocity is bounded by $U_{mf} < U <$

Figure 11.3 Drawings of X-ray photographs of bubbles in fluidized-beds: (a) crushed coal (b) magnesite (Rowe 1971). (With permission from *Fluidization.* Copyright: Academic Press Inc. (London) Ltd.)

(a)   (b)

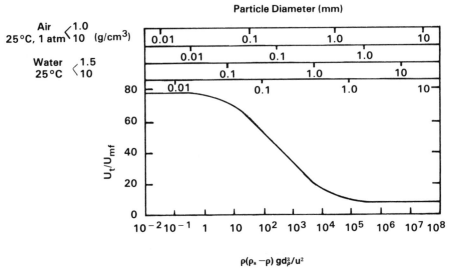

Figure 11.4   Ratio of terminal velocity to incipient fluidizing velocity with $\epsilon_{mf} = 0.4$. (Richardson 1971)

U_t. This range has been plotted as a function of particle size by Richardson (1971) for uniform spheres with $\epsilon_{mf} = 0.4$ (Eq. 11.5) as shown in Figure 11.4. It is seen that for large particles ($> \sim 1$ mm), the allowable range of U is only about 10 U_{mf}. For beds of mixed particle sizes, U is often larger than U_t for the smallest particles, so there is continuous elutriation and a cyclone recycle system is used for particle recycle. The rate of elutriation of fines (Kunii and Levenspiel 1969) at unsteady state is often expressed as:

$$-\frac{1}{A_t}\frac{dW_p}{dt} = k_e \frac{W_p}{W}$$ (11.7)

where k_e = elutriation rate coefficient (wt/area · time)
   W = total weight of solids in the bed
   W_p = weight of the amount of particles of diameter d_p
   A_t = cross-sectional area of bed surface

Correlations compiled by Leva (1960) for the elutriation rate coefficient are shown in Figure 11.5. The elutriation constant is obtained from the definition of k_e and data from steady state runs (see Problem 11.2):

$$k_e = \frac{\text{rate of removal of solids of size } d_p \text{ per area of bed surface}}{\text{fraction of bed consisting of size } d_p}$$ (11.7a)

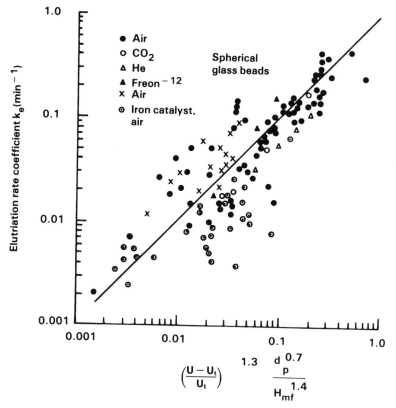

Figure 11.5 Correlation of elutriation rate coefficients for two-component systems. (Leva 1960)

11-3 TWO-PHASE MODELS

Two-phase models for bubbling beds focus on a single bubble, which is representative of all bubbles. This necessarily requires the assumptions that each bubble rises without coalescence through the bed and that all bubbles have the same diameter and behave the same as the representative bubble. Bubbles do grow as they rise and do coalesce. Therefore, these assumptions are oversimplifications and yet they bring out the most important characteristic of the bubbling bed, namely bypassing of the catalyst due to bubbles. This recognition of bypassing is the main improvement over earlier models (up to around 1960) that could not explain observed conversions below those for complete mixing. Since bubbles are considered to be free from catalyst particles and no reaction takes place in them, the gaseous reactant in the bubbles is depleted only through mass transfer from the bubble to its surroundings, which is the emulsion phase, and possible cross-flow between the bubble and emulsion phases. Various two-phase models (Davidson and Harrison 1963; Partridge and Rowe 1966; Kunii and Levenspiel 1969) all

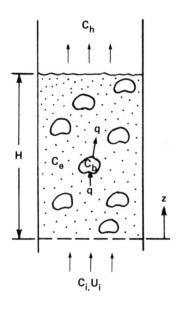

Figure 11.6 Two-phase model of Davidson and Harrison. (Davidson, Harrison, Darton, LaNauze, in *Chemical Reactor Theory: A Review,* ed. Lapidus and Amundson, © 1979. Reprinted by permission of Prentice-Hall, Inc., Englewood Cliffs, NJ.)

have these basic ingredients, but they differ in treating the interchange of mass between the two-phases. The models due to Davidson and Harrison, and Kunii and Levenspiel will be examined in some detail here.

The two-phase model of Davidson and Harrison is shown in Figure 11.6. The bubbles are assumed devoid of catalyst particles and to maintain representative characteristics such as diameter as they rise. Thus a bubble acts as a plug rising through the bed and the concentration within the plug changes only through interaction with the emulsion phase. If k_G is an overall mass transfer coefficient for the mass transport across the bubble film, and q is a volumetric rate of the cross-flow, a steady-state mass balance for a single bubble is:

$$(q + k_G S)(C_e - C_b) = U_A V_b \frac{dC_b}{dz} \tag{11.8}$$

where k_G = overall film mass transfer coefficient
q = volumetric rate of cross-flow
S = surface area of a bubble/volume
C_e = concentration of key species in emulsion phase
C_b = concentration of key species in the bubble
U_A = absolute bubble velocity
V_b = volume of the bubble

As indicated earlier, representative bubble characteristics such as volume and surface area are assumed to remain constant, i.e., independent of z. Because of very good mixing in the emulsion phase, this phase can be treated as a CSTR (sometimes, it is treated as a PFR). Therefore, Eq. 11.8 can be integrated directly to give:

$$C_b = C_e + (C_i - C_e)\exp(-Qz/U_A V_b) \tag{11.9}$$

where $Q = q + k_G S$. Since the basis of the two-phase theory is that the excess flow (flow in excess of the value needed for incipient fluidization) represented by $[(U_i - U_{mf})$ (area)] passes through the bed as bubbles and the gas flow in the emulsion phase is at its fluidization velocity U_{mf}, a mass balance at the top of the bed gives:

$$U_i C_h = U_{mf} C_{eh} + (U_i - U_{mf}) C_{bh} \tag{11.10}$$

where the subscript h refers to the value at $z = H$. Since C_e is constant at C_{eh}, Eq. 11.10 can be rewritten with the aid of Eq. 11.9:

$$C_h = C_e + \beta(C_i - C_e)\exp(-\alpha) \tag{11.11}$$

where $\beta = (U_i - U_{mf})/U_i$
$\quad \alpha = QH/U_A V_b$

A mass balance around the bed gives:

$$SU_i(C_i - C_h) = R_G(1 - \epsilon)HS$$

or

$$U_i(C_i - C_h) = (1 - \epsilon)HR_G[C_e] \tag{11.12}$$

where ϵ is the bed voidage including bubbles and the interstitial gas in the emulsion phase, and S is the bed cross-sectional area. In this regard, it is noted that R_G is based on the apparent catalyst volume and that the global rate is equal to the intrinsic rate in most cases where catalyst deactivation is unimportant. Given the global rate of the reaction, Eqs. 11.11 and 11.12 can be solved simultaneously for C_e and C_h.

It is instructive to examine the conversion behavior of a first-order reaction for which $R_G[C_e] = r_c[C_e] = kC_e$. Equations 11.11 and 11.12 yield:

$$\frac{C_h}{C_i} = \frac{1 - Z + KZ}{1 - Z + K} \tag{11.13}$$

where $Z = \beta e^{-\alpha} = (U_i - U_{mf})e^{-\alpha}/U_i$, and $K = Hk(1 - \epsilon)/U_i$. At incipient fluidization, $U_i = U_{mf}$, $Z = 0$, and there is no bubble flow. Therefore, Eq. 11.13 reduces to $C_h/C_i = 1/(1 + K)$, which is the conversion behavior of a CSTR.

An actual bed at incipient fluidization would operate more nearly in plug-flow; the results obtained are for a vigorously bubbling fluidized-bed. Two-phase models have been credited for accounting for the conversion behavior of a bubbling bed, which gives a conversion lower than that for a CSTR. This can be clearly seen by subtracting from Eq. 11.13 the concentration ratio for a CSTR $[(C_h/C_i)_c = 1/(1 + K)]$:

$$\frac{C_h}{C_i} - \left(\frac{C_h}{C_i}\right)_c = \frac{K^2 Z}{(1 + K - Z)(1 + K)} \qquad (11.14)$$

Since $0 < Z < 1$, the right hand side of this equation is always positive. Therefore, the conversion with a bubbling bed is always poorer than with a CSTR, due to bypassing of gas through the bubbles. It is seen, on the other hand, that the bed approaches the behavior of a CSTR when the value of α is large since then Z approaches zero. According to the definition of α, a large value of α corresponds to slowly rising bubbles, whose velocity is proportional to $(U_i - U_{mf})$, a large mass transfer rate from the bubble to the emulsion phase, or both. One would expect the bed to behave as a CSTR in both cases. When $(U_i - U_{mf})$ is small, bubbles are also small and it is well recognized that the bed behaves as a CSTR, and when the mass transfer rate is high, bubbles essentially act as a source of reactant and bypassing of reactant is minimized. General conversion behavior of a fluidized-bed as affected by increasing velocity emerges from the foregoing discussion. A fluidized-bed behaves more like a PFR at incipient velocity. As the velocity is increased, small bubbles appear which causes good mixing, leading to the behavior of a CSTR. As the velocity is further increased, bubbles become larger causing the bypassing of reactant, leading to the conversion behavior of Eq. 11.13, which gives a conversion smaller than that for a CSTR. It is seen then that the conversion attainable decreases with increasing velocity. Therefore, typical industrial beds operating in the bubbling regime result from a compromise between conversion and the high throughput required of an industrial bed. Controllability of conversion for highly exothermic reactions also emerges from Eq. 11.13. When K is very large, Eq. 11.13 approaches Z.

The model of Kunii and Levenspiel also focuses on a single bubble, as shown

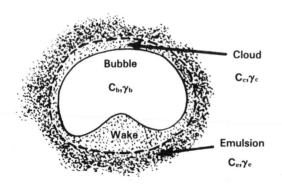

Figure 11.7 Single-bubble model of Kunii and Levenspiel (1969). (Reprinted from *Fluidization Engineering*, © Wiley, by permission of John Wiley & Sons, Inc.)

Cloud

C_c, γ_c

Bubble

C_b, γ_b

Wake

Emulsion

C_e, γ_e

in Figure 11.7. Here, γ is the ratio of solid volume to bubble volume. A representative bubble with its fixed diameter and other bubble properties is treated as a batch reactor ascending through the bed:

$$-\frac{1}{V_b}\frac{dn}{dt} = -U_A\frac{dC_b}{dz} \tag{11.15}$$

The events of the reaction in each phase and the mass transfer between phases are traced for the determination of the overall rate of the reaction:

Overall disappearance = reaction in bubble + transfer to cloud and wake.
Transfer to cloud and wake $\approx$ reaction in cloud and wake + transfer to emulsion.
Transfer to emulsion $\approx$ reaction in emulsion.

or

$$-U_A\frac{dC_b}{dz} = \gamma_b R_G[C_b] + K_{bc}(C_b - C_c) \tag{11.16}$$

$$K_{bc}(C_b - C_c) \approx \gamma_c R_G[C_c] + K_{ce}(C_c - C_e) \tag{11.17}$$

$$K_{ce}(C_c - C_e) \approx \gamma_e R_G[C_e] \tag{11.18}$$

where the K_{ij}'s are coefficients of mass transfer from phase i to phase j. Equations 11.16, 11.17, and 11.18 can be used to eliminate the unknowns, C_c and C_e. For a first-order reaction $R_G = r_c = k_r C$ and Eq. 11.16 can be expressed as:

$$-U_A\frac{dC_b}{dz} = K_o C_b$$

where the overall reaction-mass transfer coefficient K_o is obtained from Eqs. 11.16 through 11.18:

$$K_o = k_r \left[\gamma_b + \cfrac{1}{\cfrac{k_r}{K_{bc}} + \cfrac{1}{\gamma_c + \cfrac{1}{k_r/K_{ce} + 1/\gamma_e}}} \right] \tag{11.19}$$

Then, the conversion x is simply given by:

$$1 - x = \exp(-K_o H/U_A) \tag{11.20}$$

While the solution is straightforward for a first-order reaction, trial and error calculations have to be carried out for C_c and C_e for more general rate expressions. In general $\gamma_e > \gamma_c \gg \gamma_b$, and γ_b can be set to zero in Eq. 11.19.

The model of Partridge and Rowe (1966) takes into account the bubble clouds, while the bubble wakes are treated as a part of the emulsion phase.

While the two-phase models are superior to earlier simple models, there are ample opportunities for further improvements. These have to do on one hand with better ways of determining *a priori* the bubble properties appearing in the models. On the other hand, bubble growth and its effect on the performance of the fluidized-bed has to be accounted for. All the models discussed above assume a constant effective bubble diameter. The correlations for the parameters in the two-phase models are given in Section 11–5. It is a well-known experimental fact that the bubbles grow as they ascend through the bed. This growth can be ascribed to expansion, coalescence, and gas transfer from the emulsion phase. On the other hand, bubbles also tend to split. Therefore, there will be a broad distribution of bubble sizes at any point in the bed and this distribution will change with the bed height.

This section concludes with the comparison of various models by Pyle (1972), shown in Figure 11.8. For the first-order reaction considered, the differences between the models are particularly noticeable near the bottom of the bed where the effect of the small bubbles used by Partridge and Rowe is in evidence. Over much of the range covered, moreover, the predictions of the models differ by up to 100% in either concentration or bed height. What is obvious from this comparison is a need for more reliable and general correlations for bubble properties.

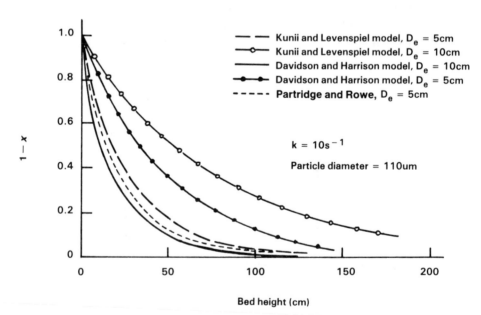

Figure 11.8 Comparison of various models and their dependence on effective bubble diameter. (Pyle 1972; reprinted with permission from Advances in Chemistry Series. Copyright 1972 American Chemical Society.)

11–4   FURTHER CONSIDERATIONS

The pressure drop versus superficial velocity diagram shown in Figure 11.2 for various flow regimes is for a particular particle size. A different diagram results for a different particle size. Geldart (1973) made a useful classification of solid particles into four characteristic groups, as shown in Figure 11.9, in order to identify the modes of fluidization according to the size and density of the particles. Bubble-correlations are often based on particles belonging to group B (Horio and Wen 1977). These correlations, when extended to group A particles, for instance, could cause the model behavior to deviate from experimental observation unless care is taken. The classification of Geldart is useful in recognizing the dependence of bubbling behavior on the particle size. When fine particles are fluidized, there is a region in which the bed expands without bubble formation. This bubble-free region disappears when larger particles are fluidized, as shown in Figure 11.10 (DeJong and Nomden 1974). For large particles, bubbling will start when the gas velocity reaches the minimum bubbling velocity.

Fluidization has become virtually synonymous with operation in the bubbling regime, corresponding typically to gas velocities of around 1 ft/s. An underlying function of fluidization has often been to afford contact between a gas and a large inventory of solid surface per unit bed volume. A question that can be raised is whether the same end can be achieved at gas velocities beyond the bubbling regime. The point advanced by Yerushalmi and co-workers (1976) is that for particles belonging to group A in Figure 11.9, fluidization can be achieved at gas velocities much higher than 1 ft/s. Since the solid feed rate determines the state of fluidization,

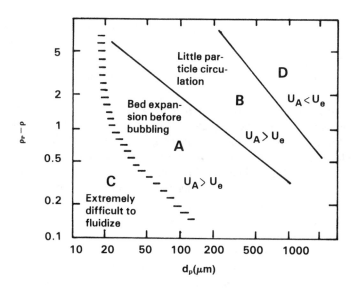

Figure 11.9   Four groups of particles proposed by Geldart (1973).

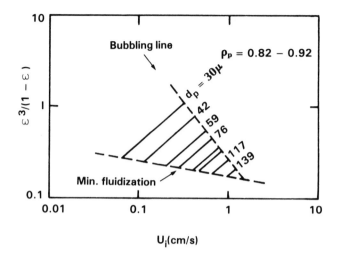

Figure 11.10   Region of bubble-free operation. (Dejong and Nomden 1974)

for satisfactory fast fluidization the solid rate has to be high. In this "fast fluidization" regime, there is no gas-bypassing since there are no bubbles. A relatively
dilute phase (low solid concentration) results from the fluidization. The fast fluidization, already practiced commercially (Reh 1971), needs further systematic studies.
Another important state of fluidization for fine particles is "turbulent fluidization"
(Kehoe and Davidson 1971; Massimilla 1973). As the gas velocity is increased, a
breakdown of the slugging regime into a state of continuous coalescence occurs
before the carryover of particles takes place. Thus, industrial beds of fine particles,
which often operate at velocities around 2 to 3 ft/s, might very well be operating
in the turbulent regime. Such a bed would operate as a nonbubbling bed at high
velocities and could be modeled as a simple CSTR, plug-flow reactor, or a combination of both, in which the gas phase is treated as plug-flow and the solid phase
as well-mixed flow. A model based on the latter approach is detailed later for a
riser cracker.

   The two-phase theory originally proposed by Toomey and Johnstone (1952)
made a significant contribution in the development of models for bubbling fluidized-beds. Although the general concept of gas bypassing in the form of bubbles
is applicable to any bubbling bed, the interpretation of the concept as applied to
the specifics of modeling is diverse. This diversity has led to a number of bubbling
fluidized-bed models that have varying degrees of complexity. The two-phase models
considered earlier use the gas bubbles as the means of relating the model parameters
to operating conditions. Thus, an average bubble size, which is often estimated
at the middle point in the bed, becomes the key parameter. The bubble size is
often considered as an adjustable parameter in the absence of bubble size information. Although this approach could be used successfully in some cases, there are
a number of problems associated with it. In the first place, use of an average
bubble size cannot adequately describe the bed since bubbles are present in varying

sizes undergoing growth and coalescence as they rise. The inadequacy becomes acute when the rate of reaction is fast and the phenomena near the distributor region dominate (Horio and Wen 1977). A logical alternative is to allow the bubble size to vary along the bed axis. Another problem with the approach of a constant bubble size is its inability to deal with the flow reversal of gas in the emulsion phase, which occurs when the gas velocity is much higher than the minimum fluidization velocity. For this flow reversal, first verified experimentally by Kunii and co-workers (1967), Fryer and Potter (1972) proposed the following equation for estimating the critical superficial gas velocity U_{cr}, above which the flow reversal will occur:

$$\frac{U_{cr}}{U_{mf}} = \left(1 + \frac{1}{\epsilon_{mf}f_w}\right)[1 - \epsilon(1 + f_w)] \tag{11.21}$$

where f_w is the ratio of wake volume to bubble volume. Under flow reversal conditions, the average gas concentration goes through a minimum in the bed.

In the following sections, a model for a bubbling fluidized-bed that allows for the variation of bubble properties along the bed axis will be considered, together with a model for a nonbubbling fluidized-bed at high velocities.

11–5 A MULTIPHASE, VARIABLE BUBBLE PROPERTY MODEL

Peters and co-workers (1982) combined the bubble assemblage concept of Kato and Wen (1969) and the three-phase model of Fryer and Potter (1972) to advance a cell (compartment) model in which an equivalent spherical diameter for a bubble is the cell size (height). The equivalent diameter is allowed to vary along the bed axis and each cell, consisting of bubble, cloud, and emulsion phases, is considered to be perfectly mixed in each phase, as shown in Figure 11.11. For each phase in a cell, a mass balance can be written, accounting for the convective mass flow into and out of the cell, and for the mass transport between phases in the same cell. A solution for the axial concentration profiles is obtained from the bottom up, starting with the inlet conditions that specify the size (height) of the first cell. The solution for the first cell is then used to calculate the next cell size (equivalently the bubble diameter) which allows solution of the mass balance equations for the second cell. These procedures are repeated until the cumulative cell size reaches the expanded bed height.

Consider the nth cell in Figure 11.11. According to the model, the mass balance for the bubble phase is:

$$\epsilon_b V_{bn} \frac{dC_{bn}}{dt} = U_{b,n-1}SC_{b,n-1} - U_{bn}SC_{bn}$$
$$+ (P_{bc})_n SC_{cn} + (K_{bc})_n V_{bn}(C_{cn} - C_{bn}) \qquad 1 \leqslant n \leqslant N \quad (11.22)$$

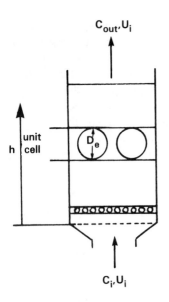

C_{out}, U_i

h | unit cell

C_i, U_i

Figure 11.11 Cell model of Peters et al. (1982). (Reprinted with permission from *Chemical Engineering Science*. Copyright by Pergamon Press, Inc.)

where the first subscript b is for the bubble phase, the second subscript n is for the cell number, and the subscript c is for the cloud phase. Here, U is the superficial gas velocity such that U_{bn} is the superficial gas velocity of the bubble phase out of the nth cell, S is the bed cross-sectional area, C is the gas concentration, V is the volume of the phase such that V_{bn} is the volume of bubble phase in the nth cell, P_{bc} is the gas cross-flow between bubble and cloud, K_{bc} is the gas interchange coefficient from the bubble to the cloud phase, and N is the total number of cells. The volume fraction of gas in the bubble, ϵ_b, is assumed constant at unity; the corresponding quantities for the cloud (ϵ_c) and emulsion phases (ϵ_e) are assumed to be constant at ϵ_{mf}. Since ϵ_b is unity by assumption, there are no catalyst particles in the bubble, and thus, no consumption term due to reaction appears in Eq. 11.22. The first two terms in the right hand side of Eq. 11.22 are the convective mass flow into and out of the nth cell. Although the third term is called mass exchange due to "cross-flow," it is not the cross-flow referred to in the Davidson and Harrison model (Eq. 11.8), but rather, it is included to satisfy the overall mass balance for the cell. This cross-flow is included in the mass balance for the emulsion phase to account for the emulsion phase degasing. The degasing coefficient for the emulsion phase, $(P_{ce})_n$, may be evaluated by:

$$(P_{ce})_n = (P_{bc})_n + U_{cn} - U_{c,n-1} \tag{11.23}$$

$$(P_{bc})_n = U_{bn} - U_{b,n-1} \tag{11.24}$$

The fourth term in Eq. 11.22 represents the mass transfer at the bubble-cloud interface. The mass balance for the cloud phase is:

$$\epsilon_c V_{cn} \frac{dC_{cn}}{dt} = U_{c,n-1} SC_{c,n-1} - U_{cn} SC_{cn}$$

$$+ (P_{ce})_n SC_{en} - (P_{bc})_n SC_{cn} + (K_{ce})_n V_{bn}(C_{en} - C_{cn}) \qquad (11.25)$$
$$+ (K_{bc})_n V_{bn}(C_{bn} - C_{cn}) - V_{cn} R_G[C_{cn}] \qquad 1 \leqslant n \leqslant N$$

where the subscript e denotes the emulsion phase. The mass balance for the emulsion phase has to be written for three different regions of the bed when flow reversal occurs at one of the cells, say n_r. The cell number n_r at which flow reversal starts is found in the cell calculations when the superficial gas velocity in the emulsion phase, calculated according to the relationship given below, becomes negative:

$$U_{en} = U_i - U_{An}\{\delta_{bn}\epsilon_b + \delta_{cn}\epsilon_c\} \qquad (11.26)$$

where the absolute bubble velocity for the nth cell, U_{An}, is determined (Davidson and Harrison 1963) from:

$$U_{An} = (U_i - U_{mf}) + 0.71[g(D_e)_n]^{1/2} \qquad (11.27)$$

Here, g is the acceleration due to gravity, D_e is the effective bubble diameter, and δ_{in} is the volume fraction of the bed in the nth cell occupied by phase i. Thus, for the emulsion phase:

$$\epsilon_e V_{en} \frac{dC_{en}}{dt} = U_{e,n-1} SC_{e,n-1} - U_{en} SC_{en} \qquad 1 \leqslant n < n_r$$
$$- (P_{ce})_n SC_{en} + (K_{ce})_n V_{bn}(C_{cn} - C_{en}) - V_{en} R_G[C_{en}] \qquad (11.28)$$

$$\epsilon_e V_{en} \frac{dC_{en}}{dt} = |U_{en}| SC_{e,n+1} + U_{e,n-1} SC_{en,n-1} \qquad n = n_r$$
$$- (P_{ce})_n SC_{en} + (K_{ce})_n V_{bn}(C_{cn} - C_{en}) - V_{en} R_G[C_{en}] \qquad (11.29)$$

$$\epsilon_e V_{en} \frac{dC_{en}}{dt} = |U_{en}| SC_{e,n+1} - |U_{e,n-1}| SC_{en} \qquad n_r < n \leqslant N$$
$$- (P_{ce})_n SC_{en} + (K_{ce})_n V_{bn}(C_{cn} - C_{en}) - V_{en} R_G[C_{en}] \qquad (11.30)$$

Cell by cell calculations can be carried out for given initial conditions and the following relationship:

$$C_{b0} = C_{c0} = C_{e0} = C_i \qquad \text{at the inlet} \qquad (11.31)$$

The gas concentration averaged over the cross-sectional area of the bed, C_{av}, is given by:

$$C_{av} = C_b \delta_b + C_c \delta_c + C_e \delta_c \qquad (11.32)$$

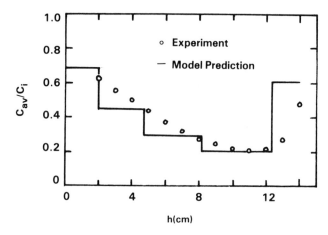

Figure 11.12 A comparison between cell model predictions and experimental data. (Peters et al. 1982; reprinted with permission from *Chemical Engineering Science.* Copyright by Pergamon Press, Inc.)

The average outlet concentration, when flow reversal occurs, is obtained from:

$$C_{out} = \frac{U_{bN}C_{bN} + U_{cN}C_{cN}}{U_{bN} + U_{cN}} \tag{11.33}$$

The inlet concentration to the Nth cell in Eq. 11.30, i.e., $C_{e,N+1}$, is given by C_{out}. If flow reversal does not occur, only Eq. 11.28 is required for the emulsion phase and the outlet concentration is given by:

$$C_{out} = \frac{U_{bN}}{U_i}C_{bN} + \frac{U_{cN}}{U_i}C_{cN} + \frac{U_{eN}}{U_i}C_{eN} \tag{11.34}$$

The steady state version of Eqs. 11.22 through 11.30 can be used for design purpose.

A comparison between cell model predictions and the experimental data of Fryer and Potter (1975) is shown in Figure 11.12 (Peters et al. 1982). The basic parameters used to calculate the model parameters for the prediction in Figure 11.12 are given in Table 11.1. These parameters are the inlet conditions, the condi-

Table 11.1   Basic Parameters for the Simulation in
Figure 11.12 (Peters et al. 1982)

$C_i = 1.0$ g/cm³	$U_{mf} = 1.7$ cm/s
$U_i = 12.65$ cm/s	$H_{mf} = 11.2$ cm
	$\epsilon_{mf} = 0.48$
$D_G = 0.18$ cm²/s	
$D = 22.9$ cm	$R_G = kC; \ k = 1.70$ s⁻¹
$N_D = 61$	

tions at the minimum fluidization (U_{mf}, ϵ_{mf}, and H_{mf}), reaction kinetic parameters, the molecular diffusivity of the gas (D_G), the bed diameter (D), and the number of orifice openings on the distributor (N_D). It is seen that the model describes the experimental data reasonably well. Perhaps more important is the ability of the model to predict the minimum concentration in the bed caused by flow reversal in the emulsion phase.

In order to use Eqs. 11.22–11.30, the model parameters in these equations have to be evaluated from the basic parameters. The relationships for the model parameters, which are given by Peters and co-workers, are summarized in Table 11.2 in the order in which they will be calculated. As indicated earlier, these calculations have to be carried out cell by cell. Since the effective bubble diameter D_e required for specifying the cell height is implicit, an iterative procedure involving

Table 11.2 Relationships for Model Parameters (Adapted from Peters et al. 1982 by permission of Pergamon Press, Inc.)

D_e (Mori and Wen 1975):

$$\frac{D_m - D_e}{D_m - (D_e)_0} = \exp(-0.3h/D)$$

$$D_m = 0.652[S(U_i - U_{mf})]^{2/5}$$

$$(D_e)_0 = \begin{array}{l} 0.347[S(U_i - U_{mf})/N_D]^{2/5}: \text{ perforated distribution plates} \\ 0.00376\,(U_i - U_{mf})^2: \text{ porous distribution plates} \end{array}$$

Range of validity:

$$0.5 < U_{mf} < 20 \text{ cm/s}; \qquad 0.006 < d_p < 0.045 \text{ cm}$$
$$U_i - U_{mf} < 48 \text{ cm/s}; \qquad D < 130 \text{ cm}$$

U_A: Eq. 11.27

δ_c/δ_b (Murray 1965):

$$\frac{\delta_c}{\delta_b} = \frac{1}{\epsilon_{mf}U_A/U_{mf}}$$

H:

$$\frac{H - H_{mf}}{H} = \frac{Y(U_i - U_{mf})}{U_A|_{h=U_{mf}/2}}$$

$$Y = 0.7585 - 0.0013(U_i - U_{mf}) + 0.0005(U_i - U_{mf})^2$$

ϵ (Kato and Wen 1969):

$$1 - \epsilon = \begin{array}{ll} \dfrac{H_{mf}}{H}(1 - \epsilon_{mf}) & \text{for } h < H_{mf} \\[2ex] \dfrac{H_{mf}}{H}(1 - \epsilon_{mf})\left\{\exp\left[-\left(\dfrac{h - H_{mf}}{H - H_{mf}}\right)\right]\right\} & \text{for } h \geqslant H_{mf} \end{array}$$

ϵ_j: $\epsilon_b = 1$, $\epsilon_c = \epsilon_e = \epsilon_{mf}$

Table 11.2 (*continued*)

N_b (Kato and Wen 1969):

$$N_b = \frac{6S(\epsilon - \epsilon_{mf})}{\pi D_e^2 (1 - \epsilon_{mf})}$$

V_j:

$$V_b = N_b \left(\frac{\pi}{6}\right) D_e^3$$

$$V_c = V_b \left(\frac{U_{mf}}{\epsilon_{mf} U_A - U_{mf}}\right)$$

$$V_e = SD_e - V_c - V_b$$

$\delta_j (j = b, c, e)$:

$$\delta_j = V_j / SD_e$$

$U_j (j = b, c, e)$:

$$U_b = U_A \delta_b \epsilon_b; \qquad U_c = \frac{\delta_c}{\delta_b} \frac{\epsilon_c}{\epsilon_b} U_b; \qquad U_e \text{ given by Eq. 11.26}$$

K_{ij} (Murray 1965; Sit and Grace 1981):

$$K_{bc} = 2\left(\frac{U_{mf}}{D_e}\right); \qquad K_{ce} = 6.78\left(\frac{D_G \epsilon_{mf} U_A}{D_e^3}\right)^{1/2}$$

P_{ij}: Eqs. 11.23 and 11.24

---

improvement of D_e calculated from the previous cell information is required. In arriving at the superficial gas velocity in the emulsion phase, the gas flow belonging to the wake phase was neglected (Yacono et al. 1979) such that the superficial gas velocity, which is constant at the inlet velocity U_i, is the sum of U_b, U_c, and U_e.

The bubble correlations given in this section are also applicable to the two-phase models. The best results are often obtained when these properties, particularly the bubble size, are evaluated at the conditions halfway up the bed. Since the bubble correlations are given in Table 11.2, only the parameters in the two-phase models are given here, which are different from those in Table 11.2. The factor α (Eq. 11.11) in the Davidson and Harrison (1977) model (Eq. 11.8) is given by:

$$\alpha = \frac{F_q U_{mf} H_{mf}}{g^{1/2} D_e^{3/2}} + \frac{F_d H_{mf} D_G^{1/2} \epsilon_{mf}}{g^{1/4} D_e^{7/4} (1 + \epsilon_{mf})} \tag{11.35}$$

where the factors F_q and F_d, which depend on D_e/D, are given in Figure 11.13.

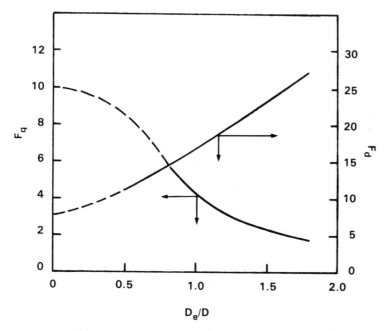

Figure 11.13 Interphase transfer factors in Eq. 11.35. (Davidson, Harrison, Darton La-Nauze, in *Chemical Reactor Theory: A Review,* ed. Lapidus and Amundson © 1977. Reprinted by permission of Prentice-Hall, Inc., Englewood Cliffs, NJ)

For the Kunii and Levenspiel model, K_{ce} is that given in Table 11.2, but they use the following for K_{bc}:

$$K_{bc} = 4.5\left(\frac{U_{mf}}{D_e}\right) + 5.85\left(\frac{D_G^{1/2}g^{1/4}}{D_e^{5/4}}\right) \tag{11.36}$$

For the volume fraction γ, they give:

$$\epsilon_b(\gamma_b + \gamma_c + \gamma_e) = (1 - \epsilon_{mf})(1 - \epsilon_b) \tag{11.37}$$

$$\gamma_c = (1 - \epsilon_{mf})\left[\frac{3U_{mf}/\epsilon_{mf}}{0.711(gD_e)^{1/2} - U_{mf}/\epsilon_{mf}} + \phi\right] \tag{11.38}$$

$$U_A \approx \frac{U_i - U_{mf}}{\epsilon_b} \tag{11.39}$$

where ϕ is estimated from Figure 11.14. The value of γ_b is usually set to zero since it is in the range of 0.001 to 0.01.

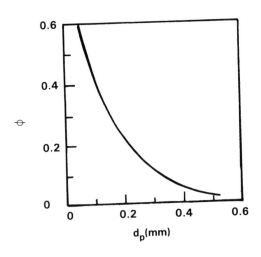

Figure 11.14 Value of ϕ in Eq. 11.38. (Kunii and Levenspiel 1969; reprinted from *Fluidization Engineering*, © Wiley, by permission of John Wiley & Sons, Inc.)

11–6 NONBUBBLING BEDS AT HIGH VELOCITIES

As indicated earlier, industrial beds of fine particles operating at high gas velocities may be in the turbulent regime or may operate as a fast fluidized-bed. For such a bed, no bubbles are present, and thus, the bed could be modeled as a simple CSTR, as a plug-flow reactor, or as a combination of both in which the gas phase is treated as plug-flow and the emulsion phase as well-mixed. Here, the latter approach taken by Weekman (1968) is considered for a riser cracker since the same approach can be applied to nonbubbling beds operating at high velocities.

Weekman's model is based on assumptions of plug-flow for the gas and of CSTR behavior for the emulsion phase with a CSTR internal age distribution of particles. In essence, this model treats the fluidized-bed as a PFR with a CSTR age distribution of the catalyst being deactivated in the bed. If the weight fraction of gas oil in the riser is f_m, the continuity equation is:

$$\frac{\partial}{\partial t}(\rho f_m) + \frac{\partial}{\partial z}(v_0 \rho_0 f_m) = -(1 - \epsilon) R_G(f_m, t) \tag{11.40}$$

where ρ_0 and v_0 are the density and velocity at the inlet, respectively. Here, the following relationship has been used:

$$v\rho = v_0 \rho_0 \tag{11.41}$$

The density can be related to the weight fraction by:

$$\rho = \frac{\rho_0}{f_m + m(1 - f_m)} \tag{11.42}$$

where m is the ratio of the product molecular weight to the reactant molecular weight. Combining Eqs. 11.40 and 11.42, and rendering the result dimensionless, one has:

$$\frac{m\rho_0}{t_m[f_m + m(1 - f_m)]^2}\frac{\partial f_m}{\partial\theta} + \frac{v_0\rho_0}{z_0}\frac{\partial f_m}{\partial x} = -(1 - \epsilon)R_G \qquad (11.43)$$

where $\theta = t/t_m$, $x = z/z_0$, and t_m is the catalyst residence time. If V_r is the reactor volume and ϵ the bed voidage, $v_0\rho_0/z_0 = F_0/V_r\epsilon$, F_0 being the mass rate of feed. Then, Eq. 11.43 can be rewritten as:

$$B\frac{\partial f_m}{\partial\theta} + \frac{\partial f_m}{\partial x} = -AR_G(f_m, \theta) \qquad (11.44)$$

where $B = \dfrac{m\rho_0\epsilon}{\bar{\rho}St_m[f_m + m(1 - f_m)]^2} = \dfrac{\text{oil transit time}}{\text{catalyst decay time }(t_m)}$ $\qquad (11.45a)$

$S = F_0/V_r\bar{\rho};$ $\bar{\rho} = $ liquid density of feed before the feed enters the riser $\qquad (11.45b)$

$A = (1 - \epsilon)/\bar{\rho}S \qquad (11.45c)$

Here, S is the space velocity based on the liquid feed, which vaporizes upon entering the riser. The catalyst decay time is much longer than the vapor residence time. Therefore, Eq. 11.44 can be approximated by:

$$\frac{\partial f_m}{\partial x} = -AR_G(f_m, \theta) \qquad (11.46)$$

An exponential decay model was used to represent the decay of activity due to coking, such that:

$$R_G = r_c\exp(-\lambda\theta) = kf(f_m)\exp(-\lambda\theta) \qquad (11.47)$$

where λ is a constant. The internal age distribution of the catalyst according to an ideal CSTR model (Levenspiel 1972) is:

$$I(\theta) = \exp(-\theta) \qquad (11.48)$$

Then, the average activity of the catalyst particles in the bed can be calculated from:

$$\bar{k} = k\int_0^\infty I(\theta)\exp(-\lambda\theta)d\theta$$
$$= k/(1 + \lambda) \qquad (11.49)$$

Hence,

$$\frac{df_m}{dx} = \frac{Ak}{1+\lambda} f(f_m)$$ (11.50)

Given the concentration dependence of the rate of the reaction $f(f_m)$, this equation can be integrated for the conversion. Second-order kinetics were used by Weekman (1968) for $f(f_m)$.

Summary

The state of fluidization is perhaps the most important factor in the design of a fluidized-bed. Particle size is a dominant factor in determining the state of fluidization. Thus, for industrial beds of fine particles belonging to group A in Figure 11.9, bubbling fluidization would be rarely found since turbulent and fast fluidization would allow better contact between the catalyst particles and the gas at much higher throughputs than bubbling fluidization could allow. For larger particles, however, bubbling fluidization would be used. For nonbubbling beds operating at high velocities, simple models, in which CSTR behavior for the emulsion phase and PFR behavior for the gas phase are assumed, seem to describe the conversion behavior adequately, as the studies of Pansing (1956) and Weekman (1968) indicate. For bubbling beds, however, multiphase treatment involving bubbles is required. The two-phase theory for bubbling beds is based on a single bubble with a fixed diameter and associated bubble properties that are assumed representative of all bubbles in the bed. While the two-phase models based on this theory account for gas bypassing, and the bubble properties can be estimated a priori, the model predictions are often in error. A better understanding of bubble properties is certainly warranted. The cell model, which allows the parameters to vary along the bed height, is an improvement over the two-phase model and can account for flow reversal in the emulsion phase. While the accuracy of the fluidized-bed reactor models, in particular the models for bubbling beds, is still insufficient for general use in the design and scale-up of the beds, these models do provide a frame within which each particular application can be formulated.

NOTATION

A	constant defined in Eq. 11.45c
C	concentration of key species
C_b	bubble-phase gas concentration
C_c	cloud-phase concentration; coke concentration
C_e	emulsion-phase gas concentration
C_h	C at the top of bed
C_i	inlet concentration

C_{jn}	concentration of gas in phase j ($j = b, c, e$) in the nth cell
C_{out}	outlet concentration
d_p	particle diameter
D	bed diameter
D_e	effective bubble diameter
D_G	molecular diffusivity
f	$r_c = kf(c)$
f_m	weight fraction of gas oil in feed
f_w	ratio of wake volume to bubble volume
F, F_o	volumetric flow rate, mass rate in Eq. 11.45b
F_d, F_g	constants in Eq. 11.35
g	acceleration due to gravity
G	mass rate per area
h	height from distributor plate
H	bed height occupied by fluidized particles
H_{mf}	H at incipient fluidization
k, k_f	intrinsic rate constants
k_e	elutriation constant defined in Eq. 11.7a
k_G	gas-phase film mass transfer coefficient in Eq. 11.8
k_r	rate constant in Eq. 11.19
K	film mass transfer coefficient in Eq. 11.1; constant defined in Eq. 11.13
k_{ij}	mass transfer coefficient from phase i to phase j
K_o	overall reaction-mass transfer coefficient defined by Eq. 11.19
m	ratio of product molecular weight to reactant molecular weight
n	number of cells as counted from bed inlet; moles of main reactant in Eq. 11.5
n_r	n at which flow reversal occurs in the emulsion phase
N	flux, oxygen concentration; last cell
N_b	number of bubbles in a cell
N_D	number of distributor holes
P	bulk pressure
ΔP_b	pressure drop across bed
P_{ce}, P_{bc}	quantities given by Eqs. 11.23 and 11.24
P_i	interfacial pressure
q	cross-flow rate in Eq. 11.8
Q	$q + k_G S$
R_G	global rate based on particle volume
R'_G	global rate based on pellet weight
S	FN_0/W in Eq. 11.3; space velocity defined by Eq. 11.45b; bed cross-sectional area
t	time
t_m	transit time of gas oil
U	superficial velocity
U_A	superficial absolute bubble velocity given by Eq. 11.27
U_e	average gas velocity in emulsion phase in Figure 11.9
U_{cr}	superficial velocity defined in Eq. 11.21

U_i	inlet superficial velocity
U_{jn}	superficial velocity of gas in phase j ($j = b, c, e$) in the nth cell
U_{mf}	incipient superficial velocity
U_t	terminal velocity
v	velocity of gas oil vapor
V_b	bubble volume
V_{jn}	volume of phase j in a cell ($j = b, c, e$) in the nth cell
w	catalyst weight
W	total weight of solids in bed
W_p	weight of solid particles of diameter d_p
x	conversion; z/z_0
z	distance from the bottom of the riser of catalytic cracker
z_0	height of the riser of catalytic cracker
Z	constant defined in Eq. 11.13

Greek Letters

α	$QH/U_A V_b$
β	$(U_i - U_{mf})/U_i$
γ	solid volume/bubble volume
δ_j	volume fraction of bed occupied by phase j
ϵ	bed voidage
ϵ_j	volume fraction of gas in phase j ($j = b, c, e$)
ϵ_{mf}	ϵ at incipient fluidization
λ	constant in Eq. 11.47
μ	fluid viscosity
ρ	fluid density
ρ_p	bulk density of particulate phase
ρ_s	solid density
ϕ	constant in Eq. 11.38
θ	t/t_m

Subscripts

b	bubble-phase
c	cloud-phase
e	emulsion-phase
i	inlet or feed
n	cell number as counted from bottom

PROBLEMS

11.1. Plot incipient and terminal velocities as a function of particle size for particle sizes in the range of 10 to 300 microns. Use the properties of air for the fluid ($\rho_s = 0.5$ g/cm³). What can you conclude regarding the range of velocities for fluidization ($U_{mf} < U < U_t$) as affected by particle size? Assume $\epsilon_{mf} = 0.4$.

11.2. Since any particle whose terminal velocity is smaller than the operating velocity is elutriated, the elutriation constant in Eq. 11.7 depends on particle size. When particles of wide size distribution $P_0(d_p)$ are fed continuously into a bed as shown in Figure 11.2P, fine particles are entrained by the gas while the remainder are discharged through an overflow pipe. Suppose that the bed weight W, the feed properties, F_f (mass rate) and $P_f(d_p)$, and elutriation constant $k_e(d_p)$ are known and that one wants to find the exit conditions: F_1, F_2, $P_1(d_p)$, $P_2(d_p)$. For steady state operation, the definition of k_e (Eq. 11.7a) gives:

$$k_e = \frac{F_2 P_2(d_p)}{S P_b(d_p)} \qquad (p11.1)$$

where the size distribution in the bed P_b is the same as P_1 under the assumption of perfect mixing.

a. Show that

$$F_1 P_1(d_p) = \frac{F_0 P_0}{1 + S k_e / F_1} \qquad (p11.2)$$

Use the fact that $F_0 P_0 = F_1 P_1 + F_2 P_2$

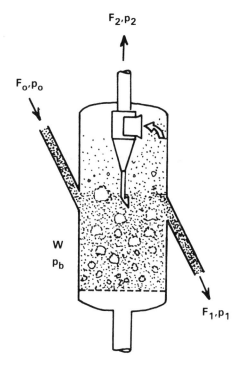

F_2, P_2

F_0, P_0

W
P_b

F_1, P_1

Figure 11.2P Fluidized-bed at steady state for the determination of size distributions. (Kunii and Levenspiel 1969; reprinted from *Fluidization Engineering,* © Wiley, by permission of John Wiley & Sons, Inc.)

b. Noting that $\int_{size}P(R)dR = 1$, integrate Eq. p11.2 to obtain

$$\frac{F_1}{F_0} = 1 - \frac{F_2}{F_0} = \int_{(dp)min}^{(dp)\,max} \frac{P_0(R)dR}{1 + Sk_e/F_1} \qquad (p11.3)$$

c. Write procedures for determining $P_1(d_p)$ and $P_2(d_p)$. Refer to Kunii and Levenspiel (1969).

11.3. For the simple Langmiur-Hinshelwood kinetics of $r_c = kC/(1 + K_eC)$, obtain the conversion using the Davidson-Harrison model. Only the parameters K_e, $Z(= \beta e^{-\alpha})$ and $K[= Hk(1 - \epsilon)/U_i]$ should appear in the expression. Plot C_h/C_i as a function of Z and K for $K_e = 10^5$ and $K_e = 10^4$. Use C_i of 5×10^{-5} g mol/cm³.

11.4. Consider a deactivation-affected reaction of first-order. Since the particle size is very small, uniform deactivation can be assumed. Further, assume that a pseudo-steady state is applicable with respect to deactivation.

a. Show that the effective rate constant at any given time can be expressed as

$$k_f = k_0\exp(-\gamma t)$$

and therefore, $R_G = k_0Cexp(-\gamma t)$. Here, γ is a constant depending on concentrations and temperature, and k_0 is the rate constant for fresh catalyst. At steady state with a continuous flow of catalyst particles, γ remains constant.

b. Show that the average activity of a catalyst (average effective rate constant) is given by:

$$\bar{k}_f = k/(1 + \gamma)$$

if one assumes complete mixing for the internal age distribution of the catalyst particles.

c. Use the Davidson-Harrison model to obtain an expression similar to Eq. 11.13. Compare the C_h/C_i ratio between deactivating and nondeactivating catalysts for γ of 5.

11.5. Write the equations necessary for the calculation of conversion according to the Kunii-Levenspiel model for a second-order reaction, for which $R_G = r_c = k_rC^2$. Note that an overall reaction-mass transfer coefficient such as the one in Eq. 11.19 cannot be expressed explicitly.

11.6. Use the data given in Table 11.1 to obtain the results shown in Figure 11.12 based on the cell model.

REFERENCES

Davidson, J.F. and D. Harrison, *Fluidized Particles,* Cambridge University Press, London (1963).

Davidson, J.F., D. Harrison, R.C. Darton and R.D. LaNauze, in *Chemical Reactor Theory,* Chap. 10, pp. 625, 630, eds. Lapida and Amundson, Prentice-Hall, Englewood Cliffs, N.J. (1977).

DeJong, J.A.H. and J.F. Nomden, Power Technol., *9,* 91 (1974).

Froment, G.F., and K.B. Bischoff, *Chemical Reactor Analysis and Design,* p. 664, Wiley, New York (1979).

Fryer, C. and O.E. Potter, Ind. Eng. Chem. Fund., *11,* 338 (1972).

Fryer, C. and O.E. Potter, International Fluidization Conference, Preprint III-1, Pacific Grove (1975).

Geldart, D., Powder Technol., *7,* 285 (1973).

Horio, M. and C.Y. Wen, AIChE Symp. Ser., *73,* No. 161, p 9 (1977).

Kato, K. and C.Y. Wen, Chem. Eng. Sci. *24,* 1351 (1969).

Kehoe, P.W.K. and J.F. Davidson, Inst. Chem. Eng. (London) Symp. Ser., *33,* 97 (1971).

Kunii, D. and O. Levenspiel, *Fluidization Engineering,* Wiley, New York (1969).

Kunii, D., K. Yoshida and I. Hiraki, Proc. Int. Symp. Fluidization, p 243, Netherlands Univ. Press (1967).

Leva, M., *Fluidization,* McGraw-Hill, New York (1960).

Levenspiel, O., *Chemical Reaction Engineering,* 2nd ed., Wiley, New York (1972).

Massimilla, L., AIChE Symp. Ser., *69,* No. 128, 11 (1973).

Mori, S. and C.Y. Wen, AIChE J., *21,* 109 (1975).

Murray, J.D., J. Fluid Mech., *21,* 465 (1965).

Pansing, W.F., AIChE J., *2,* 71 (1956).

Partridge, B.A. and P.N. Rowe, Trans. Inst. Chem. Eng. (London), *44,* T335, T349 (1966).

Peters, M.H., L.S. Fan and T.L. Sweeney, Chem. Eng. Sci., *37,* 553 (1982).

Pyle, D.L., Adv. in Chem. Ser., *109,* 106, Am. Chem. Soc., Washington (1972).

Reh, L., Chem. Eng. Prog., *67* (2), 58 (1971).

Richardson, J.F., in *Fluidization,* p 25, eds. Davidson and Harrison, Academic Press, London (1971).

Rowe, P.N., in *Fluidization,* eds. Davidson and Harrison, Academic Press, London (1971).

Sit, S.P. and J.R. Grace, Chem. Eng. Sci., *36,* 327 (1981).

Toomey, R.D. and H.F. Johnstone, Chem. Eng. Prog., *48,* 220 (1952).

Weekman, V.W., Jr., Ind. Eng. Chem. Proc. Des. Dev., *7,* 90 (1968).

Weisz, P.B. and R.D. Goodwin, J. Catalysis, *6,* 226 (1966).

Yacono, C., P.N. Rowe and H. Angelino, Chem. Eng. Sci., *34,* 171 (1979).

Yerushalmi, J., D.H. Turner and A.M. Squires, Ind. Eng. Chem. Proc. Des. Dev., *15,* 47 (1976).

Zenz, F.A. and D.F. Othmer, *Fluidization and Fluid-Particle Systems,* Reinhold Pub. Co., New York (1960).

CHAPTER 12

Multiphase Reactors

12-1 INTRODUCTION

Gas-liquid reactions catalyzed by a solid catalyst are carried out in multiphase reactors. These reactors are usually used when the gaseous reactant is too volatile to liquefy or when the liquid reactant is too nonvolatile to vaporize. Typical examples are hydrogenation, desulfurization, and hydrocracking of petroleum feeds. Many different reactor configurations can result depending on the mode of feeding and the manner in which the catalyst is placed in the reactor. The two most common multiphase reactors are slurry and trickle-bed reactors, which have been described in some detail in Chapter 7. Since other reactor configurations can be analyzed in a similar manner as for slurry and trickle-bed reactors, attention here will be focused on these reactors.

There are some similarities between fluidized-beds and slurry reactors. However, there is a distinct difference to be recognized with respect to the relative movement between particles and fluid. Unlike the fluidized-bed, there is little relative movement between particles and fluid in a slurry reactor, even though the liquid is agitated mechanically. The particles tend to move with the liquid, which considerably simplifies the modeling. Because of the high heat capacity and relatively high thermal conductivity of the liquid, uniform temperature control is easy to achieve for highly exothermic reactions. Therefore, slurry reactors should be considered as an alternative to fixed-beds for highly exothermic reactions. Because of the very small size of the catalyst particles, diffusional effects are often negligible, even though the intraparticle effective diffusivity in liquid-filled pores is of the order of 10^{-5} cm^2/s. The reactor can be run in a batch or continuous mode with respect to the catalyst. The catalyst particles are retained in the vessel in the batch mode. The most serious disadvantage of slurry reactors is the difficulty in handling the catalyst particles in the vessel. There is a problem of retaining the catalyst when operating in a batch mode and of separating the catalyst from the product when operating in a continuous mode. For instance, screens placed in the outlet lines tend to clog.

Gaseous and liquid reactants can be fed to a fixed-bed packed with catalyst pellets cocurrently or countercurrently and in the case of cocurrent feeding, the flow may be either upward or downward. Typical industrial practice is to feed both reactants downward in a cocurrent manner. A fixed-bed with this type of

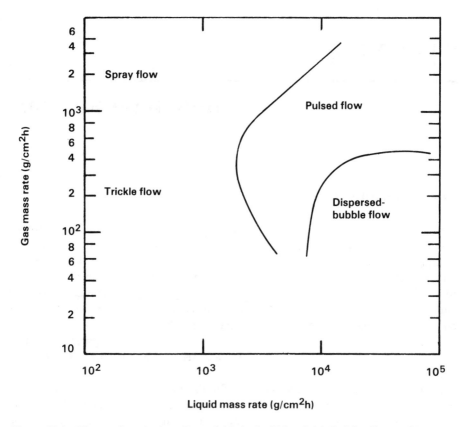

Figure 12.1 Flow regimes in downflow trickle-beds. (After J. M. Smith, *Chemical Engineering Kinetics,* © 1980; with permission of McGraw-Hill Book Co., NY.)

feeding is called a trickle-bed. The advantages are that there is no limitation on the flow rates imposed by flooding and the liquid is much more thinly and evenly distributed than with upward flow. One important characteristic of a trickle-bed is that the gas-phase is continuous throughout the bed. Downflow packed beds can have different flow regimes depending on the respective flow rates of gas and liquid as shown in Figure 12.1 (Smith 1981). The trickle-flow regime results when both flow rates are relatively low, in which the gas phase is continuous and the liquid phase is dispersed. Increasing the gas flow rate leads to pulsed flow. For high liquid throughputs, the liquid phase may be continuous and the gas phase dispersed: this is called bubble flow. The lines separating different flow regimes are approximate boundaries, and thus, should not be taken too rigorously.

With the global rates developed in Chapter 7 for slurry and trickle-bed reactors, it is now possible to proceed with the design of these reactors. Reactions of interest are:

$$A(g) + bB(l) \xrightarrow{\text{catalyst}} \text{products} \tag{12.1}$$

for which the intrinsic rate based on the species A is represented by:

$$r_A = kf(C_A, C_B)$$

Here, C_A is the liquid-phase concentration of species A, and C_B is the liquid-phase concentration of species B. The basis for r_A is unit volume of the catalyst.

12–2   SLURRY REACTORS

With good agitation and little relative movement between particles and liquid, it is a good assumption that the liquid phase is well-mixed. The situation with respect to the gas phase is less certain. With very strong agitation, the gas bubbles may coalesce and redisperse, and may have a residence time approaching that of a completely mixed fluid. With less agitation, the bubbles approach plug-flow as they rise through the liquid. Fortunately, it is only for cases of intermediate solubility of the gaseous reactant in the liquid that the residence time distribution of the bubbles is important. It has been shown (Niiyana and Smith 1976) that for either very soluble or slightly soluble gases, the residence time distribution of the gas bubbles in the liquid has no effect on reactor performance. Since most applications of importance involve either very soluble or slightly soluble gases, the gas phase will be treated as being in plug-flow.

For the slurry reactor shown in Figure 12.2, a mass balance for the gas phase (bubbles) can be written as:

$$\epsilon_b v_b \frac{d(C_{Ag})}{dz} = -\epsilon_s (R_G)_A = k_L a_L (C_{Ag}/H - C_{AL}) \tag{12.2}$$

where C_{Ag} = bulk concentration of species A in the gas phase
$\quad v_b$ = bubble velocity
$\quad k_L$ = liquid-side mass transfer coefficient
$\quad a_L$ = interfacial surface area of bubbles per unit liquid volume
$\quad \epsilon_b$ = fraction of the liquid volume occupied by bubbles
$\quad \epsilon_s$ = fraction of the liquid volume occupied by catalyst particles
$\quad C_{AL}$ = bulk concentration of species A in liquid phase

Here, it has been assumed that the resistance to mass transfer is dominated by the liquid-side of the interface so that the gas-side interfacial concentration C_{ig} is equal to the bulk concentration C_g for the gaseous species A. It has also been assumed that Henry's law holds so that $C_{Ag} = HC_{eq}$ where H is the Henry's law constant and C_{eq} is the liquid-phase concentration in equilibrium with C_{Ag}. For slightly soluble gases, the bubble velocity v_b can be assumed constant, as implied by Eq. 12.2. For very soluble gases, v_b is not constant, but it will be

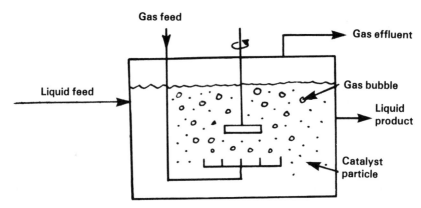

Figure 12.2 Slurry reactor. (After J. M. Smith, *Chemical Engineering Kinetics*, © 1980; with permission of McGraw-Hill Book Co., NY.)

seen that this balance equation is not necessary. Mass balances for the well-mixed liquid phase are:

$$Q_l(C_{BL_{in}} - C_{BL}) = b(R_G)_A[C_{AL}, C_{BL}]V\epsilon_s \tag{12.3}$$

$$\frac{V}{Z}\int_0^Z k_L a_L(C_{Ag}/H - C_{AL})dz = Q_l C_{AL} + (R_G)_A[C_{AL}, C_{BL}]V\epsilon_s \tag{12.4}$$

where V = liquid volume of the reactor
  Q_l = volumetric flow rate of liquid
  C_{BL} = bulk concentration of liquid reactant B
  b = stoichiometric coefficient in Eq. 12.1
  $C_{BL_{in}}$ = C_{BL} at the reactor inlet
  Z = length of the reactor

Equation 12.3 is the mass balance of liquid reactant B, which is well-mixed. In Eq. 12.4, the inlet concentration of A in the liquid phase is zero, and therefore, the total mass of A transferred from the gas to the liquid phase should be equal to the sum of species A leaving the reactor and the amount of this species consumed by the reaction. Since the liquid phase is well-mixed, both C_{AL} and C_{BL} are uniform throughout the liquid phase. Equations 12.2 through 12.4 completely describe the slurry reactor provided that the global rate is expressed in terms of bulk-liquid concentrations, i.e., $(R_G)_A[C_{AL}, C_{BL}]$.

Two types of reactions can be considered: one in which the removal of species A in the gas stream is of major interest, as in the removal of pollutants, and the other in which the treatment of liquid reactant B is of major interest, as in the desulfurization of petroleum feeds. For the former, only Eq. 12.2 is of concern. The global rate developed in Chapter 7 can be used to rewrite the equation:

$$\epsilon_b v_b \frac{dC_{Ag}}{dz} = -\left(\frac{1}{1/k_c a_c + \epsilon_s/k_L a_L}\right)(C_{Ag}/H - C_{As}) \tag{12.5}$$

where Eq. 7.14 has been used for R_G. Here, a_c is the external area of a catalyst particle per unit volume of catalyst. The pellet surface concentration of species A, C_{As}, is determined from:

$$\left(\frac{1}{1/k_c a_c + \epsilon_s/k_L a_L}\right)(C_{Ag}/H - C_{As}) = (R_G)_A = kf(C_{As}, C_{Bs}) \qquad (12.6)$$

$$(k_c)_B a_c (C_{BL} - C_{Bs}) = b(R_G)_A = bk(C_{As}, C_{Bs}) \qquad (12.7)$$

The assumptions involved are that the reaction is diffusion-free and that the concentration of liquid reactant B remains constant. Given the intrinsic kinetics, Eqs. 12.5 through 12.7 can be used to design the reactor. For a first-order reaction $(r_A = kC_A)$, Eq. 12.6 can be solved for C_{As}, which upon inserting into Eq. 12.5 gives:

$$\epsilon_b v_b \frac{dC_{Ag}}{dz} = -\left(\frac{1}{1/k_c a_c + \epsilon_s/k_L a_L + 1/k}\right)\frac{C_{Ag}}{H} \qquad (12.8)$$

When the liquid reactant is the species of primary interest, Eq. 12.2 need not be included in the design if the gaseous species is present in excess and the liquid is saturated. For instance, hydrogen used for the treatment of petroleum feeds is usually present in excess and readily soluble in the petroleum. In such a case, only Eq. 12.3 is involved in the design, which can be rewritten as:

$$Q_l(C_{BL_{in}} - C_{BL}) = k_c a_c (C_{BL} - C_{Bs})V\epsilon_s \qquad (12.9)$$

The particle surface concentration C_{Bs} is obtained from:

$$(k_c)_B a_c (C_{BL} - C_{Bs}) = bkf(C_{As}, C_{Bs}) \qquad (12.10)$$

$$(k_c)_A a_c (C_{Ag}/H - C_{As}) = kf(C_{As}, C_{Bs}) \qquad (12.11)$$

It is seen that Eqs. 12.9 through 12.11 are sufficient to design the reactor.

The general design problem posed by Eqs. 12.2 through 12.4 can be solved without difficulty. The solution of Eq. 12.2 yields:

$$(C_{Ag})_o = (C_{Ag})_{in} e^{-A} + C_{AL}(1 - e^{-A}) \qquad (12.12)$$

where $A = k_L a_L / H\epsilon_b v_b$

$(C_{Ag})_o = C_{Ag}$ at reactor outlet

$(C_{Ag})_{in} = C_{Ag}$ at reactor inlet

This equation has to be solved together with the following algebraic equations:

$$Q_L(C_{BL_{in}} - C_{BL}) = bkf(C_{As}, C_{Bs})V\epsilon_s \qquad (12.13)$$

$$\frac{V}{Z} \epsilon_b v_b [(C_{Ag})_{in} - (C_{Ag})_o] = Q_l C_{AL} + kf(C_{As}, C_{Bs}) V \epsilon_s \tag{12.14}$$

$$(k_c)_B a_c (C_{BL} - C_{Bs}) = bkf(C_{As}, C_{Bs}) \tag{12.15}$$

$$(k_c)_A a_c (C_{AL} - C_{As}) = kf(C_{As}, C_{Bs}) \tag{12.16}$$

Equations 12.13 and 12.14 follow from Eqs. 12.3 and 12.4. The integral in Eq. 12.4 can be evaluated using Eq. 12.2, as given in Eq. 12.14.

Take as an example a simple reaction for which $r_A = kf(C_A, C_B) = kC_A C_B$. Assume, for brevity, that the mass transfer coefficients are the same, i.e., $(k_c)_A = (k_c)_B$. It follows from Eqs. 12.15 and 12.16 that:

$$C_{BL} - C_{Bs} = b(C_{AL} - C_{As}) \tag{12.17}$$

This, upon inserting into Eq. 12.15, yields:

$$(k_c)_B a_c (C_{AL} - C_{As}) = kC_{As}\{C_{BL} - b(C_{AL} - C_{As})\} \tag{12.18}$$

This can be solved for C_{As} to give:

$$C_{As} = (bkC_{AL} - kC_{BL} - (k_c)_B a_c + [(bkC_{AL} - kC_{BL} - (k_c)_B a_c)^2 + 4bk(k_c)_B a_c C_{AL}]^{1/2} / 2bk \tag{12.19}$$

It has been assumed that the root of Eq. 12.18, given by Eq. 12.19, yields C_{As}, which is positive. Now that C_{As} and C_{Bs} are expressed in terms of C_{AL} and C_{BL}, these can be used in Eqs. 12.13 and 12.14 to obtain C_{BL} and $(C_{Ag})_o$ with the aid of Eq. 12.12. If the problem is to specify the liquid volume (reactor volume) V for a given desired conversion, say 60% conversion of B, Eq. 12.13 yields:

$$0.6 \, Q_l C_{BLin} = bkC_{As} C_{Bs} V \epsilon_s$$

$$= \frac{V\epsilon_s (k_c)_B a_c}{2k} \{bkC_{AL} + 0.4C_{BLin} + (k_c)_B a_c$$

$$+ [(bkC_{AL} - 0.4kC_{BLin} - (k_c)_B a_c)^2 + 4bk(k_c)_B a_c C_{AL}]^{1/2}\}$$

This can be solved for C_{AL}, which upon inserting into Eq. 12.14 gives a relationship containing V as the only unknown when Eq. 12.12 is used for $(C_{Ag})_o$. It is seen that even for relatively simple kinetics, the design requires trial and error calculations for this general case.

It has been assumed that the intraparticle resistances are negligible and that the reactor is isothermal. Even with a very small particle size (~100 micron), the diffusion resistance can become important for some fast reactions since the effective diffusivity in liquid-filled pores is of the order of 10^{-5} cm²/sec. In such cases, the rate constant k should be replaced by the effective rate constant $k\eta$, where η is the internal effectiveness factor.

12-3 TRICKLE-BED REACTORS

The design of trickle-beds is similar to that of fixed-beds, although it is more complex. In trickle-beds, reactions take place in liquid-filled pores, and therefore, the presence of both gas and liquid phases increases the transport resistances. Many complex situations can arise depending on the nature of the reaction, the distribution of liquid flow, and wetting. For extremely exothermic reactions, the pellet is not necessarily isothermal. Further, vaporization of the liquid reactant can occur. The gas can be assumed to be uniformly distributed across the diameter, but the liquid tends to flow toward the reactor wall. A reactor-to-pellet diameter ratio of 18 or more, however, ensures uniform liquid distribution (Herskowitz and Smith 1978). When the liquid velocity is relatively low, the catalyst pellets are not completely wetted with liquid and partial wetting occurs. Further, axial dispersion cannot be neglected at low velocities. Trickle-beds play an important role in petroleum processing and several reviews are available by Satterfield (1975), Goto et al. (1977), Hofmann (1978), and Herskowitz and Smith (1983). A book by Shah (1979) is also available.

In the first part of this section, attention will be restricted to the simple conditions of uniform gas and liquid distributions, a nonvolatile liquid, trickling flow, isothermal pellets, and complete wetting of the pellets by the liquid. Nonisothermal pellets will also be treated for the case of first-order reactions. In the second part, a more complex case at low liquid velocities is examined.

12-3-1 Simple Trickle-Beds

Consider a simple trickle-bed for which one can assume uniform gas and liquid distributions, a nonvolatile liquid, trickling flow, isothermal pellets, and complete wetting. For an isothermal reactor with a uniform distribution of the gas and liquid phases, there will be no radial gradients of concentration. Although axial dispersion is more important in trickle-beds than in fixed-beds because of relatively low fluid velocities, we will assume that the dispersion is negligible. Under the assumptions, one can write for the gas phase (refer to Figure 12.3):

$$v_g \frac{dC_{Ag}}{dz} = -(k_L a_L)_A (C_{Ag}/\mathrm{H} - C_{AL}) \tag{12.20}$$

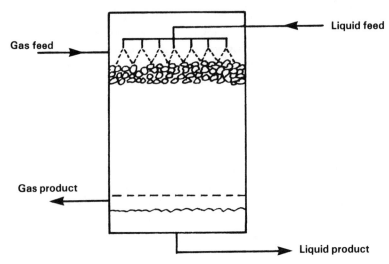

Figure 12.3 Trickle-bed reactor. (After J. M. Smith, *Chemical Engineering Kinetics*, ©
1980; with permission of McGraw-Hill Book Co., NY.)

where v_g is the superficial velocity of the gas. For reactant A in the liquid phase,
the conservation equation is:

$$v_l \frac{dC_{AL}}{dz} = -(k_c a_c)_A (C_{AL} - C_{As}) + (k_L a_L)_A (C_{Ag}/H - C_{AL}) \qquad (12.21)$$

where v_l is the superficial velocity of the liquid. Here, the gaseous species A in
the liquid phase is replenished by gas-liquid mass transfer and depleted by reaction
through liquid-solid mass transfer. For the liquid reactant B, the conservation
equation is:

$$v_l \frac{dC_{BL}}{dz} = -(k_c a_c)_B (C_{BL} - C_{Bs}) \qquad (12.22)$$

Boundary conditions are the feed concentrations. The relationships between the
surface and bulk-liquid concentrations are obtained by equating the mass transfer
rate at the solid interface to the global rate:

$$(k_c a_c)_A (C_{AL} - C_{AS}) = (R_G)_A = k \eta f(C_{AS}, C_{Bs}) \qquad (12.23)$$

$$(k_c a_c)_B (C_{BL} - C_{Bs}) = b(R_G)_A = bk \eta f(C_{AS}, C_{BS}) \qquad (12.24)$$

where η is the internal effectiveness factor excluding the effects of external resistances. Given the intrinsic rate $kf(C_A, C_B)$, Eqs. 12.20 through 12.24 can be solved
for the design. Since only one species was involved in Chapter 7 for the effectiveness
factor, here the effectiveness factor for two species is examined. The pellet conservation equations for species A and B, when combined, give:

$$bD_A \frac{d^2 C_A}{dx^2} = D_B \frac{d^2 C_B}{dx^2} \qquad (12.25)$$

where C_A and C_B are the concentrations of A and B in the pellet, and D_A and D_B are the effective diffusivities of A and B in the liquid-filled pores. Integration of this equation using the usual boundary conditions yields:

$$C_B = C_{Bs} - \left(\frac{bD_A}{D_B}\right)(C_{As} - C_A) \qquad (12.26)$$

Therefore, the intrinsic rate can be written as:

$$r_A = kf(C_A, C_B) = kf\left[C_A, C_{Bs} - \frac{bD_A}{D_B}(C_{As} - C_A) \right] \qquad (12.27)$$

$$\equiv kf_1(C_A)$$

It follows that a generalized effectiveness factor similar to Eq. 7.17 in Chapter 7 can be written as:

$$\eta(C_{As}, C_{Bs}) = \frac{1}{Lf_1(C_{As}, C_{Bs})}\left[(2D_A k) \int_{C_A^L}^{C_{As}} f_1(C_A) dC_A \right]^{1/2} \qquad (12.28)$$

where C_A^L is the pellet center concentration of species A, and f_1 is defined in Eq. 12.27. As in Chapter 10, the pellet center concentration C_A^L can be approximated by $C_{As}/\cosh \phi_G$, where ϕ_G is the generalized modulus. Equations 12.23, 12.24, and 12.28 can be solved for C_{As} and C_{Bs} in terms of C_{AL} and C_{BL}. Then the balances of Eqs. 12.20 through 12.22 are solely in terms of C_{Ag}, C_{AL}, and C_{BL}, which can readily be solved. It should be recognized in Eqs. 12.25 through 12.27 that the reaction stoichiometry is not sufficient to relate C_{Bs} to C_{As} or C_{AL} to C_{BL} since C_A is determined not only by the reaction stoichiometry but also by the mass transport rate.

Take as an example a reaction which is first-order with respect to both A and B: $r_A = kf(C_A, C_B) = kC_A C_B$. The intrinsic rate can be expressed (Eq. 12.27) as:

$$r_A = kf_1(C_A) = kC_A\left[C_{Bs} - \frac{bD_A}{D_B}(C_{As} - C_A) \right]$$

If it is assumed that the reaction is diffusion-limited so that $C_A^L = 0$, Eq. 12.28 becomes:

$$\eta(C_{As}, C_{Bs}) = \frac{\left[(2D_A k)\left(\frac{1}{2}C_{Bs} - \frac{1}{6}\frac{bD_A}{D_B}C_{As}\right) \right]^{1/2}}{LkC_{Bs}}$$

Then, Eqs. 12.23 and 12.24, for instance, become:

$$(k_c a_c)_A (C_{AL} - C_{As}) = \frac{C_{As}}{L} \left[2D_A k \left(\frac{1}{2} C_{Bs} - \frac{bD_A}{\partial D_B} C_{As} \right) \right]^{1/2}$$

$$(k_c a_c)_B (C_{BL} - C_{Bs}) = \frac{bC_{As}}{L} \left[2D_A k \left(\frac{1}{2} C_{Bs} - \frac{bD_A}{6D_B} C_{As} \right) \right]^{1/2}$$

These two equations can be solved with Eqs. 12.20 through 12.22. It is seen that even for the simplest form of rate expression, a numerical method has to be used. Considerable simplification results when the rate of the reaction depends only on species A. In such cases, the global rate developed in Chapter 7 can be used directly for the design.

As indicated earlier, the pellet can be nonisothermal for some reactions such as hydrocracking. Here, a simple case in which the gaseous reactant is present in excess and the liquid reactant is saturated with the gaseous reactant will be considered. Under these conditions, the equilibrium concentration $C_{eq} (= C_{Ag}/H)$ is the same everywhere in the reactor, and C_{AL} is also constant. Further, it is assumed that the reaction is first-order with respect to the liquid reactant. Then the rate of the reaction is represented by:

$$\begin{aligned} r_B &= k g_1(C_{AL}) C_B \\ &= k_a(T, C_{AL}) C_B \end{aligned} \tag{12.29}$$

where the apparent intrinsic rate constant k_a is now constant with respect to C_{AL} since $g_1(C_{AL})$ is constant throughout the reactor. Thus, only the conservation equations concerning species B are necessary for the design. For an adiabatic reactor, these are:

$$v_l \frac{dC_{BL}}{dz} = -(R_G)_B \tag{12.30}$$

$$(\rho C_p)_l v_l \frac{dT_L}{dz} = (-\Delta H)(R_G)_B - (Ua_L)(T_L - T_g) \tag{12.31}$$

$$(\rho C_p)_g v_g \frac{dT_g}{dz} = (Ua_L)(T_L - T_g) \tag{12.32}$$

where T_L is the bulk-liquid temperature, T_g the bulk-gas temperature, and U the overall gas-liquid heat transfer coefficient. The approximate global rate developed in Chapter 7 gives:

$$(R_G)_B = \frac{k_a(T_L) C_{BL}}{\phi(1 + \phi/B_{ML})} \left\{ 1 + \frac{\beta\gamma(1/3 + \phi/B_{hL})}{2(1 + \phi/B_{mL})^2} \right\} \tag{12.33}$$

where $\phi = L(k_a/D_B)^{1/2} > 3$
$\quad B_{mL} = L(k_c)_B/D_B$
$\quad B_{hL} = Lh/\lambda$
$\quad\quad \beta = (-\Delta H)D_B C_{BL}/\lambda T_L$
$\quad\quad \gamma = E_a/R_g T_L$

The three balance equations (Eqs. 12.30 through 12.32), need to be solved for the design with the aid of Eq. 12.33 for the global rate. While relatively simple results have been obtained for first-order reactions, full pellet-side conservation equations have to be solved for rate expressions more complex than first-order.

12–3–2   Trickle-Beds at Low Liquid Velocities

As indicated earlier, the catalyst pellets are not necessarily wetted completely when the liquid velocity is low. According to the literature data summarized by Sicardi and co-workers (1980) partial wetting can occur when the liquid mass flow rate is less than approximately 10 kg/m²s. Their results show that partial wetting always occurs when the liquid rate is less than 2 kg/m²s. Although internal partial wetting in which the pores are not completely filled with liquid can occur, the term partial wetting normally refers to the condition where the pores are full, but the external surface of the catalyst is only partially covered by liquid. At low liquid velocities, dispersion in the liquid phase cannot be neglected. Thus, even with the assumptions of a nonvolatile liquid, isothermal pellets, trickling flow, and uniform gas and liquid distributions, one has to deal with axial dispersion and partial wetting at low liquid velocities.

The fraction of the pellet surface covered by liquid, q, which is sometimes termed the wetting efficiency, has been correlated by Mills and Dudukovic (1981) as follows:

$$q = \tanh\left\{0.064\ \mathrm{Re}_l^{0.033}\ \mathrm{Fr}_l^{0.195}\ \mathrm{We}_l^{-0.171}\left(\frac{a_t d_p^2}{\epsilon_B^2}\right)^{-0.0615}\right\} \qquad (12.34)$$

where $\mathrm{Re}_l = \dfrac{v_l d_p \rho_l}{\mu_l}$

$\mathrm{Fr}_l = \dfrac{a_t L_m^2}{\rho_{Lg}^2 g} \qquad\qquad\qquad\qquad (12.35)$

$\mathrm{We}_l = \dfrac{L_m^2}{\delta_l \rho_l a_t}$

Here, d_p is the pellet diameter, ρ_l the liquid density, μ_L the liquid viscosity, a_t the external surface area of the pellet per unit volume of the bed, L_m the mass superficial velocity per unit area, g the acceleration due to gravity, δ_l the liquid surface tension, and ϵ_B the bed porosity. This correlation is based on very few data points and is less accurate at lower liquid velocities. As indicated in Chapter 7, the usual range of q is between 0.6 and 1.0. An exact value of q is not often critical in assessing the effect of the wetting efficiency on the overall effectiveness factor (Mills and Dudukovic 1979) unless the mass transfer resistances are significant. This can be readily seen from the following approximate relationship for the overall effectiveness factor for first order kinetics given in Chapter 7:

$$\eta_o = q\eta_L + (1 - q)\eta_G \qquad (7.24)$$

where the effectiveness factors for a pellet completely covered by liquid (η_L) and by gas (η_G) are given by

$$\eta_L = \frac{\tanh \phi/\phi}{1 + \phi \tanh \phi/B_{mL}} \qquad (7.25)$$

$$\eta_G = \frac{\tanh \phi/\phi}{1 + \phi \tanh \phi/HB_{mG}} \qquad (7.26)$$

If the mass transfer resistances are small such that B_{mL} and B_{mG} are large, η_L is approximately the same as η_G, and thus, η_o is independent of q as apparent from Eq. 7.24. However, an accurate value of q would be required for the overall effectiveness factor if the mass transfer resistances are significant. In light of the fact that B_{mL} is usually small, although HB_{mG} is large, better correlations than Eq. 12.34 are necessary to describe a trickle-bed at low liquid velocities. In general, partial wetting can be detrimental or beneficial depending on which phase contains the limiting reactants (Herskowitz and Smith 1983). If the limiting reactant is liquid as in hydroprocessing reactions, partial wetting reduces the rate of the reaction due to the decrease in liquid-solid mass transfer area. If the gaseous reactant is the limiting reactant as in hydrogenation reactions, partial wetting enhances the rate of the reaction because of the absence of gas-liquid mass transfer resistance for the nonwetted part of the pellets. This can readily be seen from Eqs. 7.24 through 7.26 above since $\eta_G > \eta_L$ ($B_{mL} < HB_{mG}$), and thus, the overall effectiveness factor increases with decreasing q.

The approximate overall effectiveness of Eq. 7.24 above can be used for reactor design for first-order reactions. For nonlinear kinetics, however, it is difficult to obtain an analytical expression for the overall effectiveness factor that is accurate for partially wetted pellets. Even when a relatively accurate effectiveness factor is available, the presence of different bulk and interfacial concentrations in both phases reduces the utility of the overall effectiveness factor except for limiting cases. Therefore, an internal effectiveness factor applicable to both wetted and nonwetted parts of the pellet is used here instead. For the low liquid velocities being considered, only dispersion in the liquid phase need be considered, and the gas phase may be treated as though it were in plug-flow (Levec and Smith 1976).

In fact, correlations for Peclet numbers required for reactor modeling are available almost exclusively for the liquid phase. Buffham and Rathor (1978) give the following correlation:

$$Pe_l = 0.45(Fr_l')^{0.27} \tag{12.36}$$

where

$$Pe_l = \frac{v_l d_p}{h_e E}; \qquad Fr_l' = \frac{v_l^2}{h_e^2 d_p g} \tag{12.37}$$

Here, E is the axial dispersion coefficient, and h_e is the dimensionless external holdup, which is a sum of the dynamic holdup and the external static holdup. Correlations for the holdup can be found in Shah (1979) and Gianetto et al. (1978).

Consider the design of a trickle-bed operating at low liquid velocities for the reaction given by Eq. 12.1. For the gas phase, one has:

$$v_g \frac{dC_{Ag}}{dz} = -(k_L a_L)_A (C_{Ag}/H - C_{AL})q - (k_g a_c)_A (C_{Ag} - C_{Ags})(1-q) \tag{12.38}$$

since a fraction, q, of the pellet surface is covered by liquid and a fraction, $(1-q)$, is covered by gas. Here, C_{Ags} is the concentration and k_g is the mass transfer coefficient at the gas-solid interface. For reactant A in the liquid phase, the conservation equation is:

$$-E \frac{d^2 C_{AL}}{dz^2} + v_l \frac{dC_{AL}}{dz} = -v_g \frac{dC_{Ag}}{dz} - (k_c a_c)_A (C_{AL} - C_{As})q \tag{12.39}$$

For the liquid reactant B, one has:

$$-E \frac{d^2 C_{BL}}{dz^2} + v_l \frac{dC_{BL}}{dz} = -(k_c a_c)_B (C_{BL} - C_{Bs})q \tag{12.40}$$

The boundary conditions are:

$$C_{Ag} = (C_{Ag})_f$$

$$-E \frac{dC_{BL}}{dz} = v_l [(C_{BL})_f - C_{BL}] \tag{12.41}$$

$$-E \frac{dC_{AL}}{dz} = v_l [(C_{AL})_f - C_{AL}] \qquad \text{at } z = 0$$

$$\frac{dC_{BL}}{dz} = \frac{dC_{AL}}{dz} = 0 \qquad \text{at } z = Z \tag{12.42}$$

where the subscript f denotes feed conditions. The relationships for the interfacial concentrations are obtained by equating the mass transfer rate at the interface to the global rate. Unlike the simple trickle-bed, however, one has to consider wetted and nonwetted parts separately. Thus, for reactant A:

$$(k_c a_c)_A (C_{AL} - C_{As}) = (R_G)_{Aw} = k\eta_w f(C_{As}, C_{Bs}): \quad \text{wetted part} \quad (12.43)$$

$$(k_g a_c)_A (C_{AG} - C_{Ags}) = (R_G)_{Ad} = k\eta_d f(C_{Ags}/H, C_{Bs}): \quad \text{dry part} \quad (12.44)$$

where η_w and η_d are the internal effectiveness factors based on C_{As} and C_{Ags}/H, respectively. Here, it is assumed for η_d that the concentration of A in the liquid at the mouth of the liquid-filled pore, which is exposed to gas, is at C_{Ags}/H. It is also assumed that the concentration of B at the pore mouth is at C_{Bs}. For the reactant B, one has:

$$(k_c a_c)_B (C_{BL} - C_{Bs}) = (R_G)_B = b(R_G)_A = b\eta_w f(C_{As}, C_{Bs}) \quad (12.45)$$

Here, η_w is given by Eq. 12.28, and η_d is simply η given by Eq. 12.18 with C_{As} replaced by C_{Ags}/H. The pellet center concentrations necessary for η_w and η_d can be obtained from the following approximate relations:

$$C_A^L = \frac{C_{As}}{\cosh(\phi_G)_w} \quad (12.46)$$

$$(C_A^L)_d = \frac{C_{Ags}/H}{\cosh(\phi_G)_d} \quad (12.47)$$

where $(\phi_G)_w$ and $(\phi_G)_d$ are given by:

$$(\phi_G)_w = \frac{L f_1(C_{As}, C_{Bs})}{\left[(2D_A k) \int_0^{C_{As}} f_1(C_A) dC_A \right]^{1/2}} \quad (12.48)$$

$$(\phi_G)_d = \frac{L f_1(C_{Ags}/H, C_{Bs})}{\left[(2D_A k) \int_0^{C_{Ags}} f_1(C_A) dC_A \right]^{1/2}} \quad (12.49)$$

For $(\phi_G)_w$, $f_1(C_A)$ is given by Eq. 12.27; for $(\phi_G)_d$, $f_1(C_A)$ is given by Eq. 12.27 with C_{As} replaced by C_{Ags}/H.

Design of a trickle-bed at low liquid velocities requires solving Eqs. 12.38 through 12.45. Considerable simplification results if the wetting efficiency is unity, in which case $q = 1$, η_w is η given by Eq. 12.28, and the equations for the dry part of the pellet disappear. While the design can be carried out using the conservation equations, the accuracy with which the Peclet number and the wetting effiency can be estimated would limit the utility of the result.

Summary

In this chapter, the design of the two most common multiphase reactors has been treated. While the design of slurry reactors is relatively straightforward, that of trickle-beds can be complicated when the reaction is highly exothermic. For the nonisothermal case, only a very simple case where the liquid is nonvolatile has been treated to illustrate the design procedures. Further complications can arise when liquid velocity is low. At low liquid velocities, the catalyst particles may become partially wetted and axial disperion may be significant, at least for the liquid phase. One distinct feature of multiphase reactors, when compared to fixed-bed reactors, is that the conservation equations have to be written for the gaseous reactant for both the gas and the liquid phases. The other feature is the important role of mass transfer resistances due to the presence of three phases, in particular the liquid phase. Modeling of trickle-beds in which the liquid evaporates and the pellet surface is partially wetted at low liquid velocities requires further work.

NOTATION

a_c	external area of catalyst particles per unit volume of catalyst
a_L	gas-liquid interfacial area per unit volume of liquid
a_t	external surface area of pellets per unit volume of bed
b	stoichiometric coefficient for the liquid reactant in Eq. 12.1
B_{hL}	Lh/λ
B_{mG}	$k_g L/D$
B_{mL}	$k_L L/D$; $L(k_c)_B/D_B$ in Eq. 12.33
C_A	concentration of species A
C_B	concentration of species B
C_p	molar specific heat
C_{Ag}	gas-phase bulk concentration of species A
C_{AL}	liquid-phase bulk concentration of species A
C_A^L	pellet center concentration of species A
$(C_A^L)_d$	C_A^L for dry part of pellet
C_{As}	pellet surface concentration of species A
C_{BL}	liquid-phase bulk concentration of species B
C_{Bs}	pellet surface concentration of species B
C_{Ags}	concentration of A at gas-solid interface
d_p	pellet (particle) diameter
D_i	effective diffusivity in liquid-filled pores
D_A	effective diffusivity of species A in liquid-filled pores
D_B	effective diffusivity of species B in liquid-filled pores
E	axial dispersion coefficient (m²/s) for mass
f	dependence of rate on concentrations
g	gravity
h	gas-solid film heat transfer coefficient
h_e	external holdup
H	Henry's constant

k	rate constant
k_a	apparent rate constant
k_c	mass transfer coefficient at the interface between catalyst and liquid
k_g	mass transfer coefficient at the gas-solid interface
k_L	mass transfer coefficient at the interface between gas and liquid
L	characteristic length of pellet
L_m	liquid mass rate per unit area
q	fraction of pellet surface covered by liquid (wetting efficiency)
Q_l	volumetric flow rate of liquid
r_A	intrinsic rate of reaction based on species A
r_B	intrinsic rate of reaction based on species B
$(R_G)_A$	global rate for species A
$(R_G)_B$	global rate for species B
T_g	gas temperature
T_L	liquid temperature
U	overall heat transfer coefficient between gas and liquid
v_b	bubble velocity
v_g	superficial velocity of gas
v_l	superficial velocity of liquid
V	liquid volume
z	axial coordinate
Z	reactor length

Greek Letters

β, γ	quantities defined in Eq. 12.33
ϵ_b	fraction of liquid volume occupied by bubbles
ϵ_B	bed porosity
ϵ_s	fraction of liquid volume occupied by catalyst particles
η	internal effectiveness factor
η_d	η given by Eq. 12.28 with C_{As} replaced by $C_{Ags/H}$
η_w	η given by Eq. 12.28
ϕ	Thiele modulus, $L(k/D)^{1/2}$
ϕ_G	generalized modulus given by Eq. 12.48 or 12.49
ρ	molar density
λ	effective thermal conductivity
μ	viscosity
δ	surface tension

Subscripts

A	species A
b	bubble
B	species B
d	nonwetted part of pellet

f	feed
G	gas phase
l	liquid
L	liquid phase
w	wetted part of pellet
o	reactor outlet
in	reactor inlet

PROBLEMS

12.1. Specify the liquid volume of a slurry reactor required to convert 60% of the liquid reactant at 525°K for the following reaction:

$$A(g) + 3B(l) \xrightarrow{\text{catalyst}} \text{products}$$

The gaseous reactant is present in excess for saturation at 10^{-4} mol/cm³ with a Henry's constant of 3 (mol/cm³ gas)/(mol/cm³ liquid). The inlet concentration of species B in aqueous solution is 10^{-5} mol/cm³ with a volumetric flow rate of 1 cm³/s. The liquid-solid mass transfer coefficient including the surface area to volume ratio ($k_c a_c$) is 2 s^{-1} for both A and B. The intrinsic rate is:

$$r_A = 10^{10} \exp(-20,000/R_g T) C_B C_A^{1/2} \text{ mol/(s·cm}^3 \text{ catalyst)}$$

Assume that the fraction of the liquid volume occupied by the catalyst is 0.02.

12.2. Resolve Problem 12.1 when the reaction is affected by dependent deactivation. The rate of deactivation is given by:

$$r_p = 0.01 \; r_A$$

Assume that the catalyst particles are continuously removed from the vessel, but that the internal age distribution of the particles is that of an ideal CSTR:

$$I(\theta) = \exp(-\theta); \qquad \theta = V/Q_l$$

12.3. Show for an adiabatic trickle-bed that the conservation equations can be written as:

$$v_g \frac{dC_{Ag}}{dz} = g_1 \left(C_{Ag}, C_{AL}, C_{BL} \right)$$

$$v_l \frac{dC_{AL}}{dz} = g_2 \left(C_{AL}, C_{BL} \right)$$

$$v_l \frac{dC_{BL}}{dz} = g_3 \left(C_{AL}, C_{BL} \right)$$

If axial dispersion effects are included, the conservation equations become:

$$-(E_g)_A \frac{d^2 C_{Ag}}{dz^2} + v_g \frac{dC_{Ag}}{dz} = g_1 (C_{Ag}, C_{AL}, C_{BL})$$

$$-(E_L)_A \frac{d^2 C_{AL}}{dz^2} + v_l \frac{dC_{AL}}{dz} = g_2 (C_{AL}, C_{BL})$$

$$-(E_L)_B \frac{d^2 C_{BL}}{dz^2} + v_l \frac{dC_{BL}}{dz} = g_3 (C_{AL}, C_{BL})$$

Using the results obtained in Chapter 9 regarding the equivalence between plug-flow and axial dispersion models, determine whether the plug-flow conservation equations given above are sufficient as a first-order approximation of the dispersion model.

12.4. Suppose that the reaction in Problem 12.1 is carried out in an adiabatic trickle-bed. What is the catalyst volume required for the desired conversion? Assume that the effective diffusivities are the same at 10^{-5} cm²/s and the spherical pellet diameter is 1 cm. Use the following data:

$$k_c = 0.02 \text{ cm/s for both } A \text{ and } B$$

$$(k_L a_L)_A = 0.01 \text{ s}^{-1} \qquad v_g = v_l = 5 \text{ cm/s}$$

12.5. Explain why Eqs. 12.30 and 12.33 are not sufficient to describe the trickle-bed when in fact Eq. 12.33 contains only C_{BL}.

REFERENCES

Buffham, B.A. and M.N. Rathor, Trans. Inst. Chem. Eng., 56, 266 (1978).
Gianetto, A., G. Baldi, V. Specchia and S. Sicardi, AIChE J., 24, 1087 (1978).
Goto, S., J. Levec and J.M. Smith, Cat. Rev., 15, 187 (1977).
Herskowitz, M. and J.M. Smith, AIChE J., 24, 439 (1978).
Herskowitz, M. and J.M. Smith, AIChE J., 29, 1 (1983).
Hofmann, H.P., Cat. Rev., 17, 21 (1978).
Levec, J. and J.M. Smith, AIChE J., 22, 159 (1976).
Mills, P.L. and M.P. Dudukovic, Ind. Eng. Chem. Fund., 18, 139 (1979).
Mills, P.L. and M.P. Dudukovic, AIChE J., 27, 893 (1981).
Niiyana, H. and J.M. Smith, AIChE J., 22, 961 (1976).
Satterfield, C.N., AIChE J., 21, 209 (1975).
Shah, Y.T., Gas-Liquid-Solid Reactor Design, McGraw-Hill, New York (1979).
Sicardi, S., G. Baldi, A. Gianetto and V. Specchia, Chem. Eng. Sci., 35, 67 (1980).
Smith, J.M., Chemical Engineering Kinetics, 3rd ed., McGraw-Hill, New York (1981).

CHAPTER 13

Optimization

13–1 INTRODUCTION

There are two major categories of reactor optimization. The first type, which is usually the most important for single reactions, is that of maximizing conversion for a given reactor size by properly manipulating the operating conditions, typically temperature. This problem is equivalent to minimizing reactor size for a given conversion. The second type, which is relevant to multiple reactions, is that of maximizing the yield of the desired product. The overriding factor in the latter type is the relative magnitude of activation energies associated with each reaction path (Denbigh and Turner 1971; Millman and Katz 1967).

In order for an optimization problem to exist, there must be conflicting factors. These are, for instance, the rates of formation and disappearance of the species of interest in multiple reactions. If there are no constraints, there are essentially no optimization problems for irreversible single reactions since the highest possible temperature will always lead to maximum conversion and minimum reactor size, except under some unusual circumstances. Even for reversible single reactions, the highest possible temperature will lead to the maximum conversion if the reactions are endothermic since the equilibrium conversion should increase with temperature. For single reactions, therefore, only exothermic reversible reactions are of interest since then the equilibrium conversion becomes the conflicting factor against the rate constant, which increases with increasing temperature. For such single reactions, the conversion is constrained by the equilibrium conversion. On the other hand, any reaction is of interest for optimization if the reaction is affected by catalyst deactivation since a temperature increase will not only increase conversion, but will also accelerate catalyst deactivation. Thus, the conversion averaged over a period of time can be maximized only when the temperature is properly manipulated. In the sections to follow consideration will be given to single or multiple reactions constrained by equilibrium conversion and catalyst deactivation.

There are many techniques (Beveridge and Schechter 1970; Ray and Szekely 1973) available that can be used for reactor optimization. Some that are used here include simple maximization, the calculus of variations, dynamic programming, and the maximum principle. Use of these techniques leads to some very powerful optimal policies in simple cases. These policies, however, are not always applicable to realistic heterogeneous reactors since the restrictions placed on the optimal policies are not always compatible with heterogeneous reactions inherently affected by transport resistances. When an optimization problem is posed, the assumption is that the system is characterized well enough for quantitative description. For reactions affected by catalyst deactivation, however, this is not always true. Therefore, there is a need to estimate the extent of deactivation from process measurements for the optimization. This problem will be treated in some detail.

Undoubtedly, much can be gained from the optimization of reactor operating conditions. Often neglected, however, is the fact that much more can be gained by optimizing the reactor size at the same time. This fact will be pointed out and illustrated whenever appropriate to emphasize the importance of optimizing with respect to both reactor size and operating conditions.

13-2 EQUILIBRIUM CONSTRAINTS

Consider a single reaction constrained by an equilibrium conversion. An optimization problem for a given throughput is that of choosing the inlet temperature or the temperature trajectory along the reactor that will minimize the reactor size for the desired conversion. If the maximum possible conversion is the overriding concern such that a multibed arrangement is used, one can have another optimization problem of allocating the extent of reaction to each bed so as to maximize the final conversion with certain constraints. In general, an endothermic reaction in which the equilibrium conversion increases with increasing temperature is of no interest since the highest allowable temperature is the best policy. On the other hand, one can pose an optimization problem for an exothermic reaction since an increase in temperature that results in an increase in the rate also causes a decrease in equilibrium conversion.

Suppose that an exothermic reaction is carried out in an adiabatic fixed-bed. The design equations given in Table 10.1 in Chapter 10 for the adiabatic reactor are:

$$\tau = \int_{C_{out}}^{C_{in}} \left\{ \frac{L}{\left[2D_e k_b \int_{C_c}^{C_b} f(C) dC \right]^{1/2}} - \frac{1.2 L E_a(-\Delta H)}{2 h R_g [T_{in} - (-\Delta H/\rho C_p)(C_b - C_{in})]^2} \right\} dC_b$$

(13.1)

$$k_b = k_0 \exp\{(-E_a/R_g)/[T_{in} - (-\Delta H/\rho C_p)(C_b - C_{in})]\}$$

(13.2)

One can readily show that these equations are also applicable to a single reversible reaction provided that C is the concentration of one of the reactants, the heat of reaction $(-\Delta H)$ is based on a unit mole of the main reactant, and the concentration dependence of the intrinsic rate f is properly expressed (see Problem 13.1). As detailed in Chapter 10, any forward and backward rates can be expressed in terms of the main reactant concentration, stoichiometric coefficients, and diffusivity ratios. Therefore, the net intrinsic rate can be expressed as:

$$r_c = kf(C) = kg_f(C) - k_{-1}g_b(C)$$

$$= k\left[g_f(C) - \frac{1}{K}g_b(C)\right]; \qquad K \equiv k/k_{-1} \qquad (13.3)$$

where g_f and g_b are the concentration dependences of the forward and backward rates, and k and k_{-1} are the corresponding rate constants. Thus, the expression for f is:

$$f(C) = g_f(C) - \frac{1}{K}g_b(C) \qquad (13.4)$$

where K can be expressed as:

$$K = K_0 \exp\{(Q/R_g)/[T_{in} - (-\Delta H/\rho C_p)(C_b - C_{in})]\}; \qquad Q \equiv E_b - E_a \quad (13.5)$$

Here, E_b is the activation energy for the backward rate constant k_{-1}. Consider now the optimization problem of minimizing the reactor size, or equivalently τ, by properly choosing the inlet temperature (T_{in}) for a given throughput, desired conversion, and inlet concentration. It should be clear that this problem is equivalent to that of choosing the optimal temperature for the maximum conversion for a given reactor size τ. Since τ is a function only of T_{in} when Eqs. 13.2, 13.4, and 13.5 are used in Eq. 13.1, the optimization is simply a one-parameter problem, which can be solved using the usual conditions for a minimum. That is, the optimal T_{in} is that which satisfies the following conditions:

$$\frac{\partial \tau}{\partial T_{in}} = 0 \qquad \text{and} \qquad \frac{\partial^2 \tau}{\partial^2 T_{in}} > 0 \qquad (13.6)$$

A simple search can also be used in which various values of τ are calculated for assumed values of T_{in} until the minimum is found. Here again, the pellet center concentration, C_c, is obtained from:

$$C_c = \frac{1}{\cosh \phi_G}$$

$$\phi_G = \frac{L(k_b/2D_e)^{1/2}f(C_b)}{\left[\int_0^{C_b} f(C)dC\right]^{1/2}} \qquad (13.7)$$

where f is given by Eq. 13.4.

It is always desirable to know the best possible performance even when such a performance is difficult to attain due to implementation difficulties. For the optimization being considered, the question is what the temperature trajectory along the reactor should be to minimize the reactor size. For a single reaction, the optimal temperature trajectory is that satisfying the following condition:

$$\frac{\partial R_G}{\partial T_b} = 0 \tag{13.8}$$

where R_G for use in the above equation is:

$$R_G = \frac{L}{\left[2D_e k_b \int_{C_c}^{C_b} f(C)dC \right]^{1/2}} - \frac{1.2LE_a(-\Delta H)}{2hR_g T_b^2} \tag{13.9}$$

$$k_b = k_0 \exp(-E_a/R_g T_b) \tag{13.10}$$

$$K = K_0 \exp(Q/R_g T_b) \tag{13.11}$$

This result follows directly from the results of Aris (1965) obtained for reactions free from transport effects. Obviously, the optimal temperature depends on concentrations, and therefore, the implementation of the optimal policy requires concentration measurements at every point along the reactor. Furthermore, the optimal policy at some points along the reactor may require a negative or infinite temperature (Fournier and Groves 1970) at low conversions. An alternative to this ideal policy is a multibed arrangement.

The discrete, stagewise arrangement of multibed adiabatic reactors can be optimized using the techniques of dynamic programming (Bellman 1957; Roberts 1964). Dynamic programming is based on the *principle of optimality,* which states that the optimal policy for an N-stage system must be such that the decision for the nth stage ($n = 1, \ldots , N$) is optimal for the last ($N + 1 - n$) stages of the N-stage system for any input to the nth stage. Obviously, the search for the optimal policy should start with the last bed for the multibed arrangement (N number of reactors) since the optimization for $n = N$ involves only one stage, which in turn involves the optimal choice of inlet conversion and temperature to the last bed. The principle of optimality requires that the choice of inlet conversion and temperature to the $(N - 1)^{th}$ bed should be optimal for the last two beds, but subject to the optimal policy already made for the last bed. The same is true for the last three reactors, but this time subject to the optimal policy already known for the last two. This stage by stage optimization can be carried out until the first bed is reached. The principle of optimality, therefore, considerably simplifies the optimization problem. Perhaps as important as this simplification are the relationships that result from the application of the principle.

Suppose that the optimization problem is to maximize a measure of the profit. If α is the price of product and γ is the cost incurred in producing the product per unit volume of reactor per unit time, a measure of the net profit for each bed Q_n is:

$$Q_n = \alpha v A_c (C_{n-1} - C_n) - \gamma V_n \qquad (13.12)$$

where v is the superficial fluid velocity, A_c the cross-sectional area, V the reactor volume, and C_n and C_{n-1} are the inlet concentrations to the $(n + 1)^{th}$ and n^{th} beds. The net profit function Q_n can be reduced to the following performance index:

$$P_n = (C_{n-1} - C_n) - v \frac{Z_n}{v} ; \qquad v = \frac{\gamma}{\alpha} \qquad (13.13)$$

since v is constant for a given throughput, and therefore, P_n is equivalent to Q_n. Here, Z_n is the reactor length for the nth bed. The problem is then to maximize:

$$P = \sum_{n=1}^{N} P_n$$

by properly selecting the inlet concentrations and temperatures to N beds. In order to change the inlet temperatures, there has to be a heat exchanger between two adjacent adiabatic reactors. This multibed arangement is shown in Figure 13.1. The optimization problem is to be solved subject to the following system equations:

$$\frac{d(C_b)_n}{dy} = -\tau_n R_{G_n} ; \qquad \tau_n = \frac{Z_n(1 - \epsilon_B)}{v} \qquad n = 1, \ldots, N \quad (13.14)$$

$$\frac{d(T_b)_n}{dy} = \beta \tau_n R_{G_n} ; \qquad \beta = \frac{-\Delta H}{\rho C_p} \qquad (13.15)$$

where ϵ_B is the bed voidage, $(C_b)_n$ and $(T_b)_n$ are the bulk-fluid concentration and temperature in the nth bed, and the global rate R_G is that given by Eq. 13.9. For the adiabatic reactor under consideration, $(T_b)_n$ is related to $(C_b)_n$ by:

$$(T_b)_n = T_n + \beta[C_n - (C_b)_n] \qquad (13.16)$$

which results when Eqs. 13.14 and 13.15 are combined. Therefore, $(R_G)_n$ is a function solely of $(C_b)_n$, which allows one to rewrite Eq. 13.14 in the form of Eq. 13.1 for τ_n, i.e.:

$$\tau_n = \int_{C_{n-1}}^{C_n} - \frac{d(C_b)_n}{(R_G)_n} \qquad (13.17)$$

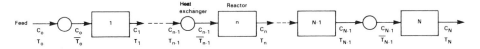

Figure 13.1   Multistage reactor/heat exchanger arrangement.

since the integrand of Eq. 13.1 is simply $-1/(R_G)_n$. Thus, the performance index for the nth bed P_n can be rewritten as:

$$P_n = \int_{C_{n-1}}^{C_n} \left[-1 + \frac{\lambda}{R_{G_n}} \right] d(C_b)_n; \qquad \lambda = \frac{\nu}{1 - \epsilon_B} \qquad (13.18)$$

The last part of the integrand follows from the fact that $\nu Z_n / \nu = \nu \tau_n / (1 - \epsilon_B)$. As discussed earlier, application of the principle of optimality begins with the last bed for the overall optimization, which can be stated as:

$$\operatorname*{Max}_{\{\bar{T}_{N-1}, C_N\}} P_N = \operatorname{Max} \int_{C_{N-1}}^{C_N} \left[-1 + \frac{\lambda}{R_{G_N}} \right] d(C_b)_N \qquad (13.19)$$

It should be noted that the feed concentration C_0 is specified such that the decision variables for the first bed are $\bar{T}_0$ and C_1 since T_1 is fixed by the relationship of Eq. 13.16 applied at the outlet of the first bed. This means that the decision variables for the Nth reactor are $\bar{T}_{N-1}$ and C_N. It should also be noted that $\bar{T}_n \neq T_n$ because of the heat exchanger between the two adjacent beds. Thus, $\bar{T}_{N-1}$ and C_N have to be chosen such that:

$$\frac{\partial P_N}{\partial C_N} = 1 - \frac{\lambda}{R_{G_N}} = 0 \qquad (13.20)$$

$$\frac{\partial P_N}{\partial \bar{T}_{N-1}} = \lambda \int_{C_{N-1}}^{C_N} \frac{-1}{R_{G_N}^2} \frac{\partial R_{G_N}}{\partial \bar{T}_{N-1}} d(C_b)_N = 0 \qquad (13.21)$$

The first condition states that:

$$R_{G_N}(C_N, T_N) = \lambda \qquad (13.22)$$

The integrand in Eq. 13.21 contains $\bar{T}_{N-1}$ (T_{in} in Eqs. 13.1 and 13.2) as the only unknown parameter, which can be differentiated with respect to $\bar{T}_{N-1}$ for the indicated integration with respect to $(C_b)_N$ (C_b in Eq. 13.1). Simultaneous solution of Eqs. 13.21 and 13.22 is illustrated in Figure 13.2. A curve satisfying Eq. 13.22 is first drawn as shown in the figure (A_N curve), which can be generated by calculating the C_N values corresponding to assumed values of T_N. For every point on the A_N curve, there is a pair of C_{N-1} and $\bar{T}_{N-1}$ satisfying the adiabatic line of Eq. 13.16, i.e.:

$$T_N = \bar{T}_{N-1} + \beta(C_{N-1} - C_N) \qquad (13.23)$$

In order to find the pair, C_{N-1} and $\bar{T}_{N-1}$, corresponding to a point (C_N, T_N) on the A_N curve, the values of C_N and T_N satisfying Eq. 13.22 are used in the above relationship to obtain C_{N-1} in terms of $\bar{T}_{N-1}$. Now that C_N is known and C_{N-1}

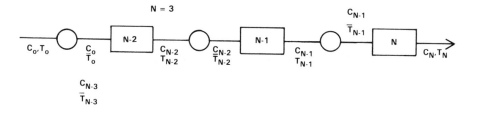

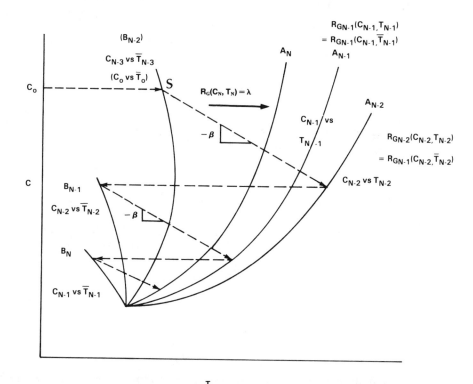

Figure 13.2 Optimal solution paths for three-stage reactor/heat exchanger on C-T plane.

is expressed in terms of $\overline{T}_{N-1}$, Eq. 13.21 can be solved for $\overline{T}_{N-1}$ as the only unknown. The pairs of C_{N-1} and $\overline{T}_{N-1}$ corresponding to the pairs of C_N and T_N on the A_N curve can be plotted as shown in Figure 13.2 (B_N curve). The optimal values of the inlet temperature and concentration to the Nth bed should be on this curve B_N since any pair of C_{N-1} and $\overline{T}_{N-1}$ on this curve satisfies both Eqs. 13.21 and 13.22 due to the adiabatic relationship of Eq. 13.23.

Consider now the last two beds. According to the principle of optimality, $\overline{T}_{N-2}$ and C_{N-1} have to be chosen in such a way that:

$$\operatorname*{Max}_{\{\overline{T}_{N-2}, C_{N-1}\}} (P_N + P_{N-1}) = \operatorname{Max} [P_{N-1} + \operatorname{Max} P_N]$$

$$= \operatorname{Max} \left[\int_{C_{N-2}}^{C_{N-1}} \left(-1 + \frac{\lambda}{R_{G_{N-1}}} \right) d(C_b)_{N-1} \right.$$

$$\left. + \operatorname{Max} \int_{C_{N-1}}^{C_N} \left(-1 + \frac{\lambda}{R_{G_N}} \right) d(C_b)_N \right] \qquad (13.24)$$

The maximization requires the following conditions to be satisfied:

$$\frac{\partial}{\partial C_{N-1}} (P_N + P_{N-1}) = 0 \qquad (13.25)$$

$$\frac{\partial}{\partial \overline{T}_{N-2}} (P_N + P_{N-1}) = 0 \qquad (13.26)$$

subject to the optimal conditions already established for the last bed. Since P_N is dependent on C_N, $\overline{T}_{N-1}$, and C_{N-1}, Eq. 13.25 becomes:

$$\frac{\partial P_N}{\partial C_{N-1}} = \frac{\partial}{\partial C_{N-1}} \int_{C_{N-1}}^{C_N} \left(-1 + \frac{\lambda}{R_{G_N}} \right) d(C_b)_N$$

$$= \left[-1 + \frac{\lambda}{R_{G_N}(C_N, T_N)} \right] \frac{\partial C_N}{\partial C_{N-1}} + \left[1 - \frac{\lambda}{R_{G_N}(C_{N-1}, \overline{T}_{N-1})} \right]$$

$$- \left[\int_{C_{N-1}}^{C_N} \frac{\lambda}{R_{G_N}^2} \frac{\partial R_{G_N}}{\partial \overline{T}_{N-1}} \frac{\partial \overline{T}_{N-1}}{\partial C_{N-1}} d(C_b)_N \right]$$

$$= \left[1 - \frac{\lambda}{R_{G_N}(C_{N-1}, \overline{T}_{N-1})} \right] \qquad (13.27)$$

The first and third terms vanish because of the optimal conditions (Eqs. 13.20 and 13.21) already determined for the last bed. It follows then that:

$$\frac{\partial}{\partial C_{N-1}} (P_N + P_{N-1}) = \left[-1 + \frac{\lambda}{R_{G_N}(C_{N-1}, T_{N-1})} \right] + \left[1 - \frac{\lambda}{R_{G_N}(C_{N-1}, \overline{T}_{N-1})} \right] = 0$$

$$(13.28)$$

which reduces to:

$$R_{G_{N-1}}(C_{N-1}, T_{N-1}) = R_{G_N}(C_{N-1}, \overline{T}_{N-1}) \qquad (13.29)$$

The optimal condition states that the rate at the outlet of the $(N - 1)^{th}$ bed must equal that at the inlet of the last (N^{th}) bed. Thus, this condition specifies the heat exchanger load (refer to Figure 13.2). Here again, a pair of C_{N-1} and T_{N-1} values corresponding to a point on the curve B_N (left hand side of Eq. 13.29) can be calculated from Eq. 13.29. The locus of C_{N-1} and T_{N-1} satisfying the equation is shown in Figure 13.2 as the curve A_{N-1}. The second optimal condition of Eq. 13.26 leads to:

$$\frac{\partial}{\partial \overline{T}_{N-2}}(P_N + P_{N-1}) = \frac{\partial P_{N-1}}{\partial \overline{T}_{N-2}} = \lambda \int_{C_{N-2}}^{C_{N-1}} \frac{-1}{R_{G_{N-1}}^2} \frac{\partial R_{G_{N-1}}}{\partial \overline{T}_{N-2}} d(C_b)_{N-1} = 0 \quad (13.30)$$

This relationship is identical to that of Eq. 13.21 with the exception of the stage subscripts. Procedures identical to those used to generate the B_N curve can be followed to obtain the locus of C_{N-2} and $\overline{T}_{N-2}$ satisfying Eqs. 13.29 and 13.30, which is the curve B_{N-1} in Figure 13.2. This curve represents the optimal inlet temperature and concentration to the $(N - 1)^{th}$ bed. The procedures discussed so far can be repeated for the optimization until the first bed is reached.

The optimal solution for a three-bed arrangement $(N = 3)$ is illustrated in Figure 13.2. First, A_N and B_N curves representing the optimal conditions for each bed and heat exchanger are constructed as detailed above and shown in Figure 13.2. Once these curves are available, the optimal solution is straightforward if it is recognized that the A_N and B_N curves are related to each other by the adiabatic operating line with a slope of $(-\beta)$ as evident from Eq. 13.23 and the A_{N-1} and B_N curves are related to each other by the constant concentration line for the heat exchanger (Eq. 13.29). All that is needed for the solution is to locate one point on any of the curves. Since the feed concentration C_0 is specified, one can locate the point corresponding to C_0 on B_1 (B_{N-2} in Figure 13.2 for C_0 versus $\overline{T}_0$) designated as S. From S, the adiabatic line is drawn to the A_{N-2} curve as shown by the dotted line in the figure; the constant concentration line is then followed to the B_{N-1} curve, and the adiabatic line is drawn again to the A_{N-1} curve followed by the constant concentration line to the B_N and then to the A_N curve via the adiabatic line. Thus, the points on the A_N and B_N curves at which these meet with the dotted lines represent the optimal inlet conditions to the three beds, corresponding to the specified feed concentration C_0. It is seen that the inlet temperature is the highest for the first bed and becomes progressively lower for subsequent beds. Further, the outlet stream from each bed should be cooled before it is fed to the next bed. The problem of including cooling costs in the performance index has been treated by Lee and Aris (1963).

It should be recognized that a shell and tube reactor/heat exchanger in which the sections of the heat exchanger are maintained at different temperatures to obtain optimal operation is entirely analogous to the multibed arrangement just considered. The same procedures should apply even though the calculations are much more complex since the adiabatic relationship no longer holds and a term for cooling has to be added to Eq. 13.15. The application of dynamic programming invariably involves numerical calculations. Nevertheless, some very useful relationships emerge, such as in Eq. 13.29 in the process of applying the principle of optimality.

13-3   CATALYST DEACTIVATION

Many optimization problems arise as a natural consequence of the fact that the catalytic activity decreases with time when the catalyst is deactivated. Given this fact, the question is how to optimize the reactor operation so as to maximize yield or a performance index that is a measure of profitability. The optimization can also involve sequences of beds to allow reaction and regeneration in a swing-cycle. Often neglected, but just as important as the optimization of operating conditions, is optimization with respect to reactor size, which was treated to some extent in Chapter 10. The usual industrial practice of operating a reactor affected by catalyst deactivation is to manipulate the reactor inlet temperature so as to maintain a constant conversion at the outlet. The reactor throughput is also manipulated to compensate for the declining catalytic activity, but usually only when the temperature cannot be raised any further due to the limitation on the maximum temperature allowed. When a shell and tube reactor/heat exchanger is used, the coolant temperature is also manipulated.

The usual practice of maintaining a constant conversion is optimal with respect to the maximum possible yield under some limiting conditions (Szepe and Levenspiel 1968; Chou et al. 1967). Consider an isothermal, plug-flow reactor. Suppose the deactivation kinetics are such that the fractional catalytic activity A can be described by:

$$\frac{dA}{dt} = -k_p(T)A; \qquad A(0) = 1 \tag{13.31}$$

Suppose further that transport resistances are negligible so that the mass balance for a single reaction in terms of conversion x can be written as:

$$\frac{dx}{dy} = \tau r_c(x, A, T); \qquad \tau = \frac{(1 - \epsilon_B)Z}{v}, \qquad y \in [0,1] \tag{13.32}$$

If the objective is to maximize the yield, i.e., outlet conversion, over the catalyst life t_f, which is fixed, the performance index P becomes:

$$P = \int_0^{t_f} x \, dt \tag{13.33}$$

The pseudo-steady state assumption has been made in writing Eq. 13.32 for a slowly deactivating catalyst. Equation 13.31 can be solved for T to give:

$$T = \frac{E_p}{R_g} \ln [-k_{p_0}A/A']; \qquad k_p = k_{p_0} \exp(-E_p/R_g), \ A' \equiv \frac{dA}{dt} \qquad (13.34)$$

Thus, the conversion can be expressed as:

$$x(A,T) = x[A, E_p \ln (-k_{p_0}A/A')/R_g]$$
$$= x(A,A') \qquad (13.35)$$

where Eq. 13.34 has been used for T. Equation 13.35 allows one to write Eq. 13.33 as:

$$p = \int_0^{t_f} x(A,A')dt \qquad (13.36)$$

According to a basic theorem of the calculus of variations (Bliss 1961), the condition for an extremum of the function given in the form of Eq. 13.36 is:

$$x(A,A') - A'x_{A'}(A,A') = \text{constant}; \qquad x_{A'} \equiv \frac{\partial x}{\partial A'} \qquad (13.37)$$

It is seen that the optimal policy would be that of maintaining the conversion constant if $A'x_{A'}$ is zero or a function solely of x since then the left hand side of Eq. 13.37 should be constant, which implies constant x. It can be shown (Chou et al. 1967) that Eq. 13.37 leads to the optimal policy of maintaining a constant conversion for an irreversible reaction in which only one rate constant and temperature dependence is involved, and that the policy is also applicable to nonisothermal reactors. The limitations under which the policy of a constant conversion is valid are embodied in Eq. 13.37: x should be a function of A and A' and not dependent explicitly on t, and $A'x_{A'}$ should be a function solely of x. The policy gives the maximum of P when an optimal temperature is chosen and hence requires a search for the optimal temperature.

The problem of maximizing the performance index of Eq. 13.33 can also be solved using the maximum principle, which will be discussed shortly. For the deactivation kinetics given by:

$$\frac{dA}{dt} = -k_p g_1(A)g_2(x) \qquad (13.38)$$

the optimal policy (Crowe and Therien 1974; Gruyaert and Crowe 1976) obtained from the maximum principle is again to maintain a constant exit conversion when the inlet temperature is manipulated without any constraint, provided that the temperature dependence in Eq. 13.32 is separable such that:

$$r_c = k(T)Ag_3(x) \tag{13.39}$$

When unconstrained temperature control is distributed over the entire reactor, the optimal policy is to maintain a constant exit conversion if the activation energy for the deactivation reaction is the same as that of the main reaction; if not, it is constant conversion over the entire reactor. It should be pointed out here that the optimal policy based on the pseudo-steady state assumption regarding Eqs. 13.32 and 13.38 is valid (Gruyaert and Crowe 1976) only when the change of temperature is felt at the outlet before another change is made at the inlet.

The optimal policies obtained so far are very powerful. However, the conditions under which the policies are valid are restrictive and the deactivation kinetics are somewhat unrealistic. For instance, the temperature dependence in rate expressions cannot be completely factored out for heterogeneous reactions, which are inherently affected by transport resistances. Consider independent deactivation. For an adiabatic reactor (Chapter 10):

$$\frac{dC_b}{dy} = -\tau R_G \tag{13.40}$$

$$\frac{dN_b}{dy} = -\tau R_p \tag{13.41}$$

If the deactivation is due to uniform poisoning, the global rates for the single main and deactivation reactions are:

$$R_G = \frac{J^{1/2}/L}{1 - \dfrac{1.2E_a(-\Delta H)}{2hR_gT_b^2}J^{1/2}} \; ; \qquad J = 2D_ek_b(1-\gamma)\int_{C_c}^{C_b} f(C)dC \tag{13.42}$$

$$R_p = (k_p)_sN_b(1-\gamma) \tag{13.43}$$

where $(k_p)_s$ is the value of k_p evaluated at the pellet surface temperature T_s. The rate at which the fraction of catalyst deactivated γ changes with time is given by:

$$\frac{d\gamma}{dt} = \frac{(k_p)_s}{Q}N_b(1-\gamma) \tag{13.44}$$

If the reaction is structure-insensitive, the activity A is simply:

$$A = 1 - \gamma \tag{13.45}$$

It is immediately seen that the condition of separable temperature dependence, which is necessary for the optimal policy of constant conversion, cannot be satisfied.

Even for this simplest case of deactivation, the condition is not met because the rate constant k_b, for instance, is not separable as apparent from Eq. 13.42, not to mention the separability of the temperature dependence of $(k_p)_s$, N_b, and γ. This does not mean that realistic problems cannot be solved. They can still be solved by the maximum principle, for instance, although the results are not as elegant as the policy of constant conversion.

The results discussed earlier for concentration dependent deactivation kinetics were based on Pontryagin's maximum principle (e.g., Koppel 1972) as extended to pseudo-steady state systems (Sirazetdinov and Degtyarev 1967). Here, a weak but more general maximum principle is considered due to Ogunye and Ray (1971). Let the concentrations (conversions) and temperatures be denoted by state variables **x** and the control variables such as temperatures (which are not state variables) by **U** and the catalytic activities by **A**. In general, under the pseudo-steady state assumption one has:

$$\frac{dx_i}{dy} = f_i(\mathbf{x}, \mathbf{A}, \mathbf{U}); \qquad x_i(0, \theta) = v_i(\theta) \qquad i = 1, 2, \ldots, r \qquad (13.46)$$

$$y \in [0,1], \qquad \theta \in [0,1]$$

$$\frac{dA_i}{d\theta} = g_j(\mathbf{x}, \mathbf{A}, \mathbf{U}); \qquad A_j(z, 0) = w_j(z) \qquad j = 1, 2, \ldots, q \qquad (13.47)$$

Suppose that the problem is to maximize the following performance index:

$$P = \int_0^1 \int_0^1 G(\mathbf{x}, \mathbf{A}, \mathbf{U}, \mathbf{v}, \mathbf{w}) \, dy \, d\theta \qquad (13.48)$$

subject to Eqs. 13.46 and 13.47 and the following upper and lower limits on the distributed (**u** and **w**) and boundary (**v**) control variables:

$$\mathbf{v}_* \leqslant \mathbf{v} \leqslant \mathbf{v}^*, \, \mathbf{U}_* \leqslant \mathbf{U} \leqslant \mathbf{U}^*, \, \mathbf{w}_* \leqslant \mathbf{w} \leqslant \mathbf{w}^* \qquad (13.49)$$

If one defines a Hamiltonian H and adjoint variables λ_i and μ_j for Eqs. 13.46 and 13.47 by:

$$H = G + \sum_i \lambda_i f_i + \sum_j \mu_j g_j \qquad (13.50)$$

it is necessary for optimal controls to satisfy the conditions in Table 13.1. The adjoint variables are then defined by:

$$\frac{d\lambda_i}{dy} = -\left[\frac{\partial G}{\partial x_i} + \sum_k \lambda_k \frac{\partial f_k}{\partial x_i} + \sum_j \mu_j \frac{\partial g_j}{\partial x_i} \right]; \qquad \lambda_i(1, \theta) = 0 \quad (13.51)$$

$$\frac{d\mu_j}{d\theta} = -\left[\frac{\partial G}{\partial y_j} + \sum_i \lambda_i \frac{\partial f_k}{\partial A_j} + \sum_k \mu_k \frac{\partial g_k}{\partial A_j} \right]; \qquad \mu_j(y, 1) = 0 \quad (13.52)$$

Table 13.1   Necessary Conditions for Optimal Controls (Ogunye and Ray 1971)

Category	Necessary conditions
$U(y,\theta)$	$\dfrac{\partial H}{\partial U_k}=0$ for all U_k and H be a maximum with respect to U_k at a constraint on U_k.
$U(y)$	$\displaystyle\int_0^1 \dfrac{\partial H}{\partial U_k}=d\theta=0$ and $\displaystyle\int_0^1 Hd\theta$ be a maximum with respect to U_k at a constraint on U_k.
$U(\theta)$	$\displaystyle\int_0^1 \dfrac{\partial H}{\partial U_k}\,dy=0$ and $\displaystyle\int_0^1 Hdy$ be a maximum at a constraint on U_k.
$v(\theta)$	$\dfrac{\partial H_1}{\partial v_i}=\displaystyle\int_0^1 \dfrac{\partial G}{\partial v_i}\,dy+\lambda_i(0,\theta)=0$ and $H_1=\displaystyle\int_0^1 Gdy+\lambda_i(0,\theta)v_j(\theta)$ be a maximum with respect to v_j at a constraint on v_j.
constant v	$\displaystyle\int_0^1 \dfrac{\partial H_1}{\partial v_i}\,dy=0$ and $\displaystyle\int_0^1 H_1 d\theta$ be a maximum with respect to v_i at a constraint on v_i.
$w(y)$	$\dfrac{\partial H_2}{\partial w_j}=\displaystyle\int_0^1 \dfrac{\partial G}{\partial w_j}\,d\theta+\mu_j(y,0)=0$ and $H_2=\displaystyle\int_0^1 Gd\theta+\mu_i(y,0)w_i(y)$ be a maximum with respect to w_j at a constraint on w_j.
constant w	$\displaystyle\int_0^1 \dfrac{\partial H_2}{\partial w_j}\,dy=0$ and $\displaystyle\int_0^1 H_2 dy$ be a maximum with respect to w_j at a constraint on w_j.

Take as a simple example an isothermal reactor in which consecutive reactions $(B_1 \xrightarrow{1} B_2 \xrightarrow{2} B_3)$ take place. Suppose that the isothermal temperature is the control variable. If the catalytic activity A is the same for both reactions, which is given by:

$$\frac{dA_1}{d\theta}=-\rho\exp\left\{-\frac{1}{pU_1}\right\}(A_1^2)=g_1;\qquad A_1(y,0)=w_1(y);\qquad \theta\in[0,1]\quad(13.53)$$

the conversion with the feed of only B_1 can be described by:

$$\frac{dx_1}{dy} = A_1\beta_1(1 - x_1)\exp\left(-\frac{1}{U_1}\right) = f_1 \qquad (13.54)$$

$$\frac{dx_2}{dy} = A_1\beta_2(x_1 - x_2)\exp\left(-\frac{p_1}{U_1}\right) = f_2 \qquad (13.55)$$

where U_1 is the dimensionless temperature R_gT/E_1, β_1 and β_2 are constants involving reactor length and preexponential factors, p is the ratio of activation energy for reaction 1 to that for the deactivation reaction, and p_1 is the ratio of activation energy for reaction 2 to that for reaction 1. Suppose that the objective is to maximize the production of B_2, i.e.:

$$P = \int_0^1 \int_0^1 (f_1 - f_2)\,dy\,d\theta \qquad (13.56)$$

subject to the following constraint:

$$U_1 \leqslant U_1^* \qquad (13.57)$$

The Hamiltonian is then given by:

$$H = f_1 - f_2 + \lambda_1 f_1 + \lambda_2 f_2 + \mu_1 g_1 \qquad (13.58)$$

and the necessary condition, which is for $U(\theta)$ in Table 13.1, is:

$$\int_0^1 \frac{\partial H}{\partial U_1}\,dy = 0 \text{ and } H \text{ be a maximum at a constraint on } U_1 \qquad (13.59)$$

where $\partial H/\partial U_1$ is given by:

$$\frac{\partial H}{\partial U_1} = (1 + \lambda_1)\frac{1}{U_1^2}A_1\beta_1\exp\left(-\frac{1}{U_1}\right)(1 - x_1)$$

$$+ (-1 + \lambda_2)\frac{p_1}{U_1^2}A_1\beta_2\exp\left(-\frac{p_1}{U_1}\right)(x_1 - x_2)$$

$$- \mu_1 p \frac{1}{pU_1^2}\exp\left(-\frac{1}{pU_1}\right)A_1^2 \qquad (13.60)$$

Since the necessary condition does not yield any analytic solution, and thus, requires a numerical search for a maximum starting with an initial guess for $U_1(\theta)$, the condition for the numerical search can be written as:

$$[U_1(\theta)]_{improved} = [U_1(\theta)]_{old} + \delta U_1(\theta) \tag{13.61}$$

$$\delta U_1(\theta) = \epsilon \int_0^1 \frac{\partial H}{\partial U_1} dy \tag{13.62}$$

where ϵ is a small, positive scaling number for convergence to the optimal $U_1(\theta)$. The last condition simply means that at every ascent to a local maximum, $\delta U_1(\theta)$ should move in the direction of the integral to reach the local maximum. When a constraint is reached, $U_1(\theta)$ should follow U_1^*. The numerical search involves solving the equations for A_1, x_1, and x_2 with an initial guess of $U_1(\theta)$ and then the equations for the adjoint variables, which are in turn used to calculate $\partial H/\partial U_1$ (Eq. 13.60) for the increment $\partial U_1(\theta)$, obtaining the improved $U_1(\theta)$ for the next iteration, and repeating the same procedures until δU_1 or $\delta P/P$ reaches a small number. The adjoint equations for the example can be written from Eqs. 13.51 and 13.52. For instance, the adjoint variable λ_1 is defined by:

$$\frac{d\lambda_1}{dy} = -\left[-A_1\beta_1(1 + \lambda_1) \exp\left(-\frac{1}{U_1} \right) \right.$$
$$\left. + A_1\beta_2(-1 + \lambda_2) \exp\left(-\frac{p_1}{U_1} \right) \right]; \qquad \lambda_1(1,\theta) = 0 \tag{13.63}$$

The weak maximum principle can certainly be used to maximize the performance index of Eq. 13.33, for instance, for the independent deactivation problem given by Eqs. 13.40 through 13.45 with the addition of a heat balance at the pellet-bulk fluid interface for the pellet surface temperature $T_s[(k_p)_s]$. Nevertheless, the optimal inlet temperature will be a function of time. Thus, the inlet temperature is to be manipulated in a prescribed manner in time without any regard to what happens to the reactor. This open-loop control is rarely practiced in actual applications because of the uncertainties regarding the model, measurements, and disturbances. Rather, closed-loop control using feedback from the process is usually practiced. This fact should not discourage one from using the maximum principle for optimization, for it will at least indicate what the best possible performance is in a relative sense. At the same time it can yield in some simple cases a very powerful optimal policy such as the constant conversion policy, which can be implemented by a feedback control scheme.

It is evident from the foregoing discussion that optimal policies are difficult to obtain for realistic heterogeneous reactors undergoing catalyst deactivation. Further, it is clear that some feedback information needs to be used for the implementation of a control scheme. While it has been assumed so far that the deactivation kinetics are known, detailed knowledge of deactivation is rarely available. This problem is particularly acute for structure-sensitive reactions in which the activity of the catalyst changes irregularly with the extent of deactivation. A number of questions can be raised, therefore, regarding optimal control of a heterogeneous reactor undergoing catalyst deactivation. A central question is whether a measure of catalyst deactivation can be estimated from process measurements and, if so, how feedback control can be realized with the measurements. A related question that was alluded to earlier is the interrelationship between reactor design and control and its use to achieve the best possible reactor performance. These questions will be considered one at a time.

Consider a single reaction in an adiabatic reactor. Assume that the global rate for fresh catalyst, which is Eq. 13.9, is known and let the "local activity factor" h be defined by:

$$h \equiv \frac{R_{G_d}}{R_{G_f}} \tag{13.64}$$

where R_{G_f} and R_{G_d} are the global rates for fresh and deactivated catalyst, respectively. Now the intrinsic rate for Eq. 13.9 is expressed as:

$$r_c = kf(C, \mathbf{K}) \tag{13.65}$$

where $\mathbf{K}$ denotes equilibrium constants that are evaluated at the bulk-fluid temperature T_b when used in Eq. 13.9. Further, T_b can be expressed in terms of the bulk-fluid concentration of the main reactant C_b and the inlet temperature and concentration for the adiabatic reactor under consideration. Therefore, R_{G_f} given by Eq. 13.9 is a function solely of C_b. Thus, for the deactivated catalyst:

$$\frac{dC_b}{dy} = -\tau R_{G_d} = -\tau h[y;t] \, R_{G_f} [C_b; T_{in}, C_{in}] \tag{13.66}$$

If this equation is integrated from the reactor inlet to the outlet, one has:

$$-\int_{C_{in}}^{C_{out}} \frac{dC_b}{\tau R_{G_f} [C_b; T_{in}]} = \int_0^1 h[y;t] dy \equiv H[t] \tag{13.67}$$

where C_{out} and C_{in} are the reactor outlet and inlet concentrations, at any given time. By definition, $H[0] = 1$ for fresh catalyst since $h = 1$ throughout the reactor at $t = 0$. The overall activity factor H is an overall measure of catalyst deactivation for the reactor at any given time and can be calculated directly from Eq. 13.67 by measuring the outlet concentration. Further, the overall activity factor can be calculated whether the deactivation is independent, dependent, or for that matter, of any kind since by definition h is simply the ratio of R_{G_f} to R_{G_d} regardless of what form R_{G_d} may take.

The overall activity factor can be used in a straightforward manner for a piecewise feedback control. Suppose that the control policy is to maintain a constant outlet conversion. Suppose further that the inlet temperature is manipulated intermittently, as in the usual operation of an adiabatic reactor, so as to maintain the constant outlet conversion. If one lets the subscript c denote the current quantities and n the new ones resulting from a change in the inlet temperature, Eq. 13.67 can be written twice to give:

$$H_c = \int_0^1 h_c dy = -\int_{C_{in}}^{C_{out}} \frac{dC_b}{\tau R_{G_f}[C_b;(T_{in})_c]}$$ (13.68)

$$H_n = \int_0^1 h_n dy = -\int_{C_{in}}^{C_d} \frac{dC_b}{\tau R_{G_f}[C_b;(T_{in})_n]}$$ (13.69)

where C_d is the desired outlet concentration corresponding to the desired conversion, which will be attained by changing the current inlet temperature $(T_{in})_c$ to the new inlet temperature $(T_{in})_n$. As shall soon be seen, the local activity factors h_c and h_n depend mainly on the fraction of the catalyst deactivated and slightly on the temperature difference between the bulk fluid and the pellet surface. If the change in the temperature difference due to a change in T_{in} is neglected for the time being, one can set:

$$h_c = h_n$$ (13.70)

for the purpose of calculating the new inlet temperature that yields the desired outlet concentration. The justification here is that the fraction of the catalyst deactivated changes negligibly while the inlet temperature is adjusted from the current value to a new one. Then, Eqs. 13.68 through 13.70 can be used to obtain:

$$-\int_{C_{in}}^{C_d} \frac{dC_b}{R_{G_f}[C_b;(T_{in})_n]} = H_c$$ (13.71)

where H_c is the known current overall activity factor calculable from Eq. 13.68 just prior to the change in the inlet temperature based on the measured current outlet concentration C_{out} and inlet temperature $(T_{in})_c$. Equation 13.71 can be solved for $(T_{in})_n$ by a trial and error method since the left hand side of the equation can be evaluated for an assumed value of $(T_{in})_n$. The value of $(T_{in})_n$ that satisfies Eq. 13.71 will yield the desired conversion, and therefore, the inlet temperature can be changed to $(T_{in})_n$ to maintain the conversion at the desired level. Since the control is piecewise, it will maintain the conversion at the desired level only for a short period of time. The conversion thereafter will gradually decrease with time until the inlet temperature is raised again. A bandwidth for the allowed decrease in conversion can be used to trigger the adjustment of the inlet temperature.

The control algorithm of Eqs. 13.68 and 13.71 is based on the assumption that the temperature difference between the bulk fluid and the pellet surface is negligible. When the temperature difference is not negligible, the discrepancy between h_c and h_n caused by the difference can be accounted for by adjusting C_d as follows:

$$\overline{C}_{d_i} = C_d + \beta(\overline{C}_{d_{i-1}} - C_{e_{i-1}}) \qquad i = 1, 2, \ldots \qquad (13.72)$$

where $\overline{C}_{d_i}$ is the adjusted C_d when a step change in the inlet temperature is made at the ith step, β is a proportionality constant with a value between zero and unity, and C_e is the measured outlet concentration. For a given bandwidth for the allowed decrease in the outlet concentration that will trigger the temperature adjustment, the piecewise control algorithm can be stated as follows:

$$-\int_{C_{in}}^{\overline{C}_{d_i}} \frac{dC_b}{\tau R_{G_f}[C_b; (T_{in})_i]} = (H_c)_{i-1} \qquad i = 1, 2, \ldots \qquad (13.73)$$

where $(H_c)_{i-1}$ is calculated from:

$$(H_c)_{i-1} = -\int_{C_{in}}^{C_{e_{i-1}}} \frac{dC_b}{\tau R_{G_f}[C_b; (T_{in})_{i-1}]}; \qquad (H_c)_0 = 1 \qquad (13.74)$$

with the measured outlet concentration and inlet temperature, and $\overline{C}_{d_i}$ is given by Eq. 13.72. Each time the outlet concentration decreases to the allowed bandwidth, the new inlet temperature $(T_{in})_i$ is calculated for the manipulation of the inlet temperature. The reactor behavior (Hong and Lee 1983) that results when the adiabatic reactor detailed in Table 10.5 is controlled according to the algorithm of Eqs. 13.72 through 13.74 is shown in Figure 13.3. The reactor response for $\tau = 20$s, which was simulated according to the procedures in Chapter 10 for the reaction affected by uniform independent deactivation, was used as the process response for the manipulation of the inlet temperature. The bandwidth was set so that temperature adjustments were made when percent conversion fell to 60% from the initial conversion of 70%. The final time was set as the time at which the final bandwidth becomes one-tenth of the initial bandwidth. As shown in Figure 13.3, the outlet temperature, which is the maximum reactor temperature in this case, is well below the maximum temperature allowed. Due to the deactivation, both the conversion and the outlet temperature decrease with time and the conversion eventually reaches the lower bound given by the bandwidth, triggering an adjustment of the inlet temperature. The new inlet temperature $(T_{in})_1$ calculated from the algorithm is seen to bring the conversion back to the desired level. When the outlet temperature reaches the allowed maximum, which occurs at around $t = 2 \times 10^5$s, the control policy is simply to increase T_{in} such that $T_{out} = T_{max}$ since the conversion can no longer be brought back to the desired level due to the temperature constraint. Consequently, the bandwidth decreases with time and eventually becomes less than one-tenth of the original bandwidth, resulting in the reactor shutdown for catalyst regeneration according to the constraint imposed. In practice, however, the conversion may be allowed to decrease below 60% when the maximum temperature is reached. In the example, the time average conversion is at the middle of the bandwidth, which is 65%.

The control algorithm of Eqs. 13.72 through 13.74 does not require any

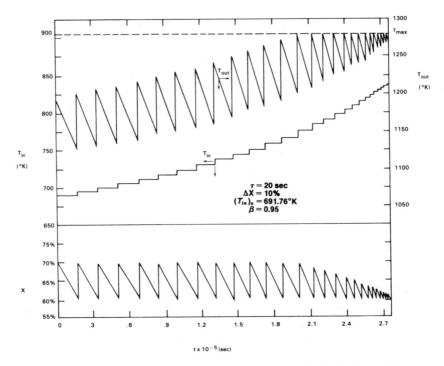

Figure 13.3   Reactor behavior when subjected to piecewise feedback control for constant conversion.

knowledge of catalyst deactivation and is applicable to any type of deactivation. It can also be extended (Hong and Lee 1983) to multiple reactions and nonadiabatic reactors, although the application to nonadiabatic reactors requires measurements of coolant temperature along the reactor. It should be noted here that reactor transients can exhibit "wrong way" behavior (Crider and Frost 1966) in which the transients initially move away from the "right" steady state, but eventually reach the right steady state in a few residence times. Since the control algorithm is piecewise and intermittently used for the temperature manipulation in intervals of hours rather than seconds due to the time scale of deactivation, which is much larger than the residence time under the assumption of pseudo-steady state, the right steady state values would be used in the control algorithm by the time the temperature change is made. Further, wrong way behavior is caused by a decrease in temperature rather than an increase (Mehta et al. 1981).

As indicated earlier, the usual optimization practice is to design first without due consideration of catalyst deactivation and then optimize the reactor performance. It is intuitively clear that much better reactor performance can be attained if the reactor size and operating conditions are both optimized during the design phase. Suppose that one wishes to maximize the yield of a single reaction per unit reactor volume by properly specifying the reactor size τ and inlet temperature $T_{in}(t)$. Then the optimization problem is:

$$\underset{\{\tau, T_{in}(t)\}}{\text{Max}} F = \frac{1}{\tau} \int_0^{t_f} x \, dt \qquad (13.75)$$

subject to mass and heat balances for the reactor and $T_{in} \leq (T_{in})_{max}$ with a fixed t_f. This two-parameter optimization problem can be solved one at a time by first searching for the maximum for a given τ and then finding the value of τ that yields the maximum of F. The results obtained for a constant conversion policy for the optimal $T_{in}(t)$ are shown in Figure 13.4. The algorithm of Eqs. 13.72 through 13.74 was used for the policy of constant conversion, which depends on τ and initial T_{in}. Under the constant conversion policy, therefore, the problem for a given τ is that of finding an initial conversion (or initial inlet temperature) that yields a maximum. As evident from Figure 13.4, much better reactor performance is attained when the reactor is optimally sized, τ of 22s yielding the maximum. Any other choice of τ would have resulted in poorer performance even if the reactor were operated optimally. This example clearly demonstrates the need for optimizing both reactor size and operating conditions.

So far optimization problems have been considered in which the regeneration

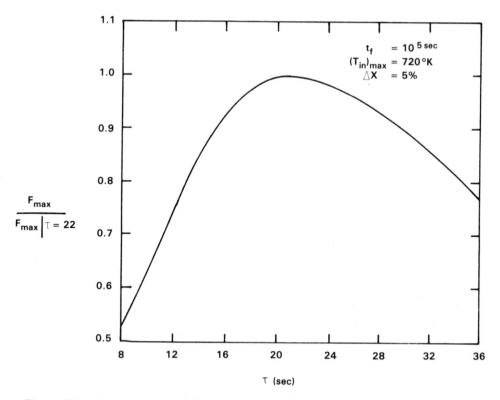

Figure 13.4 Maximum values of the performance index of Eq. 13.75 as a function of reactor size.

time is negligible compared to the operation time as in, for example, reforming and hydrocracking reactions. If the regeneration and operation times are of similar magnitude, there would be a unique operation/regeneration cycle optimal for a single fixed-bed or a train of fixed-beds. This optimization problem was considered by Weekman (1968) for the fixed-bed catalytic cracking of gas oil. The simplified model for catalytic cracking discussed in Chapters 8 and 11 was used:

$$F \xrightarrow{k_1} G \xrightarrow{k_2} C$$

in which F is gas oil, G is gasoline, and C is coke. As detailed in Chapter 8, the rate affected by the coking is expressed as:

$$r_1 = k_1 A y_1^2 \tag{13.76}$$

where A is the fractional activity remaining after coking, and y_1 is the weight fraction of gas oil. The fractional activity is correlated as follows:

$$A = e^{-\alpha t} \tag{13.77}$$

where α is a decay constant dependent on the catalyst and feedstock, and t is time on stream. For an isothermal plug-flow reactor in which the rates are affected only by coking, Eqs. 13.76 and 13.77 can be used in the reactor conservation equation to obtain the following time-averaged conversion (Weekman 1968) of gas oil:

$$\bar{x} = \frac{1}{\lambda} \ln \left[\frac{1 + A_0}{1 + A_0 \exp(-\lambda)} \right] \tag{13.78}$$

where

$$A_0 = K_0/S; \qquad K_0 = \rho_0 \epsilon_B k_1 / \rho_L \tag{13.79}$$

$$\lambda = \alpha t_o \tag{13.80}$$

Here, S is the liquid hourly space velocity (vol/vol · time), ρ_0 the initial vapor density, ρ_L the liquid charge density at room temperature, and t_o the duration of reactor operation. The performance criterion used was a measure of the approach to operation at 100% conversion without coking, termed the reactor efficiency and given by:

$$E_r = \frac{N F_0 t_o \bar{x}}{F_0 t_t} = \frac{\text{total actual product}}{\text{total ideal product}} \tag{13.81}$$

where F_0 is the feed rate, and t_t is the total time for N cycles of reaction of length t_o and regeneration of length t_r. Thus, the reactor efficiency becomes:

$$E_r = \frac{1}{\alpha(t_o + t_r)} \ln\left[\frac{1 + A_0}{1 + A_0 \exp(-\alpha t_o)}\right] \qquad (13.82)$$

where Eq. 13.78 has been used in 13.81.

Two different cases can be considered for the optimal operation: regeneration independent of operation and regeneration dependent on operation. For the first case, the reactor efficiency is simply a function solely of t_o. An inspection of the expression of Eq. 13.82 should reveal that there is a maximum for a given value of t_r. The dependence of E_r on t_o is shown in Figure 13.5 for various regeneration times. It is seen that the maxima are not materially affected by the regeneration time, although the optimal cycle length increases with increasing regeneration time as would be expected.

When regeneration and operation times are dependent on each other, the

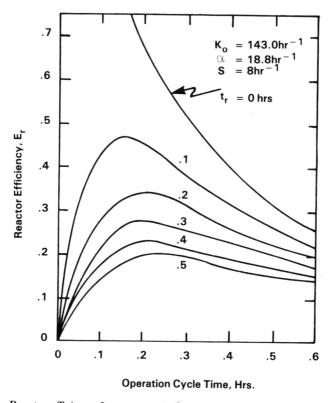

Figure 13.5 Reactor efficiency for regeneration independent of operation. (Weekman 1968; reprinted with permission from *Industrial and Engineering Chemistry, Process Design and Development.* Copyright by American Chemical Society.)

manner in which the regeneration is carried out has to be determined for the interrelationship between regeneration and operation. As shown by Voorhies (1945), coke formation in catalytic cracking can be well correlated by a negative order, i.e., self-inhibiting reaction:

$$\frac{dC_c}{dt} = k_c C_c^{-n} \tag{13.83}$$

which, upon integration with $C_c = 0$ at $t = 0$, yields the Voorhies relationship:

$$C_c = a t_o^b \tag{13.84}$$

where

$$a = [(1+n)k_c]^b \tag{13.85}$$

and

$$b = \frac{1}{1+n} \tag{13.86}$$

Further, if the coke burning is diffusion-free, one has:

$$-\frac{dC_c}{dt_r} = k_r C_{O_2} C_c^m \tag{13.87}$$

where m is unity for monolayer coke and less than unity for multilayer coke as detailed in Chapter 5. Solving Eq. 13.87 for constant C_{O_2} with $C_c = a t_o^b$ at $t_r = 0$ and with $C_c = 0$ as $t_r \to \infty$ yields:

$$t_r = \frac{a^{1-m} t_o^{(1-m)/(n+1)}}{k_r C_{O_2}(1-m)} ; \qquad m \neq 1 \tag{13.88}$$

It can be observed from Eq. 13.82 that if $E_r \to 0$ as $t_o \to 0$, an optimum E_r will be found as t_o is increased since as $t_o \to \infty$, $E_r \to 0$. Substituting Eq. 13.88 into 13.82 and taking the limit as $t_o \to 0$ by l'Hôpital's rule gives:

$$\operatorname*{Lim}_{t_o \to 0} E_r = \operatorname*{Lim}_{t_o \to 0} \left(\frac{A_0}{1+A_0}\right)\left[1 + \frac{a^{1-m} t_o^{-(m+n)/(n+1)}}{(1+n)k_r C_{O_2}}\right]^{-1} \tag{13.89}$$

For the reactor efficiency to approach zero as $t_o \to 0$, the exponent on t_o must be negative. Thus, if $n > 0$ and $m \geq 0$, an optimum will exist in E_r as t_o is increased. This result shows that for zero or any positive order coke burning an optimum operation/regeneration cycle will exist, provided the Voorhies parameter b is between zero and unity. Shown in Figure 13.6 is such a relationship, illustrated

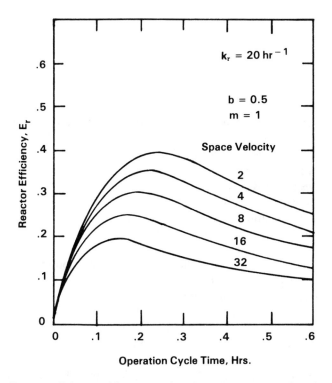

Figure 13.6   Reactor efficiency with regeneration dependent on operation. (Weekman 1968; reprinted with permission from *Industrial and Engineering Chemistry, Process Design and Development.* Copyright by American Chemical Society.)

for $m = 1$, $b = 0.5$, and a final coke content at the end of regeneration of 0.03 wt% coke on catalyst. Considerable improvement is seen to result from the optimization. Here again, the reactor size (inverse of space velocity) is seen to play an important role.

Summary

Equilibrium constraints and catalyst deactivation lead naturally to optimization problems. Simple maximization, dynamic programming, maximum principles, and other techniques can be used to solve the optimization problems. Optimization is usually carried out with respect to operating conditions. As pointed out repeatedly in this chapter, it should also be done with respect to reactor size. Perhaps the most powerful optimal policy for a reactor affected by deactivation is that of a constant conversion. However, it is not usually applicable to realistic systems and more work is needed for simple and yet general policies applicable to realistic systems. Means for estimating the extent of deactivation from process measurements and their use for optimization is another area that needs further work since detailed knowledge of deactivation is usually unavailable. An extensive review of reactor optimization in the presence of catalyst deactivation is given by Kovarik and Butt (1982).

NOTATION

a	constant given by Eq. 13.85
A	fraction of catalyst activity remaining after deactivation
A_c	cross-sectional area
A_0	quantity defined in Eq. 13.79
A'	dA/dt
$\mathbf{A}$	vector of A
b	constant given by Eq. 13.86
C	concentration of main reactant
C_b	bulk-fluid concentration of main reactant
$(C_b)_n$	C_b in nth bed
C_c	coke concentration; pellet center concentration given in Eq. 13.7
C_d	desired concentration of main reactant
C_e	reactor outlet concentration
C_n	concentration of main reactant to the nth stage
C_0	feed concentration of main reactant
C_p	concentration of poisoning species
C_{d_0}	C_d at time zero
$\overline{C}_{d_i}$	modified C_d at the ith control action, given by Eq. 13.72
C_{in}	inlet concentration of main reactant
C_{out}	outlet concentration of main reactant
D_e	effective diffusivity
E_a	activation energy for key reaction, forward reaction
E_b	activation energy for backward reaction
E_p	activation energy for poisoning reaction
E_r	reactor efficiency defined by Eq. 13.81
$f(C)$	concentration dependence of the rate of reaction
f_i	function given in Eq. 13.46
F	performance index defined in Eq. 13.75
F_0	feed rate
g_b	concentration dependence of rate of the backward reaction
g_f	concentration dependence of rate of the forward reaction
g_j	function given in Eq. 13.47
g_1, g_2, g_3	functions given in Eqs. 13.53, 13.38, and 13.39, respectively
G	function given in Eq. 13.48
h	local activity factor defined by Eq. 13.64; film heat transfer coefficient
H	overall activity factor defined by Eq. 13.67; Hamiltonian defined by Eq. 13.50
H_c	current H value given by Eq. 13.68
$(H_c)_i$	value of H_c calculated at the ith control, given by Eq. 13.73
H_n	new H value given by Eq. 13.69
$(-\Delta H)$	heat of reaction per mole of main reactant

J	quantity given in Eq. 13.42
k	rate constant of main reaction
k_b	k elevated at T_b
k_c	rate constant for coking reaction
k_0	preexponential factor of k
k_p	rate constant for poisoning reaction
k_{p_0}	preexponential factor for k_p
k_r	rate constant for coke regeneration reaction
$(k_p)_s$	k_p evaluated at pellet surface temperature
k_1	rate constant in Eq. 13.76
K	equilibrium constant; k/k_{-1}
K_0	preexponential factor of K; constant given in Eq. 13.79
$\mathbf{K}$	vector of K
L	pellet characteristic length
m	exponent in Eq. 13.87
n	negative order of coking reaction
N	total number of beds or stages; number of operation/regeneration cycles
N_b	bulk-fluid concentration of poisoning reaction
p	E_a/E_p
P	performance index given by Eq. 13.33; performance index given by Eq. 13.48
P_n	performance index for n^{th} stage given by Eq. 13.18
Q	poison capacity of catalyst in moles poisoning species per pellet volume; $E_b - E_a$
Q_n	measure of net profit for n^{th} stage given by Eq. 13.12
r_c	intrinsic rate of main reaction
r_1	rate given by Eq. 13.76
R_g	gas constant
R_p	global rate of deactivation
R_G	global rate of main reaction
R_{G_d}	global rate of deactivated catalyst
R_{G_f}	global rate of fresh catalyst
R_{G_n}	R_G for n^{th} stage
S	space velocity, vol/(vol · time)
t	time
t_f	final time at which catalyst is regenerated; catalyst life
t_o	duration of operation in operation/regeneration cycle
t_r	length of time for catalyst regeneration
t_t	$t_o + t_r$
T	temperature
T_b	bulk-fluid temperature
$(T_b)_n$	T_b in nth bed
T_n	outlet temperature for n^{th} stage
T_{n-1}	inlet temperature to n^{th} stage after cooling
T_{in}	reactor inlet temperature

$(T_{in})_c$	current value of T_{in}
$(T_{in})_n$	new value of T_{in}
U_1	$R_g T / E_1$ in Eq. 13.53
$\mathbf{U}$	vector of control variables
U^*	maximum value of control variable
U_*	minimum value of control variable
v	superficial fluid velocity
v_i	inlet value of state variable x_i
v^*	maximum value of v_i
v_*	minimum value of v_i
V	reactor volume
w_j	initial distribution of activity
w^*	maximum value of w_j
w_*	minimum value of w_j
x	conversion
x_i	i^{th} state variable
$\mathbf{x}$	vector of state variables
$\bar{x}$	average conversion
$x_{A'}$	$\partial x / \partial A'$
y	dimensionless reactor length
y_1	weight fraction of gas oil
Z	total reactor length

Greek Letters

α	product price per mole; decay constant in Eq. 13.77
β	$(-\Delta H)/\rho C_p$; proportionality constant in Eq. 13.72
β_1, β_2	constants in Eqs. 13.54 and 13.55
γ	costs per unit volume of reactor; fraction of catalyst deactivated
ϵ	small nonnegative constant in Eq. 13.62
ϵ_B	bed voidage
θ	dimensionless time
λ	$v/(1 - \epsilon_B)$; αt_o
λ_i	adjoint variable in Eq. 13.51
μ_j	adjoint variable in Eq. 13.52
ρ	constant in Eq. 13.53; density
ρ_0	initial vapor density
ρ_L	liquid charge density at room temperature
ν	γ/α
τ	$(1 - \epsilon_B)Z/v$
ϕ_G	generalized Thiele modulus given by Eq. 13.8

Subscripts

b	bulk fluid
c	current

n	new; n^{th} stage
N	N^{th} stage
s	pellet surface
in	reactor inlet
out	reactor outlet

PROBLEMS

13.1. Consider the following second-order reversible reaction:

$$A \overset{k}{\underset{k_{-1}}{\rightleftharpoons}} B$$

$$kg_f = kC^2; \qquad k_{-1}g_b = k_{-1}C_B^2$$

where C is the concentration of the main reactant A. Show that combining the pellet conservation equations for A and B yields:

$$C_B = (C_B)_b + C_b - C$$

if one assumes equal diffusivities. Show further for a plug-flow reactor that:

$$C_{in} - C_b = (C_B)_b$$

if one assumes no B in the feed, where C_{in} is the inlet concentration of A, and the subscript b denotes the bulk-fluid concentration. Show that $f(C)$ in Eq. 13.4 for this problem is:

$$f(C) = C^2 - \frac{1}{K}(C_{in} - C)^2; \qquad K = k/k_{-1}$$

Obtain an expression that can be solved for the optimal T_{in} which maximizes C_{out} using Eq. 13.1, given C_{in} and τ. Assume that the pellet center concentration C_c is negligible.

13.2. Construct a diagram corresponding to Figure 13.2 for a two-bed adiabatic reactor and specify optimal inlet conditions to each bed using dynamic programming. For the SO_2 oxidation under consideration, assume that the global rate is the same as the intrinsic rate given (Collina et al. 1971) by:

$$r_c = \frac{k_1 p_{O_2} p_{SO_2} \left(1 - \dfrac{p_{SO_3}}{p_{SO_3} p_{O_2}^{1/2} K_p}\right)}{22.414(1 + K_2 p_{SO_2} + K_3 p_{SO_3})^2}$$

where $r_c =$ mol SO_2/g cat hr
  $k_1 = \exp(12.160 - 5473/T)$

$$K_2 = \exp(-9.953 + 8619/T)$$
$$K_3 = \exp(-71.745 + 52596/T)$$
$$K_p = \exp(11300/T - 10.68)$$

Use the following conditions of Lee and Aris (1963):

Feed: 7.8 mole % SO_2, 10.8 mole % O_2, 81.4 mole % N_2 at 1 atm and 37°C

Mean specific heat = 0.221 cal/g°C, $(-\Delta H) = 21.4$ kcal/mole, $\lambda = 0.00068$ mol/g cat hr

13.3. Obtain the temperature trajectory, $T(t)$, which is optimal with respect to the performance criterion of Eq. 13.33. Assume for an isothermal plug-flow reactor that the conservation equation is given by:

$$\frac{dC_b}{dy} = -\tau A k C_b$$

and the activity A is given by:

$$\frac{dA}{dt} = -k_p A$$

Show first that under the condition of constant conversion, the activity is given by:

$$A(t) = (1 - \beta t)^p$$

where p is the ratio of activation energy for k to that for k_p, and β is a constant. Assume that the rate constants are given by:

$$k_p = A_p \exp(-E_p/R_g T)$$
$$k = A_a \exp(-E_a/R_g T)$$

Find an expression for the optimal constant conversion.

13.4. For the example problem in the text (Eqs. 13.53 through 13.57), write the complete equations necessary for the optimization. Based on these equations, give detailed procedures readily usable for computer programming.

13.5. Extend the method of estimating the extent of deactivation to simple parallel and consecutive reactions. Specifically, derive an expression for H_c (Eq. 13.68) when the local activity factor h is the same for both reactions. Repeat the case in which h is different for different reaction paths.

13.6. Construct a diagram similar to Figure 13.6 using the following conditions:

$$K_0 = 143 \text{ hr}^{-1}, \ b = 0.5, \ m = 1$$
$$k_r = 20 \text{ hr}^{-1}, \ C_{O_2} = 10^{-4} \text{ mole/cm}^3, \ \alpha = 18.8 \text{ hr}^{-1}$$

Assume that the final coke weight after burnoff is 0.03% and that the initial coke content is 5%.

REFERENCES

Aris, R., *Introduction to the Analysis of Chemical Reactors,* Prentice-Hall, Englewood Cliffs, N.J. (1965).

Bellman, R., *Dynamic Programming,* Princeton University Press, Princeton, N.J. (1957).

Beveridge, G.S.B. and R.S. Schechter, *Optimization Theory and Practice,* McGraw-Hill, New York (1970).

Bliss, G.A., *Lectures on the Calculus of Variations,* University of Chicago Press, Chicago (1961).

Chou, A., W.H. Ray and R. Aris, Trans. Inst. Chem. Eng., *45,* 53 (1967).

Collina, A., D. Corbetta and A. Cappelli, Eur. Symp., *Use of Computers in the Design of Chemical Plants,* Firenze (1971).

Crider, J.E. and A. S. Frost, AIChE J., *12,* 514 (1966).

Crowe, C.M. and N. Therien, Can. J. of Chem. Eng., *52,* 822 (1974).

Denbigh, K.G. and J.C.R. Turner, *Chemical Reactor Theory,* 2nd ed., Cambridge University Press (1971).

Fournier, C.D. and F.R. Groves, Chem. Eng., *77,* 157 (1970).

Gruyaert, F. and C.M. Crowe, AIChE J., *22,* 985 (1976).

Hong, J.C. and H.H. Lee, AIChE J. (in press).

Koppel, L.B., *Introduction to Control Theory with Application to Process Control,* Prentice-Hall, Englewood Cliffs, N.J. (1972).

Kovarik, F.S. and J.B. Butt, Cat. Rev., *24,* 441 (1982).

Lee, K.Y. and R. Aris, Ind. Eng. Chem. Proc. Des. Dev., *2,* 300 (1963).

Mehta, P.S., W.N. Sams and D. Luss, AIChE J., *27,* 234 (1981).

Millman, M.C. and S. Katz, Ind. Eng. Chem. Proc. Des. Dev., *6,* 447 (1967).

Ogunye, A.F. and W.H. Ray, AIChE J., *17,* 43 (1971).

Ray, W.H. and J. Szekely, *Process Optimization,* Wiley, New York (1973).

Roberts, S.M., *Dynamic Programming in Chemical Engineering and Process Control,* Academic Press, New York (1964).

Sirazetdinov, T.K. and G.L. Degtyarev, Automation Remote Control, *28,* 1642 (1967).

Szepe, S. and D. Levenspiel, Chem. Eng. Sic., *23,* 881 (1968).

Voorhies, A., Ind. Eng. Chem., *37,* 318 (1945).

Weekman, V.W., Jr., Ind. Eng. Chem. Proc. Des. Dev., *7,* 90, 253 (1968).

PART IV

Transport Properties

CHAPTER 14

Transport Properties and Experimental Methods

Throughout this book various transport properties and transfer coefficients have been used. These include effective diffusivity and thermal conductivity for mass and heat transport in catalyst pellets, film transfer coefficients for mass and heat transfer across the pellet-bulk fluid interface, transport properties for the degree of dispersion of mass and heat in the reactor, and heat transfer coefficients for heat exchange between the cooling medium and the reactor. In this chapter these transport properties and transfer coefficients are treated in detail, including experimental methods for obtaining these properties.

14–1 PELLET TRANSPORT PROPERTIES

This section considers the effective diffusivity and thermal conductivity for the transport of mass and heat in the pellets. In Chapter 4, the effective diffusivity for a key species was defined by the following flux relationship:

$$N = -D_e \frac{dC}{dz} \qquad (14.1)$$

where the subscript 1 in Eq. 4.7 is now eliminated for the key species. An explicit expression for D_e is impossible to obtain from the general flux expressions of Eq. 4.13 for a multicomponent system even when the pellet is assumed to be isobaric and surface diffusion is neglected, as has been assumed in all treatments so far. It is often satisfactory, however, to calculate a theoretical value of the diffusivity based on a pseudo-binary system in which the key species is one species and the rest of the system is another. This theoretical estimate of the diffusivity may be used for comparison with the diffusivity obtained experimentally using the definition of Eq. 14.1. For a binary system, the effective diffusivity D_e can be expressed as:

$$\frac{1}{D_e} = \frac{1 + (m^{1/2} - 1)x}{D_{e,12}} + \frac{1}{D_{e,k}} \qquad (14.2)$$

which is Eq. 4.16 for the key species. Here, $D_{e,\,12}$ and $D_{e,k}$ are effective molecular and Knudsen diffusivities, x is the mole fraction of the key species, and m is the ratio of the molecular weight of the key species to that of the other species. The effective diffusivities, which account for the obstruction of movement of gas molecules in a porous medium, are related to the obstruction-free molecular and Knudsen diffusivities by:

$$D_{e,\,12} = D_{m,\,12} f_e(\epsilon/\kappa) \tag{14.3}$$

$$D_{e,k} = D_k f_e(\epsilon/\kappa) \tag{14.4}$$

where $f_e(\epsilon/\kappa)$ is a correction factor for the effective values of the diffusivities in the porous medium. Here, ϵ and κ are the porosity and tortuosity of the medium.

For an accurate calculation of the molecular diffusivity $D_{m,\,12}$, the Chapman-Enskog formula (Bird et al. 1960) can be used:

$$D_{m,\,12} = 0.0018583 \left\{ \left[T^3 \left(\frac{1}{M_1} + \frac{1}{M_2} \right) \right]^{1/2} \Big/ [P \, \sigma_{12}^2 \, \Omega_{12}] \right\} (\text{cm}^2/\text{sec}) \tag{14.5}$$

where the total pressure P is in atm, T is in °K, and σ_{12} is in Å and Ω_{12} are the parameters calculable from Lennard-Jones parameters: $\sigma_{12} = (\sigma_1 + \sigma_2)/2$ and $\epsilon_{12} = (\epsilon_1 \epsilon_2)^{1/2}$. These parameters are given in Tables 14.1 and 14.2. When the pore radius is very small, collisions will occur primarily between gas molecules and the pore wall, rather than between molecules. For such Knudsen diffusion in a long capillary, the diffusivity obtainable from the simple kinetic theory of gases is:

$$D_k = \frac{2}{3} \bar{r} \bar{v} \tag{14.6}$$

where $\bar{r}$ is the pore radius (capillary radius), and $\bar{v}$ is the mean molecular velocity given by:

$$\bar{v} = \left(\frac{8 R_g T}{\pi M} \right)^{1/2}$$

where M is the molecular weight, and R_g is the gas constant. The Knudsen diffusivity, therefore, is given by:

$$D_k = 9.7 \times 10^3 \, \bar{r}(T/M)^{1/2} \quad (\text{cm}^2/\text{sec}) \tag{14.7}$$

For Knudsen diffusion in a porous medium, an average pore radius may be used for $\bar{r}$ (in cm) in Eq. 14.7. It is instructive in this regard to compare the relative magnitude of molecular and Knudsen diffusivities with the help of an approximate expression for the former based on the simple kinetic theory, given by:

$$D_{m,\,12} = \frac{1}{3} \bar{v} \lambda \qquad \left(\lambda \propto \frac{1}{P} \right) \tag{14.8}$$

Table 14.1 Intermolecular Force Parameters (after Bird, Stewart, and Lightfoot, *Transport Phenomena*, © 1960, Wiley. Reprinted by permission of John Wiley & Sons, Inc.)

Substance	Molecular Weight M	σ (Å)	ϵ/k (°K)
Light elements:			
H_2	2.016	2.915	38.0
He	4.003	2.576	10.2
Noble gases:			
Ne	20.183	2.789	35.7
Ar	39.944	3.418	124.
Kr	83.80	3.498	225.
Xe	131.3	4.055	229.
Simple polyatomic substances:			
Air	28.97	3.617	97.0
N_2	28.02	3.681	91.5
O_2	32.00	3.433	113.
O_3	48.00	—	—
CO	28.01	3.590	110.
CO_2	44.01	3.996	190.
NO	30.01	3.470	119.
N_2O	44.02	3.879	220.
SO_2	64.07	4.290	252.
F_2	38.00	3.653	112.
Cl_2	70.91	4.115	357.
Br_2	159.83	4.268	520.
I_2	253.82	4.982	550.

Lennard-Jones Parameters

Substance	Molecular Weight M	σ (Å)	ϵ/k (°K)
Hydrocarbons:			
CH_4	16.04	3.822	137.
C_2H_2	26.04	4.221	185.
C_2H_4	28.05	4.232	205.
C_2H_6	30.07	4.418	230.
C_3H_6	42.08	—	—
C_3H_8	44.09	5.061	254.
$n\text{-}C_4H_{10}$	58.12	—	—
$i\text{-}C_4H_{10}$	58.12	5.341	313.
$n\text{-}C_6H_{12}$	72.15	5.769	345.
$n\text{-}C_6H_{14}$	86.17	5.909	413.
$n\text{-}C_7H_{16}$	100.20	—	—
$n\text{-}C_8H_{18}$	114.22	7.451	320.
$n\text{-}C_9H_{20}$	128.25	—	—
Cyclohexane	84.16	6.093	324.
C_6H_6	78.11	5.270	440.
Other organic compounds:			
CH_4	16.04	3.822	137.
CH_3Cl	50.49	3.375	855.
CH_2Cl_2	84.94	4.759	406.
$CHCl_3$	119.39	5.430	327.
CCl_4	153.84	5.881	327.
C_2N_2	52.04	4.38	339.
COS	60.08	4.13	335.
CS_2	76.14	4.438	488.

Table 14.2   Functions for Prediction of Transport Properties of Gases at Low Densities (after Bird, Stewart, and Lightfoot, *Transport Phenomena,* © 1960, Wiley. Reprinted by permission of John Wiley & Sons, Inc.)

kT/ϵ or kT/ϵ_{AB}	$\Omega_\mu = \Omega_k$ (For Viscosity and Thermal Conductivity)	$\Omega_{D,12}$ (For Mass Diffusivity)	kT/ϵ or kT/ϵ_{AB}	$\Omega_\mu = \Omega_k$ (For Viscosity and Thermal Conductivity)	$\Omega_{D,12}$ (For Mass Diffusivity)
			2.50	1.093	0.9996
			2.60	1.081	0.9878
0.30	2.785	2.662	2.70	1.069	0.9770
0.35	2.628	2.476	2.80	1.058	0.9672
0.40	2.492	2.318	2.90	1.048	0.9576
0.45	2.368	2.184			
0.50	2.257	2.066	3.00	1.039	0.9490
0.55	2.156	1.966	3.10	1.030	0.9406
0.60	2.065	1.877	3.20	1.022	0.9328
0.65	1.982	1.798	3.30	1.014	0.9256
0.70	1.908	1.729	3.40	1.007	0.9186
0.75	1.841	1.667	3.50	0.9999	0.9120
0.80	1.780	1.612	3.60	0.9932	0.9058
0.85	1.725	1.562	3.70	0.9870	0.8998
0.90	1.675	1.517	3.80	0.9811	0.8942
0.95	1.629	1.476	3.90	0.9755	0.8888
1.00	1.587	1.439	4.00	0.9700	0.8836
1.05	1.549	1.406	4.10	0.9649	0.8788
1.10	1.514	1.375	4.20	0.9600	0.8740
1.15	1.482	1.346	4.30	0.9553	0.8694
1.20	1.452	1.320	4.40	0.9507	0.8652
1.25	1.424	1.296	4.50	0.9464	0.8610
1.30	1.399	1.273	4.60	0.9422	0.8568
1.35	1.375	1.253	4.70	0.9382	0.8530
1.40	1.353	1.233	4.80	0.9343	0.8492
1.45	1.333	1.215	4.90	0.9305	0.8456
1.50	1.314	1.198	5.0	0.9269	0.8422
1.55	1.296	1.182	6.0	0.8963	0.8124
1.60	1.279	1.167	7.0	0.8727	0.7896
1.65	1.264	1.153	8.0	0.8538	0.7712
1.70	1.248	1.140	9.0	0.8379	0.7556
1.75	1.234	1.128	10.0	0.8242	0.7424
1.80	1.221	1.116	20.0	0.7432	0.6640
1.85	1.209	1.105	30.0	0.7005	0.6232
1.90	1.197	1.094	40.0	0.6718	0.5960
1.95	1.186	1.084	50.0	0.6504	0.5756
2.00	1.175	1.075	60.0	0.6335	0.5596
2.10	1.156	1.056	70.0	0.6194	0.5464
2.20	1.138	1.041	80.0	0.6076	0.5352
2.30	1.122	1.026	90.0	0.5973	0.5256
2.40	1.107	1.012	100.0	0.5882	0.5170

where λ is the mean free path. This relationship is for rigid spheres of equal mass and radius for the two species. Taking the ratio of the two diffusivities, one gets from Eqs. 14.6 and 14.8:

$$D_k/D_{m,12} = 2\bar{r}/\lambda \qquad (14.9)$$

It is seen from Eq. 14.9 that the magnitude of the average pore radius relative to the mean free path determines the predominant mechanism of diffusion. Since λ is of the order of 1000 Å for gases at atmospheric pressure, diffusion in a porous medium will be predominantly by the Knudsen mechanism if the average pore radius $\bar{r}$ is much less than 1000 Å.

Theoretical calculations for the effective molecular and Knudsen diffusivities, and therefore, of the effective diffusivity D_e can be made from Eqs. 14.5 and 14.7 provided that the factor f_e representing the pore characteristics is known. Various models have been developed for the calculation of f_e. Two such models are considered here. The first is that due to Feng and Stewart (1973). The simplest form of f_e (ϵ/κ) that they obtained is:

$$f_e = \frac{\epsilon}{\kappa} = \frac{\epsilon}{3} \qquad (14.10)$$

Assumptions made are isotropic pores with no dead-ended branches, independence of ϵ and κ with pore size, and lumping of an integral over the entire pore structure into a single term. The second model to consider is the random-pore model due to Wakao and Smith (1962), shown in Figure 14.1. This model was developed for a bidisperse porous medium, but is equally applicable to a monodisperse medium. The nature of the interconnections of macro and micro void regions for a bidisperse pellet shown in the figure is the essence of the model. Transport in the pellet is assumed to occur by a combination of diffusion through the macro regions (of void fraction ϵ_M), the micro regions (of void fraction ϵ_μ), and a series contribution involving both regions but dominated by the diffusion through the micro regions. A probabilistic argument is used to assign the probability of access of gaseous molecules to these three regions equal to respective available area, which are ϵ_M^2 for the macro region, $(1 - \epsilon_M)^2$ for the micro region, and $2\epsilon_M(1 - \epsilon_M)$ for the series region. Since the micropore diffusivity D_μ is based on the micro void, the ratio of $\epsilon_\mu^2/(1 - \epsilon_M)^2$ is used for the (micro void/particle) area ratio, yielding:

$$D_e = \epsilon_M^2 D_M + (1 - \epsilon_M)^2 \frac{\epsilon_\mu^2}{(1 - \epsilon_M)^2} D_\mu + 2[2\epsilon_M(1 - \epsilon_M)] \frac{\epsilon_\mu^2}{(1 - \epsilon_M)^2} D_\mu \qquad (14.11)$$

$$= \epsilon_M^2 D_M + \frac{\epsilon_\mu^2(1 + 3\epsilon_M)}{1 - \epsilon_M} D_\mu$$

where D_M and D_μ are the macropore and micropore diffusivities, respectively. For a monodisperse pellet, the above relationship reduces to either $\epsilon_M^2 D_M$ or $\epsilon_\mu^2 D_\mu$.

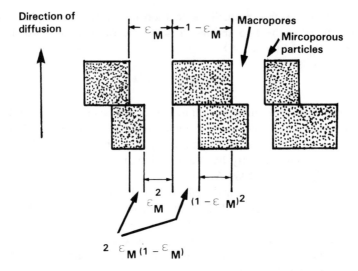

Figure 14.1   Random-pore model. (Wakao and Smith 1962; reprinted with permission from *Chemical Engineering Science.* Copyright by Pergamon Press, Inc.)

Therefore, the factor $f_e(\epsilon/\kappa)$ for the random pore model is:

$$f_e(\epsilon/\kappa) = \epsilon^2 \qquad (14.12)$$

Theoretical values of effective diffusivities can now be calculated with Eqs. 14.5 and 14.7 along with the factor f_e given by either Eq. 14.10 or Eq. 14.12. It must be pointed out, however, that the diffusivities thus calculated do not often compare satisfactorily with experimental values. Often the tortuosity is adjusted in such cases to force the theoretical values to conform to the experimental values. Detailed treatment of the tortuosity factor can be found in the book by Satterfield (1970).

Experimental determination of the effective diffusivity D_e is usually carried out in an apparatus, shown in Figure 14.2, known as a Wicke and Kallenbach chamber (1941). The chamber is operated at steady state and nearly constant pressure as is usually the case in reactors. Rewriting the definition of Eq. 14.1 for the flux, one has:

$$N = -\frac{P}{R_g T} D_e \frac{dy}{dz}$$

which, upon integration, yields:

$$N = -\frac{P}{R_g T} D_e \frac{(y_2 - y_1)}{\Delta z} \qquad (14.13)$$

where y is the mole fraction of the key species, and Δz is the length of the pellet. If the flow rates and concentrations are measured, N can be calculated for the

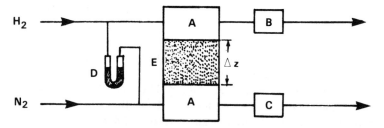

Figure 14.2 Constant-pressure apparatus for measuring diffusion rates in porous catalysts. A: mixing chambers; B: detector for composition of N_2 in H_2 stream; C: detector for composition of H_2 in N_2 stream; D: pressure equalization gauge; E: catalyst pellet. (After J.M. Smith, *Chemical Engineering Kinetics*, © 1980; with permission of McGraw-Hill Book Co., NY.)

key species, say H_2, (see Figure 14.2), in which case the subscript 2 is for the bottom of the chamber and 1 for the top. Then this flux, the measured concentrations and pellet length are substituted in Eq. 14.13 to obtain an experimental value of the effective diffusivity. It is noted that the effective diffusivities determined from the H_2-N_2 pair, for instance, can be used to calculate the diffusivitites for any other pairs for the *same* pellet (see Problem 14.1). The dynamic method (Dogu and Smith 1975) has also been used for the measurement of D_e using an apparatus similar to the Wicke-Kallenbach chamber. Their method involves introducing a pulse of one of the gases and analyzing the measured response based on a moment analysis. The dead-ended pores can affect the diffusion in this method. Experimental results show that the dynamic method yields similar (Baiker et al. 1982) or smaller values of D_e (McGreavy and Siddiqui 1980) when compared with the results of the static method. The latter discrepancy was attributed to pore size distribution. A flow-through diffusion cell has been proposed by Frost (1981). In contrast to the Wicke-Kallenbach cell, it can be operated at high temperature and pressure and does not require special sample preparation.

The effective diffusivity can also be determined from reaction conditions provided that the intrinsic kinetics are known. In the region of strong diffusion effects, the product of the effectiveness factor and the generalized Thiele modulus is unity, i.e., $\eta\phi_G = 1$, where ϕ_G is given by:

$$\phi_G = Lr_c \left[2D_e \int_0^{C_b} r_c(C)dc \right]^{-1/2} \tag{14.14}$$

Here, L is the characteristic length of the pellet, and r_c is the intrinsic rate of the reaction. Since η can be determined from the ratio of R_G/r_c where R_G is the observed rate, one has from Eq. 14.14:

$$D_e = \frac{(LR_G)^2}{2 \int_0^{C_b} r_c(C)dc} \tag{14.15}$$

It is seen that given the intrinsic kinetics, the actual rate of the reaction in the diffusion-limited region can be used to determine the effective diffusivity.

The theory of diffusion in liquids is not as well developed as that in gases. In fact, no satisfactory theory appears to exist. The main interest here is the diffusion of a dissolved gaseous reactant in the pores of support catalysts filled with liquid reactant, as in a slurry or trickle-bed reactor. The mean free path in liquids is so small that Knudsen diffusion is not significant. Thus, the diffusion rate is unaffected by the pore diameter and pressure. The effective diffusivity may be defined in the same manner as in Eq. 14.1 with the diffusivity D_{AB}, for instance, representing the effective diffusivity of species A of interest in pores filled with a solution of A and B. As expected, values of D_{AB} are much less than those for gases and are of the order of 1×10^{-5} cm²/sec. For gases dissolved in liquids, the diffusivity can be an order of magnitude larger. Several correlations (Reid et al. 1977) are available for estimating diffusivities in liquids at low concentrations (infinite dilution), but these should be used with caution.

Effective Thermal Conductivity

It is not surprising, in view of the minor role played by internal heat transport, that relatively little attention has been paid to the effective thermal conductivity inside pellets. In general, a solid is a much better conducting material than a gas. It is natural, therefore, that the conductivity of gas is the controlling factor in the overall (effective) thermal conductivity. This manifests itself in the strong dependence of the effective thermal conductivity on the void fraction of the pellets. The thermal conductivity of solid alumina is of the order of 10^{-3} (cal cm^{-1} sec^{-1} °K^{-1}) while that of helium, which has the highest thermal conductivity of gases, is of the order of 10^{-4} (cal cm^{-1} sec$^-$ °K^{-1}). In the range of typical reaction temperatures and pressures, the experimental data obtained by Masamune and Smith (1963) and Sehr (1958) indicate that the effective thermal conductivities fall in a very small band around 3×10^{-4} (cal cm^{-1} sec^{-1} °K^{-1}). It is thus seen that the gas conductivity is indeed the controlling factor and that a simple-minded approach may be sufficient for theoretical calculations of the effective thermal conductivity. Such an approach by Woodside and Messmer (1961) yielded the following relationship:

$$\lambda_e = \lambda_s \left(\frac{\lambda_g}{\lambda_s}\right)^\epsilon \tag{14.16}$$

where λ_e, λ_s, and λ_g are the effective thermal conductivity of the pellet, and the thermal conductivities of solid and gas, respectively, and ϵ is the fractional void. A more rigorous approach can be found in the article by Butt (1965).

The theory (Bird et al. 1960) of gas conductivity is well developed and the Chapman-Enskog theory for a monoatomic gas at low density gives:

$$\lambda_g = 1.989 \times 10^{-4} \frac{(T/M)^{1/2}}{\sigma^2 \Omega_k} \qquad (\text{cal cm}^{-1} \text{sec}^{-1} \text{ }^{\circ}\text{K}^{-1}) \qquad (14.17)$$

$$T \text{ in } ^{\circ}\text{K}, \sigma \text{ in } \text{Å}$$

where the values of σ and Ω_k are given in Tables 14.1 and 14.2. For a polyatomic gas, one can use the following:

$$\lambda_g = \left(\overline{C}_p + \frac{5}{4} \frac{R_g}{M} \right) \mu \qquad (14.18)$$

where $\overline{C}_p$ is the mass heat capacity, and μ is the viscosity. For the conductivity of gas mixtures, the following can be used:

$$(\lambda_g)_{mix} = \sum_{i=1}^{n} \frac{y_i \lambda_i}{\sum\limits_{i=1}^{n} y_i \phi_{ij}} \qquad (14.19)$$

$$\phi_{ij} = \frac{1}{8^{1/2}} \left(1 + \frac{M_i}{M_j} \right)^{-1/2} \left[1 + \left(\frac{\mu_i}{\mu_j} \right)^{1/2} \left(\frac{M_j}{M_i} \right)^{1/4} \right]^2 \qquad (14.20)$$

14–2   MASS AND HEAT TRANSFER COEFFICIENTS

Average transport coefficients for transfer between the bulk-fluid and particle surface can be correlated in terms of dimensionless groups that characterize the flow conditions. It is common practice to correlate experimental data in terms of j-factors. Usually, the mass transfer coefficient is obtained from the j factor for mass; the heat transfer coefficient is obtained from j factor analogy. There have been many experimental studies of mass transfer in fixed-beds and summaries and analyses of the results are available (Whitaker 1972; Dwivedi and Upadhay 1977). For Reynolds numbers greater than 10, the following relationship (Dwivedi and Upadhay 1977) between j_D and the Reynolds number represents available data:

$$j_D = \frac{0.458}{\epsilon_B} \left(\frac{d_p G}{\mu} \right)^{-0.407} \qquad (14.21)$$

where G = mass (superficial) velocity based on the cross-sectional area of the empty reactor ($= v\rho$)

μ = viscosity of fluid

ϵ_B = void fraction of the bed

d_p = diameter of a spherical catalyst pellet (For other shapes, an approximate value of d_p is that of a sphere with the same external area as the nonspherical pellet.)

ρ = fluid density

v = superficial velocity

The j-factor for mass is defined as:

$$j_D = \frac{k_g}{v}\left(\frac{\mu}{\rho D}\right)^{2/3} = \frac{k_g}{v}(\text{Sc})^{2/3} \qquad (14.22)$$

where D is the molecular diffusivity of the component being transferred, and k_g is the mass transfer coefficient. For a Reynolds number less than 10, a correlation based on a boundary layer model (Carberry 1960) may be used:

$$j_D = 1.15\left(\frac{d_p v}{v\epsilon_B}\right)^{-0.5} \qquad (14.23)$$

where v is the kinematic viscosity. This correlation may be used down to a Reynolds number of 0.5. As the Reynolds number gets smaller, the applicability of j-factors becomes increasingly uncertain.

While literature correlations sometimes make a distinction between j_H and j_D, the validity of the difference is uncertain in the absence of radiation. Hence, the original j-factor analogy may be used to obtain heat transfer coefficients:

$$j_D = j_H = \frac{h}{\rho v C_p}\left(\frac{C_p \mu}{\lambda}\right)^{2/3} = \frac{h}{\rho v C_p}(\text{Pr})^{2/3} \qquad (14.24)$$

where λ is the thermal conductivity of fluid, and h is the heat transfer coefficient. Proper values of the mass and heat transfer coefficients that should be used when axial dispersion is included in the reactor model are given by Wakao and Funazkri (1978) and Wakao et al. (1979).

For fluidized-beds, Kunii and Levenspiel (1969) have summarized available mass and heat transfer data in the form of graphs of Sherwood and Nusselt numbers versus Reynolds number. Typical of the results are those of Chu et al. (1953) which may be expressed in terms of j-factors as:

$$j_D \text{ or } j_H = 1.77\left[\frac{d_p G}{\mu(1-\epsilon_B)}\right]^{-0.44} \qquad (14.25)$$

for the range $30 < d_p G/\mu(1-\epsilon_B) < 5000$. The transfer coefficients for multi-phase reactors have already been treated in Chapter 7.

Of the two transfer coefficients for gas-phase catalytic reactions, the heat transfer coefficient is much more important than the mass transfer coefficient since the external mass transfer resistance can be neglected under realistic conditions. The heat transfer coefficient under reaction conditions may be determined using a single-pellet reactor. At steady state, one has:

$$(-\Delta H)R_G L = h(T_s - T_b)$$

where R_G = measured rate of reaction
 L = pellet volume/external surface area
 T_s = pellet surface temperature
 T_b = bulk-fluid temperature
 $(-\Delta H)$ = heat liberated per mole reactant

If the temperatures are measured, then the heat transfer coefficient can be determined from:

$$h = \frac{(-\Delta H)R_G L}{T_s - T_b}$$

In view of the good agreement obtained between measured and theoretical temperature differences across the pellet-bulk fluid film (Butt et al. 1977), the temperature measurements may give accurate values of h under reaction conditions.

14-3 REACTOR TRANSPORT PROPERTIES

In this section, consideration is given to effective thermal conductivities and diffusivities, in both the radial and axial directions, for the dispersion of mass and heat in a fixed-bed. The flux relationships for the mass and heat transport in the bed may be considered to define these transport properties. Also considered are the heat transfer coefficients at the tube wall.

Effective Thermal Conductivities ($\rho C_p K$)

Many models and correlations have been developed for the radial effective thermal conductivity (see the review by Kulkarani and Doraiswamy 1980). The thermal dispersion coefficient K used in Chapter 9 is related to the effective conductivity by $\lambda = \rho C_p K$. For nonadiabatic fixed-beds, the main heat conduction is in the radial direction, and thus, the radial conductivity is much more important than the axial conductivity. The axial conductivity represents the conduction superimposed on the bulk flow, which is quite small relative to the heat transport by the bulk flow. Therefore, most work has been directed to the radial conductivity.

The radial heat transport is complex, involving conduction, convection, and radiation between voids and solid and between solid particles. Possible modes of heat transfer in the radial direction are shown in Figure 14.3. Different physical models result depending on whether various resistances to the heat transport are in series, parallel, or a combination of both. Here an additive model is considered, which assumes that the radial effective thermal conductivity consists of static (conduction and radiation) and dynamic contributions, the latter caused by fluid motion. These two contributions are considered to be additive:

$$\lambda_{er} = \lambda_s^\circ + \lambda_d^\circ \tag{14.26}$$

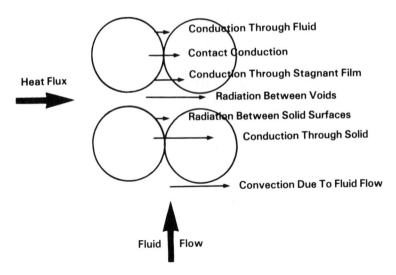

Figure 14.3 Various modes of heat transport in a packed bed. (Reprinted from Kulkarani and Doraiswamy 1980, by courtesy of Marcel Dekker, Inc.)

where λ_s° and λ_s° are the static and dynamic contributions. The static contribution is considered to consist of conduction and radiation, which are again additive:

$$\lambda_s^\circ = \lambda_s^c + \lambda_s^r \tag{14.27}$$

where λ_s^c and λ_s^r are the contributions due to conduction and radiation, respectively. The Kunii-Smith model (1960), when the radiation term is neglected, reduces to:

$$\frac{\lambda_s^c}{\lambda_g} = \epsilon_B + (1 - \epsilon_B)\frac{\beta}{\delta + 2\lambda_g/(3\lambda_s)} \tag{14.28}$$

where λ_g = fluid thermal conductivity
   λ_s = solid thermal conductivity
   ϵ_B = bed porosity
   β = ratio of the effective distance between particles to particle diameter
     $(= L/d_p)$
   δ = stagnant fluid thickness

Here, the thickness δ can be read from curve A of Figure 14.4 for dense packing with β of 0.895, and from curve B for loose packing with β of 1. An expression for the radiant contribution (Kulkarani and Doraiswamy 1980), which is equivalent to the radiation part of the Kunii-Smith model, is:

$$\lambda_s^r = \epsilon_B d_p h_b + \frac{1 - \epsilon_B}{1/\lambda_s + 1/(d_p h_b)} \tag{14.29}$$

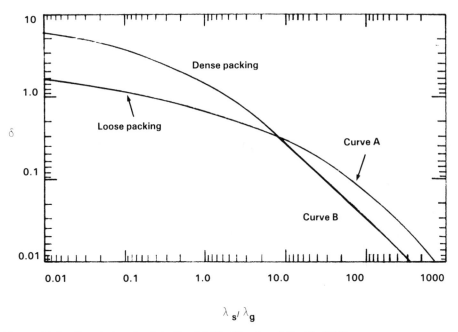

Figure 14.4 Correlation for δ in Eq. 14.28. (Kunii and Smith 1960)

where d_p is the particle diameter and the radiant heat transfer coefficient h_b is given by:

$$h_b = 1.952 \times 10^{-7} \, gT^3 \qquad \text{kcal/hr m}^2 {}^\circ\text{C} \qquad (14.30)$$

Here, g can be assumed to be equal to the emissivity e. For the dynamic contribution, the relationship suggested by DeWasch and Froment (1972) may be used:

$$\lambda_d^\circ = \frac{0.0025}{1 + 46(d_p/d_t)^2} \left(\frac{d_p G}{\mu} \right) \qquad \text{kcal/hr m}{}^\circ\text{K} \qquad (14.31)$$

where d_t is tube diameter. Use of Eqs. 14.28, 14.29, and 14.31 in Eq. 14.26 then yields the effective thermal conductivity in the radial direction. The effective thermal conductivity increases almost linearly with increasing particle Reynolds number $(d_p G/\mu)$ and ranges from 0.2 to 2 kcal/hr m°K for Reynolds number ranging from almost zero to 800.

Because of the very minor role that conduction has in axial heat transport, not much work has been done and a relatively simple correlation should suffice for the purpose of estimating the approximate magnitude of the axial effective thermal conductivity. For this purpose, the correlation of Yagi and co-workers (1960) may be used:

$$\frac{\lambda_{ea}}{\lambda_g} = \frac{\lambda_s^\circ}{\lambda_g} + \alpha \text{PrRe} \qquad (14.32)$$

where λ_{ea} = axial effective thermal conductivity

λ_s^o = static contribution, which is the same as that for radial thermal conductivity (Eq. 14.27)

$Re = d_p G/\mu$

α = factor varying with solid thermal conductivity, 0.7 for steel and 0.8 for glass

Pr = Prandtl number of fluid

Here again, the static and dynamic contributions have been assumed to be additive.

Wall Heat Transfer Coefficients

In Chapters 9 and 10, two heat transfer coefficients were used: the overall heat transfer coefficient for a one-dimensional model (U) and the wall heat transfer coefficient for two-dimensional model (h_w). The overall heat transfer coefficient can be written as:

$$\frac{1}{U} = \frac{1}{h_e} + \frac{1}{h_c} \tag{14.33}$$

where h_e is the heat transfer coefficient on the bed side, and h_c is that on the cooling or heating medium side. Here, the conduction resistance at the wall was neglected. The surface area ratios were also set to unity since, in general, the wall thickness is so small that the ratios are close to unity. The correlations for h_c can be found in any book on heat transfer. Here again, additive models for the heat transfer coefficients will be used.

For the heat transfer coefficient h_e, the correlation of DeWasch and Froment (1972) may be used for an accurate estimate:

$$(Bi)_h = (Bi)_h^o + 0.024 \left(\frac{d_p G}{\mu} \right) \tag{14.34}$$

where $(Bi)_h = h_e d_p / \lambda_g$

$(Bi)_h^o = h_e^o d_p / \lambda_g$

For the static contribution (h_e^o), the following can be used:

$$h_e^o = 6.15 \left(\frac{\lambda_s^o}{d_t} \right) \tag{14.35}$$

DeWasch and Froment also give the following relationship for the wall heat transfer coefficient for the two-dimensional model (h_w):

$$h_w = h_w^o + 0.0115 \left(\frac{d_t}{d_p} \right) \left(\frac{d_p G}{\mu} \right) \tag{14.36}$$

where the static contribution h_w^o is given by:

$$h_w^o = \frac{20\lambda_s^o}{d_t}$$

(14.37)

The overall heat transfer coefficient is based on an average bed temperature. If there exists a significant radial temperature profile and yet the one-dimensional model is used through a choice of the overall heat transfer coefficient, the following equation can be used to account for the effect of the radial profile (Crider and Foss 1965):

$$h_e = \frac{h_w}{1 + (Bi)_h/6.2}$$

(14.38)

Assuming a parabolic profile, for instance, for the temperature, the factor appearing in the denominator would be 8 instead of 6.2. This relationship also gives an indication as to whether the one-dimensional model is adequate. If $(Bi)_h$ is small, say less than 1, one may conclude that the one-dimensional model is adequate since then h_e is quite close to h_w.

It has long been recognized that the wall heat transfer coefficients (h_e and h_w) as well as the effective thermal conductivity are not constant but depend on the reactor length. In general, these coefficients decrease with increasing reactor length. On the other hand, it is extremely difficult to determine these coefficients as a function of reactor length, and therefore, constant values are used for design purposes. Most of the correlations presented so far are based on the data obtained from experimental reactors in the absence of a reaction for which the ratio of reactor length to tube radius is relatively small. If this ratio is large enough, ($100 \sim 200$) which is true for typical industrial reactors, these coefficients will approach asymptotic values, which are independent of the reactor length. Li and Finlayson (1977) obtained these asymptotic values based on the available literature data. Their results are summarized in Table 14.3.

Effective Diffusivities

Unlike the diffusion in catalyst pellets, molecular as well as eddy (turbulent) diffusion causes the mass dispersion in fixed-beds. The effective molecular diffusivity may be obtained by simply multiplying the gas molecular diffusivity by the factor ϵ_B/κ, where the tortuosity factor κ is often taken as 1.5. The theory on eddy diffusivity is not well established. Therefore, the effective diffusivities are often correlated in the following form:

$$\frac{1}{Pe} = \frac{1}{(Pe)_m} + \frac{1}{(Pe)_t}$$

(14.39)

Table 14.3   Asymptotic Values of Coefficients (Li and Finlayson 1977; reprinted with permission from *Chemical Engineering Science*. Copyright by Pergamon Press, Inc.)

Overall Heat Transfer Coefficient	*Wall Heat Transfer Coefficient*	
$\dfrac{Ud_t}{\lambda_g} \exp(6d_p/d_t) = 2.03 \left(\dfrac{d_p G}{\mu}\right)^{0.8}$	$\dfrac{h_w d_p}{\lambda_g} = 0.17 \left(\dfrac{d_p G}{\mu}\right)^{0.79}$	Spherical packing
$20 < \dfrac{d_p G}{\mu} < 7600$ and $0.05 < \dfrac{d_p}{d_t} < 0.3$		
$\dfrac{Ud_t}{\lambda_g} \exp(6d_p/d_t) = 1.26 \left(\dfrac{d_p G}{\mu}\right)^{0.95}$	$\dfrac{h_w d_p}{\lambda_g} = 0.16 \left(\dfrac{d_p G}{\mu}\right)^{0.93}$	Cylindrical packing
$20 < \dfrac{d_p G}{\mu} < 800$ and $0.03 < \dfrac{d_p}{d_t} < 0.2$		
where $d_p = 6 V_p/S_p$		

Effective thermal conductivity (radial)

$$\left(\frac{h_w d_p}{\lambda_{er}}\right)\left(\frac{\epsilon_B}{1 - \epsilon_B}\right) = 0.27$$

$$500 < \frac{d_p G}{\mu(1 - \epsilon_B)} < 6000 \text{ and } 0.05 < \frac{d_p}{d_t} < 0.15$$

where $(Pe)_m$ and $(Pe)_t$ are the Peclet numbers based on particle diameter for molecular and turbulent (eddy) diffusion, respectively, and $Pe(= d_p v/D_e)$ is the Peclet number for the combined mass dispersion. In terms of Reynolds and Schmidt numbers, the Peclet number for molecular diffusion can be written as:

$$(Pe)_m = (ReSc)\left(\frac{\kappa}{\epsilon_B}\right) \qquad Re = \frac{d_p G}{\mu}, \qquad Sc = \frac{\mu}{\rho D_m} \qquad (14.40)$$

when D_m is the molecular diffusivity for the gas species of interest. The majority of correlation efforts have been centered around the Peclet number for eddy dispersion. It has been recognized that the overall Peclet number is dominated by the turbulent Peclet number when the particle Reynolds number is greater than 100, its value approaching approximately 11 for radial dispersion and 2 for axial dispersion.

For the radial effective diffusivity, Froment (1967) found that at high Reynolds numbers the empirical correlation factor of Fahien and Smith (1955) given by $(Pe)_r/[1 + 19.4(d_p/d_t)^2]$ approaches 11. Using this as the turbulent contribution, Eq. 14.39 may be written as:

$$\frac{1}{(Pe)_r} = \frac{\epsilon_B}{1.5ReSc} + \frac{1}{11[1 + 19.4(d_p/d_t)^2]}; \qquad (Pe)_r = \frac{v d_p}{D_{er}} \qquad (14.41)$$

Table 14.4   Approximate Ranges of Transport Properties and Coefficients (Reprinted from Kulkarani and Doraiswamy 1980, by courtesy of Marcel Dekker, Inc.)

Radial effect thermal conductivity (λ_{er})	1–10 (kcal/m²hr°C)
	0.2–2 for $0 < \text{Re} < 800$
Axial effective thermal conductivity $\left(\dfrac{\lambda_{ea}}{\lambda_g}\right)$ Gas thermal conductivity	1–300
Heat transfer coefficient for a one-dimensional model (h_e)	15–75 (kal/m hr°C)
Wall coefficient for a two-dimensional model (h_w)	100–250
Radial Peclet number for mass dispersion ($(\text{Pe})_r$)	6–20
	(Typically 11 for Re > 100)
Axial Peclet number for mass dispersion ($(\text{Pe})_a$)	0.01–10
	(Typically 2 for Re > 10)

where κ has been set to 1.5. The stochastic model of Gunn (1969) for radial dispersion in a packed bed relates the turbulent Peclet number to the probability of a molecule at the entrance traveling a certain distance. This model also yields a value of 11 for the Peclet number at high Reynolds number. As for axial diffusion, the axial effective diffusivity can be obtained from:

$$\frac{1}{(\text{Pe})_a} = \frac{\epsilon_B}{1.5 \text{ReSc}} + \frac{1}{2}; \qquad (\text{Pe})_a = \frac{v d_p}{D_{ea}} \tag{14.42}$$

since the turbulent Peclet number approaches 2 at high Reynolds number. An empirical correlation developed by Wen and Fan (1975) is:

$$\frac{1}{(\text{Pe})_a} = \frac{0.3}{(\text{ReSc})} + \frac{0.5}{1 + 3.8/(\text{ReSc})} \tag{14.43}$$

$$0.008 < \text{Re} < 400 \ \text{ and } \ 0.28 < \text{Sc} < 2.2$$

This correlation, which is similar to that proposed by Bischoff and Levenspiel (1962), accounts for a maximum that exists in the indicated range of Reynolds number.

Typical ranges of transport properties and coefficients are summarized in Table 14.4.

14–4   DETERMINATION OF TRANSPORT PROPERTIES AND COEFFICIENTS

Fixed-beds packed with inert pellets are usually used for the determination of transport properties and transfer coefficients. The solution of a steady-state mass or heat balance equation is compared with concentration or temperature measurements for the determination. For thermal properties, for instance, radial temperature profile (Coberly and Marshall 1951; Michelsen 1979), axial temperatures (Yagi and Wakao 1959) or centerline temperatures (Olbrich et al. 1966) are measured. These properties can also be determined from a moment analysis of the transient response at the bed outlet obtained by introducing a pulse at the inlet. While this procedure is more convenient experimentally, possible dependence of the properties on transient conditions has not yet been fully resolved. For this reason, a steady-state experiment may be preferred for the determination. The correlations given earlier are only for approximate calculations of the properties. For any systematic study of a fixed-bed, these properties should be determined for the reactor under consideration using the experimental methods described in this section.

Perhaps the simplest to determine is the effective wall heat transfer coefficient (based on tube wall temperature) used in conjunction with the one-dimensional model. A heat balance for a bed packed with inert pellets is given by:

$$\frac{d_t G C_p}{4} \frac{dT}{dz} = h_e (T_w - T) \tag{14.44}$$

where T_w is the constant wall temperature. Integrating this equation over the reactor length (Z) gives:

$$h_e = \frac{d_t G C_p}{4Z} \ln \frac{T_w - T_{in}}{T_w - T_e} \tag{14.45}$$

where the mean outlet temperature T_e is obtained from the radial profile at the outlet by:

$$T_e = \frac{4 \int_0^{d_t/2} r T dr}{d_t^2} \tag{14.46}$$

Effective wall heat transfer coefficients determined in this way by DeWasch and Froment (1972) are shown in Figure 14.5. It is seen that the effective transfer coefficients are higher at the inlet and approach a limiting value toward the outlet. In general, this limiting value is reached at a smaller length for a smaller ratio of d_t/d_p. Note that h_e is a lumped quantity averaged over the reactor length, and therefore, is a constant for a given reactor.

For the determination of effective thermal conductivity and wall heat transfer coefficient, the following heat balance equation is solved:

$$G C_p \frac{\partial T}{\partial z} = \lambda_{er} \left(\frac{\partial^2 T}{\partial r^2} + \frac{1}{r} \frac{\partial T}{\partial r} \right) \tag{14.47}$$

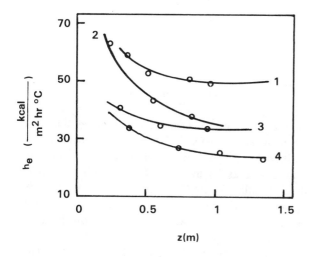

Figure 14.5 Dependence of effective wall heat transfer coefficient on reactor length for various conditions. (Dewasch and Froment 1972; reprinted with permission from *Chemical Engineering Science.* Copyright by Pergamon Press, Inc.)

with

$$T = T_{in}, z = 0$$

$$\frac{\partial T}{\partial r} = 0, r = 0$$

$$h_w(T_w - T) = \lambda_{er} \left(\frac{\partial T}{\partial r}\right), r = \frac{d_t}{2} \tag{14.48}$$

whose solution (Coberly and Marshall 1951) is given by:

$$\frac{T_w - T}{T_w - T_{in}} = 2 \sum_{n=1}^{\infty} \frac{J_0(b_n \xi)e^{-\beta b_n z}}{b_n J_1(b_n) [1 + (b_n m)^2]} \tag{14.49}$$

where

$$\xi = \frac{2r}{d_t}$$

$$\beta = \frac{4\lambda_{er}}{GC_p d_t^2} \tag{14.50}$$

$$m = \frac{2\lambda_{er}}{h_w d_t}$$

and J_0 and J_1 are Bessel functions of the first kind of order zero and one, respectively. Here, the eigenvalues b_n are defined by the relationship:

$$\frac{J_0(b_n)}{J_1(b_n)} = b_n m \tag{14.51}$$

When the value of βz is greater than 0.2, the series in Eq. 14.49 converges so rapidly that only the first term is significant. Therefore, for $\beta z > 0.2$, the equation can be reduced to:

$$\ln \frac{T_w - T}{T_w - T_{in}} = -\beta b_1^2 z + \ln \frac{J_0(b_1 \xi)}{b_1[1 + (b_1 m)^2] J_1(b_1)} \qquad (14.52)$$

If the temperature along the tube centerline is measured, for instance, and used in Eq. 14.52, a plot of the left hand side versus z should give a straight line, yielding βb_1^2 as its slope from which λ_{er} can be calculated, and also yielding the logarithmic term in the right hand side as its intercept from which h_w can be calculated. Various methods of calculating λ_e and h_w result depending on how Eq. 14.49 or Eq. 14.52 is used, given certain temperature measurements. DeWasch and Froment, for instance, suggest the selection of λ_{er} and h_w based on the minimization of the function $F = \Sigma(T_{i\,calc} - T_{i\,exp})^2$ using a radial profile.

One way of writing a heat balance for the two-dimensional model is to replace the boundary condition of Eq. 14.48 by:

$$T = T_w \text{ at } r = R$$

In such a case, the solution becomes:

$$\frac{T_w - T}{T_w - T_{in}} = 2 \sum_1^\infty \frac{J_0(\lambda_n \xi)}{\lambda_n J_1(\lambda_n)} e^{-\beta \lambda_n^2 z} \qquad (14.53)$$

where λ_n are the roots of $J_0(x) = 0$. Similar analysis as for the determination of λ_{er} and h_w given above applies.

The effective diffusivity in the radial direction can be determined from concentration measurements and the solution of the following mass balance:

$$\frac{\partial C}{\partial \bar{z}} = \frac{1}{P_{mr}} \frac{1}{\bar{r}} \frac{\partial}{\partial \bar{r}} \left(\bar{r} \frac{\partial C}{\partial \bar{r}} \right) \qquad (14.54)$$

$$\bar{r} = \frac{r}{R}, \bar{z} = \frac{z}{Z}, P_{mr} = \frac{vR^2}{D_{er}Z}$$

This equation is essentially equivalent to Eq. 14.47. However, the boundary condition at the wall is $\partial C / \partial \bar{r} = 0$. The solution (Olbrich et al. 1966) is given by:

$$\frac{C - C_\infty}{C_\infty} = 2 \sum_{i=1}^\infty \frac{J_1(\eta_i) J_0(\eta_i \bar{r})}{\eta_i J_0^2(\eta_i)} e^{-(\eta_i^2 \bar{z}/P_{mr})} \qquad (14.55)$$

where C_∞ is the concentration approached asymptotically in the bed of infinite length, and η_i are the roots of the equation $J_1(\eta_i) = 0$. For a bed of sufficient length, the exit concentration may be taken as C_∞.

Effective axial transport properties can be determined using an adiabatic reactor. Steady state mass and heat balances result in second-order ordinary differential equations when the axial dispersion is taken into consideration, solutions of which can readily be obtained. Based on these solutions and temperature or concentration measurements, the effective transport properties can be calculated in a manner similar to the procedures used for the radial transport properties. As indicated earlier, a transient experiment can also be used for the determination. Here, experimental and analytical procedures are illustrated for the determination of the effective axial transport property for mass. An unsteady state mass balance for an adiabatic reactor can be written as:

$$\frac{\partial \bar{C}}{\partial \bar{t}} + \frac{\partial \bar{C}}{\partial \bar{z}} - \frac{1}{P_{ma}} \frac{\partial^2 \bar{C}}{\partial \bar{z}^2} = 0 \qquad (14.56)$$

where $\bar{z} = z/Z$, $\bar{t} = t/\tau$, $\tau = Z/v$ and $P_{ma} = vZ/D_{ea}$. The concentration deviation from steady state is given by:

$$\bar{C} = \frac{(C - C_s)}{C_s}$$

where C_s is the steady state concentration. Taking the Laplace transform of Eq. 14.56 and solving the resulting second-order ordinary differential equation gives a solution in the following form:

$$\bar{C} = \bar{C}(s, z; P_{ma}) \qquad (14.57)$$

When a concentration pulse is introduced to the bed and the response at the outlet is measured, this outlet concentration, when multiplied by t^n and integrated over $t \in (0, \infty)$, yields the nth-order moment:

$$M_n = \int_0^\infty \bar{t}^n \bar{C}_1(\bar{t}) d\bar{t}; \qquad \bar{C}_1 = (\bar{t}, \bar{z} = 1) \qquad (14.58)$$

The moment can also be obtained from:

$$M_n = \lim_{s \to 0} \frac{d^n \bar{C}_1(s)}{ds^n} (-1)^n \qquad (14.59)$$

Since $\bar{C}_1(s) = \bar{C}(s, 1; P_{ma})$, the moment obtained from Eq. 14.59 will contain P_{ma} as the only unknown. When the moment is determined by graphical integration of the outlet response, P_{ma} can be calculated by equating Eq. 14.58 to Eq. 14.59. In this case, only the first moment ($n = 1$) is required for the determination, which will yield fairly accurate results since the higher order moments are not involved. No satisfactory comparisons have been made between the properties determined from steady state and transient experiments. For design purposes, therefore, the properties determined from steady-state experiments may be preferred.

Summary

Transport properties and coefficients have been presented in this chapter, including theories, correlations, and experimental methods. While correlations are useful for estimating the transport properties and coefficients, it is desirable for any detailed study of a reactor to determine these properties experimentally using the methods presented in this chapter.

NOTATION

b_n	solutions of Eq. 14.51
$(Bi)_h$	Biot number for heat given in Eq. 14.34
C	concentration
C_b	bulk-fluid concentration
C_p	specific heat content
C_∞	concentration at the exit of a sufficiently long bed
d_p	particle diameter
d_t	tube diameter
D_e	effective diffusivity
$D_{e,12}$	effective molecular diffusivity
$D_{e,K}$	effective Knudsen diffusivity
D_{ea}	effective diffusivity in the axial direction in a fixed-bed
D_{er}	effective diffusivity in the radial direction in a fixed-bed
D_k	Knudsen diffusivity
$D_{m,12}$	molecular diffusivity
D_M	macropore diffusivity
D_μ	micropore diffusivity
f_e	correction factor for pore geometry given in Eqs. 14.3 and 14.4
G	mass rate per unit area
h	film heat transfer coefficient
h_b	radiant heat transfer coefficient
h_c	coolant-side heat transfer coefficient
h_e	tube-side heat transfer coefficient in a one-dimensional reactor model
h_w	wall heat transfer coefficient in a two-dimensional reactor model
$(-\Delta H)$	heat liberated by reaction per mole
j_D, j_H	j factor for mass and heat, respectively
J_0, J_1	Bessel functions of the first kind of order zero and one, respectively
k	Boltzmann constant
k_g	film mass transfer coefficient
L	ratio of volume to external surface area of pellet; effective distance between particles

m	M_1/M_2; quantity defined in Eq. 14.50
M	molecular weight
M_n	n^{th} moment
N	molar flux
P	total pressure
Pe	Peclet number
$(\text{Pe})_a$	axial, mass Peclet number given in Eq. 14.42
$(\text{Pe})_m$	Peclet number for molecular diffusion given in Eq. 14.40
$(\text{Pe})_r$	radial, mass Peclet number given in Eq. 14.41
$(\text{Pe})_t$	Peclet number for turbulent diffusion in Eq. 14.39
P_{ma}	vZ/D_{ea}
P_{mr}	$vR^2/D_{er}Z$
Pr	Prandtl number for fluid, $C_p\mu/\lambda$
r	radial coordinate
$\bar{r}$	r/R; average pore radius
r_c	rate of reaction
R	radius
Re	Reynolds number d_pG/μ
R_g	gas constant
R_G	global rate
s	Laplace transform variable
s_p	surface area of a cylinder
Sc	Schmidt number, $\mu/\rho D_m$
t	time
$\bar{t}$	t/τ
T	temperature
T_b	bulk-fluid temperature
T_e	reactor outlet temperature
T_s	pellet surface temperature
T_w	tube wall temperature
T_{in}	reactor inlet temperature
U	overall heat transfer coefficient based on h_e
v	superficial fluid velocity
$\bar{v}$	average molecular velocity
V_p	volume of a cylinder
y	mole fraction
z	pellet coordinate, axial coordinate
$\bar{z}$	z/Z
Z	reactor length

Greek Letters

α	constant factor in Eq. 14.32
β	quantity defined in Eq. 14.50; quantity defined in Eq. 14.28
δ	stagnant fluid thickness

ϵ	porosity
ϵ_B	bed porosity
ϵ_M	macropore porosity
ϵ_μ	micropore porosity
κ	tortuosity factor
η	effectiveness factor
λ	thermal conductivity; mean free path of a molecule
λ_e	effective thermal conductivity
λ_g	gas thermal conductivity
λ_s	solid thermal conductivity
λ_{ea}	effective thermal conductivity in axial direction
λ_{er}	effective thermal conductivity in radial direction
λ_d°	dynamic contribution to λ_{er} in Eq. 14.26
λ_s°	static contribution to λ_{er} in Eq. 14.26
λ_s^c	conduction component of λ_s° in Eq. 14.27
λ_s^r	radiation component of λ_s° in Eq. 14.27
μ	viscosity
ν	kinematic viscosity
ρ	density
σ, σ_{12}	parameters given in Table 14.1
Ω_k, Ω_{12}	parameters given in Table 14.2
τ	Z/v
ϕ_G	generalized Thiele modulus
ϕ	quantity defined by Eq. 14.20

PROBLEMS

14.1. The following results have been obtained using a Wicke and Kallenbach diffusion cell:

$$(D_{H_2})_e = 0.07 \text{ cm}^2/\text{s}$$

$$(D_{N_2})_e = 0.04 \text{ cm}^2/\text{s}$$

The same catalyst pellets are to be used for combustion of C_6H_8 in excess air. Find the effective diffusivity of C_6H_8, $(D_{C_6H_8})_e$. Assume for simplifications that temperature, pressure, and the constants in the Lennard-Jones potential function remain the same. (Ans. 0.017 cm²/s)

14.2. One set of data obtained by Coberly and Marshall (1951) for a bed ($d_t = $ 5in) packed with inert pellets is given below for the centerline temperatures.

z (in)	5.75	11.5	17.25	23.0	34.5	46.0
T (°F)	115	126	156	160	187	197

For this run, $G = 475$ lb/ft²hr, $T_w = 220°F$, and $T_{in} = 100°F$. Determine from the data the effective thermal conductivity in the radial direction λ_{er} and the wall heat transfer coefficient h_w. Use the physical properties of air.

14.3. Consider an adiabatic fixed-bed. Suppose the bed is packed with inert pellets.

    a. Write the unsteady-state, dimensionless heat balance for the bed such that the Peclet number appears as the only constant in the equation.

    b. Rewrite the equation in terms of temperature deviation from the steady state value.

    c. Suppose that at time zero, a temperature impulse of unit magnitude is introduced at the inlet and that the outlet response is recorded. What is the expression for the first moment? Is this equal to the area under the outlet response curve?

REFERENCES

Baiker, A., M. New and W. Richarz, Chem. Eng. Sci., *37*, 643 (1982).

Bird, R.B., W.E. Stewart and E.N. Lightfoot, *Transport Phenomena*, Wiley, New York (1960).

Bischoff, K.B. and O. Levenspiel, Chem. Eng. Sci., *17*, 245 (1962).

Butt, J.B., AIChE J., *11*, 106 (1965).

Butt, J.B., D.H. Downing and J.W. Lee, Ind. Eng. Chem. Fund., *16*, 270 (1977).

Carberry, J.J., AIChE J., *6*, 460 (1960).

Chu, J.C., J. Kaiil and W.A. Wetterath, Chem. Eng. Prog., *49*, 141 (1953).

Coberly, C.A. and W.R. Marshall, Jr., Chem. Eng. Prog. *47*, 141 (1951).

Crider, J.E. and A.S. Foss, AIChE J., *11*, 102 (1965).

DeWasch, A.P. and G.F. Froment, Chem. Eng. Sci., *27*, 567 (1972).

Dogu, G. and J.M. Smith, AIChE J., *21*, 58 (1975).

Dwivedi, P.N. and S.N. Upadhay, Ind. Eng. Chem. Proc. Des. Dev., *16*, 157 (1977).

Fahien, R.W. and J.M. Smith, AIChE J., *1*, 25 (1955).

Feng, C.F. and W.E. Stewart, Ind. Eng. Chem. Fund., *12*, 143 (1973).

Froment, G.F., Ind. Eng. Chem. *59*, 27 (1967).

Frost, A.C., AIChE J., *27*, 813 (1981).

Gunn, D.J., Trans. Inst. Chem. Eng., *47*, T351 (1969).

Kulkarani, B.D. and L.K. Doraiswamy, Cat. Rev., *22*, 431 (1980).

Kunii, D. and O. Levenspiel, *Fluidization Engineering*, Wiley, New York (1969).

Kunii, D. and J.M. Smith, AIChE J., *6*, 71 (1960).

Li, C.H. and B.A. Finlayson, Chem. Eng. Sci., *32*, 1055 (1977).

Masamune, S. and J.M. Smith, J. Chem. Eng. Data, *8*, 54 (1963).

McGreavy, C. and M.A. Siddiqui, Chem. Eng. Sci., *35*, 3 (1980).

Michelsen, M.L., Chem. Eng. J., *18*, 67 (1979).

Olbrich, W.E., J.B. Agnew and O.E. Potter, Trans. Inst. Chem. Eng., *46*, T207 (1966).

Reid, R.C., J.M. Prausnitz and T.K. Sherwood, *The Properties of Gases and Liquids*, Chap. 11, McGraw-Hill, New York (1977).

Satterfield, C.N., *Mass Transfer in Heterogeneous Catalysis*, MIT Press, Cambridge, Mass. (1970).

Sehr, R.A., Chem. Eng. Sci., *2*, 145 (1958).

Wakao, N. and J.M. Smith, Chem. Eng. Sci., *17*, 825 (1962).

Wakao, N. and T. Funazkri, Chem. Eng. Sci., *33*, 1375 (1978).

Wakao, N., S. Kaguel and T. Funazkri, Chem. Eng. Sci., *34*, 325 (1979).

Wen, C.Y. and L.T. Fan, *Models for Flow Systems and Chemical Reactors*, Marcel-Dekker, New York (1975).

Whitaker, S., AIChE J., *18*, 361 (1972).

Wicke, E. and R. Kallenbach, Kolloid-Z, *17*, 135 (1941).

Wolff, H.J., K.H. Radeke and D. Gelbin, Chem. Eng. Sci., *34*, 101 (1979).

Woodside, W. and J.H. Messmer, J. Appl. Phys., *32*, 1688 (1961).

Yagi, S. and N. Wakao, AIChE J., *5*, 79 (1959).

Yagi, S., D. Kunii and N. Wakao, AIChE J., *6*, 543 (1960).

SUBJECT INDEX

Acidity
  of catalyst, 6
  Brønsted and Lewis acid, 6
  correlations, in terms of, 31, 32
  determination of, 27
Activation energy
  homogeneous vs. heterogeneous, 5
  diffusional effect on, 104
Active site, 55
  of acidic catalysts, 6, 27
Activity factor, 165
Activity, on-line estimation of, 465
Adsorption
  apparatus for, 25
  chemical, 20, 21, 24
  energetics of, 22
  heat of, 43
  hysteresis of physical, 43
  inhibition, 62
  in surface reaction, 55
  isotherms, 23, 35
  multilayer, 35
  physical, 20, 21, 40, 41
  types of physical, 40
Arrhenius number, 108
Aspect ratios
  definitions of, 289
  role of, in design, 291
Atom migration
  mechanism of, 202
  kinetics of sintering for, 211
Autothermic reactor, 309
Average pore radius, 44
Axial dispersion
  boundary conditions in the model of, 288
  coefficients, 499
  CSTR analogy, 326
  in fixed-bed, 287
  in trickle-bed, 443

Bernoulli equation, 294
Biot numbers, 107
Blowout, 312
Bubble phase, 405
Bubble properties, 419
BET (Brunauer-Emmett-Teller)
  theory, 35
  surface area, 39

Calculus of variations, 459
Carbonium ion, 6
Catalyst
  characterization of, 35
  classification of, 5
  definition of, 3
  support for, 10
  uses of, 8
Catalyst deactivation. *See* Chemical deactivation; Physical deactivation
Catalyst preparation
  impregnation, 10
    single pore model of, 14
    equilibration time for, 17
    parameter determination of, 18
  precipitation, 10
    calcination, 11
    drying, 11
Catalyst regeneration, 180, 201, 470
Catalytic activity
  correlations of, 27
  distribution of, 18, 126, 141
Catalytic converter, 312, 314
Cell model
  for fixed-bed, 297
  for fluidized-bed, 415
Chapman-Enskog formulas, 484, 491
Chemical adsorption (chemisorption). *See* Adsorption

AUTHOR INDEX

515